Springer Series in Materials Science

Volume 307

The Springer Series in Materials Science covers the complete spectrum of materials research and technology, including fundamental principles, physical properties, materials theory and design. Recognizing the increasing importance of materials science in future device technologies, the book titles in this series reflect the state-of-the-art in understanding and controlling the structure and properties of all important classes of materials.

More information about this series at http://www.springer.com/series/856

S. Noor Mohammad

Synthesis of Nanomaterials

Mechanisms, Kinetics and Materials Properties

Springer

S. Noor Mohammad
Washington, DC, USA

ISSN 0933-033X ISSN 2196-2812 (electronic)
Springer Series in Materials Science
ISBN 978-3-030-57584-7 ISBN 978-3-030-57585-4 (eBook)
https://doi.org/10.1007/978-3-030-57585-4

This Springer imprint is published by the registered company Springer Nature Switzerland AG
The registered company address is: Gewerbestrasse 11, 6330 Cham, Switzerland

Preface

This book *Synthesis of Nanomaterials: Mechanisms, Kinetics and Materials Properties* is aimed at the senior undergraduate and graduate students in the disciplines of science and engineering, particularly physics, chemistry, materials science and engineering, chemical engineering, electrical engineering, mechanical engineering, bioengineering, and biology. Researchers in the areas of nanomaterials and nanosciences from industry, academia, and government agencies may particularly find the book useful. However a basic level of competency in physics, chemistry and mathematics would be required for this. The first few of the eighteen chapters of the book provide background of the basic nanomaterials synthesis science. It should greatly serve as a ready reference for the understanding of the fundamental chemistry and physics of the latest advances in nanomaterials synthesis. It should provide a broad perspective and scientific depth for the readers who intend to comprehend the rapidly developing new sciences in the interdisciplinary area of nanomaterials synthesis.

The book contains a great deal of information departed from the ones based on existing conventional knowledge of nanomaterials science. I believe that the book may hence evoke controversy. It may be seen as challenging the consensus built over the years. System of any kind and in any context tries to preserve the status quo; it tries to disregard the new ones by all means possible. Having spent 50+ years in the field of chemistry and physics of materials, and having 250 research articles written and published in major journals in this area, this author feels obliged to share his knowledge, analyses, and conclusions, some of which different from the norm. His goal is to raise the level of awareness among the scientists and engineers and to initiate the discussions that, in turn, may entail major changes, as well as revisions of the existing, conventional knowledge in favor of the new and corrected ones. If his ideas and knowledge different from the conventional ones are found correct, and this book is considered pioneering—distinctly different from the ones based largely on existing knowledge—the beneficiaries will be all of us—ourselves, our scientists, our researchers, our children, our grand children, and our society, as a whole.

The author owes a debt of gratitude to numerous scientists who worked with him, corrected him whenever necessary, provided him guidance, and shared their knowledge with him. They include Dr. Sudhir Kumar De, Dr. Zhi-Fang Fan, Dr. Haseeb F. Rashid, Prof. Charles E. Rogers, Late Prof. Osman. K. Mawardi, Prof. M. A. Sobhan, Dr. Arif Khan, and Dr. Saeed Ganji. Born and brought up in a remote village in India, the author has an extremely humble background. The most favorite dreams of his parents about him turned out to be wrong because of two men—Mr. Nagendra Nath Das at Golgram High School and Prof. H. N. Bose at the Indian Institute of Technology, Kharagpur, both in India. They were his teachers, mentors, guides, and supervisors. Almost everything that he has got in life has been made possible because of these two great teachers. Dr. Arif Khan and his wife Roksana Parvin have done everything during the past 20 years to keep him alive and active in research. They have been truly guides in his pursuit for new knowledge and new science. David Silverio Hernandez Pineda has always provided technical assistance needed for this research. And, lastly his daughter Lina has been a constant source of encouragement for him to continue his work. The publishing of the book would not be easy without the help and cooperation of Dr. Zachary Evenson, the Physics Editor at the Springer Nature Publishing Company. He is grateful to all of them.

Lastly, about 20 years ago, Mr. Maoqi He—at that time a highly experienced senior researcher and a member of the author's research team—became most instrumental in initiating nanomaterials research in the author's laboratory. He contributed immensely to the author's endeavor in nanoscience and nanotechnology. Truthfully, he led in 1999, probably for the first time, to the discovery of self-catalytic nanomaterials growth [Appl. Phys. Lett. 77, 3731 (2000)] in our laboratory. He deserves full credit for this. With all love, affection, and appreciation, this book is dedicated to him. The author has made sincere effort to make the book free from errors. He fears it has still some errors, for which he apologizes.

Washington, DC, USA S. Noor Mohammad

Contents

1	**Introduction**		**1**
	References		9
2	**Nanomaterials Synthesis Routes**		**13**
	2.1	Introductory Comments	13
	2.2	Basics	14
	2.3	Control Over Nanomaterials Growths	15
	2.4	CVD Route for Synthesis	15
	2.5	MBE Route for Synthesis	17
	2.6	Pulsed Laser Deposition Route for Synthesis	18
	2.7	Solvothermal Route for Synthesis	19
	2.8	ARC Discharge Route for Synthesis	19
	2.9	Differences Between CVD and PVD Techniques	21
	2.10	Sol-Gel Route for Synthesis	21
	2.11	Chemical Beam Epitaxy	23
	2.12	Conclusions	24
	References		25
3	**Catalyst Nanoparticles**		**27**
	3.1	Background	27
		3.1.1 Steps for Nanoparticle Formation	28
		3.1.2 METANOs and SUBSANOs for Growths	30
	3.2	METANO and METANO Synthesis	30
		3.2.1 Synthesis Methods	32
		3.2.2 Synthesis Approaches	34
		3.2.3 Catalytic Reactivity	35
	3.3	SUBSANO and SUBSANO Synthesis	37
		3.3.1 SUBSANO Synthesis by Surface Treatment	37
		3.3.2 SUBSANOs Generated by Stress	38
		3.3.3 SUBSANOs Created by Droplets	40

3.4 Nanoparticle Surface Composition Called R_L Species 40
3.4.1 Definition 40
3.4.2 R_L Species Composition and Characteristics 41
3.5 Types of Nanoparticles 42
3.5.1 Type 1 Nanoparticles 42
3.5.2 Type 2 Nanoparticles 42
3.5.3 Type 3 Nanoparticles 43
3.5.4 Type 4 Nanoparticles 43
3.6 Effects of Surface, Interface, Size, and Density of Nanoparticles 44
3.7 Bimetallic Nanoparticles 44
3.7.1 Structural Diversity of BNPs 44
3.7.2 BNP Properties Different from the Corresponding Bulk 45
3.7.3 BNP Based R_L Species 46
3.8 Ostwald Ripening of Nanoparticles 46
3.9 Functionalization of FECA Nanoparticles 47
3.10 Lifetime of Feca Metal Nanoparticles 48
References 48

4 Pre-synthesis and Synthesis Events 53
4.1 Basics 53
4.2 Pre-synthesis Process: Formation of Catalyst Support 54
4.3 Crucial Synthesis Stages 55
4.4 Key Synthesis Events 56
4.4.1 Event 1 56
4.4.2 Event 2 58
4.4.3 Surface Activation Barrier 58
4.4.4 Reaction and Interaction on Nanoparticle Surface . . . 59
4.4.5 Anisotropic Growth and Nucleation for This Growth 59
4.4.6 Adhesive Properties of the Nanoparticle Seed 60
4.4.7 Contact Angle 62
4.5 The Nanoparticle Core and Shell Structures for Synthesis 64
4.6 Nanoparticle Periphery with Shell and Hill for Synthesis 66
4.7 Smaller Nanoparticles Generally Yield SWCNTs 67
References 67

5 The VLS Mechanism 69
5.1 Historical Background 69
5.2 Important Requirements 70
5.3 The Eutectic Phase 71
5.3.1 Formation of Droplet 71
5.3.2 Incubation Time 72

5.3.3 Thermodynamic and Kinetic Conditions for Incubation Time 73
5.3.4 Binary Phase Diagrams 74
5.3.5 Important Requirements for the Formation of Droplets 77
5.4 FECA Metal Selection 78
5.4.1 Possible Selection Criteria 78
5.4.2 Illustrative Examples 79
5.5 Growth Dynamics 79
5.6 Temperature Dependency 80
5.6.1 Inconsistency in Growth Characteristics 80
5.6.2 Several VLS Growth Rates Compared 80
5.6.3 VLS Growth at Eutectic Temperature 83
5.7 Failures of the VLS Mechanism to Mediate Some Nanowire Growths 83
5.7.1 Lack of Atomic-Scale Control Over Growth 83
5.7.2 Silicon Nanowire Growth Rate as Function of Various Growth Parameters 85
5.7.3 InAs Nanowire Growth Rate as Function of Various Growth Parameters 86
5.7.4 Defects in Nanowires by the VLS Mechanism 87
5.7.5 Impact of MET Contamination in Nanowire 88
5.8 Failures of the VLS Mechanism to Mediate Carbon Nanotube Growths 89
5.8.1 Location, Shape, and Size of Catalyst Particle During Growth 89
5.8.2 Crystallographic Relationship During Growth 89
5.8.3 Presumed Stages of Growth 89
5.8.4 Possible Mechanism for Growth 90
5.9 Criteria for Nanomaterial Growths by the VLS Mechanism 90
5.9.1 Criterion 1 91
5.9.2 Criterion 2 92
5.9.3 Criterion 3 93
5.9.4 Criterion 4 94
5.9.5 Criterion 5 96
References 97

6 Vapor–Solid–Solid Growth Mechanism 101
6.1 Basics 101
6.2 Illustrations of Nanowire Growth by the VSS Mechanism 102
6.2.1 Au-Mediated Low-Temperature ZnO Nanowire Growths 102
6.2.2 Cu-Mediated Low-Temperature Ge Nanowire Growths 103

6.2.3 Nanowire Growths via Solid FECA Material 103
6.2.4 Temperature-Dependent Variation of Nanowire and of MET Catalyzing This Nanowire 104
6.3 Illustrations of Carbon Nanofiber Growth by the VSS Mechanism 105
6.4 Illustrations of Carbon Nanotube Growth by the VSS Mechanism 106
6.5 Strengths of the VSS Mechanism 107
6.5.1 Superior Crystal-Phase Control 107
6.5.2 Abrupt Interface Composition 107
6.6 Controversy and Weaknesses of the VSS Mechanism 108
6.7 Rate-Limiting Steps During the VSS Growths 109
6.8 Comparison of VSS Growth Rates with the VLS Growth Rates 110
6.9 Temperature-Dependent Growths 113
6.9.1 Temperature-Dependent Growth Rate of Nanowires 113
6.9.2 Temperature-Dependent Growth Rate of Carbon Nanotubes 114
References 116

7 **Vapor–Solid Growth Mechanism** 121
7.1 Basics 121
7.2 Illustrations of Nanowire Growth by the VS Mechanism 123
7.2.1 Nanowire Growths on Non-stoichiometric SiO_z and GeO_z Wafer Surfaces 123
7.2.2 Nanowire Growths on Activated Amorphous Carbon 123
7.2.3 Nanowire Growths on Au Cluster Surface 124
7.2.4 Chemical Vapor Deposition of Tungsten Oxide Nanowires 124
7.2.5 Comparison of Nanowires by the VS and the VSS Mechanisms 126
7.3 Illustrations of Carbon Nanotube Growths by the VS Mechanism 127
7.3.1 CNT Growths on Nanosized Diamond 127
7.3.2 Basics of CNT Growths on Nanosized Diamond 127
7.3.3 MWCNT Growths on Defective Surfaces 127
7.3.4 SWCNT Growths on SiO_2 Surfaces 128
7.4 Illustrations of Nanobelt Growth by the VS Mechanism 129
7.5 Nanomaterial Growth Rates by the VS Mechanism 130
7.6 The Role of SUBSANO in Nanomaterial Growths 132
7.7 Growths by Water-Assisted Means 134
7.7.1 Basics 134

7.7.2 Effect of Precursor Flow 134
7.7.3 Effect of Temperature on CNT Growth 135
References 136

8 Solution and Supercritical Fluid-Based Growth Mechanisms 139
8.1 Background 139
8.2 Various Features of Solution and Supercritical Fluid-Liquid-Solid Mechanisms 141
8.2.1 Basics 141
8.2.2 Environment for the Synthesis of III-V Nanowires 142
8.2.3 Supercritical Fluid Conditions 143
8.2.4 First Realization of Supercritical Fluid Condition . . . 144
8.2.5 Inference 144
8.3 Solution- and Supercritical-Fluid-Based Growth Techniques 144
8.4 SFLS Nanowire Characteristics 145
8.4.1 Nanowire Size Distribution 145
8.4.2 Example of Ge Products from Low-Temperature Reactions 145
8.4.3 SFLS Growth of Si and Ge Nanowires 146
8.5 Strengths 147
8.5.1 Growth of Highly Promising Ge Nanowires 147
8.5.2 High Crystallinity of Nanowires 148
8.5.3 Other Advantages 148
8.6 Weaknesses 150
8.6.1 Challenges 150
8.6.2 Low-Temperature Growths 150
8.6.3 Solution-Phase Mechanisms Inferior to Vapor-Phase Mechanisms 151
8.6.4 Illustrative Demonstrations of SFLS and SoLS Mechanisms 151
8.7 SoSS (SFSS) Mechanism 153
8.7.1 Background 153
8.7.2 Specifics of the SoSS and the SFSS Mechanisms . . . 153
8.7.3 Interesting Feature of Chalcogenite Nanoparticles . . . 153
8.7.4 Nanowire Quality Dependent on Precursor 154
References 154

9 Solid–Liquid–Solid Growth Mechanism 159
9.1 Various Features of the Synthesis 159
9.1.1 Basics 159
9.1.2 General Hypothesis for Growths 161
9.1.3 Why Nanowire Growths Do Not Take Place at $T < T_{sls}$ 162

9.2 Why SLS Nanowires Are Often Amorphous or Have Amorphous Shell? . . . 163
9.3 Illustrations of the Catalyst and Temperature for the SLS Growth . . . 164
9.4 Illustrations of Oxide and Nitride Nanowires Growth by the SLS Mechanism . . . 167
9.4.1 SiO_z Nanowire Growths . . . 167
9.4.2 In_2O_3 Nanowire Growths . . . 168
9.4.3 Si_3N_4 Nanowire Growths . . . 168
References . . . 171

10 Oxide-Assisted Growth Mechanism . . . 173
10.1 Introduction . . . 173
10.1.1 Basics . . . 173
10.1.2 High-Temperature Reaction, Formation of Clusters, and Their Impacts . . . 174
10.1.3 Experimental Demonstrations Of The OAG Mechanism . . . 177
10.2 Advantages and Disadvantages of the OAG Mechanism . . . 179
10.2.1 Advantages . . . 179
10.2.2 Disadvantages . . . 180
10.3 Analysis of the Observed Oxide-Assisted Growths . . . 181
10.4 Formation of the Oxide Sheath . . . 182
10.5 Role of Metal in Oxide-Assisted Growth . . . 183
10.6 Role of Sulfur in Oxide-Assisted Growth . . . 184
References . . . 185

11 Self-catalytic Growth (SCG) Mechanism . . . 187
11.1 Introduction . . . 187
11.2 Approaches to Obtain Met-Free Nanoparticle Seeds . . . 188
11.3 Examples of Met-Free Nanowire Growths . . . 189
11.3.1 Selective Area Epitaxy . . . 190
11.3.2 CVD Growth of III–V Nitride Nanowires . . . 190
11.3.3 Vapor Deposition of InAs Nanowires . . . 191
11.3.4 Patterned Growth of InAs Nanowires . . . 193
11.3.5 Boron Nitride Nanotube Growths . . . 194
11.4 Understanding the Self-catalytic Nanowire Growth . . . 194
11.4.1 Tips, Hillocks, and Roughness Crucial for Growths . . . 194
11.4.2 Grains and Grain Boundaries Formed Prior to Growths . . . 195
11.5 Role of Oxide in Nanowire Growth . . . 195
11.6 Various Stages of Nanowire Growth . . . 196
11.7 Novelty of the SCG Mechanism . . . 197

11.8 Differences Between Selective Area Epitaxy and Self-catalytic Growth . . . 198
11.9 Nanoparticles Crucial for Nanowire Growths . . . 199
11.9.1 Porous SiO_z Film Formed on Substrate . . . 199
11.9.2 Buffer Layer Formed on Substrate . . . 200
11.9.3 Both X and X_mY_n Nanoparticles Can Lead to X_mY_n Nanowire Growths . . . 201
11.9.4 Superiority of SEG Mechanism to Other Mechanisms . . . 202
References . . . 203

12 VQS Mechanism for Nanomaterials Syntheses . . . 207
12.1 Forwarding Note . . . 207
12.2 The Concept of Quasiliquid (Quasisolid) Nanoparticle Surface . . . 208
12.2.1 Quasiliquid (Quasisolid) Medium Defined . . . 208
12.2.2 Structure and Morphology of Nanoparticle . . . 209
12.2.3 Nanoparticle Surfaces Influenced by Various Parameters . . . 210
12.2.4 Illustrations of Nanoparticle Surfaces . . . 210
12.3 Surface Coarsening of Nanoparticle Surface . . . 212
12.3.1 Elements of Surface Coarsening; Tamman and Heutting Temperatures . . . 212
12.3.2 Basics of Surface Coarsening . . . 215
12.3.3 Illustrations of Surface Roughness . . . 215
12.4 Surface Looseness and the Porosity ρ_c . . . 216
12.4.1 Effect of Annealing and Temperature on the Opening of Mask . . . 216
12.4.2 Migration of the R_S Source Species and of the Droplets . . . 217
12.4.3 Illustrations of Nanopores Generated on Nanoparticle Surface . . . 218
12.5 Melting (Semi-melting) of Nanoparticle Surface . . . 218
12.5.1 Basics . . . 218
12.5.2 A Simple Model of Surface Melting . . . 219
12.5.3 Illustrations of Surface Melting . . . 220
12.5.4 Possible Causes of Surface Melting . . . 220
12.6 Nanoparticle Structure and Morphology . . . 221
12.7 Phase Transition(s) and Phase Separation(s) . . . 223
12.8 Creation of High-Energy Sites (HETs) . . . 224
12.9 NP1 and NP2 Nanoparticles . . . 225
12.9.1 Basic Definitions . . . 225
12.9.2 Illustrations . . . 225
12.10 Evidences of Phase Transitions . . . 226

12.11 Evidences of Phase Transitions and Co-existence of Multiple Phases . . . 226
12.12 Evidence of Phase Separation . . . 228
12.13 Experimental Evidences of the Benefits of Surface Treatments . . . 228
12.13.1 Surface Treatment Yields Surface Amorphicity and Surface Roughness . . . 228
12.13.2 Surface Treatment Yields Surface Porosity . . . 230
12.13.3 Optimal Level of Defect and Amorphicity Essential for Nucleation and Growth . . . 231
12.14 Distinctive Features of Nanomaterials Syntheses by the VQS Mechanism . . . 232
12.14.1 Transformation from Vapor Phase to Solid Phase . . . 232
12.14.2 Formation of Intermediate Quasiliquid (Quasisolid) Phase . . . 233
12.14.3 The Need of Streamlining for Porosity and Growth . . . 234
12.14.4 Role of Catalyst Support and Dipole Moment in FECA Surface Functionalization . . . 236
12.14.5 Illustrations of the Role of Catalyst Support in FECA Nanoparticle Surface Functionalization . . . 237
12.14.6 Amphoteric Characteristics of Catalyst Support Should Be Preferred . . . 238
12.14.7 Catalyst Support Should Enhance FECA Nanoparticle Surface Disturbance and Polarity . . . 238
12.15 Nanomaterial Growths by Low-Melting Point Metals . . . 239
12.16 Nanomaterials Tips . . . 240
12.17 Concluding Remarks . . . 242
12.17.1 Surface Energy . . . 243
12.17.2 Nanomaterials Nucleation . . . 243
12.17.3 Superiority of the VQS Mechanism . . . 245
References . . . 246

13 Growths on METANO Surface by the VQS Mechanism . . . 253
13.1 Forwarding Note . . . 253
13.2 Basic Concepts . . . 254
13.2.1 Formation of R_L Species . . . 254
13.2.2 Phase Transformations and Generations . . . 254
13.2.3 Possible Events During the Pre-nucleation Stage of Growth . . . 255
13.3 Illustrative Demonstration of the R_L Species . . . 256
13.3.1 Non-eutectic R_L Species Created by Some Oxide-Assisted Growth Experiments . . . 256

13.3.2 Eutectic R_L Species Created by Some Non-oxide-Assisted Growth Experiments . . . 259
13.3.3 Non-eutectic R_L Species Created by Some Non-oxide-Assisted Growth Experiments . . . 259
13.4 The Role of Surface Energy in the Met-Mediated Growths . . . 260
13.4.1 Surface Energy Defined . . . 260
13.4.2 METANO Surface Characteristics . . . 260
13.4.3 Barrier to the Exchange of Materials on the METANO Surface . . . 261
13.5 Model for the Role of Surface Energy in the Met-Mediated Growth . . . 262
13.5.1 Model for the Exchange of Materials . . . 262
13.5.2 Alternative Model for the Exchange of Materials . . . 263
13.6 Analyses of the Role of Surface Energy in the Metano-Mediated Growth . . . 265
13.6.1 Reduced Solubility of METANO Surface . . . 265
13.6.2 Exchange of Materials on METANO Surface . . . 269
13.7 Why Are Au-Mediated Si and Ge Nanowire Growths So Successful? . . . 271
13.7.1 Possible Reasons of Au Being Suitable for the VLS Growths of Si and Ge Nanowires . . . 271
13.7.2 Reasons for Other Metals not Being Very Suitable for the VLS Growths of Si and Ge Nanowires . . . 271
13.7.3 Surface Energy, Activation Energy, and Exchange of Materials on METANO Surface . . . 272
13.8 Carbon Solubility in Metano . . . 273
13.8.1 Reduced Solubility . . . 273
13.8.2 Effective Barrier to Exchange of Materials . . . 274
13.8.3 Important Revelation . . . 276
13.9 Why CNT Growth Rates with Fe, Co, and Ni Are Very High . . . 276
13.9.1 Experimental Demonstration . . . 276
13.9.2 Possible Causes of Said Observations . . . 278
13.9.3 Possible Causes of Discrepancy Based on Calculated Results . . . 279
13.9.4 Implication of Higher Solubility of C in Fe, Co, and Ni for SWCNT Growths . . . 279
13.9.5 Implication of Lower Solubility of C in Fe for MWCNT Growths . . . 280
13.10 Limit of Growth Rate . . . 281
13.11 Conclusions . . . 282
References . . . 284

14 Growths on SUBSANO Surface by the VQS Mechanism 289
14.1 Forwarding Note and Basic Concepts 289
14.1.1 Forwarding Note 289
14.1.2 A Critical Look at SUBSANOs 290
14.2 Illustrative Demonstrations of the R_L Species 290
14.2.1 Nanomaterials Growth on Solid Solution (Clustered) Islands 291
14.2.2 Nanomaterials Growth on Coarsened Substrates 291
14.2.3 Nanomaterials Growth on Metallic Substrates 292
14.2.4 Nanomaterials Growth on Nonmetallic Substrates . . . 292
14.3 Nanomaterials Growths on Surfaces 293
14.3.1 Basics 293
14.3.2 Illustrations of SFSs 295
14.3.3 SFS Types and Defects Generation in These Types of SFSs 298
14.3.4 Heterointerfaces and Charge Transfers in SFSs 299
14.3.5 Influence of the Layer Thicknesses of Type-I and Type-II SFSs on Catalytic Activity 301
14.3.6 Influence of the Metallic SUBSANO Layer Thickness of Type-III SFSs on Catalytic Activity . . . 302
14.4 High Catalytic Activities of Catalyst Surfaces 306
14.4.1 Key Catalyst Activities 306
14.4.2 Catalyst Surface Characteristics for Effective Catalytic Activities 307
14.4.3 Knudsen Diffusion, Interstitial Diffusion, and Substitutional Diffusion 311
14.4.4 Low-Temperature Decomposition of Gaseous Precursors 312
14.4.5 Is Catalyst Poisoning Real? 313
14.4.6 Catalyst Template Effects for Supersaturation 313
14.4.7 Membrane Template Effects 314
14.5 Conclusions 315
References 316

15 Simple Theoretical Model for Growth by the VQS Mechanism . . . 321
15.1 Forwarding Note 321
15.2 FECA Nanoparticle Porosity for Nanowire Growths 322
15.3 FECA Nanoparticle Porosity for Nanotube Growths 323
15.4 Theoretical Models for Porosity, Pore Radius, and HET in Terms of Amorphicity 324
15.4.1 Surface Amorphicity 324
15.4.2 Pore Radius, Porosity, and HET Reactivity Defined in Terms of Effective Amorphicity 325
15.5 Knudsen Diffusivity 326

15.5.1 Formulation of Knudsen Diffusivity 326
15.5.2 Experimental Support for the Knudsen Diffusivity . . . 327
15.6 Molecular and Knudsen Diffusion 328
15.7 Knudsen Permeability 329
15.8 Knudsen Diffusion Through Rough R_L Species 329
15.9 Nanomaterial Growth Rate 330
15.10 Diffusivity and Permeability for Growths 331
15.10.1 Calculated Results for Diffusivity and Permeability 331
15.10.2 Surface Roughness Effects on Diffusivity 332
15.10.3 Surface Amorphicity Effects on Diffusivity 334
15.11 Carbon Nanotube Growth Rates 338
15.11.1 Variation of CNT growth rate with CNT diameter 338
15.11.2 Temperature-Dependent Variation of CNT Growth Rate 340
15.11.3 CNT Growth Rates Dependent on Growth Duration 341
15.11.4 CNT Growth Rates Dependent on Precursor Flow Rate 342
15.11.5 Inference 343
References 344

16 The General, Versatile Growth Mechanism 347
16.1 Generality of Growth Mechanism 347
16.1.1 Basics of Material Phases 348
16.1.2 Multiple Phases Participating in Nanomaterials Growths 348
16.1.3 General Pathway for Nanomaterials Growths 349
16.2 Important Hallmarks 351
16.2.1 Factors Influencing Nanomaterials Growths 351
16.2.2 Optimal Phase for SUBSANO-Mediated Growths . . . 351
16.2.3 Optimal Phase for METANO-Mediated Growths . . . 352
16.2.4 Uniqueness of the Nanoparticle Phase 353
16.2.5 Characteristics of SECINI Formed During Growths 354
16.3 Illustrations of Hallmarks 354
16.4 The VLS Mechanism is a Special Case of the VQS Mechanism 356
16.5 The VSS Mechanism, for Low-Temperature ($T < T_E$) Growth, is a Special Case of the VQS Mechanism 357
16.5.1 Basics 357
16.5.2 Illustration with ZnO Nanowire Growths 357
16.5.3 Illustration with GaN and InN Nanowire Growths 359

16.5.4 Role of Intermediate Phases in Growths . . . 359
16.5.5 Influence of Dopants, Contaminants and Stresses on Growths . . . 360
16.5.6 Discrepancy in Growth Rates Explained . . . 362
16.6 The VSS Mechanism, for High-Temperature ($T > T_E$) Growth, is a Special Case of the VQS Mechanism . . . 364
16.7 The VS Mechanism is a Special Case of the VQS Mechanism . . . 367
16.7.1 Exceptional Roles of Surface Porosity and HETs in Growths . . . 367
16.7.2 Exceptional Role of Surface Disorder in Growths . . . 369
16.7.3 Exceptional Role of Oxygen Contaminants in Growths . . . 369
16.7.4 The Role of Contaminant Assemblages in Growths . . . 370
16.7.5 Impact of Substrate Scratching on Growths . . . 370
16.8 Nanobelt Growth is by the VQS Mechanism, not by the VS or the VLS Mechanism . . . 373
16.8.1 Preliminary Note . . . 373
16.8.2 Nanobelt Synthesis and Conflicts and Contradictions in This Synthesis . . . 373
16.8.3 Illustrative Demonstration of Nanobelts Being Growths by the VQS Mechanism . . . 374
16.8.4 The R_L Species Responsible for Nanobelt Growths . . . 380
16.9 SFLS and SoLS Mechanisms Are the Special Cases of the VQS Mechanism . . . 383
16.9.1 Catalyst-Mediated Si and Ge Nanowires Grown by the SFLS Mechanism . . . 383
16.9.2 Catalyst-Free Si and Ge Nanowires Grown by the SFLS Mechanism . . . 385
16.10 SLS Mechanism is a Special Case of the VQS Mechanism . . . 386
16.10.1 Demonstration with Si Nanowire Growths . . . 386
16.10.2 Demonstration with Indium Oxide Nanowire Growth . . . 387
16.10.3 Demonstration with Silicon Nitride Nanowire Growths . . . 387
16.10.4 Demonstration with Si Covered by a Layer of Ni . . . 387
16.10.5 Inference . . . 388
16.11 OAG Mechanism is a Special Case of the VQS Mechanism . . . 388
16.12 SCG Mechanism is a Special Case of the VQS Mechanism . . . 390

16.12.1 SCG Mechanism Based on Substrate Surface Disorder 390
16.12.2 SCG Mechanism Based on Substrate Surface Roughness 391
16.12.3 SCG Mechanism Based on Substrate Surface Porosity 391
16.12.4 Inferences 392
16.13 Boron Nitride Nanotube Growths Arc by the VQS Mechanism 392
16.13.1 BNNTs and CNTs Compared 392
16.13.2 Key Features of BNNT Growths 395
16.13.3 BNNT Growths by Using Borazine 397
16.13.4 BNNT Growths by Using B–N–O Powder 397
16.13.5 BNNT Growths by Using Amorphous Boron 398
16.13.6 Effects of Processing Parameters on BNNT Growths 398
16.13.7 Effects of Laser Irradiation on BNNT Growths 399
16.14 Conclusions 399
16.14.1 Key Conclusions 399
16.14.2 Sticking of the Source Species and of the Precursors of the Source Species 401
16.14.3 Impact of Pressure on Nanoparticle Surface 402
16.14.4 Steps Involved in Growth Kinetics 403
16.14.5 Competitive Role of Temperature and Pressure During Growths 404
References 405

17 Conclusions 413
17.1 General Conclusions 413
17.2 Concerns About the VSS and the VS Mechanisms Reaffirmed 418
17.3 Concerns About Catalyst Droplets During Growths Affirmed 419
17.4 The Key Elements of the Book 419
17.5 Design Rules and Guidelines 422
References 426

Appendix A: Effective Amorphicity of FECA Nanoparticle Surface 427

Index 435

Acronyms and Abbreviations

AES	Auger electron spectroscopy
AFM	Atomic force microscopy
BNNT	Boron nitride nanotube
CBE	Chemical beam epitaxy
CNF	Carbon nanofiber
CNT	Carbon nanotube
CVD	Chemical vapor deposition
DFT	Density functional theory
D_{nano}	The diameter of FECA nanoparticle
D_{NW}	The diameter of nanowire
EDS, EDX	Energy dispersive X-ray spectroscopy
EELS	Electron energy loss spectroscopy
EMNO	Engineered metal nano (or engineered nano metal)
ESNO	Engineered substrate nano (or engineered nano substrate)
ETEM	Environmental transmission electron microscopy
FECA	Foreign element catalytic agent
FECANO	FECA nanoparticle
FESEM	Field emission scanning electron microscopy
FTIR	Fourier transform infrared
FWCNT	Few walled carbon nanotube
FWHM	Full width at half maximum
HAADF	High-angle annular dark-field imaging
HET	High energy site
HRTEM	High-resolution transmission electron microscopy
HXO	High-temperature oxide
ITO	Indium tin oxide
L/S	Liquid/solid
LED	Light emitting diode
LEED	Low-energy electron diffraction
LXO	Low-temperature oxide

MBE	Molecular beam epitaxy
MET	Metal
METANO	Metal nanoparticle
MOCVD	Metalorganic chemical vapor deposition
MOVPE	Metalorganic vapor-phase epitaxy
MWCNT	Multiwalled carbon nanotube
NMET	Non-metal
OAG	Oxide-assisted growth
OMVPE	Organometallic vapor phase epitaxy
PL	Photoluminescence
PLD	Pulse laser deposition
PVD	Physical vapor deposition
QL/S	Quasiliquid/solid
QS/S	Qasisolid/solid
r_{as}	Radius of the R_S species
rf	Radiofrequency
RHEED	Reflection high-energy electron diffraction
RIE	Reactive ion etching
R_L	Nanoparticle material Species
RMS	Root mean square
r_{nano}	Radius of FECA nanoparticle
r_{NW}	Radius of nanowire
R_S species	Nanomaterial Source species
SAED, SAD	Selected area (electron) diffraction
sccm	Standard Cubic Centimeters per Minute
SCG	Self-catalytic growth
SECINI	**S**et of **E**nergy, **C**onfiguration (composition), **I**ntermediate phases, **N**anointeractions (all of interatomic, intermolecular, and internuclear interactions), and **I**nteratomic/intra-atomic bonding
SEM	Scanning electron microscopy
SFSS	Supercritical fluid-Solid-Solid
SFSs	The surfaces of functional substrates
SLS	Solid-Liquid-Solid
SoLS	Solution-Liquid-Solid
SoSS	Solution-Solid-Solid
STEM	Scanning tunneling electron microscopy
SUBSANO	Substrate nanoparticle, metallic or non-metallic
SWCNT	Single walled carbon nanotube
TDS	Time-domain spectroscopy
T_E	Eutectic temperature
TEELs	Temporary energy exposure limits
TEM	Transmission electron microscopy
T_M	Melting temperature
UHV	Ultrahigh vacuum
VLS	Vapor-liquid-solid

VS	Vapor-solid
VSS	Vapor-solid-solid
XPS	X-ray photospectroscopy
XRD	X-ray diffraction
α phase	Source species phase; the phase in which source species is produced
Θ_M	Dipole moment

Chapter 1
Introduction

Abstract In the introduction of the book on the kinetics, mechanism and synthesis of nanomaterials (nanocrystals), it has been noted that there can be no nanomaterial technology and no nanomaterial research if there is no nanomaterial. And there is no nanomaterial for manufacturing if there is no viable mechanism for controlled large-scale high-quality nanomaterials growths. The nanomaterials of interests are nanowires, carbon nanotubes, BN nanotubes, graphene, carbon nanofibers, nanodots, quantum dots, nanorings, semiconductor nanobelts, etc. Unfortunately, we still lack knowledge of how to better control nanomaterial growths or to adequately optimize their characteristics. So, understanding the growth mechanism is important for developing a controlled synthetic route for one-dimensional and quasi-two-dimensional structures of desired morphology, composition, shape, size, and properties. This book provides a forum to thoroughly review strengths and weaknesses, success and failure, and the potential and prospects of various growth mechanisms. The goal is also to gain deeper understanding of the fundamentals of these mechanisms, the causes of their limitations, and the conflicts and controversies haunting them, and to explore new avenues leading to broader, genuine, and, if possible, universal, appeal.

Nanomaterials (nanocrystals), and nanotechnology based on these nanomaterials [1], are fascinating, not only among scientists, engineers, fashion designers, and architects, but also among the common human beings. They can be both colloidal and non-colloidal nanomaterials. They can have great potential for practical applications. The enormous focus of nanomaterial research to find means to realize this potential has been derived from progress in their synthesis. Simply there can be no nanomaterial research if there is no nanomaterial. Although not conclusive, the said progress has already led to their narrow size distributions of just a few percent, rational shape-engineering, compositional modulation, electronic doping, and tailored surface chemistries. The immediate consequence of these has been the realization of photovoltaic and light-emitting devices competitive to those from other state-of-the-art materials. They have also been found promising for near- and mid-infrared technologies. On the fundamental side, new insights are being achieved of the nanomaterial synthesis, chemical transformations, and self-organizations. New

S. N. Mohammad, *Synthesis of Nanomaterials*, Springer Series in Materials Science 307, https://doi.org/10.1007/978-3-030-57585-4_1

phenomena are being discovered, as well, in various areas of nanomaterials. Nevertheless, the experimental advances are often poorly understood as theoretical framework to explain the mechanism of these processes is still poor. As a result, we still lack knowledge of how to better control their growth or to adequately optimize their characteristics.

To reiterate, nanomaterials possess extraordinary properties and functionalities. They have tremendous potentials in areas such as automotive, electronics, photonics, aerospace, healthcare, and biomedical applications. The products, systems, and tools made from them are smaller, better, lighter, and faster. And hence, many different kinds of nanomaterials, the synthesis routes of these nanomaterials, advanced characterization techniques of these nanomaterials, and computational models and theories of these nanomaterials are being studied. They are classified according to their morphology and composition. Based on their morphology, they may be nanowires [2–7], nanotubes [8–12], nanobelts, nanosheets, nanodots (quantum dots) [13, 14], nanofibers [15, 16], graphene [17–19], etc. (see Fig. 1.1). Nanowires, nanotubes, and nanofibers generally have circular cross section. Nanowires and nanotubes are rigid, but nanofibers are flexible. Nanotubes are hollow from inside. Nanobelts [20] and nanoribbons [21] are also solid, but have rectangular cross section, instead of circular cross section. All of them may be elemental, binary, ternary, or quaternary quasi-one-dimensional or quasi-two-dimensional materials, referred hereafter to as nanomaterials (nanocrystals). They may, for example, be carbon nanotubes (CNTs), boron nitride nanotubes (BNNTs), carbon nanofibers (CNFs), and semiconductor (Si, Ge, InP, GaAs, GaN, etc.) nanotubes. Carbon nanotubes may be single-walled carbon nanotubes (SWCNTs), double-walled carbon nanotubes (DWCNTs), few-walled carbon nanotubes (FWCNTs), and multiwalled carbon nanotubes (MWCNTs) [7–10]. Boron nitride nanotubes [11] and semiconductor nanotubes [12] may also be single-walled, double-walled, or multiwalled nanotubes. Nanodots are actually miniaturized nanowires, and nanorings are miniaturized nanotubes, both about 2–100 nm in diameter and 2–40 nm in height. Perhaps the most important forms of nanodots are quantum dots (QDs), which are tiny (generally 10 nm or less in dimension) man-made semiconductor particles. They possess optical and electronic properties quite different from those of bulk materials. And these are determined by their size and shape. If excited by light or electricity, they may emit light of specific wavelengths. If they have smaller size (e.g., radius of 2–3 nm), they may emit shorter wavelength light yielding violet, blue, or green color. In contrast, if they have larger sizes (e.g., radius of 5–6 nm), they emit longer wavelengths yielding yellow, orange, or red color.

Graphene [17–19] is a single layer of carbon atoms shown schematically in Fig. 1.2. It has a tightly bound hexagonal honeycomb lattice. It is the thinnest nanomaterial known to human beings. Only one-atomic-layer thick, it is the lightest material (about 1 m^2 in size weighing ~0.77 mg) and the strongest nanomaterial (about 100–300 times stronger than steel) with a tensile strength of 130 GPa and a Young's modulus of 1 TPa or 150,000,000 psi. It is also the best conductor of heat at room temperature. It has a conductivity of about $(4.84 \pm 0.44) \times 10^3$ to $(5.30 \pm 0.48) \times 10^3$ W m^{-1} K^{-1}. It is the best conductor of electricity ever known. It has electron mobility

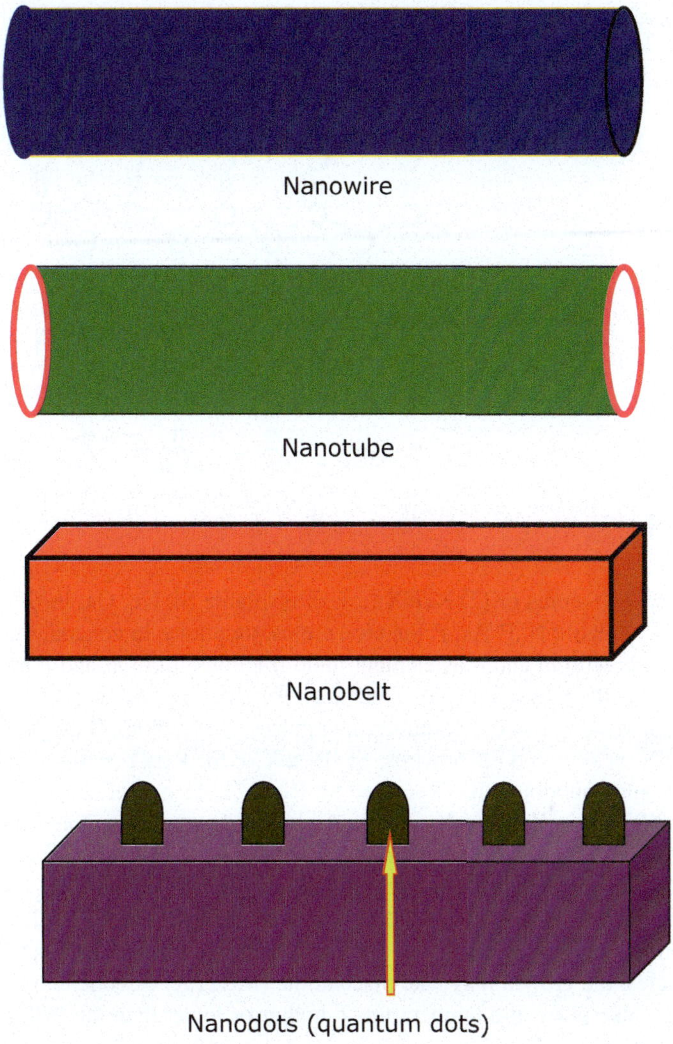

Fig. 1.1 Schematic diagrams of nanowire of circular cross section, nanotube of circular cross section, nanobelt of rectangular cross section, and nanodot of circular cross section

higher than even 200,000 $cm^2\ V^{-1}\ s^{-1}$. It is exceptionally promising for the manufacture of high-frequency electronic, biological, chemical, optical, and magnetic sensors, ultrawide bandwidth photodetectors, and energy storage and generation.

New strategies for nanomaterial synthesis are of fundamental importance for the advancement of nanoscience and nanotechnology. Following these strategies, nanomaterials are derived from their source species, in general, referred to as the R_S species. They are promising for highly downscaled, new, and multifunctional technology processing. If composed of (1) just $R_S{\equiv}X$ (e.g., nonvolatile Si, Ge, C,

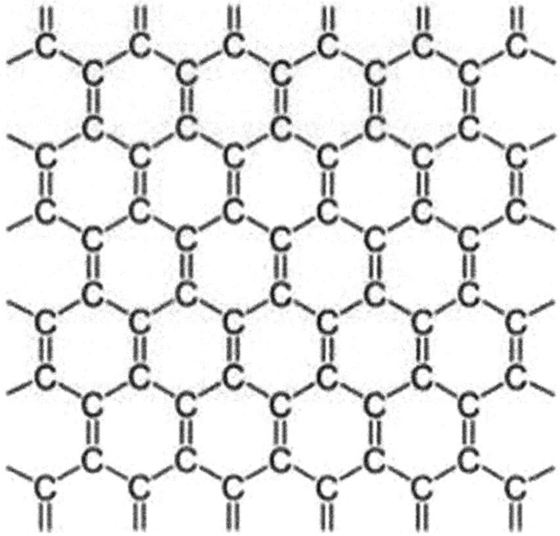

Fig. 1.2 Quasi-two-dimensional lattice structure of graphene

Ga, In, Zn, etc.) atoms or (2) $R_S{\equiv}X$ (e.g., nonvolatile Si, Ge, Ga, In, Zn, etc.) and $R_S{\equiv}Y$ (e.g., volatile N, P, As, P, Se, etc.) atoms, they may, in general, be the X_mY_n nanomaterials with m and n as element mole fractions. Both m and n are integers, though one of them may be zero. They may, for example, be carbon ($R_S{\equiv}X = C$, $R_S{\equiv}Y = 0$, $m = 1$, $n = 0$) nanotubes, nanofibers or nanodots, GaN ($R_S{\equiv}X = Ga$, $R_S{\equiv}Y = N$, $m = n = 1$) nanowires, Si ($R_S{\equiv}X = Si$, $R_S{\equiv}Y = 0$, $m = 1$, $n = 0$) nanowires and nanodots, SiC ($R_S{\equiv}X = Si$, $R_S{\equiv}Y = C$, $m = 1$, $n = 1$) nanowires, and ZnSe ($R_S{\equiv}X = Zn$, $R_S{\equiv}Y = Se$, $m = 1$, $n = 1$) nanodots. They are currently synthesized at a wide range of temperature T with or without the mediation of foreign element catalytic agents (FECAs) produced or aggregated as nanoparticles. The said nanomaterials are grown on metallic or nonmetallic FECA nanoparticles (diameter D_{nano}) composed of metal (MET) or nonmetal (NMET). If grown on metallic FECA nanoparticles, the growth may take place at the (MET, X) eutectic temperature T_E; it may also take place at a temperature T higher or lower than the (MET, X) alloy eutectic temperature T_E.

During the past years, many different mechanisms have been employed to synthesize nanomaterials (nanocrystals). The focus of attention for this study has however been the synthesis, and the mechanisms (listed in Table 1.1) for this synthesis, which produce high-quality materials, particularly the materials promising for nanotechnology developments. Most of these mechanisms are based on thermal diffusion of the source species through droplet (diffusivity D_{RS}) or on the diffusion through solid (Knudsen diffusivity D_{KND}). They involve three macroscopic states (referred to as macrostates), namely $\wp 1$, $\wp 2$, and $\wp 3$. If $\wp 1$ is the solid microstate, $\wp 2$ is the liquid microstate, $\wp 3$ is the quasiliquid or quasisolid microstate, $\wp 4$ is the solution or fluid microstate, $\wp 5$ is the supercritical fluid microstate, $\wp 6$ is the solid microstate, then for example, the VLS (vapor–liquid–solid) mechanism [22–27] involves $\wp 1$

Table 1.1 List of most well-known mechanisms for the syntheses of nanomaterials and of the $\wp 1$ (vapor), $\wp 2$ (liquid), $\wp 3$ (quasiliquid, quasisolid), $\wp 4$ (solution), $\wp 5$ (supercritical fluid), and $\wp 6$ (solid) macrostates defining these mechanisms

Group	Mechanism	Perceived phases involved in the growth mechanism	Proposed phases involved in the growth mechanism
A-1	VLS	Vapor–liquid–solid ($\wp 1$-$\wp 2$-$\wp 6$)	Vapor–quasiliquid (quasisolid)–solid ($\wp 1$-$\wp 3$-$\wp 6$)
A-2	SLS	Solid–liquid–solid ($\wp 6$-$\wp 2$-$\wp 6$)	Solid–quasiliquid (quasisolid)–solid ($\wp 6$-$\wp 3$-$\wp 6$)
A-3	SoLS	Solution–liquid–solid ($\wp 4$-$\wp 2$-$\wp 6$)	Solution–quasiliquid (quasisolid)–solid ($\wp 4$-$\wp 3$-$\wp 6$)
A-4	SFLS	Supercritical fluid–liquid–solid ($\wp 5$-$\wp 2$-$\wp 6$)	Supercritical fluid–quasiliquid (quasisolid)–solid ($\wp 5$-$\wp 3$-$\wp 6$)
A-5	OAG	Oxide vapor–solid ($\wp 1$-$\wp 6$)	Oxide vapor–quasiliquid (quasisolid)–solid ($\wp 1$-$\wp 3$-$\wp 6$)
A-6	SCG	Vapor–solid–solid ($\wp 1$-$\wp 6$-$\wp 6$)	Vapor–quasiliquid (quasisolid)–solid ($\wp 1$-$\wp 3$-$\wp 6$)
B-1	VSS	Vapor–solid–solid ($\wp 1$-$\wp 6$-$\wp 6$)	Vapor–quasiliquid (quasisolid)–solid ($\wp 1$-$\wp 3$-$\wp 6$)
B-2	SoSS	Supercritical fluid–solid–solid ($\wp 5$-$\wp 6$-$\wp 6$)	Supercritical fluid–quasiliquid (quasisolid)–solid ($\wp 5$-$\wp 3$-$\wp 6$)
B-3	SSS	Solution–solid–solid ($\wp 4$-$\wp 6$-$\wp 6$)	Solution–quasiliquid (quasisolid)–solid ($\wp 4$-$\wp 3$-$\wp 6$)
C-1	VS	Vapor–solid ($\wp 1$-$\wp 6$)	Vapor–quasiliquid (quasisolid)–solid ($\wp 1$-$\wp 3$-$\wp 6$)

(vapor), ℘2 (liquid), and ℘6 (solid) macrostates; the VQS (vapor–quasiliquid–solid and vapor–quasisolid–solid) mechanism [28] involves ℘1 (vapor), ℘3 (quasiliquid or quasisolid), and ℘6 (solid) macrostates; the VSS (vapor–solid–solid) mechanism [29–31] involves ℘1 (vapor), ℘6 (solid), and ℘6 (solid) macrostates. The VS (vapor–solid) mechanism [32, 33] involves, on the other hand, ℘1 (vapor) and ℘6 (solid) macrostates. The most well-known mechanisms are the VLS mechanism, VSS mechanism, VS (vapor–solid) mechanism, SFLS (supercritical fluid–liquid–solid) and SoLS (solution–liquid–solid) mechanisms [34, 35], SLS (solid–liquid–solid) mechanism [36], OAG (oxide-assisted growth) mechanism [37, 38], SCG (self-catalytic growth) mechanism [39–41], etc. These mechanisms expressed in terms of microstates are listed in Table 1.1. It would be interesting to note that all of these mechanisms based on the proposed microstates exhibit significant commonality.

Most of the mechanisms stated above, however, appear to suffer from the lack of understanding and also from narrow scope of applications. For example, the VLS mechanism has rarely been found to be useful for the synthesis of carbon nanotubes and carbon nanofibers. The fundamentals of the VS and VSS mechanisms for nanowire and nanotube growths are not fully understood. The OAG mechanism has often been found to yield defective nanowires. Experiments for nanowire, nanotube, and nanodot growths are performed almost under identical conditions; they make use of similar parameters. Yet it is not clear why some experiments produce nanowires, some other experiments produce nanotubes, and some other experiments produce nanodots. There is widely held idea that the R_S ($R_S \equiv X$, $R_S \equiv Y$) species undergo bulk diffusion through nanoparticle for nanomaterial growth [42]. But there are intriguing issues underlying this diffusion, which have not been resolved. A thorough examination of various mechanisms has not been performed, and the commonality of these mechanisms has not been explored. General mechanistic platform for all nanomaterials growths could not therefore be established. Bakkers and Verheijen [43] grew InP nanocrystals on Si substrate employing FECA≡Au. They found that the nanocrystals become nanowires at $T < 500$ °C, but nanotubes at $T > 500$ °C. Cao et al. [44] found that, by manipulating the H_2O content, carbon nanostructures can be tailored to produce carbon nanotubes (CNTs) of various inner diameters and also carbon nanowires (CNWs) of various diameters. The very causes of these have not been addressed. The size-dependent melting [45–49] of nanoparticles is very real; yet the impact of this melting on nanomaterial synthesis has not been carefully elucidated.

The growth mechanism is notably the general phenomenon [50, 51] dictating the realization of one-dimensional and quasi-two-dimensional nanomaterial morphology. The synthesis routes (see Fig. 1.3) are the experimentally employed chemical and/or physical processes incorporating growth mechanism for the synthesis of nanomaterial structures at the liquid/solid (L/S) interface. A novel growth mechanism should (a) explain how one-dimensional growth occurs, (b) provide a kinetic and thermodynamic rationale for this growth, and (c) dictate predictable applicability of the growth of a wide variety of nanomaterials. As stated earlier, during the past years, many different nanomaterials have been fabricated by many

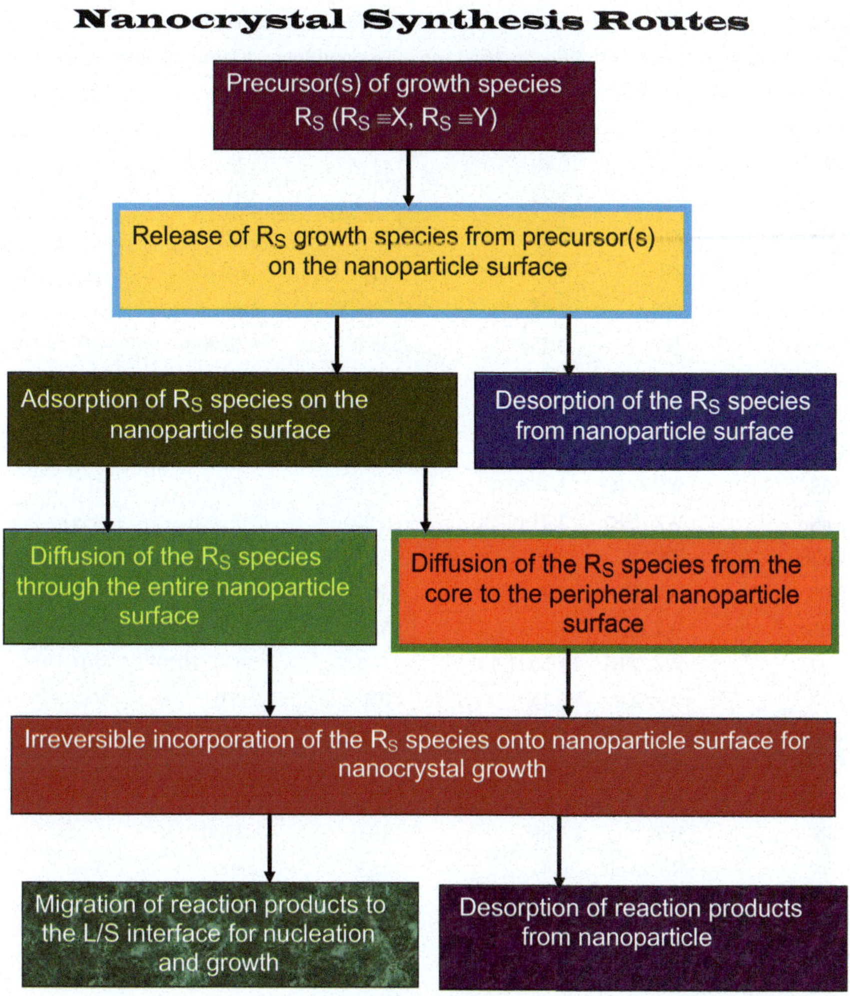

Fig. 1.3 Schematic diagram of nanomaterial (nanocrystal) synthesis route

different means. Yet our understanding of the basic processes underlying these fabrications remains premature and even probably immature. Growth of many nanomaterials has been experimentally accomplished, though without satisfactory elucidation of the underlying mechanism. Nevertheless, understanding the growth mechanism is very important, and even very vital, for developing a controlled synthetic route for quasi-one-dimensional structures of desired morphology, composition, shape, size, and properties. Our goal in this study is to thoroughly review strengths and weaknesses, success and failure, and the potential and prospects of various growth mechanisms. Most importantly, it will be to gain deeper understanding of the fundamentals of these mechanisms, the causes of their limitations, and the

Table 1.2 Room-temperature physical properties for some bulk semiconductors

Semiconductor	Crystal structure	Energy band gap (eV)	Exciton Bohr radius (nm)	Carrier mobility $cm^2/(V\ s)$	
				Electron	Hole
Si	Diamond	1.12	4.3	1750	450
Ge	Diamond	0.66	11.5	2300–3800	1820–2400
GaAs	ZB	1.43	12.4	9340	450
GaP	ZB	2.26	–	189	140
InP	ZB	1.35	11.0	6460	180
AlP	ZB	2.45	–	80	–
InAs	ZB	0.356	35.0	33,000	450
InSb	ZB	0.18	60.0	77,000	1100
AlSb	ZB	1.58	–	200–400	550
GaSb	Sphalerite	0.726	20.5	3000	1000
ZnS	WZ (ZB)	3.75 (372)	2.20	140–170	72
ZnSe	WZ (ZB)	2.71 (2.58)	3.80	(1500)	(355)
ZnTe	ZB	2.27	6.7	600	100
CdS	WZ (ZB)	250 (2.46)	30	390 (70–85)	48
CdSe	WZ (ZB)	1.751 (1.675)	5.4	900	50
CdTe	WZ (ZB)	1.50 (1.44)	(7.5)	650 (1050)	(104)
BN	WZ, ZB, sphalerite	6.10	0.8	200	500
GaN	WZ	3.39	2.3	1250	850
AlN	WZ	6.20	1.5	200	14
InN	WZ	1.89	10.0	3200	–
PbS	Rock salt	0.41	18.0	800	1000
PbSe	Rock salt	0.278	46.0	1500	1500
PbTe	Rock salt	0.23	150.0	1600	750

conflicts and controversies haunting them, and explore avenues leading to broader, if possible universal, appeal. These are all central to assessing experimental parameters dictating the shape, size, properties, and monodispersity of nanomaterials, and the ease with which the synthesis of these nanomaterials is tailored. These nanomaterials comprising semiconductors (see Table 1.2) and carbon allotropes (Table 1.3), in particular, would be the focus of our attention. The purpose would be to introduce and expose various issues pertaining to nanomaterial synthesis and the mechanisms of this synthesis to researchers in the field. To find the said mechanisms applicable to the syntheses of most, if not all nanomaterials, particularly nanowires, nanotubes, nanodots, nanofibers, nanobelts, quantum dots, and graphene, would be truly monumental. If and when found, they would vastly advance our ability to produce nanomaterials suitable for highly scaled-down and greatly promising new technology developments.

Table 1.3 List of some known allotropes of carbon

No	Allotrope	Salient characteristics
1	Diamond	Diamond is the hardest known natural mineral, which makes it an excellent abrasive. Each carbon atom in diamond is covalently bonded to four other carbon atoms in a tetrahedron
2	Graphite	It is a pure form of carbon. Unlike diamond, graphite is an electrical conductor. It is the most stable form of carbon under standard conditions
3	Amorphous carbon	It has no long-range pattern of atomic positions and no definite structure; it consists of small irregular crystallites
4	Buckyball	It is spherical fullerene—a molecule composed of 60 carbon atoms formed in the shape of a hollow ball
5	Carbon nanotube	It is cylindrical carbon molecule possessing extraordinary strength and unique electrical properties
6	Glassy carbon	It is non-graphitizing carbon. Unlike many other non-graphitizing carbons, glassy carbon is impermeable to gases and is chemically very inert
7	Graphene	It is a single-layer graphite with extraordinary electrical, thermal, and physical properties
8	Carbon nanofoam	It is made of a low-density cluster assembly of carbon atoms struck together in a loose three-dimensional web
9	Carbon onion	This is a spherical or concentric carbon shell structure of small diameter, which is generally smaller than 10 nm. It possesses high electrical conductivity

References

1. M.V. Kovalenko, L. Manna, A. Cabot, Z. Hens, D.V. Talapin, C.R. Kagan, V.I. Klimov, A.L. Rogach, P. Reiss, D.J. Milliron, P. Guyot-Sionnnest, G. Konstantatos, W.J. Parak, T. Hyeon, B.A. Korgel, C.B. Murray, W. Heiss, Prospects of nanoscience with nanocrystals. ACS Nano **9**(2), 1012–1057 (2015)
2. M. Meyyappan, M.K. Sunkara, *Inorganic Nanowires: Applications, Properties, and Characterization* (CRC Press, Boca Raton, FL, 2009)
3. J. Shi, X. Wang, Functional semiconductor nanowires via vapor deposition. J. Vac. Sci. Technol. B **29**, 060801 (2011)
4. M. Kirkham, The role of the catalysis in the growth of one-dimensional nanostructures. Doctoral thesis, Georgia Institute of Technology, Atlanta (2009)
5. J. Ramanujam, D. Shiri, A. Verma, Silicon nanowire growth and properties: a review. Mater. Express **1**, 105–126 (2011)
6. M.M. Khan, Novel applications of nanowires in different aspects of life. A short review. J. Ind. Res. Tech. **1**, 135–146 (2011)
7. B.K. Teo, X.H. Sun, Silicon-based low-dimensional nanomaterials and nanodevices. Chem. Rev. **107**, 1454–1532 (2007)
8. A. Moisala, A.G. Nasibulin, E.I. Kauppinen, The role of metal nanoparticles in the catalytic production of single-walled carbon nanotubes—a review. J. Phys. Condens. Matter **15**, S3011–S3035 (2003)
9. M. Terrones, Science and technology of the twenty first century: synthesis, properties, and applications of carbon nanotubes. Annu. Rev. Mater. Res. **33**, 419–501 (2003)

10. V. Jourdain, C. Bichara, Current understanding of the growth of carbon nanotubes in catalytic chemical vapour deposition. Carbon **58**, 2–39 (2013)
11. S. Kalay, Z. Yilmaz, O. Sen, M. Emanet, E. Kazanc, M. Çulha, Synthesis of boron nitride nanotubes and their applications. Beilstein J. Nanotechnol. **6**, 84–102 (2015)
12. Y. Sun, J. Hu, Z. Chen, Y. Bando, D. Golberg, Prospective important semiconducting nanotubes: synthesis, properties and applications. J. Mater. Chem. **19**, 7592–7605 (2009)
13. F. Wang, V.N. Richards, S.P. Shields, W.E. Buhro, Kinetics and mechanics of aggregative nanocrystal growth. Chem. Mater. **26**, 5–21 (2014)
14. A. Al-Ahmadi (ed.), *Quantum Dots—A Variety of Applications* (InTech, Rijeka, Croatia, 2012)
15. N.M. Rodriguez, A review of catalytically grown carbon nanofibers. J. Mater. Res. **8**, 3233–3250 (1993)
16. H. Song, W. Shen, Carbon nanofibers: synthesis and applications. J. Nanosci. Nanotechnol. **14**, 1799–1810 (2014)
17. M. Batzill, The surface science of grapheme: metal interfaces, CVD synthesis, nanoribbons, chemical modifications, and defects. Surf. Sci. Rep. **67**, 83–115 (2012)
18. S.N. Mohammad, VQS (vapor-quasiliquid-solid, vapor-quasisolid-solid) mechanism lays down general platform for the syntheses of graphene by chemical vapor deposition. J. Appl. Phys. **120**, 214305 (2016)
19. R. Muñoz, C. Gómez-Aleixandre, Review of CVD synthesis of graphene. Chem. Vapor. Depos. **19**, 297–322 (2013)
20. M.S. Alqahtani, N.M.A. Hadia, S.H. Mohamed, Synthesis, optical, structural, and electrical properties of single-crystalline CdS nanobelts. Appl. Phys. A **123**, 298 (2017)
21. A. Narita, Z. Chen, Q. Chen, K. Mullen, Solution and on-surface synthesis of structurally defined graphene nanoribbons as a new family of semiconductors. Chem. Sci. **10**, 964 (2019)
22. R.S. Wagner, W.C. Ellis, Vapor-liquid-solid mechanism of single crystal growth. Appl. Phys. Lett. **4**, 89 (1964)
23. S.N. Mohammad, Analysis of the vapor-liquid-solid mechanism for nanowire growth and a model for this mechanism. Nano Lett. **8**, 1532–1538 (2008)
24. D. Wang, Y.-L. Chang, Q. Wang, J. Cao, D.B. Farmer, R.G. Gordon, H. Dai, Surface chemistry and electrical properties of germanium nanowires. J. Am. Chem. Soc. **126**, 11602–11611 (2004)
25. Y. Kim, H.J. Joyce, Q. Gao, H.H. Tan, C. Jagadish, M. Paladugu, J. Zou, A.A. Suvoroval, Influence of nanowire density on the shape and optical properties of ternary InGaAs nanowires. Nano Lett. **6**, 599–604 (2006)
26. K.A. Dick, A review of nanowire growth promoted by alloys and non-alloying elements with emphasis on Au-assisted III-V nanowires. Prog. Crystal Growth Character. Mater. **54**, 138–173 (2008)
27. P.C. McIntyre, A.F. Morral, Semiconductor nanowires: to grow or not to grow? Mater. Today Nano **9**, 100058 (2020)
28. S.N. Mohammad, For nanowire growth, vapor-solid-solid (vapor-solid) mechanism is actually vapor-quasisolid-solid (vapor-quasiliquid-solid) mechanism. J. Chem. Phys. **131**, 224702 (2009)
29. Y. Tatsumi, M. Hirata, M. Shigi, Characteristics of whisker growth in amorphous silicon. Jpn. J. Appl. Phys. **18**, 2199–2206 (1979)
30. O. Lotty, S. Biswas, T. Ghoshal, C. Glynn, C. O'Dwyer, N. Petkov, M.A. Morris, J.D. Holmes, Containing the catalyst: diameter controlled Ge nanowire growth. J. Mater. Chem. C **1**, 4450 (2013)
31. J.L. Lensch-Falk, E.R. Hemesath, D.E. Perea, L.J. Lauhon, Alternative catalysts for VSS growth of silicon and germanium nanowires. J. Mater. Chem. **19**, 849–857 (2009)
32. Y.S. Won, Y.S. Kim, O. Kryliouk, T.J. Anderson, Growth mechanism of catalyst- and template-free group III-nitride nanorods. J. Crystal Growth **310**, 3735–3740 (2008)
33. S. Ambrosini, M. Fanetti, V. Grillo, A. Franciosi, S. Rubini, Vapor-liquid-solid and vapor-solid growth of self-catalyzed GaAs nanowires. AIP Adv. **1**, 042142 (2011)
34. F. Wang, A. Dong, W.E. Buhro, Solution–liquid–solid synthesis, properties, and applications of one-dimensional colloidal semiconductor nanorods and nanowires. Chem. Rev. **116**, 10888–10933 (2016)

35. T. Hanrath, Germanium nanowires: synthesis, characterization, and utilization. Doctoral thesis, University of Texas, Austin (2004)
36. J.-H. Lee, R.E. Geer, Templated Si-based nanowires via solid-liquid-solid (SLS) and vapor-liquid-solid (VLS) growth: novel growth mode, synthesis, morphology control, characteristics, and electrical transport, in *Cutting Edge Technology*, ed. by D. Vasileska (InTech, Croatia, 2010). ISBN: 978-953-7619-93-0
37. S.T. Lee, R.Q. Zhang, Y. Lifshitz, Oxide-assisted growth of silicon and related nanowires: growth mechanism, structure and properties, in *The Chemistry of Nanomaterials: Synthesis, Properties and Applications*, ed. by C.N. Rao, A. Muller, A.K. Cheetham (Wiley VCH GmbH, Weinheim, Germany, 2004)
38. S.N. Mohammad, Investigation of the oxide-assisted growth mechanism for nanowire growth and a model for this mechanism. J. Vac. Sci. Technol. B **26**, 1993–2007 (2008)
39. S.N. Mohammad, Why self-catalyzed nanowires are most suitable for large-scale hierarchical integrated designs of nanowire nanoelectronics. J. Appl. Phys. **110**, 084310 (2011)
40. S.N. Mohammad, Self-catalysis: a contamination-free, substrate-free growth mechanism for single-crystal nanowire and nanotube growth by chemical vapor deposition. J. Chem. Phys. **125**, 094705 (2006)
41. S.N. Mohammad, Self-catalytic solution for single-crystal nanowire and nanotube growth. J. Chem. Phys. **127**, 244702 (2007)
42. J.H. Lee, Z.M. Wang, E.S. Kim, N.Y. Kim, H.S. Park, G.J. Salamo, Various quantum and nanostructures by III-V droplet epitaxy on GaAs substrate. Nanoscale Res. Lett. **5**, 308–314 (2010)
43. E.P.A.M. Bakkers, M.A. Verheijen, Synthesis of InP nanotubes. J. Am. Chem. Soc. **125**, 3440–3441 (2003)
44. L.-M. Cao, Y.-S. Chen, C.-L. Yang, Y.-Q. Song, J. Yang, J.-P. Jia, Selective fabrication of carbon nanowires, carbon nanotubes, and graphene by catalytic chemical liquid deposition. Mater. Res. Bull. **55**, 229–236 (2014)
45. K. Dick, T. Dhanasekaran, Z. Zhang, D. Meisel, Size-dependent melting of silica-encapsulated gold nanoparticles. J. Am. Chem. Soc. **124**, 2312–2317 (2002)
46. P. Buffat, J.P. Borel, Size effect on the melting temperature of gold particles. Phys. Rev. A **13**, 2287–2298 (1976)
47. E. Sutter, P. Sutter, Phase diagram of nanoscale alloy particles used for vapor–liquid–solid growth of semiconductor nanowires. Nano Lett. **8**, 411–414 (2008)
48. K.K. Nanda, Size-dependent melting of nanoparticles: hundred years of thermodynamic model. Pramana **72**, 617–628 (2009)
49. F. Gao, Z. Gu, Melting temperature of metallic nanoparticles, in *Handbook of Nanoparticles*, ed. by M. Aliofkhazraei (Springer, Cham, 2016), pp. 661–690
50. S.N. Mohammad, General hypothesis for nanowire synthesis. I. Extended principles and evidential (experimental and theoretical) demonstration. J. Appl. Phys. **110**, 054311 (2011)
51. S.N. Mohammad, General hypothesis for nanowire synthesis. II: Universality. J. Appl. Phys. **110**, 054312 (2011)

Chapter 2
Nanomaterials Synthesis Routes

Abstract There are many different routes for the synthesis of one-dimensional and quasi-two-dimensional structures of nanomaterials. These include the chemical vapor deposition (CVD), the physical vapor deposition, the molecular beam epitaxy (MBE), the pulsed laser deposition, the solvothermal synthesis, the arc discharge, the sol–gel synthesis, and the chemical beam epitaxy. Salient features of all these routes have been described. The strengths and weaknesses of them have been elucidated. The differences between the chemical vapor deposition and the physical vapor deposition have been presented. Among various routes, the CVD and the MBE routes are most widely employed. Important variants of both of them have been narrated. The differences between the CVD and the MBE routes have been detailed. Various means to perform controlled synthesis employing the CVD and the MBE routes have been explained.

2.1 Introductory Comments

There are two general approaches to the synthesis of X_mY_n nanomaterials. These are bottom-up approaches and top-down approaches. The bottom-up approaches lead to the miniaturization of materials components (down to atomic level), for which there occurs self-assembly and hence the realization of nanostructures. During self-assembly, the physical forces operate at nanoscale. These forces combine basic units into larger stable structures. Typical examples are the formation of quantum dots by epitaxy and formation of nanoparticles from colloidal dispersion. The top-down approaches, on the other hand, make use of larger (macroscopic) initial structures to be processed into nanostructures, generally by applying externally controlled processes. Typical examples are etching through the mask, ball milling, and application of severe plastic deformation.

S. N. Mohammad, *Synthesis of Nanomaterials*, Springer Series in Materials Science 307, https://doi.org/10.1007/978-3-030-57585-4_2

2.2 Basics

The potential of realizing X_mY_n nanomaterials for applications in electronics, optics, medicine, sensor, solar cell, and others requires that nanomaterials be manufactured and assembled, and that they have size, shape, composition, and structure tightly controlled within a narrow distribution range. A synthesis route must be capable of realizing them together with quasi one-dimensional heterostructures through controlled doping and interfacing. Both the longitudinal heterostructures and coaxial heterostructures would be important for integrated circuit technology and for opto-electronic devices such as LEDs, laser diodes, and quantum cascade lasers. Vapor-phase synthesis is probably the most extensively explored approach for the formation of these nanostructures. Such synthesis requires that the initial starting reactants for the formation of nanostructures be the gas-phase species. It is also important that the compositions and concentrations of the gaseous reactants be judiciously chosen to ensure that the secondary nucleation events be suppressed. Many different routes have been followed to achieve this goal. These include chemical vapor deposition (CVD), physical vapor deposition (PVD), molecular beam epitaxy(MBE), pulse laser deposition (PLD) [1], plasma deposition [2], arc discharge [3], solvothermal technique [4], hydrothermal technique [5], etc. Other less known techniques are inert gas condensation, ion sputtering, ball milling, aerosol-based process, cryochemical synthesis, spray drying, etc. Well-known variants of both the CVD and MBE processes listed in Table 2.1 have been tried. Among various CVD variants, the thermal CVD [6] and the plasma-enhanced CVD (PECVD) [7] have been the most widely used synthesis routes for growths.

Table 2.1 Variants of CVD and MBE techniques for the synthesis of nanomaterials (nanocrystals, nanostructures)

No	Variants of CVD	Variants of MBE
1	Atmospheric pressure chemical vapor deposition	Chemical beam epitaxy
2	Metal-organic chemical vapor deposition	Solid-source MBE
3	Low -pressure chemical vapor deposition	Gas-source MBE
4	Laser chemical vapor deposition	Hydride MBE
5	Photochemical vapor deposition	Metal-organic MBE
6	Chemical vapor infiltration	Focused ion beam implanter MBE
7	Plasma-assisted chemical vapor deposition	Hybrid MBE
8	Plasma-enhanced chemical vapor deposition (PECVD)	–

2.3 Control Over Nanomaterials Growths

Considering their utmost importance, we emphasize and repeat that X_mY_n nanomaterials growths require control over nanomaterial nucleation. They require as well appropriate choice of the precursor(s) of the nanomaterial source species, namely the R_S ($R_S{\equiv}X$ and $R_S{\equiv}Y$) species, control over growth conditions (temperature, pressure, ambient carrier gas, etc.), and pre-planned optimization of the surface characteristics of FECA nanoparticles. The nanoparticles may be islands and/or clusters on a metallic or non-metallic substrate. They may as well be metallic nanoparticles produced on a wafer. Nanoparticles may be about 0.5–100 nm in diameter [8]. They may be called SUBSANO if they are created on substrate (wafer), but METANO if they are derived from metallic FECA [9]. Both the SUBSANOs and METANOs may have coarsened surface with pits and hillocks and sharp edges produced on them.

2.4 CVD Route for Synthesis

CVD is a process in which X_mY_n nanomaterials are deposited on a substrate or a nanoparticle surface from the vapor -phase source species $R_S{\equiv}X$ and $R_S{\equiv}Y$ [10–12]. It is a sequential process comprising seven phases as detailed [13] in Table 2.2. These source species are released by decomposing their precursors on the said surface. As the precursor gases pass over the nanoparticle (or substrate) surface at elevated growth temperature, there occurs chemical reaction that ultimately leads to the formation of solid nanomaterial on the nanoparticle (substrate) surface. Most

Table 2.2 Sequential phases of the chemical vapor deposition

Phase 1	Diffusion of gaseous reactants onto a surface. If the reactants are not available in the vapor phase, but available in the solid or liquid phase, they can be converted to vapor-phase reactants using a suitable carrier gas such as hydrogen
Phase 2	Adsorption of the reacting species onto surface sites after migration on the surface
Phase 3	Surface chemical reaction between the reactants, which may be catalyzed by the surface maintained at a certain desired temperature, at a desired pressure under the ambient of process gases, and at an appropriate gas flow rate (s); typical temperature range is 500–1200 °C
Phase 4	Desorption of the reaction by-products
Phase 5	Migration of the by-products away from the active surface area
Phase 6	Incorporation of the condensed, solid product into the microstructure of the growing nanomaterials
Phase 7	The grown nanomaterial products are bonded to the surface irrespective of it is a catalyst surface or a substrate surface. The nature of the bond between the nanomaterial product and the underlying substrate (catalyst nanoparticle) is an intrinsic property of the deposition system

often, the deposition is thermally driven. But it may be driven by photo- and plasma-assisted processes, as well. The chemical reaction playing the role in the deposition is carried out under non-equilibrium conditions. Also, the phase and morphology of the nanomaterials depend on the precursor of the source species. The growth can be conformal and large area growth with the advantage of reproducibility. The purity of the grown materials can also be very high. The temperature of the nanoparticle (substrate) surface is critical for growth as the reaction of the said surface can be different at different temperature.

Unlike techniques such as MBE, wet chemical techniques, and hydrothermal techniques, the CVD has many advantages. We cite a few of these advantages. (1) It (e.g., CVD) is the most popular technique to produce, for example, CNTs, for which thermal decomposition of precursor (s) takes place on the nanoparticle (substrate) surface. (2) The X_mY_n nanomaterials synthesis by CVD takes place at moderately high temperature, under high vacuum deposition conditions, and on rationally designed catalyst, substrate or support. The X_mY_n nanomaterials, thus produced, exhibit high crystal integrity, clean and clear facet structure, and controlled composition/impurity. (3) The vapor deposition is generally carried out in conventional thin-film deposition systems. The existing knowledge of thin-film growth is therefore utilized to understand the CVD grown nanomaterials characteristics. (4) With control over product composition, dimension, location, and morphology, CVD grown nanomaterials can be obtained on properly engineered FECA nanoparticles, and even on substrates and substrate supports. (5) The broad range of vapor deposition conditions for CVD can allow the realization of a variety of nanomaterials from numerous material sources available at our disposal. (6) The CVD provides reasonably good control of the impurity concentration and distribution in the products. It also realizes wide range in thickness of the products. (7) The CVD is a near equilibrium process with a broad choice of the reactants species, large-scale multiwafer processing, and in situ etching of the substrates. (8) CVD permits the use of almost any kind of inorganic materials as well as some kinds of organic materials. (9) The CVD process is more environment friendly than processes such as electroplating. The CVD has a few disadvantages, as well. To mention a few, (a) it has numerous control variables. (b) It makes use of many flammable, corrosive, and toxic gases. (c) It has restriction on the kind of substrates that can be coated. (d) It generates stresses in films deposited on materials with different thermal expansion coefficients.

As stated earlier, CVD is a relatively straightforward technique for X_mY_n nanomaterials growth. It is very convenient for growths as it enables the precursors of the desired X_mY_n nanomaterial to be supplied in an oxygen-free vapor form, such as germane for Ge nanowire growth and silane for Si nanowire growth. The window of temperature required for growth constitutes both the advantages and disadvantages. If used for high-temperature growth, the range of FECA materials to be used to produce the X_mY_n nanomaterials, for example, by the VLS mechanism is greater. The growth thus enjoys a greater degree of freedom in altering growth conditions such as temperature and pressure. The high-temperature growth though entails higher surface diffusion of the R_S species and higher Ostwald ripening. Larger nanoparticles resulting from Ostwald ripening will cause non-uniform diameter distributions.

If, on the other hand, used for low-temperature growth, the X_mY_n nanomaterials produced by the CVD would have uniform diameters and narrower diameter distributions. Also, doping would be more readily achieved at lower temperatures, which would allow more facile tuning of the electrical properties of the X_mY_n nanomaterials thus produced. Importantly, the low-temperature CVD growths would be more compatible with Si processing, which is desired for the realization of semiconductor nanotechnology.

2.5 MBE Route for Synthesis

Molecular beam epitaxy (MBE) is essentially an ultrahigh precision, ultraclean evaporation technique. It has a set of in situ tools, such as Auger electron spectroscopy (AES) and/or reflection high-energy electron diffraction (RHEED) for characterization of the deposited layers during growth. In other words, it is an atomic layer by atomic layer nanocrystal growth technique [14, 15]. The term "beam" simply means that the evaporated species do not interact with each other or with any other gases present in vacuum in the chamber. Thus, they reach the wafer surface without any interruption due to the long mean free paths of the beams.

It may be gas-source MBE if it uses gaseous sources for growth, metal-organic MBE if it uses metal-organic compounds for growth, chemical beam epitaxy if it uses chemical beam for growth, or solid-source MBE if it uses solid source for growth. A notable feature of this technique is that it makes use of reaction of molecular or atomic beams with a heated FECA or substrate to realize this growth. There occurs a unidirectional kinematic flow of atoms or molecules with no collisions among them during growth. Performed in an UHV environment (typically below 10^{-10} Torr), it is suitable for achieving extremely low background impurity levels in the nanomaterials thus produced. Hence, an advantage of MBE is again that under UHV conditions, the beams possess mean free paths much longer than the typical 20–30 cm source-to-substrate distance. And hence, they do not interact with each other in the gas phase. A critical issue to be watched during MBE growth is the surface conditions and surface morphology of the FECA nanoparticle or the substrate on which nanomaterials are grown. These surface conditions must be such that they promote reaction of atomic and/or molecular *beams* with each other and/or with the FECA nanoparticle (substrate) surface, relying on *kinetic processes* such as adsorption, desorption, dissociation, decomposition, migration, reaction, and incorporation. These surface conditions must also facilitate the realization of extremely high purity, highly abrupt heterostructures, precise control of composition of these heterostructures, doping, and interfaces. They must enable desired growth direction and precise lateral uniformity. All these can cause very abrupt changes in dopant profiles as it can interrupt the dopant fluxes with shutters located next to the effusion cells. It differs from CVD in that the growth rate is not dependent on the substrate orientation or the substrate temperature. This growth rate is however diameter-dependent. Unlike CVD, the MBE growth rate is comparably small, about 1–10 nm/min.

The versatility of the MBE process for X_mY_n nanomaterials growths was evident from the MBE growth of both carbon nanotubes and carbon nanowires by Liu et al. [16]. The growths of these nanotubes and nanowires were determined by taking into consideration the peculiarity in the nanoparticle crystallinities and the growth conditions. For example, the larger nanowire diameter was realized under lower pressure if the nanoparticle was a Ni nanoparticle exhibiting amorphous surface. Larger nanowire diameter was obtained also for growth on larger nanoparticle. The larger the nanowire diameter, the smaller was the growth rate. Nanowire growth at pre-defined positions on the nanoparticle (substrate) could be realized as the fluxes could be accurately controlled.

2.6 Pulsed Laser Deposition Route for Synthesis

Pulsed laser deposition (PLD) is a simple technique for the deposition of X_mY_n nanomaterials [17, 18]. For this, the surface of the target material is irradiated with a sufficiently intense pulsed laser beam yielding a vapor flux of the desired material. When collected onto a substrate (nanoparticle), it produces a X_mY_n nanomaterial (e.g., nanowire). Both the target and the substrate (nanoparticle) are placed into a vacuum chamber to ensure minimum interaction of them with the ambient atmosphere. However, the focused pulsed laser beam enters the chamber through a suitable window. Also, the substrate temperature is controlled to heat the substrate; inert gases are introduced as well and the pressure is maintained. The most notable advantage of PLD is the stoichiometric ejection of atoms from the target material. This is attributed to the fast, transient nature of the ablation process and the high laser fluence. They together create a surface temperature of approximately 5000 K within a few nanoseconds. Short pulses of the fluence levels, thus generated, are much higher than the ablation threshold. Because of this, all components of the target are ablated equally, irrespective of their binding energies. This does not happen in evaporation and sputtering, where the volatility of the species affects the atom ejection rate.

Most of the PLD systems currently employed for growth make use of ultraviolet excimer lasers emitting in the ultraviolet, such as KrF (248 nm), ArF (193 nm), and F_2 (157 nm) range. The UV radiation is most suitable for this purpose as it has very short penetration depth in materials. And only the atoms closest to the surface are ejected. Ejection of subsurface atoms by longer wavelength radiation is undesirable as these atoms come out as clusters. A notable advantage of the PLD is that an inert gas atmosphere (e.g., Ar atmosphere) prevailing at the substrate surface helps tune the energy of the species arriving at this substrate. The said inert gas atmosphere increases the scattering rate of the ejected species and slows them down. By increasing the inert gas pressure, the ejected species can be allowed to collide with each other in order to form clusters of atoms on the substrate (nanoparticle) surface. The size of these clusters can also be controlled. Pulsed laser deposition is particularly useful for the formation of quantum dots. This is evident from the GaN quantum dots growths by Goodwin et al. [19] and ZnSe and CdS quantum dots growths by Melnik et al.

[20]. Pulsed laser deposition suits also the growth of nanowires [21] and boron nitride nanotubes [22].

2.7 Solvothermal Route for Synthesis

Solvothermal route for nanomaterials synthesis is a powerful tool as it controls well the morphology of the X_mY_n nanomaterials to be grown [23]. This synthesis is performed in sealed containers at a temperature exceeding the boiling point of the solvents. It is achieved generally by the increase of pressure. The increase of both the temperature and pressure facilitates the solubility and the reactivity of the reagents. The temperature is particularly very effective in growth as it can greatly dictate the solvent properties such as viscosity, heat capacity, and dielectric constant. Also, the reagents can be more active at high temperature with the solvent solubility changing with the change in temperature. To reiterate, the solvent properties such as polarity, dielectric constants, and viscosity, are very important for the solvothermal growth. They offer high diffusivity leading to the increase in mobility of the dissolved ions and better mixing of the reagents. The interactions of solvent with reactants, intermediates, and products can duly alter the reaction equilibrium, kinetics, and products. A few good examples of the solvent are water, dimethylformamide, polyols, oleylamine, and their mixtures. They have excellent compatibility with many reactants.

We cite some examples. Zou et al. [24] discussed general hydrothermal/solvothermal route for the synthesis of semiconductor nanowires. Tang et al. [25] found that solvothermal synthesis is very important for the low-temperature synthesis of nanowires. Using polymer matrix, and appropriate pressure generated by solvothermal reactions, these nanowires could grow as long as 100 μm and be well crystallized. Tang et al. [25] noted that the solvothermal route could be suitable, for example, for the growth of CdS–CdSe core–sheath nanowires and very long CdS nanowires. Yang et al. [26] found that the solvothermal route could be suitable also for the growth of nanotubes. They produced arrays of ZnO nanotubes on zinc foil. XRD (X-ray diffraction) analysis and the SEM images indicated that the structure of the ZnO nanotubes was single crystalline with a wurtzite structure.

2.8 ARC Discharge Route for Synthesis

Arc discharge is a powerful two-electrode technique [27] for the synthesis of X_mY_n nanomaterials, particularly carbon nanotubes and carbon nanofibers. It is characterized by low voltages (in the range 10–50 V) and high currents (in the range 1–100 Amp) condition. The high current heats up the electrodes causing thermionic emission to occur. And this thermionic emission is required to produce an arc discharge. In this technique, the source material is vaporized through an arc discharge in areas between the two electrodes. The vaporization is duly followed by condensation, nucleation, and growth. This growth can be of metal oxide nanoparticles, carbon nanotubes, carbon nanofibers, fullerenes, or graphene. The primary components of the arc discharge are the said two electrodes, namely anode made of source material and cathode made of carbon or tungsten. There must be background gases; for example, gas for plasma action or chemical reaction, a gas for quenching or condensation, and a gas to source the arc current. The arc voltage drop is essentially independent of the arc current. At a given pressure, the arc voltage drop varies inversely with the molecular weight of the gas. This voltage drop however varies inversely with the gas pressure.

The solid metal electrode may serve as the source material. It may otherwise be a mixture of source materials containing, for example, carbon. The background gas may be argon or helium to produce the inert gas plasma The background gas may however be methane, oxygen, hydrogen, ethanol, or diborane to serve as reactive species. So, to reiterate, the inert gas is ionized to generate plasma; others are used as reactant species. The reactive species undergo reaction(s) under high vacuum and high-temperature environment. Although the temperature around the electrode can be very high, it drops off very rapidly due to the temperature gradient being very steep. The quenching therefore easily takes place. And due to this quenching, the evaporated source material becomes supersaturated and eventually nucleated leading to growth. There are several variants of the arc discharge methods. These include high pressure, high temperature, and submerged arc discharges.

The arc discharge deposition led originally to the growth of carbon nanotubes [28, 29]. For this, a graphitic rod acted as a negative cathode. The positive anode was placed a few millimeters away from the negative cathode. When evaporated, the arc discharge yielded carbon products, which were deposited around the chamber walls or over the cathode electrode substrate. Carbon nanotubes and carbon nanofibers could thus be synthesized in the arc discharge alternating current or direct current environment. The direct current environment yielded largely the MWCNTs. However, the metal catalysts (such as Ni, Fe, and Co) coated with pure graphite on the graphite anode, and keeping this graphite of the cathode intact, produced largely SWCNTs under essentially the same current environment. The arc discharge method could also produce BNNTs [30]. The electric discharge for this had cooled copper cathode and an h-BN rod inserted into a hollow tungsten anode. The deposition on the cathode contained BNNTs. Some arc discharge with graphite as cathode and HfB_2 as anode, or Cu as cathode and tantalum tube filled with BN powder as anode, also produced BNNTs. The BNNT yields were nevertheless low, and the BNNT purity was poor.

2.9 Differences Between CVD and PVD Techniques

PVD technique primarily uses physical means to deposit X_mY_n nanomaterials [31]. A number of steps are performed for this under high-temperature vacuum conditions. The first of them is the transformation of a solid precursor material into vapor, typically by the use of high-power electric or laser tools. The vapor atoms are then transferred to a chamber containing a substrate or a nanoparticle. Source material atoms are adsorbed on the substrate or nanoparticle yielding the solid nanomaterial (nanocrystal). PVD is particularly useful for the synthesis of extremely hard, corrosion-resistant materials. Obviously, these materials have high-temperature tolerance and superior ablation resistance. PVD is an environmentally friendly process for nanomaterials synthesis. It has nevertheless a few drawbacks including high cost. This cost emanates from intense heating and cooling required for synthesis. It is relatively slow and not ideal for the growth of materials exhibiting non-visible surfaces.

There are a number of differences between the CVD and the PVD techniques for growths. In the CVD technique, a precursor gas flows into a chamber, over a substrate or nanoparticle surface maintained at the growth temperature. Deposition of X_mY_n nanomaterial takes place on this surface due to the chemical reaction in vapor phase. The deposition process is therefore an atomistic process in which the R_S ($R_S \equiv X$ and $R_S \equiv X$) species to be deposited are atoms or molecules or a combination of both. PVD technique is different from CVD technique. Unlike the CVD technique, the PVD technique makes use of evaporation, sputtering, and other physical processes to produce vapors of materials. This technique does not therefore resort to chemical processes. The PVD technique suffers therefore from low deposition rate and low -pressure requirements. It needs additionally subsequent annealing to complete the deposition. No doubt, CVD and PVD can be used separately. There can though be benefits if they are used together to create PECVD method.

2.10 Sol-Gel Route for Synthesis

Sol-gel process is a method [32] for producing solid X_mY_n nanomaterials from small molecules. These small molecules (precursors) are converted into a colloidal solution (sol) and then into an integrated network (gel). To achieve it, several steps must be involved. These include (1) hydrolysis, (2) condensation/polymerization of monomers produced by hydrolysis, (3) growth of particles from condensed/polymerized monomers, and (4) finally the gel formation by some drying process. These steps are governed by several different experimental parameters such as pH, temperature, concentration of the reactants, and presence of additives.

It is one of the well-established synthetic approaches [33] to synthesize novel metal oxide and mixed oxide nanoparticles. This approach has control over the textural and surface properties of the materials. The formation of these metal oxides however involves consecutive steps different from the ones stated above. For example, the metal precursor(s) initially undergoes rapid hydrolysis. The metal hydroxide solution, thus produced, is then subjected to condensation. Three-dimensional gels are thus formed. The gels are next dried by a suitable drying process, and depending on the mode of drying, the products resulting from this process are converted to xerogel or aerogel.

The main advantages of the sol-gel method are (1) low processing temperatures, (2) complex shapes of the products, and (3) the absence of impurities in the end products. Appropriate choice of compounds must though be made to achieve the goal. The main disadvantages of the sol-gel method are, on the other hand, relatively long period of time for processing flow and the difficulties in phase separation. Such difficulties arise particularly in the synthesis of hybrid coating.

Traditional precursors for sol-gel coatings are alkoxysilanes such as tetraethyl orthosilicate (TEOS) and tetramethyl orthosilicate (TMOS). These are toxic compounds. To overcome this difficulty, less toxic and more environmental-friendly precursors such as precursor solutions of alkoxide (e.g., tetrabutylorthotitanate) and diethanolamine dissolved in ethanol and then mixed with water (with or without additives) in a certain suitable proportion were tried to prepare TiO_2. A precursor solution of tetraethylorthosilicate $Si(OC_2H_5)_4$, H_2O, C_2H_5OH, and HCl mixed at a certain molar ratio was similarly tried to produce SiO_2 coating. Several different compounds (inhibitors, pigments, etc.) were tried to improve the physicochemical properties of the coating thus produced. Two main techniques commonly used to apply a sol-gel coating on the surface of a metallic substrate are dip-coating and spin-coating.

There are two different sol-gel methods. These are aqueous sol-gel method and non-aqueous sol-gel method. If water is used as reaction medium to obtain the product, it is the aqueous sol-gel method. If organic solvent is used as reaction medium to obtain the product, it is, on the other hand, the non-aqueous sol-gel method. The formation of metal oxides requires oxygen. The advantage of aqueous sol-gel method is that this oxygen is readily supplied by its water solvent. The formation of metal oxides also requires metals, which are provided by precursors such asmetal acetates, metal nitrates, metal sulfates, metal chlorides, and metal alkoxides. Due to high reaction affinity of alkoxides for water, metal alkoxides are though widely used as the precursors for the production of metal oxide nanoparticles. The aqueous sol-gel method has some drawbacks, as well. For example, hydrolysis, condensation, and drying may simultaneously take place in this method. To control nanoparticle morphology and the reproducibility of the final product is then jeopardized. The problem may particularly be severe during the preparation of nanooxides, but not so much during the preparation of metal oxides.

Non-aqueous sol-gel method is free from some of the major drawbacks found in aqueous sol-gel method. We cite some examples. In non-aqueous sol-gel method, oxygen required for the formation of metal oxide is supplied not from the solvents, but from alcohols, ketones, aldehydes, etc. It is supplied as well by metal precursors. Further, the organic solvents, mentioned above, serve as versatile tools to tune some key components, such as morphology, surface properties, nanoparticle size, and composition of the final oxide materials. The non-aqueous sol-gel method, unlike the aqueous sol-gel method, is particularly useful for the production of monoxides. The metaloxide nanoparticles by the non-aqueous sol-gel method can be produced either by the surfactant-controlled approach or by the solvent-controlled approach. Among them, the surfactant-controlled approach would lead to direct transformation of metal precursor into the respective metal oxide at higher temperature. It would also provide control over the nanoparticle shape and the agglomeration of these nanoparticles.

The sol-gel route is indeed widely followed to fabricate nanodots and nanoparticles [33]. It has also been used to synthesize nanowires and nanotubes. Using zinc nitrate and urea as precursors, zinc nitrate as zinc ions source, and urea as a basic medium for hydrolysis, ZnO nanowires and nanotubes were synthesized by Wu et al. [34]. An improved sol-gel template method was employed for this purpose. X-ray diffraction and selected area electronic diffraction demonstrate that they possessed hexagonal wurtzite polycrystalline crystal structures. Silicon nanotubes were fabricated by Yoo et al. [35] through facile surface sol-gel reaction. These nanotubes showed excellent electrochemical performances when used as anodes for lithium rechargeable batteries. These nanotubes also showed good cycle stability.

2.11 Chemical Beam Epitaxy

Chemical beam epitaxy (CBE) was first reported by Tsang [36, 37]. The goal of this technique was to combine the advantages of MOCVD (metal-organic chemical vapor deposition) and MBE. It is performed in an ultrahigh vacuum system. So, the ballistic beams of molecular precursors impinge on the heated substrate surface and decompose into constituent atoms before they are incorporated into lattice sites of a growing epitaxial film. To be more specific, the precursors are thermally cracked into molecules. And these molecules obtained before transporting into the growth chamber are ballistic in nature. This implies that CBE can have a great variety of precursors chosen to provide the flexibility in both selection of reaction pathways and the implementation of process control. This is done however at the cost of getting a far more complex surface chemistry encountered in CBE than in organometallic vapor -phase epitaxy (OMVPE) and MBE. Note that both OMVPE and MBE utilize ballistic beams obtained from elemental sources. Unlike OMVPE, the CBE growth is conducted without the use of a carrier gas. Precursor gases flow under ultrahigh vacuum conditions and, as a result, precursor molecules in the gas phase rarely react or interact with each other. The chamber pressure is in the range of 1×10^{-7} to 5×10^{-4} mbar. Also, in situ analytical techniques that require an electron beam (ion beam) probe, such as reflection high-energy electron diffraction (RHEED), can be part of the CBE.

Jensen et al. [38] made use of CBE to grow InAs nanowires on Au nanoparticle deposited on InAs (111)B substrates. The nanowires exhibited high aspect ratios and high homogeneity both in length and width. Nanowire growth rate as a function of diameter, density, and time indicated ~80% of the growth due to In diffusing into the (111)B substrate surface. There was a decrease in the growth rate of nanowire length with increase in nanowire diameter. This could happen only because the supply of the R_S species for growth during the pro-nucleation stage of growth was constant. Persson et al. [39] investigated GaAs nanowire growth on (111) B-oriented substrates as a function of temperatures and source pressures. They found two factors important for nanowire growth. These are sufficiently long diffusion length of the group-III source species (a) on the two-dimensional substrate surface and (b) at the side facets. Both of them led to the rod-shaped nanowire growths. Persson et al. also found that the growth conditions suppressing growth rate on adjacent surfaces can enhance the nanowire growth. They suggested that the growth was based on surface-selective growth with a solid seed particle rather than conventional vapor–liquid–solid (VLS) growth responsible for growth.

2.12 Conclusions

Various synthesis routes for the synthesis of X_mY_n nanomaterials have been presented. Among them, variants of both CVD and MBE are widely employed for the synthesis of nanowires and carbon nanotubes. As stated in Table 2.3, they have, however, their strengths and weaknesses. The primary disadvantage of the physical and chemical methods is that they are extremely expensive, though sophisticated. Besides physical and chemical methods, mechanical methods are also employed for syntheses. These include mechanical grinding, high-energy ball milling, mechanical

Table 2.3 Comparison of the MBE and the CVD routes of synthesis

No	MBE system and MBE growth	CVD system and CVD growth
1	It can control the film thickness down to fractions of monolayers	It is less precise; it can still control the film thickness down to fractions of nanometers
2	It functions under ultrahigh vacuum condition; pressure should be ~10^{-8} Torr	It can function even at a gas pressure as low as ~1 Torr
3	It is a physical deposition process	It is a chemical deposition process
4	It is used primarily in laboratories for the process optimization	It is preferred and used for large-scale production
5	It is a more sophisticated system; it is hence more expensive for growth	It is a less sophisticated system; it is generally less expensive for growth
6	Substrate is usually maintained at lower temperature during growth	Substrate is usually maintained at higher temperature during growth
7	The growth rate is very low, typically lower than 1 μm/h	The growth rate is high, typically higher than 1 μm/h
8	The system enables to use in situ real-time characterization tools such as RHEED to monitor the behavior of growth	The system is not well equipped for in situ real-time characterization with tools such as RHEED that monitors the behavior of growth

alloying, and reactive milling. The advantages of these techniques are that they are simple and require low-cost equipment, provided that a coarse feedstock powder can be made, and the powder can be processed. Recently, biological and green synthesis techniques, that use biological systems for the synthesis of nanoparticles, have been proposed. These include nanoparticle synthesis by (a) plant extracts, (b) bacteria, (c) fungi, (d) yeast, and (e) biological particles. An advantage of the biological methods is that they may be cost-effective and free from adverse processing conditions as they allow the synthesis at physiological pH, temperature , and pressure.

References

1. D.B. Chrisey, G.K. Hubler (eds.), *Pulsed Laser Deposition of Thin Films* (Wiley, New York, 1994)
2. U.R. Kortshagen, R.M. Sankaran, R.N. Pereira, S.L. Girshick, J.J. Wu, E.S. Aydil, Nonthermal plasma synthesis of nanocrystals: fundamental principles, materials, and applications. Chem. Rev. **116**, 11061–11127 (2016)
3. S.N. Mohammad, Arc discharge and laser ablation processes for nanotube synthesis: exploration, understanding and rediscovery. Rev. Nanosci. And Nanotechnol. **2**, 309–345 (2013)
4. S.H. Feng, G.H. Li, Hydrothermal and solvothermal syntheses, in *Modern Inorganic Synthetic Chemistry*, ed. by R. Xu, W. Pang, Q. Huo, 2nd edn. (Elsevier, Amsterdam, 2017), p. 73–104
5. J.A. Darr, J. Zhang, N.M. Makwana, X. Weng, Continuous hydrothermal synthesis of inorganic nanoparticles: applications and future directions. Chem. Rev. **117**, 11125–11238 (2017)
6. H.J. Fan, P. Werner, M. Zacharias, Semiconductor nanowires : from self-organization to patterned growth. Small **2**, 700–717 (2006)
7. M.S. Bell, K.B.K. Teo, R.G. Lacerda, W.I. Milne, D.B. Hash, M. Meyyappan, Carbon nanotubes by plasma-enhanced chemical vapor deposition. Pure Appl. Chem. **78**, 1117–1125 (2006)
8. S.N. Mohammad, Systematic investigation of the growth mechanisms for the synthesis of the conventional, doped, and bamboo-shaped nanotubes, primarily the carbon nanotubes. Carbon **75**, 133–148 (2014)
9. S.N. Mohammad, Some possible rules governing the syntheses and characteristics of nanotubes, particularly carbon nanotubes. Carbon **71**, 34–46 (2014)
10. S. Sivaram, *Chemical Vapor Deposition: Thermal and Plasma Deposition of Electronic Materials* (Springer, New York, 1995)
11. K.A. Shah, B.A. Talli, Synthesis of carbon nanotubes by catalytic chemical vapor deposition: a review on carbon sources, catalysts and substrates. Mater. Sci. Semicond. Process. **41**, 67–82 (2016)
12. K.L. Choy, *Chemical Vapor Deposition (CVD): Advances, Technology and Applications* (CRC Press, Boca Raton, 2019)
13. S. Sivaram, *Chemical Vapor Deposition* (Springer Science, New York, 1995)
14. M.A. Herman, H. Sitter, *Molecular Beam Epitaxy: Fundamentals and Current Status* (Springer, Berlin, Heidenberg, Germany, 2012)
15. M. Henini (ed.), *Molecular Beam Epitaxy : From Research to Mass Production*, 2nd edn. (Elsevier, Amsterdam, 2018)
16. R.-M. Liu, J.-M. Ting, J.C.A. Huang, C.-P. Liu, Growth of carbon nanotubes and nanowires using selected catalysts. Thin Solid Films **420–421**, 145–150 (2002)
17. Y. Lin (ed.), *Advanced Nano Deposition Methods* (Wiley, New York, 2016)
18. R. Eason, *Pulsed Laser Deposition of Thin Films: Applications-Led Growth of Functional Materials* (Wiley, New York, 2006)
19. T.J. Goodwin, V.J. Leppert, S.H. Rishbud, Synthesis of gallium nitride quantum dots through reactive laser ablation. Appl. Phys. Lett. **70**, 3122 (1997)

20. N.N. Melnik, A.V. Simakin, G.A. Shafeev, V.V. Voronov, A.G. Vitukhnovsky, Formation of ZnSe and CdS quantum dots via laser ablation in liquids. Chem. Phys. Lett. **366**, 357–360 (2002)
21. A.M. Morales, C.M. Lieber, A laser ablation method for the synthesis of crystalline semiconductor nanowires. Science **279**, 208–211 (1998)
22. Z. Wang, Y. Shimizu, T. Sasaki, K. Kawaguchi, K. Kamura, N. Koshizaki, Catalyst-free fabrication of single crystalline boron nanobelts by laser ablation. Chem. Phys. Lett. **368**, 663–667 (2003)
23. S.H. Feng, G.H. Lu, Hydrothermal and solvothermal syntheses, Chap. 4, in *Modern Inorganic Synthetic Chemistry*, ed. by R. Xu, W. Pang, Q. Huo, 2nd edn. (Elsevier, Amsterdam, 2017) p. 73–104
24. G. Zou, H. Li, Y. Zhang, K. Xiong, Y. Qian, Solvothermal/hydrothermal route to semiconductor nanowires. Nanotechnology **17**, S313 (2006)
25. K.-B. Tang, Y.-T. Qian, J.-H. Zeng, X.-G. Yang, Solvothermal route to semiconductor nanowires. Adv. Mater. **15**, 448–450 (2003)
26. J.H. Yang, J.H. Zheng, H.J. Zhai, L.L. Yang, L. Liu, M. Gao, Solvothermal growth of highly oriented wurtzite-structured ZnO nanotube arrays on zinc foil. Crystal Res Technol. **44**, 619–623 (2009)
27. H. Knight, *The Arc Discharge* (Chapman and Hall, London, 1960)
28. S. Iijima, Helical microtubules of graphitic carbon. Nature **354**, 56 (1991)
29. Y. Ando, X. Zhao, Synthesis of carbon nanotubes by arc-discharge method. New Diamond Front. Carbon Technol. **16**, 123–137 (2006)
30. Y.-W. Yeh, Y. Raitses, B.E. Koel, N. Yao, Stable synthesis of few-layered boron nitride nanotubes by anodic arc discharge. Sci. Rep. **7**, 3075 (2017)
31. J.E. Mahan, *Physical Vapor Deposition of Thin Films* (Wiley-VCH, Weinheim, Germany, 2000)
32. C.J. Brinker, G.W. Scherer, *Sol-Gel Science: The Physics and Chemistry of Sol-Gel Processing* (Academic Press, New York, 1990)
33. R. Sui, P. Charpentier, Synthesis of metal oxide nanostructures by direct sol-gel chemistry in supercritical fluids. Chem. Rev. **112**, 3057–3082 (2012)
34. G.S. Wu, T. Xie, X.Y. Yuan, Y. Li, L. Yang, Y.H. Xiao, D. Zhang, Controlled synthesis of ZnO nanowires or nanotubes via sol–gel template process. Solid-St. Commun. **134**, 485–489 (2005)
35. J.-K. Yoo, J. Kim, Y.S. Jung, K. Kang, Scalable fabrication of silicon nanotubes and their application to energy storage. Adv. Mater. **24**, 5452–5456 (2012)
36. W.T. Tsang, Chemical beam epitaxy of InP and GaAs. Appl. Phys. Lett. **45**(11), 1234–1236 (1984)
37. W.T. Tsang, From chemical vapor epitaxy to chemical beam epitaxy. J. Cryst. Growth **95**, 121–131 (1989)
38. L.E. Jensen, M.T. Björk, S. Jeppesen, A.I. Persson, B.J. Ohlsson, L. Samuelson, Role of surface diffusion in chemical beam epitaxy of InAs nanowires. Nano Lett. **4**(10), 1961–1964 (2004)
39. A.I. Persson, B.J. Ohlsson, S. Jeppesen, L. Samuelson, Growth mechanisms for GaAs nanowires grown in CBE . J. Cryst. Growth **272**, 167–174 (2004)

Chapter 3
Catalyst Nanoparticles

Abstract One-dimensional and quasi-two-dimensional structures of nanomaterials are produced on substrate or support. But the nanomaterials growths are catalyzed by nanoparticles. The catalysts are mostly the foreign element catalytic agents (FECAs). Catalyst nanoparticles are therefore FECA nanoparticles (FECANOs). Various steps to form these FECANOs have been reported. Classification of them into metal nanoparticles (METANOs) and substrate (both MET and NMET) nanoparticles (SUBSANOs) has been made. The means to synthesize METANOs and SUBSANOs have been described. Four different types of nanoparticles (both METANOs and SUBSANOs) have been illustrated and explained. Materials composition of the nanoparticle surface, called the R_L species, has been defined. Various features of the R_L species have been depicted. Structural diversity, property, novelty, applicability, and importance of the bimetallic nanoparticles have been discussed. At the end, the Ostwald ripening, the nanoparticle functionalization, and the lifetime of nanoparticles have been described.

3.1 Background

Catalyst nanoparticles, also called foreign element catalytic agent (FECA) nanoparticles, as stated in previous chapter(s), are the nanometer-sized solid particles engineered at atomic or molecular scale in a way that they possess novel, superior, and unique physicochemical properties not attainable in conventional bulk solids. FECA nanoparticles can be metal nanoparticles, e.g., METANOs, or substrate nanoparticles, e.g., SUBSANOs. They are produced by several different means [1, 2], such as gas condensation, attrition, chemical precipitation, ion implantation, pyrolysis, sol-gel processes, and hydrothermal processes [3]. Colloidal FECA nanoparticles [4] may have low melting temperature T_M; they may be dispersed in a solution, or deposited on a substrate or support. They can be produced from their precursors sometimes with the assistance of the radiation chemistry [5]. There can be many different precursors. Precursors of some well-known FECA nanoparticles are listed in Table 3.1.

S. N. Mohammad, *Synthesis of Nanomaterials*, Springer Series in Materials Science 307, https://doi.org/10.1007/978-3-030-57585-4_3

Table 3.1 Precursors of some well-known FECA nanoparticles

Nanoparticle	Precursors
Fe	$Fe_2(SO_4)_3$, $Fe(NO_3)_3$, $(NH_4)Fe(SO_4)_2$, Ferrocene, $Fe(CO)_5$, Fe_2O
Mo	$(NH_4)_6Mo_7O_{24}$, $Mo(acac)_2$, $Mo(CO)_6$, $MoO_2(acac)_2$
Ni	$Ni(NO_3)_2$, $Ni(acac)_2$, [Me2NN]Ni(NO), bis(triisopropylphosphine) nickel, $Pr_3P)_2Ni(CO)_2$, $Pr_3P)_2NiCl$
Co	$Co(NO_3)_2$, $CoAl_2O_4$
Bi	$Bi[N(SiMe_3)_2]_3$, $Na[N(SiMe_3)_2]$, $(CH_3CO_2)_2Bi(C_6H_5)_3$, $Bi(C_6H_5)_3$, $(CH_3OC_6H_4)_3Bi$, $Bi(N(SiMe_3)_2)_3$
Au	AuCl(oleylamine), AuCl(octadecylamine), AuBr(oleylamine), (bis(trimethylsilyl)amido)(triethylphosphine)gold (I), $(Au(N(SiMe_3)_2)(PEt_3))$, $Au(PPh_3)(NO)_3$, Au acetate, [TOA][AuSRX], $[TOA][Au(SR)_2]$
Al	$[AlCl_3]$, [TMA, $Al(CH_3)_3$], [AIP, $Al(O^iPr)_3$], $[Al(acac)_3]$, $[Al(OCH(CH_3)CH_2CH_3)_3]$
Ag	$[RAg(PPh_3)_n$ (R = Cl, Br, or NO_3 and $n = 1$ or 3)], $NO_3Ag(PPh_3)$, $AgNO_3$
Pt	$Pt(acac)_2$, $PtCl_2$, $PtCl_4$, $H_2PtCl_6{\cdot}H_2O$, $Pt(NH_3)_4(NO_3)_2$, $Pt(NH_3)_4(Ac)_2$, $(NH_4)_2PtCl_4$, and H_2PtCl_6

Me is the abbreviation of the methyl group, and acac is the abbreviation of acetylacetone, which is an organic compound with the formula $CH_3COCH_2COCH_3$. Pr_3P is the abbreviation of bis(triisopropylphosphine); PPh_3 is the abbreviation of triphenylphosphine, $P(C_6H_5)_3$; PEt3 is the abbreviation of triethylphosphine, $P(CH_2CH_3)_3$

3.1.1 *Steps for Nanoparticle Formation*

There are several steps for nanoparticle formation. All of these steps must lead to the realization of nanoparticles of desired shape, size, composition, and morphology. The nanoparticles may be formed directly on clean substrate surface (e.g., Si, SiO_2, quartz, etc.). They may otherwise be formed on support material (e.g., Al_2O_3, MgO, etc.) deposited on the substrate. As shown in Fig. 3.1, there can thus be nanoparticles composed of many different materials. Once they are formed, the nanoparticles are placed inside a growth chamber prior to nanomaterials growth. To form METANOs, a metallic FECA film is produced directly on a substrate, or on a support material which is again produced on the substrate. Rapid thermal annealing of the metallic FECA film is subsequently carried out at some suitable temperature. The metallic FECA film is, as a result, fractured creating METANOs. To form SUBSANOs, the substrate (or the support formed on this substrate) can be surface treated. It may, for example, be bombarded by some ions (molecules), or treated with some other means [6–8] at a suitable temperature T. Alternative means include surface etching with pure water or surface scratching, for example, with a diamond blade. They also include surface treatments with plasma or with some chemicals (for example, aqua regia). Islands (clusters) thus produced on the substrate (or on the support produced on this substrate) serve as SUBSANOs. For the sake of convenience, METANO material of engineered [e.g., disturbed (disordered), coarsened, amorphous, porous] surface

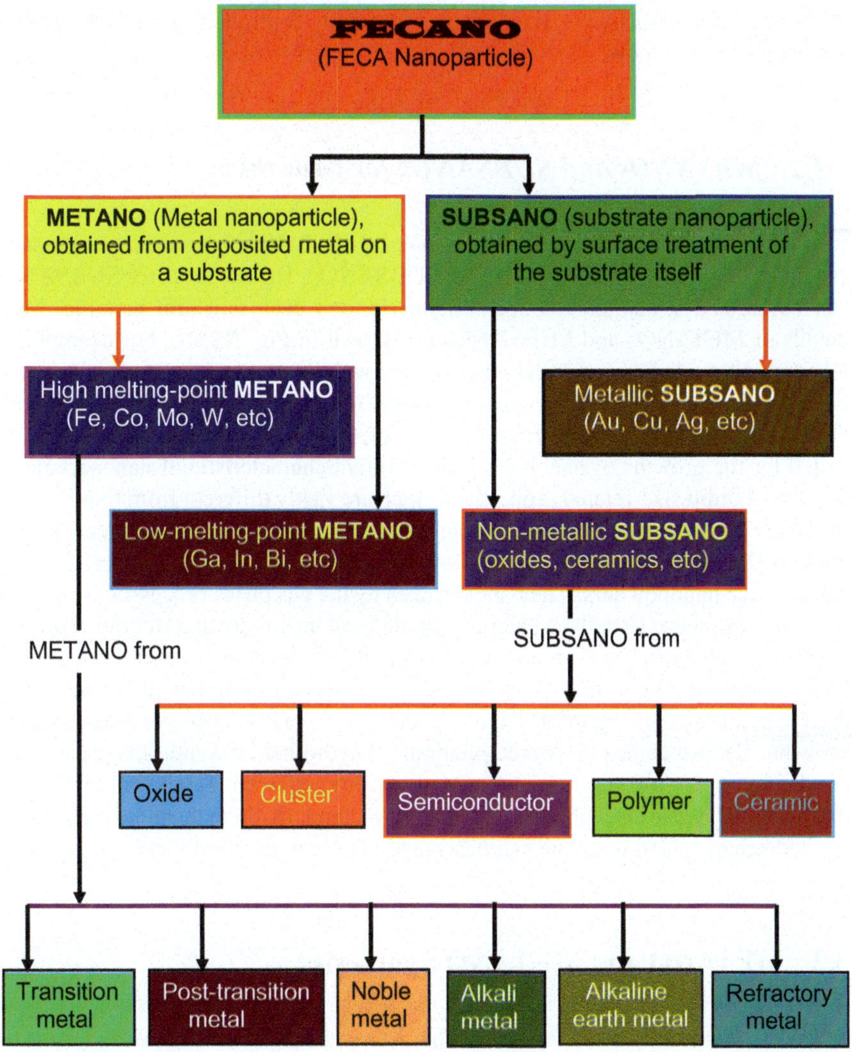

Fig. 3.1 Classification of FECA nanoparticles (FECANOs) and the list of materials classified as FECA nanoparticles

resulting from surface treatment may be called engineered metal nano (EMNO). It may be a cluster or ionic species. It may have a surface different from that of the untreated METANO. For the sake of convenience, SUBSANO material of engineered [e.g., disturbed (disordered), coarsened, amorphous, porous] surface resulting from surface treatment may similarly be called engineered substrate nano or engineered nanosubstrate (ESNO). It may very well be a cluster, a ionic species, a non-stoichiometric species, or a solid solution. It may have a surface different from that of the untreated SUBSANO. It may have judiciously controlled surface amorphicity as detailed in Appendix.

3.1.2 METANOs and SUBSANOs for Growths

Recall that metal nanoparticles on a substrate are METANOs, but substrate nanoparticles on a substrate (MET or NMET) are SUBSANOs. The substrate for SUBSANOs can be dielectric, semiconducting, polymeric, and even thin-film metallic. More details of METANOs and SUBSANOs are shown in Fig. 3.2a, b. For example, as elaborated later in Chap. 12, they can be NP1 or NP2 type nanoparticles. They can be circular if suited for the growths of nanowires, nanotubes, nanodots (quantum dots), or nanorings (quantum rings). They can, on the other hand, be rectangular if suited for the growths of nanobelts. An important characteristic of nanoparticles is that they exhibit size-related properties, which are vastly different from those of bulk materials. They are small enough to exhibit electron confinement and hence quantum effects. They can possess unexpected optical properties. When agglomerated, they can produce nanopowders. The nanoparticles by the gas phase processes are created from the gas phase. Usually, a chemical or physical means are used to utilize a vapor to produce nanoparticles via homogeneous nucleation. The nanoparticles can initially be in a liquid or solid state. During the past years, many different methods have been used to synthesize nanoparticles and nanocomposites as catalysts for nanomaterials growths. These include (1) co-precipitation, (2) hydrothermal synthesis, (3) inert gas condensation, (4) ion sputtering; (5) ion scattering, microemulsion, and microwave treatment; (6) pulse laser ablation, (7) sol-gel method, (8) sonochemical method, (9) spark discharge, (10) template synthesis, and (11) biological synthesis.

3.2 METANO and METANO Synthesis

METANOs possess a wide variety of characteristics [9]. These include (1) surface area-to-volume ratio very large as compared to that in the corresponding bulk; (2) large surface energies; (3) the transition between molecular and metallic states yielding specific electronic structure and local density of states; (4) plasmon excitation; (5) quantum confinement; (6) short-range ordering; (7) increased number of kinks; (8) a large number of low-coordination sites such as corners and edges; (9) a large number of dangling bonds, and consequently some specific and chemical properties, and also the ability to store excess electrons.

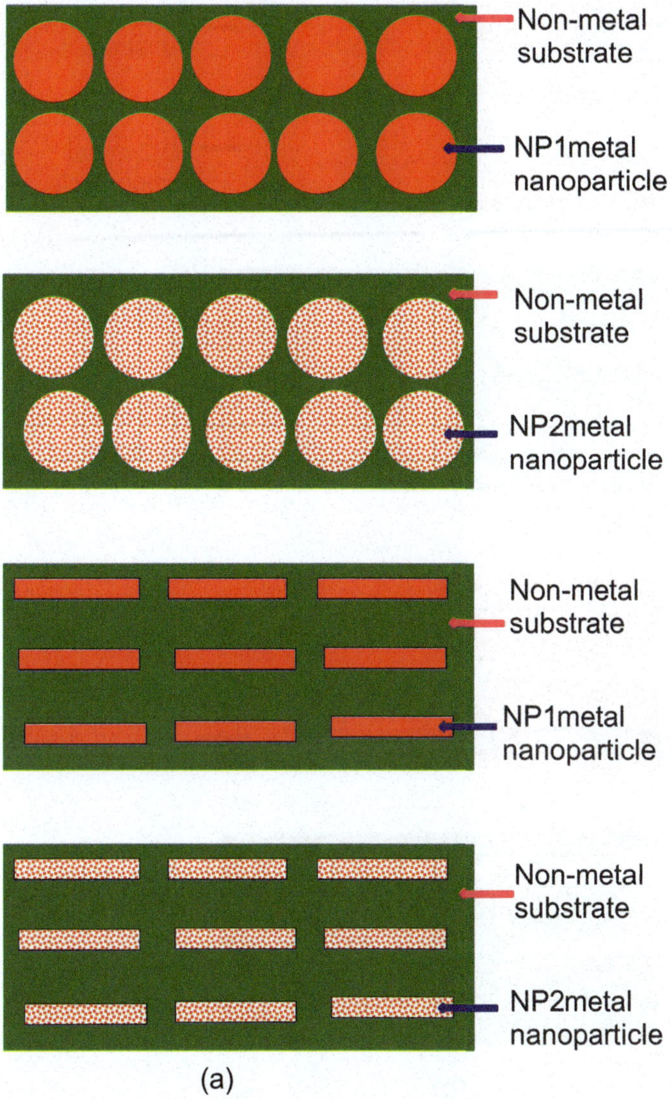

Fig. 3.2 Schematic diagrams of **a** METANOs and **b** SUBSANOs formed on substrates. Generally, circular (cylindrical) nanoparticles (METANOs and SUBSANOs) yield nanowires and nanotubes, and rectangular nanoparticles (METANOs and SUBSANOs) yield nanobelts. While red corresponds to metal, green corresponds to non-metal

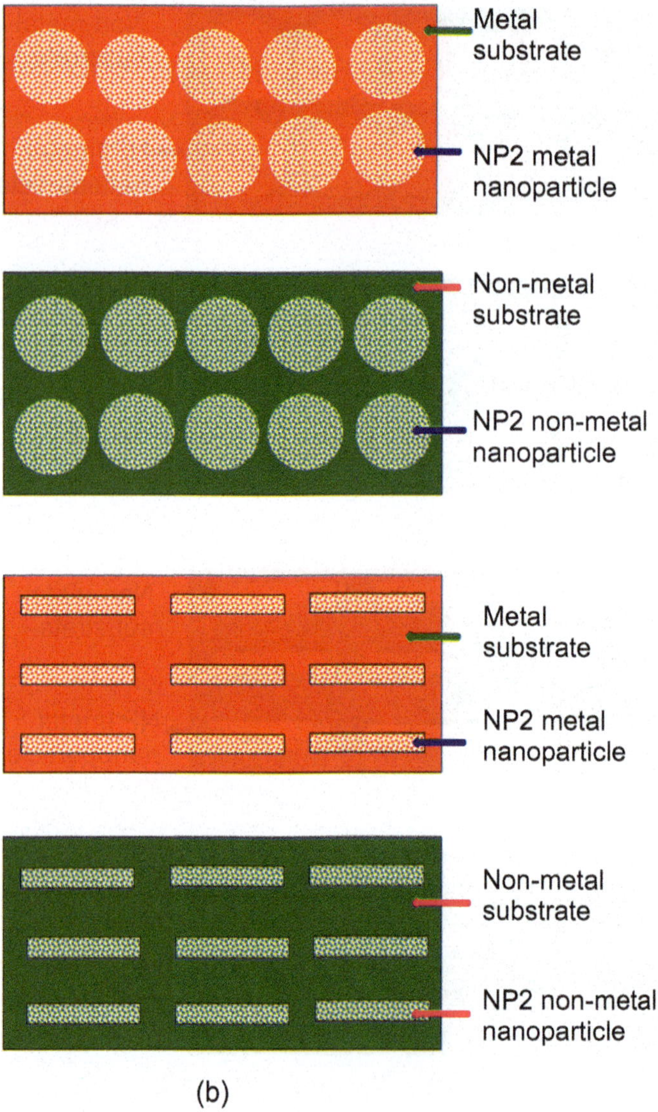

Fig. 3.2 (continued)

3.2.1 Synthesis Methods

Methods such as (1) spray pyrolysis, (2) liquid infiltration, (3) rapid solidification, (4) high-energy ball milling, (5) chemical vapor deposition, (6) physical vapor deposition, (7) sol-gel processes, and (8) colloidal processes have been employed to synthesize metal nanoparticles and nanocomposites. These metal nanoparticle and nanocomposites are synthesized by one of the two following methods: the top-down methods and the bottom-up methods. It is shown in Fig. 3.3. The crucial step of the top-down approach, as stated earlier, is the use of an external force to fracture and fragment the bulk material into smaller components. Mechanical, chemical, or some other energy sources can be used to accomplish this goal. The bottom-up approach is reverse to the top-down approach in which smaller particles (atomic or molecular species) are used to form nanoparticles, for example, by chemical reactions. While the top-down approach is the physical approach, the bottom-up approach is the chemical approach. Both of these approaches can be applied in liquid, solid, gas, supercritical fluid, or vacuum medium. The goal is always to realize a product of the

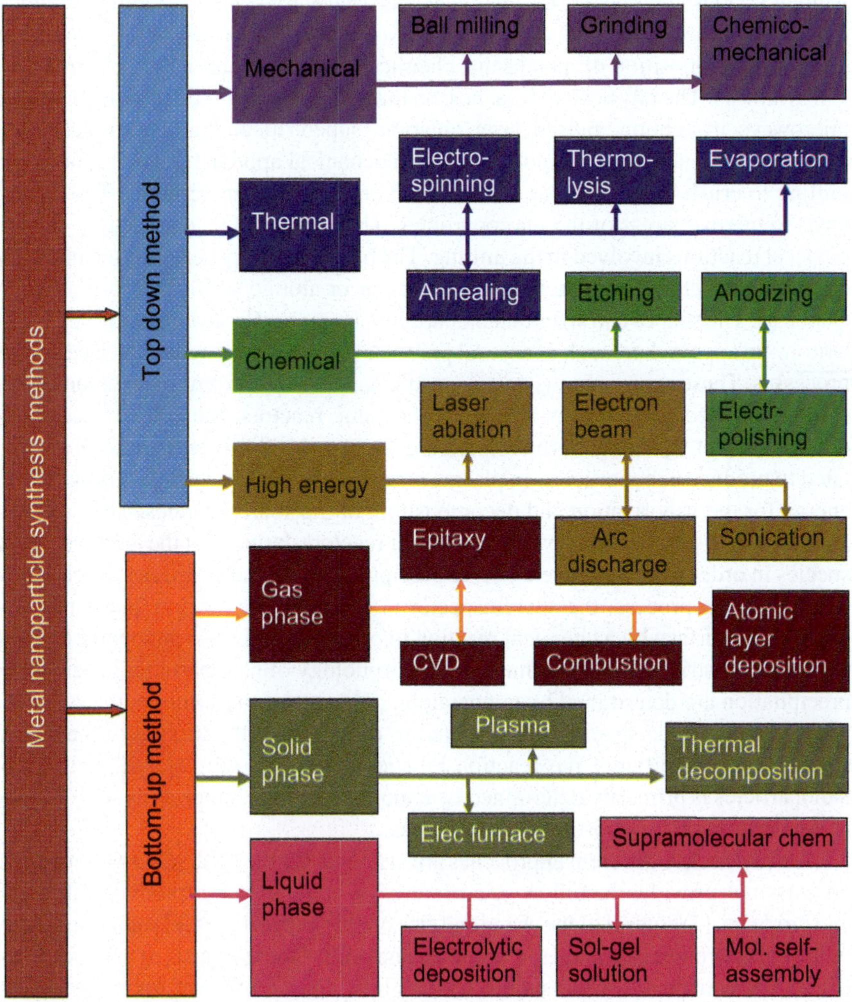

Fig. 3.3 Figurative presentation of various possible nanoparticle synthesis methods

desired shape (spheres, cubes, rectangles, octahedra, rods, wires, plates, bipyramid, etc.), size (for example, 1–100 nm), and architecture (alloy, core–shell, rattle, and dendrite, etc.). The simplicity of the approach though matters. Under experimental conditions, spherical nanoparticles of approximately 1, 3, 10, and even 20 nm sizes can be prepared reproducibly, for example, from gold, platinum, silver, palladium, etc.

3.2.2 Synthesis Approaches

Among various approaches, the chemical approaches for synthesis include chemical reduction of metal salts, the alcohol reduction, the polyol process, microemulsion, thermal decomposition of metal salts, chemical vapor deposition, and electrochemical synthesis. The physical approaches, on the other hand, include plasma treatment, microwave irradiation, pulsed laser ablation, supercritical fluidity, sonochemical reduction, and gamma-ray irradiation. The mechanical approach involves, however, milling to crush microparticles into metallic or ceramic nanoparticles. The milling leads to thermal stress of the microparticles. There can also be chemical or chemo-physical reactions involved in the milling. The bottom-up approaches often make use of the physicochemical principles of molecular or atomic self-organization for the synthesis. The shapes and sizes of nanoparticles are more effectively controlled by the bottom-up approaches, such as aerosol processes, precipitation reactions, and sol-gel processes. These approaches yield, as a result, nanoparticles from atoms or molecules exhibiting selected, complex structures. In flame reactors, the source species are decomposed at relatively high temperature (~1200–2200 °C) and are then precipitated to produce nanoparticles. In plasma reactors, plasma (ionized gas) provides the energy for the vaporization and decomposition of the source species. They are then precipitated to produce nanoparticles. In laser reactors, lasers heat the gaseous source species in order to decompose it before precipitation into nanoparticles. In chemical gas deposition process, the source species is vaporized in vacuum and condensed on a heated surface by a chemical reaction in order to yield solid nanoparticles. The shape, size distribution, crystallinity, and morphology of nanoparticles produced by precipitation are determined by temperature, pH value of the solution, the sequence in which the source species are subjected to evaporation, the mixing of the source species, and importantly the reaction kinetics. Note that the catalytic activity of nanoparticles is primarily determined by nanoparticle size. Importantly their surface areas are inversely proportional to their sizes.

Although many different approaches are available for the METANO synthesis, the one(s) which provides the following advantages is preferred: (1) reproducibility of the end products, (2) control of the shape and size of the nanoparticles, (3) monodispersity of the nanoparticles (4) easiness to conduct and realize the synthesis of nanoparticles, (5) cost effectiveness of the synthesis, (6) lower consumption of toxic precursors, but higher consumption of water and other environmentally benign solvents (e.g., ethanol) for the synthesis, (7) the use of the least number of reagents for synthesis,

(8) the use of a reaction temperature close to room temperature, (9) the use of as few synthesis steps as possible, and (10) minimal quantities of by-products and waste generated during the synthesis.

3.2.3 Catalytic Reactivity

The most important elements of METANOs is perhaps the reactivity and ability to catalyze nanomaterials growths. Notably, these are derived from their shape, size, surface, thickness, morphology, and composition. If realized from the fragmentation of thin films, the thickness of the thin film determines the thickness and the sizes of the nanoparticles. Lee et al. [10] studied the effect of Co and Ni film thickness on the sizes of nanoparticles obtained from them. They studied these nanoparticles after they had gone through the H_2 plasma pre-treatment for 10 min at 700 °C under a pressure of 1000 mTorr. The pre-treatment process acted as annealing step needed for nucleation. The variation of nanoparticle size as function of Co and Ni thin films, as observed by Lee et al. [10], is shown in Figs. 3.4 and 3.5, respectively. One can see from these figures that the nanoparticle sizes increase almost linearly with increase in the film thickness.

Atoms at the surface of nanoparticles have dangling bonds. These are unsatisfied bonds exposed at the surface and hence under the influence of an inwardly directed force. The bond distance between the surface atoms and the subsurface atoms is

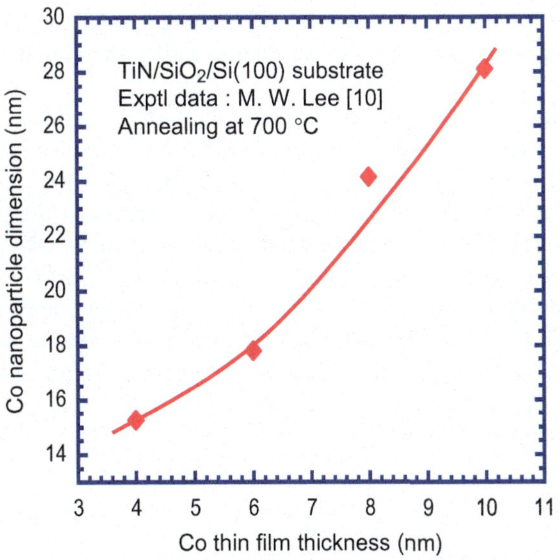

Fig. 3.4 Size of cobalt nanoparticles varied with the Co film thicknesses; Co nanoparticles were realized from Co thin film. The plots are obtained with the experimental data by Lee et el. [10]

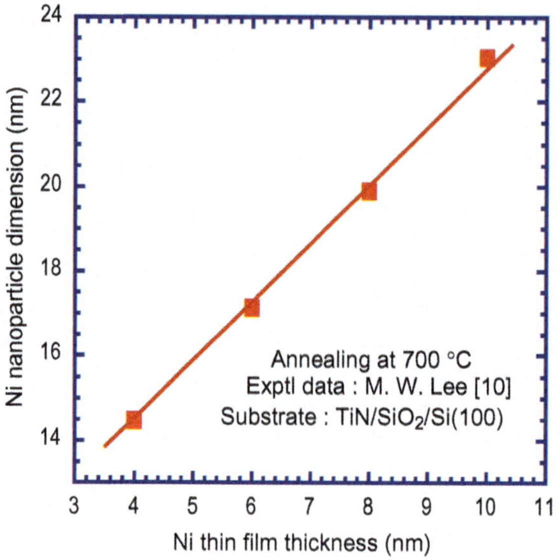

Fig. 3.5 Size of nickel nanoparticles varied with the Ni film thicknesses; Nickel nanoparticles were realized from nickel thin film. The plots are obtained with the experimental data by Lee et al. [10]

generally smaller than that between the atoms in the bulk (core). These atoms in the bulk (core) are surrounded in all directions by their neighbors. So, the electronic environment of the bulk (core) atoms is distinctly different from that of the surface atoms. The surface atoms have higher surface energy and higher tendency to react with some other atoms or molecules in order to satisfy their bonds. They are more reactive than the bulk atoms.

Geonmonond et al. [11] reported the dependence of the number of surface atoms on the total number of atoms of a nanoparticle. This dependence was obtained by Xia et al. [12] by ab initio molecular dynamics simulations. One may note from Fig. 3.6 that the number of surface atoms increases with increase in the total number of atoms of a nanoparticle. However, the smaller the nanoparticle with smaller number of total atoms, the larger is the percentage of its atoms on its surface. The percentage of surface atoms is 92, 76, 63, 52, 45, and 39 for a nanoparticle of total number of atoms of 13, 55, 147, 309, 561, and 923, respectively. Note that almost all the atoms of a nanoparticle of total number of 13 atoms are the surface atoms. As surface atoms represent the catalytic reactivity of a nanoparticle, Fig. 3.6 suggests that the smaller the size of a nanoparticle, the higher is its catalytic reactivity.

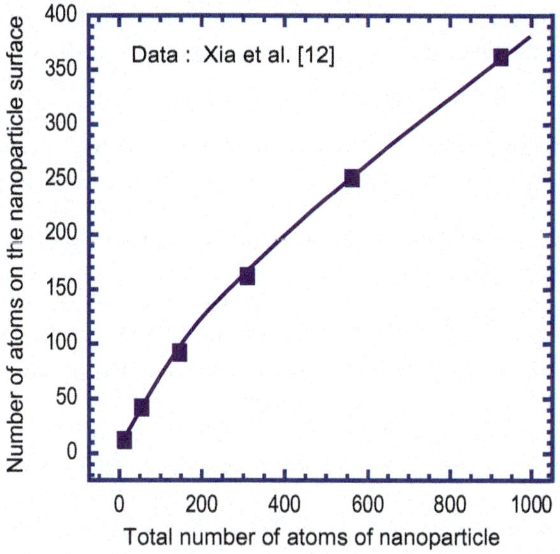

Fig. 3.6 Number of surface atoms as function of the total number of atoms of a nanoparticle. The plot is obtained with calculated data by Xia et al. [12]

3.3 SUBSANO and SUBSANO Synthesis

Recall that SUBSANOs are islands of disturbed (disordered) lattice structures formed on a substrate surface. This substrate is generally non-metallic. It can though be metallic, as well. Many different techniques and approaches have, during the past years, been employed to synthesize SUBSANOs.

3.3.1 SUBSANO Synthesis by Surface Treatment

SUBSANOs were created by RIE of SiO_2/Si substrate [13]. This RIE yielded islands of amorphous, coarsened surface lattice structures. ZnO nanowire array could be grown exclusively in the islands formed by the RIE of the SiO_2/Si substrate. An advantage of nanomaterials growth on SUBSANO surface is that it is facile and suitable for subsequent transfer-free fabrication of electronic and optoelectronic devices.

SiGe islands, Ge dots, and Ge based SUBSANOs were produced by ion implantation and thin-film deposition for the growths of silica nanowires and carbon nanotubes [14]. Ge nanostructures were the starting material for the growths of these SUBSANOs. These Ge nanostructures were produced on a thin Si buffer layer. To accommodate the stress resulting from the lattice mismatch between Si and Ge, the Si and Ge together created islands on the top of the thin wetting layer. The Ge layer also created self-assembled Ge dots. The SiGe islands had heights of 20–50 nm, but the Ge dots had diameters of 20–250 nm and heights of 10–25 nm. The wafers containing islands and dots were next implanted with carbon at a dose of 3×10^{16} cm^{-2}. The island and dot structures were transformed, as a result, into amorphous, disturbed, disordered, coarsened nanostructures, which we call SUBSANOs. Ge based SUBSANOs were produced on a 30-nm-thick SiO_2 film. They were then ion implanted at a dose of 5×10^{15} cm^{-2} and energy of 20 keV. The wafers containing Ge SUBSANOs were finally annealed in N_2 atmosphere at 600 °C for 40 min to create the final structures of the Ge SUBSANOs. A HF vapor etch removed the SiO_2 layer exposing the Ge nanocrystals formed during N_2 annealing. AFM measurements showed that the SUBSANOs had heights between 1.3 and 2.9 nm, and the mean SUBSANO density was 460 ± 30 particles/μm. They had as well amorphous, disturbed, disordered, coarsened nanostructures.

The kinetic processes by using plasma-assisted molecular beam epitaxy (PAMBE) are useful for the synthesis of thin-film substrates containing SUBSANO. Highly efficient radio-frequency (rf) plasma source can produce sufficient atomic species (e.g., atomic nitrogen for the synthesis of GaN thin films) together with a low density of ions of low kinetic energy. An advantage of this is the absence of any significant ion-induced desorption. Depending on the density of ions, SUBSANOs may be created on the substrate surface during the pre-nucleation stage of growth. They may have various shapes, sizes, density, and compositions. There may also be a time factor yielding these SUBSANOs. Calarco et al. [15] prepared GaN nanowires on SUBSANOs produced on Si(111) substrates. The nucleation process of GaN nanowires was found to depend on time. The nanowire density, as a result, increased with time and then saturated [15]. Tchernycheva et al. [16] also produced free-stranding GaN nanowires on SUBSANOs created on Si(111) substrate by plasma-assisted molecular beam epitaxy during the pre-nucleation stage of growth. The nanowires were cylindrical with a hexagonal cross section; they had diameters down to 20 nm. Mata et al. [17] investigated the growth of GaN nanowires by means of plasma-assisted molecular beam epitaxy directly on Si(111) as a function of temperature. Based on statistical analysis of the SEM images of GaN nanowires grown at different temperatures, it was concluded that density, diameter, length, and length dispersion of nanowires were strongly dependent on growth temperature [17]. They were depended as well on the growth kinetics. The growth kinetics was determined by the flux ratio of the atomic species (e.g., J_{Ga}/J_N of Ga and N, respectively, for GaN thin growth). The appropriate flux ratio (e.g., J_{Ga}/J_N) is one of the most important factors determining the growth mode and the degree of crystallinity of the SUBSANO (nanoparticle) products.

3.3.2 *SUBSANOs Generated by Stress*

The nanomaterials can be produced directly in localized regions (e.g., SUBSANOs) of a substrate without the use of any metal catalyst at moderate temperatures. The stress gradient between the substrate and a thin film of dissimilar material produced on this substrate may give rise to these regions, e.g., SUBSANOs on this substrate. This stress gradient may be generated due to the different thermal expansion coefficients of the two materials (e.g., silicon substrate and SiO_2 produced on this substrate). Such a stress gradient results, upon heating, in a tensile stress in the thin film (e.g. oxide) produced on the substrate and a compressive stress in the substrate itself. As the stresses become large at higher temperatures, there occurs fracture in various pockets at the higher temperature of growth. This is needed for relieving the compressive stress locally. SUBSANOs exhibiting disordered lattice, nanopores, pits, and hillocks are consequently generated on the thin film surface oxide surface. Flux of the substrate atoms may be released through the nanopores to the growth chamber. There is a certain temperature optimal for the generation of SUBSANOs of certain preferred dimensions and densities.

We cite some examples. The thermal expansion coefficient of SiO_2 is almost an order of magnitude smaller than that of Si. SiO_2 produced on Si thus experiences a significant tensile stress; a compressive stress is also generated in the underlying Si interfacial region. And the stress increases with increase in temperature. Prokes and Arnold [18] noted that the tensile stress in the thin SiO_2 layer increased with heating, and SUBSANOs began to form, thereby relaxing the compressive stress in these local regions. Si atom transport occurred from the compressed regions to the stress-free SUBSANOs. Accumulated Si atoms in SUBSANOs underwent nucleation yielding whiskers and/or nanowires.

Bertness et al. [19–21] analyzed the catalyzed-free spontaneous growth of GaN nanowires via MBE under a number of different growth parameters, such as relative flux of N species to Ga, type of N species present, AlN buffer layer thickness, and substrate orientation. They concluded that the nucleation of GaN nanowires was dictated by the formation of nanocolumns in the AlN buffer layer. We call them SUBSANOs. GaN nanowires were actually grown on them; these nanocolumns were grown on Si substrates. Prior to growth, a thin AlN buffer layer was produced on the Si substrate. The AlN buffer layer thickness varied from 30 to 120 nm. GaN nanowires were produced on the AlN buffer layer. Thermal expansion coefficient of Si is $2.616 \times 10^{-6}\ K^{-1}$. Thermal expansion coefficient of AlN is, however, $5.616 \times 10^{-6}\ K^{-1}$. Due to a large difference in thermal expansion coefficients of AlN and Si, AlN buffer layer produced on Si substrate experienced a significant tensile stress; a compressive stress was, on the other hand, generated in the underlying Si interfacial region. And the stress increased with increase in temperature. At a temperature of 800–850 °C, it was large enough to create fracture at various pockets of the AlN/Si interface. It was needed for relieving the compressive stress locally. SUBSANOs exhibiting disordered lattice, nanopores, and pits and hillocks were, as a result, generated on the substrate surface.

3.3.3 SUBSANOs Created by Droplets

A convenient means for catalyst-free growth of nanomaterials, for example, by MOVPE or MOCVD, is to create droplets of low-melting-point metals, such as In, and Ga on the substrate surface. Note that the low-melting-point metal droplets, with or without being alloyed with some other elements, are the SUBSANOs created on the substrate surface. Park et al. [22] utilized In droplets to produce In(As)P alloy nanowires at a growth rate of 2.4 ± 0.2 nm/s on InP(111)B substrate. They deposited liquid droplets of In (melting point 156.6 °C) on the substrate surface at 375 °C by feeding 8.4×10^{-7} mol/s of trimethylindium before MOCVD growth of nanowires. Based on SAED and EDS data, these droplets were eventually transformed into In-rich In-As droplets. These droplets were obviously molten (semimolten) solid solutions. Ji et al. [23] grew GaSb nanowires on InAs stems formed on Si(111) substrates. To achieve this growth, they preserved Ga droplets on thin InAs nanowire stems. Elemental mapping indicated that the droplets were converted eventually to Ga-rich In-Ga droplets. They had though no As or Sb atoms. EDS point analysis indicated that the droplets had 96.13 atomic % of Ga; 3.8 atomic % of In; 0.0 atomic % of As; and 0.07 atomic % of Sb.

3.4 Nanoparticle Surface Composition Called R_L Species

3.4.1 Definition

We propose that, depending on growth conditions and the environment of growth, there can be a species, called the R_L species, critical for nanomaterials growths. Formed on nanoparticle (METANO and SUBSANO) surface, this species may have many different shapes and sizes; it may as well have many different compositions. The shape and size of the nanoparticle on which it is formed may be different from its shape and size. The shape and size of the nanomaterial formed on it may though resemble the shape and size of the R_L species. In fact, the shape and size of the nanomaterial formed on it may be the same as its shape and size. It can be true for growths on both the METANO and the SUBSANO surfaces. And this R_L species may be molten, semimolten, or solid with surface-disorder (surface amorphicity)-induced charaacteristics. To emphasize and reiterate, we call this materials species the R_{LD} species or simply the R_L species. This species may specifically be expressed as $R_L \equiv (\beta_1, \beta_2, \beta_3, \beta_4$, etc.), where $\beta_1, \beta_2, \beta_3, \beta_4$, etc., are the material components of the R_L species. If composed, for example, of just β_1 and β_2, the R_L species may alternatively be expressed as $R_L \equiv (\beta_1)_z(\beta_2)_{1-z}$, where z is the mole fraction of β_1 in the R_L species. The components (constituents) of the R_L species may be the ones stated in the following:

1. Atoms, molecules, compounds, and/or ions of the bulk of the nanoparticle;
2. Atoms, molecules, compounds, and/or ions of the precursor (s) of the R_S species;
3. Atoms, molecules, compounds, and/or ions of the contaminants introduced into the chamber;
4. Atoms, molecules, compounds, and/or ions migrated from the substrate or support to the nanoparticle surface. They may be present in the R_L species during growth.

Again, the R_L species of a FECANO (both METANO and SUBSANO) may be solid, quasiliquid (quasisolid) or solid. If we define its compositional material state by ξ_m, then the solid R_L species have $\xi_m = 0$, the liquid (molten) R_L species has $\xi_m = 1$, and the quasiliquid (quasisolid) R_L species has $0 < \xi_m < 1$.

3.4.2 R_L Species Composition and Characteristics

Depending on (a) growth parameters, viz precursor pressure, growth temperature, carrier gas, etc., during growth, (b) the sticking coefficient and the temperature and pressure-dependent characteristics of the $R_S{\equiv}X$ and $R_S{\equiv}Y$ species adsorbed onto the nanoparticle surface, (c) the nature of the FECA material, and of course (d) the relative percentage of β_1, β_2, β_3, β_4, etc., present in the R_L species, there can be many different chemical compositions, lattice structure, surface morphology, surface coarsening (roughness), surface characteristics, and reactivity of the R_L species. The R_L species can, for example, be eutectic alloy, non-eutectic alloy, cluster, solid solution, or some ionic species created on the nanoparticle surface. Recall that, among them, the eutectic alloy is very distinct. It has two or more materials components and a well-defined eutectic composition at a eutectic temperature T_E, which is the lowest possible melting temperature over all of the mixing ratios of the component species. Achieving eutectic composition of an alloy at the eutectic temperature T_E is the hallmark of the VLS mechanism. The chemical compositions, lattice structure, surface morphology, surface coarsening, surface melting, dipole moment, high energy sites, reactivity, etc., of the R_L species may be determined by a well-defined **S**et of **E**nergy, **C**onfiguration (composition), **I**ntermediate phases, **N**anointeractions (all of interatomic, intraatomic, and internuclear interactions), and **I**nteratomic/intraatomic bonding, hereafter abbreviated and referred to as SECINI. This SECINI has an optimal value called SECINI0, which corresponds to the most porous state of the R_L species suitable for the uninterrupted, smoothest diffusion of the $R_S{\equiv}X$ and $R_S{\equiv}Y$ source species through this R_L species. This state can be eutectic or non-eutectic. If eutectic and porous, it has $\xi_m = 1$. If non-eutectic, it can be a porous cluster or solid solution exhibiting an effective surface amorphicity $\alpha_{amoreff} = \alpha_{amoreff0}$ (see Appendix) and $0 < \xi_m < 1$. The said composition, lattice structure, surface morphology, surface coarsening, surface melting, dipole moment, high energy sites, and surface reactivity corresponding to a SECINI are not arbitrary. They are judiciously well controlled, pre-determined, and interrelated; they all depend on the effective amorphicity $\alpha_{amoreff}$ of the R_L species surface. They together may dictate the very mechanism of the nanomaterials growths.

3.5 Types of Nanoparticles

3.5.1 Type 1 Nanoparticles

Metallocenes are organometallic compounds exhibiting a sandwich-like spatial arrangement. They consist of a transition metal (iron, titanium, zirconium, etc.) situated in between two cyclic organic compounds. Solid organometallocenes constitute type-1 nanoparticles. Good examples of these nanoparticles are ferrocene, cobaltocene, and nickelocene, which are also widely used as catalysts for CNT syntheses. Metal nanoparticles released from them in situ during growth efficiently catalyze hydrocarbon decomposition. Both MWCNTs and SWCNTs have been produced by the pyrolysis of metallocenes and phthalocyanines in a reducing atmosphere. Most importantly, pyrolysis of organometallics alone or in a mixture with hydrocarbons can yield aligned nanotube bundles that possess field-emission and hydrogen-storage properties. It has been noted that the pyrolysis of organometallics, if carried out in the presence of thiophene, can yield Y-junction nanotubes. Carbon nanotubes produced by employing organometallics can be useful for the synthesis of nanowires and inorganic compound nanotubes such as those from BN, GaN, SiC, and Si_3N_4.

3.5.2 Type 2 Nanoparticles

Nanoparticles made of metal such as iron, nickel, and cobalt constitute type-2 nanoparticles. They are most commonly used for the CVD syntheses of CNTs and CNFs [24, 25]. CNTs have been grown also by using nanoparticles ($D_{nano} > 3$ nm) of metals such as Pd, Pt, Au, Mn, W, Ti, Mg, Al, In, Na, K, and Cs. These are all monometallic nanoparticles. They may also be from transition metals, such as Fe, Ni, Co, Pd, Pt, and Ru; noble metals, such as Cu, Ag, and Au; and early transition metals, such as Mn, Cr, and Mo. But SWCNTs have been grown by using (a) monometallic nanoparticles, particularly those from transition metals and noble metals, (b) elements of the carbon family, such as diamond; (c) semiconductors such as Si, Ge, SiGe; (d) lanthanides such as Gd and Eu; and (e) other mixed compounds, such as $FeSi_2$, SiC, SiO_2, Al_2O_3, TiO_2, Er_2O_3, and ZnO.

3.5.3 *Type 3 Nanoparticles*

Catalytic potential of monometallic nanoparticles is restricted by the lack of flexibility of their properties. This is overcome by modifying these properties through the construction of bimetallic nanoparticles, which consist of two distinct metals. Polymetallic nanoparticles have also been used to serve the same purpose. These manoparticles, so far realized, have fascinating properties, such as enhanced plasmonic properties, for example, in Au-based alloy nanoparticles, enhanced magnetic properties in magnetic materials, or enhanced optical and magnetic properties in multifunctional nanoparticles. They can also have enhanced stability and dispersion. Interestingly, they can have selective biological application, as well. Bimetallic nanoparticles containing binary mixtures of active catalysts such as Ni, Fe, and Co have lately been found to display a higher activity than those from individual FECA monometallic elements. Nanoparticles from a mixture of inactive and weakly active elements, such as Mo and Mg with Fe or Co have particularly been found to be superior to these individual FECA monometallic elements.

3.5.4 *Type 4 Nanoparticles*

Nanoparticles, such as those of In, Bi, and Sn, for low-temperature solution-based nanowire growths by the SFLS (supercritical fluid–liquid–solid) and SoLS (solution–liquid–solid) mechanisms, have been achieved by thermolysis of precursors. These precursors, for which Me≡methyl, include $Bi[N(SiMe_3)_2]_3$, $Sn[N(SiMe_3)_2]_3$, or $In[N\text{-}(SiMe_3)_2]_3$ in poly(styrene$_{0.86}$-co-vinylpyrrolidinone$_{0.14}$) or poly(vinylpyrrolidinone) solutions. Among them, the Bi nanoparticles have emerged to be most promising for the solution-based growth of a wide variety of nanowires. More recently, other high-quality, narrowly dispersed Bi nanoparticles have been realized for the SoLS growth of high-quality semiconductor nanowires. In situ generations of In nanoparticles were obtained at 340 °C by reducing $InCl_3$ with $NaBH_4$ in a mixture of tri-*n*-octylphosphine (TOP) and tri-*n*-octylphosphine oxide (TOPO). Sn nanoparticles of an average diameter of 8 nm were obtained by thermal decomposition of $Sn[N(SiMe_3)_2]_2$ at 180 °C in diphenyl ether solutions of poly(1-hexadecene)$_{0.67}$-co-(1-vinylpyrrolidinone)$_{0.33}$. Ga nanoparticles were synthesized from highly monodisperse Ga nanoparticles by injecting $Ga_2(NMe_2)_6$ into a hot mixture of 1-octadecene (ODE) and di-*n*-octylamine (DOA) at a temperature of 240–310 °C.

Nanoparticles from Au have primarily been used for the synthesis of nanowires, such as those of Si, Ge, GaAs, InP, ZnSe, and CdTe [26–28]. Au, together with Ag and Al, has been found useful particularly for the growths of Si and Ge nanowires. An interesting feature of the binary phase diagrams, for example, of (Au, Si), (Au, Ge), (Al, Si), and (Ag, Si) alloys are that they are simple eutectic. Investigations on nanoparticles from metals such as Ag, Al, Bi, Cd, Co, Cu, Dy, Fe, Ga, Gd, In,

Mg, Mn, Ni, Pb, Pd, Pr, Pt, Ru, Sn, Te, Ti, and Zn suggest that some of them may effectively replace the nanoparticle of Au for various nanowire syntheses [26].

3.6 Effects of Surface, Interface, Size, and Density of Nanoparticles

Importantly, surface and interface effects greatly impact the chemical stability and catalytic activity of nanoparticles. The chemical stability of nanoparticle is generally modified by interaction with the support or substrate on which it is formed. We cite an example. An XPS analysis by Lu [29] indicated that under identical reducing conditions, Fe (III) nanoparticles formed on silicon oxide were reduced to Fe (0). It was not, however, successful if Fe (III) was formed on silicon nitride. The metal nanoparticles have different interactions with different oxide substrates. For example, Pd nanoparticles formed on propylamine-functionalized support have enhanced catalytic activity, but these nanoparticles on ethylenediamine and diethylenetriamine supports have reduced catalytic activity [30]. The nanoparticle size and density dictate the diameter and density of nanomaterials formed on this nanoparticle.

High density of nanoparticles is required for achieving high density of nanomaterials. Note that, the higher the density, the smaller is the dimension of the nanoparticles formed on a substrate. Because of this, the nanoparticle dimensions have a limit, for example, for the VLS growth. For this growth, very small nanoparticles may not lead to any nucleation. The exact composition of the R_L species on the nanoparticle surface is very important for nanomaterials growth. In fact, surface functionalization is critical to producing nanomaterials of specific morphology and desired properties.

3.7 Bimetallic Nanoparticles

3.7.1 Structural Diversity of BNPs

Unlike monometallic nanoparticles, bimetallic nanoparticles (BNPs) have been found to exhibit synergistic effects [31–35]. In fact, BNPs can exhibit not only the combination of the properties related to the two component elements, but also new properties due to synergy between them. They possess unique size-dependent optical, electronic, thermal, and catalytic properties. They may, in general, be $\mathrm{METANO}\equiv(\mathrm{MET1})_{1-z}(\mathrm{MET2})_z$, where MET1 and MET2 are the component FECA metal elements, and z is the MET2 mol fraction of the BNPs. Depending on the structure and composition, BNPs can be broadly classified as mixed structured BNPs and segregated structured BNPs. Based on atomic ordering, they may though be alloyed, intermetallic, subclustered, and core–shell BNPs. The atomic ordering of the mixed structured BNPs can be random or ordered. The mixed structure with random arrangement is the alloyed structure, but the mixed structure with ordered arrangement is the intermetallic structure. Two metals are randomly mixed in an alloyed BNP, but two metals are orderly mixed in an intermetallic BNP. One metal dissolved in another metal in solid state creates a solid solution. Solid solution BNPs can be substitutional BNPs or interstitial BNPs. They are substitutional BNPs if the size of the two metal atoms is comparable; they are otherwise interstitial BNPs. The smaller atoms position themselves between the larger atoms to generate interstitial BNPs. While the segregated BNPs are composed of two separate components with a shared interface, the core–shell BNPs are composed of core of one type of metal surrounded with shell of another type of metal. Many different techniques have been employed for the synthesis of BNPs. These include electrochemical co-reduction, chemical reduction, thermal decomposition, radiolytic synthesis, sputtering, sol-gel synthesis, microemulsion, hydrothermal synthesis, and sonochemical synthesis. The preparation conditions govern the miscibility of the two metal components and hence the consequent structure of BNP. The component metal elements and the size of these elements determine, on the other hand, the BNP properties. The nature of the atomic ordering in the structures, the atom mole fractions, and the compositions, all play an important role in the overall catalytic properties of the BNPs. Several factors that influence the structure of BNPs are the relative strengths of metal-metal bonds, the surface energy of the bulk materials, relative atomic sizes, charge transfer, and specific electronic and magnetic effects.

3.7.2 BNP Properties Different from the Corresponding Bulk

That nanoparticle can have chemical and physical properties different from those of the corresponding bulk of identical composition is quite evident from BNPs. We cite an example. Bimetallic Pt_zRu_{1-z}, $(z = 0.166)$ nanoparticles, 1.5 nm in diameter and synthesized from molecular precursors, assumed a face-centered cubic (fcc) closest-packed structure, but the bulk alloy of the same composition had a hexagonal-closest packed (hcp) structure. It happened probably because a large fraction of atoms of BNPs reside on the surface, and surface atoms contribute significantly to the increment of Gibbs free energy. Also a large fraction of atoms on the surface and at interfaces of BNPs affect their phase behaviors. Hence, it is doubtful if the bulk phase diagrams for bimetallic bulk materials can be extended to the nanometer-sized regime. Surface segregation of BNPs is quite common. This surface segregation becomes easier when the differences in atomic sizes, surface energies, and strain energies of the two component elements of a BNP are large. Note that surface segregation of one or both of the two elements can cause reconstruction of the BNP, for example, to core–shell, sandwich and onion-like structures.

3.7.3 BNP Based R_L Species

R_L≡[(MET1)$_{1-z}$(MET2)$_z$, X] species, formed from BNP≡(MET1)$_{1-z}$(MET2)$_z$ and R_S≡X, is solid at temperature T smaller than the (BNP, X) eutectic temperature T_E, but liquid at temperature $T = T_E$. It may be liquid also at $T > T_E$. The eutectic temperature T_E, for instance, of the R_L≡(Ag_2Au, Si) alloy is 710 °C. Chou et al. [36] could use Ag_2Au to efficiently grow Si nanowires at a temperature $T < 710$ °C.

BNP properties generally resemble those of the component metal elements [37] of the BNP. Connell et al. [38] observed that Ge nanowire growth rate G_R was lower with MET≡Cu, but higher with MET≡Au. Depending on the mole fraction, z, this growth rate with $Au_{1-z}Cu_z$ BNPs was intermediate to those with Cu and Au nanoparticles under identical growth conditions. Mukhopadhyay et al. [37, 39] demonstrated that BNPs promote higher yields than pure monometallic nanoparticles even at lower temperatures. Sato et al. [40] studied the possible role of Ti nanoparticles and $Ti_{1-z}Co_z$ BNPs in the CNT synthesis. They observed that $Ti_{1-z}Co_z$ alloy incorporates more carbon than Co element in the R_L≡(BNP, C) species. And due to increased heat of reaction with carbon, $Ti_{1-z}Co_z$ was more efficient than Co in yielding CNTs. They noted that the melting temperature T_M of the R_L≡($Ti_{1-z}Co_z$, C) alloy (solid solution) was generally lower than that of the R_L≡(Co, C) alloy (solid solution). Also, there was increase in CNT growth rate with increase in z of the $Ti_{1-z}Co_z$ BNP, and that this growth rate reached a peak and then decreased subsequently with further increase in z. Biswas et al. [41] obtained descent Ge nanowire growth rate with Au nanoparticle. This growth rate was though increased fivefold by using $Au_{0.80}Ag_{0.20}$ nanoparticle and twofold by using $Au_{0.90}Ag_{0.10}$ nanoparticle. These happened even at temperatures $T < T_E$.

3.8 Ostwald Ripening of Nanoparticles

The atoms/molecules of a nanoparticle are less tightly bound to the nanoparticle at the surface than in the bulk. The smaller the nanoparticle, the weaker is the atom–atom and atom–molecule bonds at the surface. In contrast, the larger the nanoparticle, the stronger is the atom–atom and atom–molecule bonds at the surface. As a consequence, atoms and molecules of smaller nanoparticle dissolve more easily than those of larger nanoparticle. The smaller nanoparticles, as a result, eventually disappear, while the larger nanoparticles become still larger. Generally, the R_L≡(MET, X) species is a droplet at $T = T_E$ for VLS growth. Unless adversely affected by growth condition, this droplet does not change in volume and size during nanomaterial growths. The diameter of nanomaterial grown by this droplet may hence be as large as that of the droplet. It is at least determined by the diameter of droplet. Depending on (1) the conditions of growth, (2) FECA metal, and (3) the R_S≡X species, the FECA metal may, however, diffuse out of the R_L≡(MET, X) droplet during nanomaterial growth. There is van der Waals attraction among various metal elements existing in a medium. Due to this attraction, metal elements as well may diffuse out of smaller droplets, but diffuse in larger droplets [42]. The driving force for this is the difference in chemical potentials of the droplets, which is caused by the difference in their curvatures. Smaller droplets may thus disappear eventually during growth leading to an increase in the average size of the remaining droplets. There may hence be variability in nanomaterial diameter and nanomaterial growth rate [43].

Experimental observations by Hannon et al. [44] indicate that Au is highly mobile on Si surface at a temperature $T \approx 600$ °C, and that the $R_L \equiv$(Au, Si) droplets on this surface undergo Ostwald ripening (coarsening). Based on discussions made above, while the smaller ones of them disappear, a few larger ones of them are enlarged in volume. An immediate consequence of it may be tapering, and in some cases, larger diameter of nanowires grown from them. Au-catalyzed nanowire growths at lower temperature and higher pressure may, however, have lower impact of Au migration. The migration of Au may even be thwarted by high enough pressure.

Kodambaka et al. [45] found that the Au migration may be marginalized even at relatively high temperatures by adding oxygen during growth. Hofmann et al. [46] suggested that there may be reshaping of the nanoparticle due to dynamic variation of the composition of the R_L species during growth. They may have high tendency to coalesce. There may also be Ostwald ripening like process driven by interfacial energy minimization. Such a process can prompt, for example, crystalline Si to dissolve back into the $R_L \equiv$(MET, X) alloy droplet.

3.9 Functionalization of FECA Nanoparticles

As discussed earlier, nanoparticles can be synthesized by many different techniques. The main problem with nanoparticles, as also indicated earlier, is their tendency to undergo agglomeration which leads to an increase in size and decrease in the energy associated with their large surface areas. Many different strategies have been employed in the past to circumvent this problem. Using these strategies, nanoparticles have, for example, been functionalized with a variety of ligands such as small molecules, surfactants, dendrimers, polymers, and biomolecules. Noticeably, nanoparticles conjugated with biomolecules can impart desired properties such as specific recognition or biocompatibility.

If functionalized, metal nanoparticles are generally very stable in solution. In fact, functionalized metal nanoparticles are far more stable than non-functionalized metal nanoparticles in solution. When placed in liquid solutions, the metal nanoparticles come close to each other and to one another. They are then affected by van der Waals forces. In the absence of agents opposing these forces, the nanoparticles aggregate. And the immediate consequence of this is the decrease in their catalytic activities. This is a serious problem, which is overcome by organometallic functionalization. Ligands are coordinated to the metal center to prevent aggregation by this functionalization. Some of these ligands can even alter the properties and sizes of the nanoparticles. Nanoparticles can also be functionalized with polymers or oligomers, which provide a protective shell layer in order to prevent them from interacting with the neighboring nanoparticles and thus sterically stabilize these nanoparticles.

3.10 Lifetime of Feca Metal Nanoparticles

Depending on growth conditions, catalyst used for nanomaterials growth may decay during growth. This decay during CNT growth may depend on many parameters including the pressure of the gaseous carbon precursor and the growth temperature T. For example, as noted by Futuba et al. [47], an increase in the pressure of C_2H_4 precursor during Fe-catalyzed and water-assisted thermal CVD of SWCNTs accompanied a gradual decrease in catalyst lifetime. Another investigation by Chen et al. [48] on the catalyst lifetime evolution as function of changing precursor pressure attributed the catalyst decay to carbon feedstock. This was substantiated by a decrease and then increase in Fe-catalyst lifetime due to increase in the pressure of precursors such as C_2H_4 and C_4H_{10}. Picher et al. [49] made use of another precursor, C_2H_5OH, to show that the lifetime of both Ni and Co nanoparticles decreased with increase in precursor pressure. This lifetime decreased also with increase in temperature. Correlation between CNT growth rate and catalyst lifetime was also studied. It was found by Vinten et al. [50] that an increase in catalyst lifetime led to a decrease in CNT growth rate. In fact, it was found to be true also in elaborate investigations by Chen et al. [48]. These researchers studied the growth rate and lifetime of over 300 SWCNTs synthesized by thermal CVD making use of carbon precursors (such as C_2H_2, C_2H_4, C_4H_{10}, and C_3H_8), various carbon concentrations, and various growth temperatures. Kodambaka et al. [45] focused on nanowire growth, rather than on nanotube growth, for similar investigation. They demonstrated that the exposure to oxygen, even at low levels, during Au-catalyzed Si nanowire growths using silane, reduces the migration of Au away from the catalyst droplets. The droplet volume therefore remained unaltered for longer periods of time permitting the growth of untapered nanowires.

References

1. B.R. Cuenya, Synthesis and catalytic properties of metal nanoparticles: size, shape, support, composition, and oxidation state effects. Thin Solid Films **518**, 3227–3250 (2010)
2. V. Johanek, M. Laurin, A.W. Grant, B. Kasemo, C.R. Henry, J. Libuda, Fluctuations and bistabilities on catalyst nanoparticles. Science **304**, 1639–1644 (2004)
3. J.A. Darr, J. Zhang, N.M. Makwana, X. Weng, Continuous hydrothermal synthesis of inorganic nanoparticles: applications and future directions. Chem. Rev. **117**, 11125–11238 (2017)
4. J. Chen, H. Jahaveri, B. Sulaiman, Y. Dahman, Synthesis, characterization and applications of nanoparticles, Chap. 1, in *Fabrication and Self-assembly of Nanobiomaterials*, ed. by A.M. Grumezescu (Elsevier, Amsterdam, 2018)

5. S.S. Joshi, S.F. Patil, S. Mahumuni, Radiation induced synthesis and characterization of copper nanoparticles. Nanostruct. Mater. **10**, 1135–1144 (1998)
6. M. Meyyappan, A review of plasma enhanced chemical vapor deposition of carbon nanotubes. J. Phys. D Appl. Phys. **42**, 213001 (2009)
7. J.-H. Lin, Z.-Y. Zeng, Metal-catalyst-free growth of carbon nanotubes/carbon nanofibers on carbon blacks using chemical vapor deposition. RSC Adv. **4**, 40251–40258 (2014)
8. L.L. Tang, W.-J. Ong, S.-P. Chai, A.R. Mohamed, Growth of carbon nanotubes over non-metallic based catalysts: a review on the recent developments. Catal. Today **217**, 1–12 (2013)
9. T.S. Rodrigues, A.G.M. da Silva, P.H.C. Camargo, Nanocatalysis by noble metal nanoparticles: controlled synthesis for the optimization and understanding of activities. Mater. Chem. A **7**, 5857–5874 (2019)
10. M.W. Lee, M.A.S.M. Haniff, A.S. Teh, D.C.S. Bien, S.K. Chen, Effect of Co and Ni nanoparticles formation on carbon nanotubes growth via PECVD. J. Experiment. Nanosci. **10**, 1232–1241 (2015)
11. R.S. Geonmonond, A.G.M. Da Silva, P.H.C. Camargo, Controlled synthesis of noble metal nanomaterials: motivation, principles, and opportunities in nanocatalysis. Annal. Brazil. Acad. Sci. **90**, 719–744 (2018)
12. Y. Xia, Y. Xiong, B. Lim, S.E. Skrabalak, Shape-controlled synthesis of metal nanocrystals: simple chemistry meets complex physics? Angew. Chem. Int. Ed. Engl. **48**(1), 60–103 (2009)
13. L. Xu, X. Li, Z. Zhan, L. Wang, S. Feng, X. Chai, W. Lu, J. Shen, Z. Weng, J. Sun, Catalyst-free, selective growth of ZnO nanowires on SiO_2 by chemical vapor deposition for transfer-free fabrication of UV photodetectors. ACS Appl. Mater. Interfaces **7**(36), 20264–20271 (2015)
14. T. Uchino, J.L. Hutchison, G.N. Ayre, D.C. Smith, K. de Groot, P. Ashburn, Metal-catalyst-free growth of silica nanowires and carbon nanotubes using Ge nanostructures. Jpn. J. Appl. Phys. **50**, 2011 (2011)
15. R. Calarco, R.J. Meijers, R.K. Debnath, T. Stoical, E. Sutter, H. Luth, Nucleation and growth of GaN nanowires on Si(111) performed by molecular beam epitaxy. Nano Lett. **7**(8), 2248–2251 (2007)
16. M. Tchernycheva, C. Sartel, G. Cirlin, L. Travers, G. Patriarche, J.-C. Harmand, L.S. Dang, J. Renard, B. Gayral, L. Nevou, F. Julien, Growth of GaN free-standing nanowires by plasma-assisted molecular beam epitaxy: structural and optical characterization. Nanotechnology **18**(38), 385306 (2007)
17. R. Mata, K. Hestroffer, J. Budagosky, A. Cros, C. Bougerol, H. Renevier, B. Daudin, Nucleation of GaN nanowires grown by plasma-assisted molecular beam epitaxy: the effect of temperature. J. Cryst. Growth **334**(1), 177–180 (2011)
18. S.M. Prokes, S. Arnold, Stress-driven formation of Si nanowires. Appl. Phys. Lett. **86**, 193105 (2005)
19. K.A. Bertness, A. Roshko, L.M. Mansfield, T.E. Harvey, N.A. Sanford, Nucleation conditions for catalyst-free GaN nanowires. J. Cryst. Growth **300**(1), 94–99 (2007)
20. K.A. Bertness, A. Roshko, L.M. Mansfield, T.E. Harvey, N.A. Sanford, Mechanism for spontaneous growth of GaN nanowires with molecular beam epitaxy. J. Cryst. Growth **310**, 3154–3158 (2008)
21. K.A. Bertness, N.A. Sanford, A.V. Davydov, GaN nanowires grown by molecular beam epitaxy. IEEE J. Select. Top. Quantum Electron. **17**, 847–858 (2011)
22. J.H. Park, M. Pozuelo, B.P.D. Setiawan, C.-H. Chung, Self-catalyzed growth and characterization of In(As) nano-wires on InP(111)B using metal-organic chemical vapor deposition. Nanoscale Res. Lett. **11**, 208 (2016)
23. X. Ji, X. Yang, T. Yang, Self-catalyzed growth of vertical GaSb nanowires on InAs stems by metalorganic chemical vapor deposition. Nanoscale Res. Lett. **12**, 428 (2017)
24. S. Esconjaureguia, C.M. Whelan, K. Maex, The reasons why metals catalyze the nucleation and growth of carbon nanotubes and other carbon nanomorphologies. Carbon **47**, 659–669 (2009)
25. M. Kumar, Y. Ando, Chemical vapor deposition of carbon nanotubes: a review on growth mechanism and mass production. J. Nanosci. Nanotechnol. **10**, 3739–3758 (2010)

26. V. Schmidt, J.V. Wittemann, U. Gösele, Growth, thermodynamics, and electrical properties of silicon nanowires. Chem. Rev. **110**, 361–388 (2010)
27. N. Wang, Y. Cai, R.Q. Zhang, Growth of nanowires. Mater. Sci. Eng. R **60**, 1–51 (2008)
28. F. Ishikawa, I. Buyanova, *Novel Compound Semiconductor Nanowires: Materials, Devices, and Applications* (Pan Stanford Publishing, Abingdon, Singapore, 2016)
29. J.Q. Lu, Nanocatalysts with tunable properties derived from polystyrene-b-poly(vinyl pyridine) block copolymer templates for achieving controllable carbon nanotube synthesis. J. Phys. Chem. C **112**(28), 10344–10351 (2008)
30. F. Parra da Silva, J.L. Fiorio, L.M. Rossi, Turning the catalytic activity and selectivity of Pd nanoparticles using ligand-modified supports and surfaces. ACS Omega **2**, 6014–6022 (2017)
31. A.K.M. Fazle Kibria, Y.H. Mo, M.H. Yun, M.J. Kim, K.S. Nahm, Effect of bimetallic catalyst composition and growth parameters on the growth density and diameter of carbon nanotubes. Korean J. Chem. Eng. **18**, 208–214 (2001)
32. J.W. Kaiser, Structural and catalytic analysis of gold–palladium composite nanoalloys. Doctoral thesis, Humboldt-Universitat, Berlin (2013)
33. B.D. Adams, G. Wu, S. Nigro, A. Chen, Facile synthesis of Pd-Cd nanostructures with high capacity for hydrogen storage. J. Am. Chem. Soc. **131**, 6930–6931 (2009)
34. A.K. Singh, Q. Xu, Synergistic catalysis over bimetallic alloy nanoparticles. ChemCatChem **5**, 652–676 (2013)
35. T. Guo, P. Nikolaev, A. Thess, D.T. Colbert, R.E. Smalley, Catalytic growth of single-walled manotubes by laser vaporization. Chem. Phys. Lett. **243**, 49–54 (1995)
36. Y.-C. Chou, C.-Y. Wen, M.C. Reuter, D. Su, E.A. Stach, F.M. Ross, Controlling the growth of Si/Ge nanowires and heterojunctions using silver-gold alloy catalysts. ACS Nano **6**, 6407–6415 (2012)
37. K. Mukhopadhyay, A. Koshino, T. Sugai, N. Tanaka, H. Shinohara, Z. Konya, J.B. Nagy, Bulk production of quasi-aligned carbon nanotube bundles by the catalytic chemical vapor deposition (CCVD) method. Chem. Phys. Lett. **303**, 117–124 (1999)
38. J.G. Connell, Z.Y. Al Balushi, K. Sohn, J. Huang, L.J. Lauhon, Growth of Ge nanowires from Au–Cu alloy nanoparticle catalysts synthesized from aqueous solution. J. Phys. Chem. Lett. **1**, 3360–3365 (2010)
39. K. Mukhopadhyay, A. Koshio, N. Tanaka, H. Shinohara, A Simple and novel way to synthesize aligned nanotube bundles at low temperature. Jpn. J. Appl. Phys. **37**, L1257 (1998)
40. S. Sato, A. Kawabata, D. Kondo, M. Nihei, Y. Awano, Carbon nanotube growth from titanium–cobalt bimetallic particles as catalyst. Chem. Phys. Lett. **402**, 149–154 (2005)
41. S. Biswas, C. O'Regan, N. Petkov, M.A. Morris, J.D. Holmes, Manipulating the growth kinetics of vapor-liquid-solid propagated Ge nanowire. Nano Lett. **13**, 4044–4052 (2013)
42. W. Ostwald, Studien über die bildung und umwandlung fester körper (Studies on the formation and transformation of solid bodies). Zeit. Physikalische Chem. **22**, 289–330 (1897)
43. N.O. Weiss, X. Duan, A guide for nanowire growth. Proc. Natl. Acad. Sci. U.S.A. **110**, 15171–15172 (2013)
44. J.B. Hannon, S. Kodambaka, F.M. Ross, R.M. Tromp, The influence of the surface migration of gold on the growth of silicon nanowires. Nature **440**, 69–71 (2006)
45. S. Kodambaka, J.B. Hannon, R.M. Tromp, F.M. Ross, Control of Si nanowire growth by oxygen. Nano Lett. **6**, 1292–1296 (2006)
46. S. Hofmann, R. Sharma, C.T. Wirth, F. Cervantes-Sodi, C. Dukati, T. Kasama, R.E. Dunnin-Borkowaski, J. Drucker, P. Bennett, J. Robertson, Ledge-flow-controlled catalyst interface dynamics during Si nanowire growth. Nat. Mater. **7**, 372–375 (2008)
47. D.N. Futaba, K. Hata, T. Yamada, K. Mizuno, M. Yumura, S. Iijima, Kinetics of water-assisted single-walled carbon nanotube synthesis revealed by a time-evolution analysis. Phys. Rev. Lett. **95**, 056104 (2005)
48. G.H. Chen, R.C. Davis, H. Kimura, S. Sakurai, M. Yumura, D.N. Futaba, K. Hata, The relationship between the growth rate and the lifetime in carbon nanotube synthesis. Nanoscale **7**, 8873–8878 (2015)

49. M. Picher, E. Anglaret, R. Arenal, V. Jourdain, Self-deactivation of single-walled carbon nanotube growth studied by in situ Raman measurements. Nano Lett. **9**, 542–547 (2009)
50. P. Vinten, J. Lefebvre, P. Finnie, Kinetic critical temperature and optimized chemical vapor deposition growth of carbon nanotubes. Chem. Phys. Lett. **469**, 293–297 (2009)

Chapter 4
Pre-synthesis and Synthesis Events

Abstract Important elements of the pre-synthesis and synthesis events of one-dimensional and quasi-two-dimensional nanomaterials synthesis have been presented. While describing the pre-synthesis process, e.g., the process prior to nanomaterial growths, the formation of the FECA (both METANO and SUBSANO) nanoparticles on a substrate or support has been emphasized. All important characteristics of the FECA nanoparticles on the substrate or support have been laid down. It has been noted that the pre-nucleation stages for the vapor-phase growths, solution-phase growths, and the solid-phase growths are quite different. Crucial synthesis stages for the synthesis of nanomaterials via the vapor-phase, solution-phase, and the solid-phase processes have then been put forth. These stages for nanowire growths and nanotube growths have been specifically listed. While laying down important attributes of the synthesis process, the hallmarks of this synthesis process, such as surface activation barrier, reaction and interaction on the nanoparticle surface, anisotropic growth and nucleation for this growth, adhesive property of the nanoparticle seed, e.g., R_L species, and the contact angle between the substrate and the nanoparticle catalyst droplet, have been unraveled and described. The nanoparticle core and shell structures for the synthesis of nanotubes have also been elucidated.

4.1 Basics

The low-dimensional materials possess nanometer and subnanometer-size dimensions. They are dominated by surface phenomena. The design and synthesis of them therefore require surface engineering both in the pre-synthesis and in the pro-synthesis processes. Among them, the pre-synthesis processes involve surface treatment and surface functionalization. Both of them introduce novel functional components exhibiting heteroatoms and defects to the growth front, and thus enhance catalytic activity of this growth front. Continuing with the presentations made in Chap. 3, we describe in the following the pre-synthesis and the pro-synthesis processes. The pro-synthesis processes are the events promoting the synthesis and leading to the nucleation, crystallization and growth.

S. N. Mohammad, *Synthesis of Nanomaterials*, Springer Series in Materials Science 307, https://doi.org/10.1007/978-3-030-57585-4_4

4.2 Pre-synthesis Process: Formation of Catalyst Support

The FECA (both METANO and SUBSANO) nanoparticles formed on a substrate are usually small and thin as compared to the substrate. As noted earlier [1] and will be discussed in the following chapters, FECA nanoparticles yet play crucial role in nanomaterial growths. In order to accomplish this goal, they must not undergo decay during nanomaterial growth. Instead, they should have characteristics most suited to growth; they should promote the realization of superior nanomaterials. They should also possess the most desired activity, stability, selectivity, and regeneration ability. They should not be mobile. Unfortunately, FECA nanoparticles formed directly on a substrate may lack these characteristics. And hence, a thin-film layer is formed on the substrate followed by FECA nanoparticle deposited on it. The intermediate layer thus formed is called the catalyst support. This catalyst support may be granular, powdered, colloidal, co-precipitated, extruded, pelleted, spherical, wirelike, honeycombed, and skeletal. It may either be inert or active in reactions. It should fulfill the following requirements:

1. FECA nanoparticles formed on it should possess narrow size distribution.
2. FECA nanoparticles formed on it should not undergo excessive sintering during growth. This implies that it should significantly immobilize the FECA nanoparticles during growth.
3. The size of the catalyst support should be so tuned and optimized that it permits the fluidization of the FECA nanoparticles when fluidized bed reactors are used for continuous nanomaterial synthesis.
4. It should define the macroscopic shape of the end product.
5. It should increase the dispersion of the active phases of the FECA nanoparticles, better control nanoparticle structures, convert these structures to be porous, and improve their mechanical strength.
6. There will be a challenge for it to insure that FECA nanoparticles do not undergo agglomeration.
7. The catalyst support should have such structure, composition, and morphology that the catalytic activity and the lifetime of the FECA nanoparticles are both influenced, and even dictated by it.
8. The catalytic activity and the lifetime of the FECA nanoparticles should therefore be different for a different catalyst support. They may be best achieved if the catalyst support properties match (mismatch) those of the substrate, particularly lattice structure, surface morphology, textural properties, and nanopore size well.
9. The said matching (mismatching) would result from good catalyst–substrate interactions. To illustrate just one, we argue that such an interaction must determine if, for example, the CNT growth would be a tip growth or base growth.
10. Al_2O_3, SiO_2, and MgO are common support materials. Together with substrates such as graphite, quartz, silicon, and silicon carbide, they should serve the job of a facilitator, a promoter, and/or supporter in yielding good nanomaterials.

4.3 Crucial Synthesis Stages

We emphasize that there are at least two stages, namely pre-nucleation stage and pro-nucleation stage, for nanomaterial synthesis. The pre-nucleation stage is the stage preparatory for nanomaterial growth. But the pro-nucleation stage is the stage promoting nanomaterial growth. While the key events of the pre-nucleation stage for the vapor-phase growth mechanisms (e.g., VLS, VSS, VS, VQS, OAG, and SCG mechanisms) may be different from those of the solid-phase growth mechanism(s) (e.g., SLS mechanism) and the solution-phase (liquid-phase) growth mechanisms (e.g., SFLS and SoLS mechanisms), the key events of the pro-nucleation stage may essentially be the same for all of these mechanisms. Various stages of nanowire growth are shown schematically in Fig. 4.1. But various stages of nanotube growth are shown schematically in Fig. 4.2. The pre-nucleation stage for nanotube growth by the vapor-phase mechanism involves the landing of the precursor(s) of the R_S species everywhere on the nanoparticle surface and release of the R_S species from their precursor(s). The pro-nucleation stage for this nanotube growth involves, on the other hand, the segregation of the R_S species from the core to the peripheral surface of nanoparticle. A shell is thus created. This shell is amorphous (semi-amorphous and amorphous-like); it surrounds the core and serves as the R_L species, which was defined in Chap. 3, Sect. 3.3. The amorphicity of the shell is controlled (see Appendix). The pro-nucleation stage—a stage subsequent to the pre-nucleation stage—involves adsorption of the R_S species onto the R_L species of the shell, diffusion of the R_S species through the R_L species to the quasiliquid/solid (QL/S) interface, and nucleation of the R_S species into nanotube. The pre-nucleation and pro-nucleation stages for nanowire growth are essentially the same as those for nanotube growth, except that the segregation of the R_S species from the core to the peripheral surface of the nanoparticle is absent and hence there is no shell formed at the peripheral surface of nanoparticle. The solution-phase mechanisms (e.g., SFLS and SoLS mechanisms) are analogous to the vapor-phase mechanisms. The main difference between them is that the pre-nucleation stage for growths by the vapor-phase mechanism is carried out in vapor phase, but the pre-nucleation stage for growths by the solution-phase (liquid-phase) mechanisms is carried out in solution phase. Also, unlike the precursors of the R_S species for most of the vapor-phase growths, the precursors of the R_S species for solution-phase growths are generally metallo-organic compounds. The pre-nucleation stage for nanomaterial growths by the solid-phase mechanism may however involve the R_S ($R_S \equiv X$ and $R_S \equiv Y$) species generated on the nanoparticle surface from component elements of the substrate (support) or the component elements of the nanoparticle formed on this substrate (support).

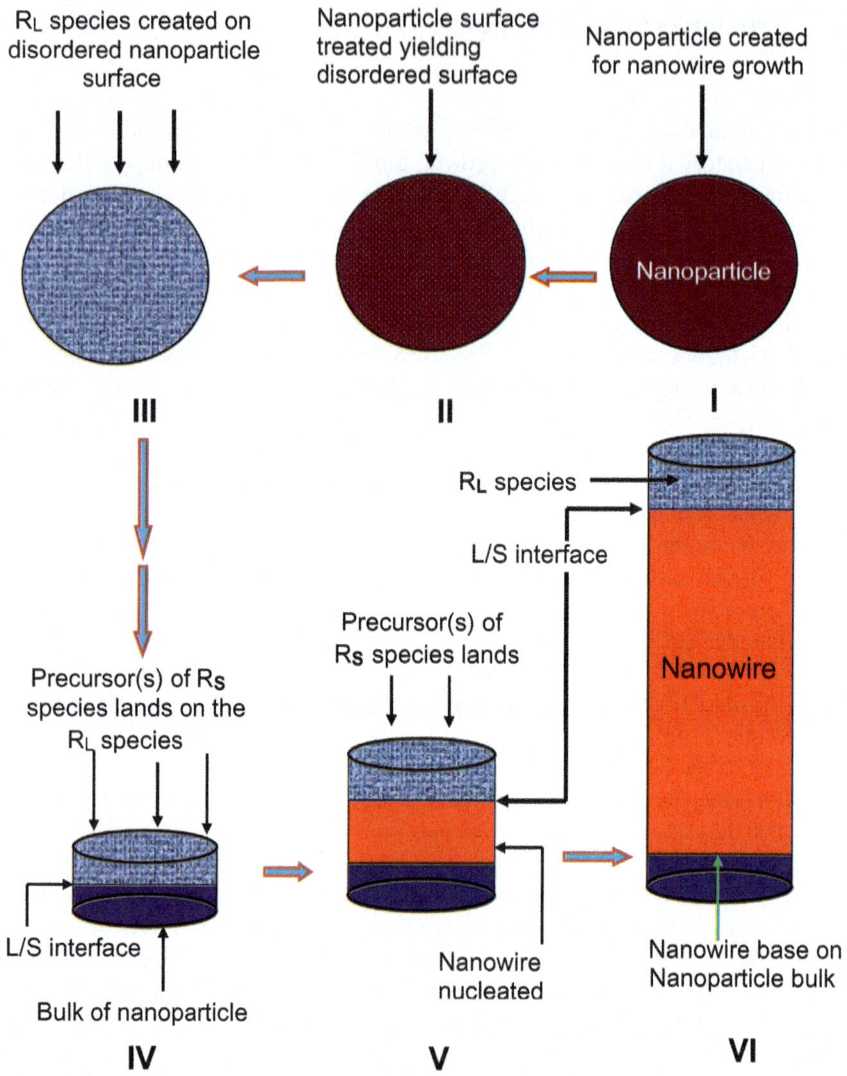

Fig. 4.1 Schematic diagrams of various steps of nanowire nucleation and growth preferably by the CVD technique

4.4 Key Synthesis Events

4.4.1 Event 1

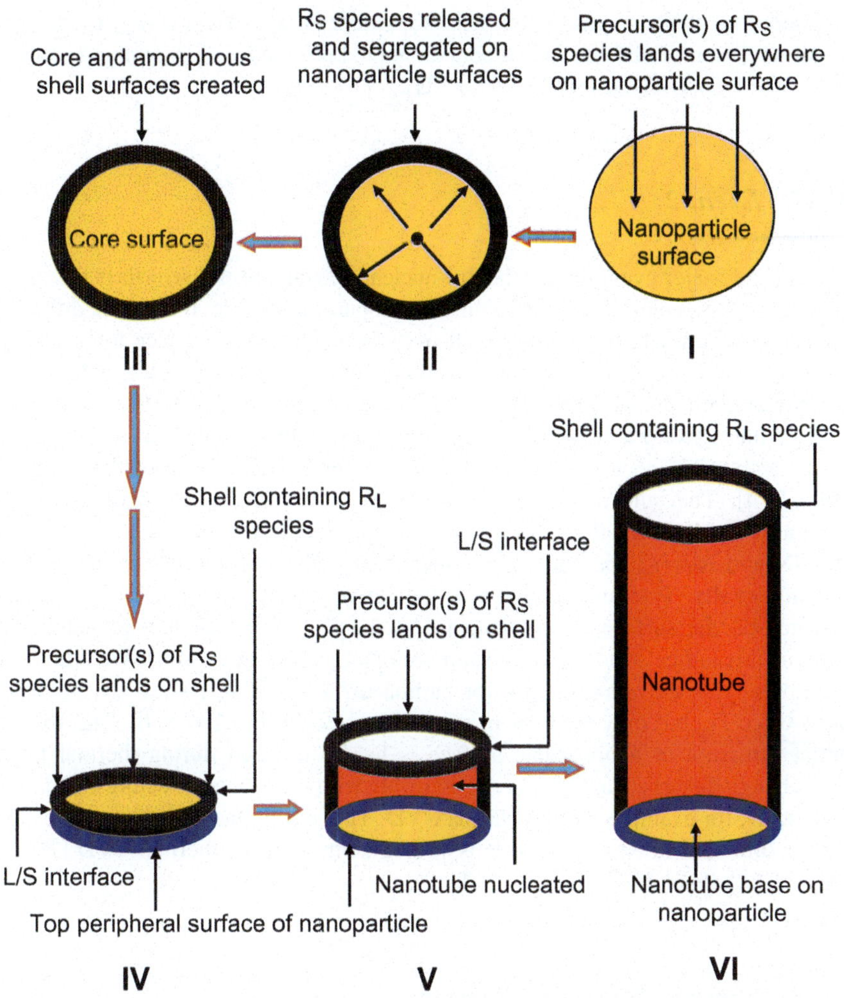

Fig. 4.2 Schematic diagrams of various steps of nanotube nucleation and growth preferably by the CVD technique

At least two events are central to the nanomaterial growths by the vapor-phase mechanism. The first major events during the pre-nucleation stage of the growth are the attraction, landing, and adhesion of the precursor(s) of the R_S ($R_S{\equiv}X$ and $R_S{\equiv}Y$) species on the nanoparticle (SUBSANO and METANO) surface and the release of the active R_S ($R_S{\equiv}X$ and $R_S{\equiv}Y$) species due to decomposition of this precursor(s) on the nanoparticle surface. Instead of being generated from precursor(s), the R_S ($R_S{\equiv}X$ and $R_S{\equiv}Y$) species may be generated on the nanoparticle surface from the component elements of the substrate (support). Under identical conditions, the stability of a precursor is determined by the strength of its chemical bonds, and hence this release is different for different precursors. This is apparent from different energy barriers for the dissociation of different precursors on Ni(111) step edges [2]; while the barrier is ~1.30 eV for C_2H_2, it is 0.90 eV for CH_4.

4.4.2 *Event 2*

The second major event during the pre-nucleation stage of the growth is the diffusion of the R_S species thus released on the nanoparticle surface. If it is bulk diffusion for nanowire growth, it takes place all over the nanoparticle surface for nanowire growth. If it is bulk diffusion for nanotube growth, it takes place only in the peripheral shell of the nanoparticle for nanotube growth [3, 4]. The bulk diffusion of the active R_S species for nanotube growth follows the surface diffusion of the active R_S species from the bulk to the peripheral surface of the nanoparticle [3, 4]. The bulk diffusion of the active R_S ($R_S{\equiv}X$ and $R_S{\equiv}Y$) species is generally much faster than that of the nanoparticle atoms; while the active R_S ($R_S{\equiv}X$ and $R_S{\equiv}Y$) species diffuse interstitially, the nanoparticle atoms diffuse substitutionally. The bulk diffusion of the active R_S ($R_S{\equiv}X$ and $R_S{\equiv}Y$) species depends on the surface lattice structure of the R_L species of nanoparticle. It is faster if this lattice structure is loose, more open, and hence porous. We cite some examples. The activation energy for carbon diffusion is 1.53–1.57 eV in close-packed fcc Fe, but only 0.83 eV in the relatively open bcc Fe [5, 6]. The surface lattice structure of nanoparticle surface is hardly affected during thermal CVD . It is though disturbed and disordered plausibly with some open structure containing nanopores, for example, during plasma CVD. This is why the activation energy for carbon diffusion in Ni nanoparticle is 116.75 kJ/mol during thermal CVD [7], but only 22.19 kJ/mol during plasma CVD [2].

4.4.3 *Surface Activation Barrier*

In general, the activation barriers for surface diffusion are lower than those for bulk diffusion. Yazyev and Pasquarello [8] performed first principle calculations of bulk and surface diffusion of carbon through Ni, Pd, Pt, Cu, Ag, and Au, and found higher activation barrier for bulk diffusion than for surface and subsurface diffusions. The calculated activation barriers for surface diffusion of carbon through Ni(111) and Ni(110) surfaces were found to be 0.4–0.5 eV [2], respectively. These are strikingly comparable to the observed activation barrier of 0.3 eV for carbon through polycrystalline Ni [9]. Note that, unlike crystalline Ni, the polycrystalline Ni is comprised of many individual grains or crystallites, and hence is porous. Based on the data presented above, the bulk diffusion through nanoparticle may be marginally small if the nanoparticle surface is solid and non-porous, moderately large if the nanoparticle surface is quasisolid (quasiliquid) and (slightly, or even highly) porous, but large if the nanoparticle surface is liquid.

4.4.4 Reaction and Interaction on Nanoparticle Surface

During the pre-nucleation stage of growth, precursor(s) of the $R_S{\equiv}X$ and $R_S{\equiv}Y$ species, together with one or more carrier gases, is flown into the growth chamber at suitable temperature and pressure. This precursor(s), landing on the nanoparticle (METANO or SUBSANO) surface, for example, for thermal or plasma-enhanced CVD undergoes a series of reactions (interactions) with the nanoparticle. These reactions may be stepwise reactions. The reactions (interactions) may vary and lead to the formation of alloy, solid solution, ionic species, etc. They may otherwise disturb the lattice of the nanoparticle surface creating islands (clusters) of materials species called the R_L species. They are called, as well, seeds. The R_L species may be solid with a network of nanopores generated in it. They may be molten (semimolten) during growth. So, they may also be called droplets. Pre-nucleation stage of nanowire growth in solution using high-temperature supercritical fluids and organic-monolayer-protected metal nanocrystals preferably requires pressurization of the solvents. Nanowire growth is tuned by proper size selection of the organic-monolayer-protected metal nanocrystals prior to injection to the reaction chamber. Temperature selection of the supercritical fluid is essential to provide the solvating medium the need to maintain nanoparticle size distribution prior to growth. The high precursor solubility and the high particle concentration in the supercritical fluid environment are keys to the synthesis of large number of nanowires.

4.4.5 Anisotropic Growth and Nucleation for This Growth

Universal to one-dimensional and quasi-two-dimensional growths is the growth anisotropy. This is the growth taking place along one crystallographic direction. This is a preferential growth of the crystalline structure influenced by several factors such as: (a) Different facets in the crystal have different growth rates, (b) imperfections are present in some specific crystal directions, and (c) there is preferential accumulation of impurities and ion in some specific facets. Control of the crystallization process is central to developing nanomaterials with atomic precision. Nucleation occurs when a small nucleus begins to form in the liquid; the nucleus then grows as the atoms and/or molecules of the R_S growth species diffused into the R_L species are gradually attached to it. The crucial point to understand is the balance between the free energy available from the driving force and the energy consumed in forming new interface(s). The critical point is also the liquid, quasiliquid, or quasisolid state of the R_L species that promotes the formation of the said small nucleus. The atomic arrangement changes during solidification from a random or short-range order to a long-range order or crystal structure. Decrease in free energy provides the driving force for solidification. This decrease in free energy is accompanied by reorganization in the crystal lattice structure. Let us be more specific. The nucleation process may consist of several partial steps [10]. First, a small fraction of the R_S species dissolved in the R_L species at a temperature $T = T_r$ come close to each other until some part (portion and segment) of the R_L species is saturated up to the equilibrium concentration, suppose, at a temperature $T = T_r$. Subsequently, supersaturation begins if the temperature of that part (segment) of the R_L species is lowered to T_s in the low-temperature zone of the growth chamber. This supersaturated portion of the R_L species is brought into contact with the solid (e.g., substrate) just underneath the nanoparticle. This marks the beginning of the growth process with transport of the R_S species to the liquid/solid interface.

In equilibrium, there is no net flux from one phase to another phase of the R_S ($R_S \equiv X$ and $R_S \equiv Y$) species and of the molecules formed from these R_S species. This flux is equal to zero. This means there must be a deviation from equilibrium in order for crystal to grow. This deviation must generate the driving force for crystallization. In fact, it would be a prerequisite for a solid phase to appear in a saturated solution. And it would be the difference ΔG between the chemical potentials of the crystallizing compound in a supersaturated solution and in the crystal or solid. To be slightly more detailed, the chemical potential μ_i is used to describe phase equilibrium and the driving force for crystallization. And it is given by the first partial derivative ΔG of Gibbs free energy G with respect to the number of particles n_i of component i. Note that the Gibbs free energy is a thermodynamic potential. It is the enthalpy minus the product of the entropy and the absolute temperature of a system. It is generally used to calculate the maximum of the reversible work to be performed by a thermodynamic system at a constant temperature and pressure. It is at its minimum in equilibrium. Also, in equilibrium, each component of the chemical potential is equal in all phases.

The supersaturation is best achieved in the area (location) of the R_L species where the system possesses excess free energy and is not in equilibrium. This area resembles the transition state area of a chemical reaction. It is though different from the transition state area which is too fleeting to detect. The supersaturation area has a finite lifetime, and it is detectable. Maliakkal et al. [11] noted the process of supersaturation occurs in two steps: formation of a critical nucleus at the liquid–solid interface and then the layer growth across this interface. Considering that metal droplet is small, the number of atoms participating in the supersaturation of liquid may be finite. These atoms may be consumed during the formation of a layer.

4.4.6 Adhesive Properties of the Nanoparticle Seed

Nanoparticles serving as seeds for growth do exhibit strong adhesive property toward the R_S species adatoms landing on it or created on its surface. And this arises from the electric field generated at the nanoparticle surface due to large accumulated charge. The accumulation is due to anisotropy in the nanoparticle surface structure and the dipole moment, which result from charge redistribution during the pre-nucleation stage of growth. During this stage of growth, there may, for example, be annealing influenced at least partly by the presence of FECANO (METANO or SUBSANO) and nanoparticle/nanomaterial (or nanoparticle/substrate, nanoparticle/support) heterointerface. The nanoparticle dimension is small. Hence, any disturbance in the nanoparticle surface during annealing, and also due to lattice mismatch and mismatch of the thermal expansion coefficient of the two materials at the heterointerface, easily gives rise to charge redistribution on the nanoparticle surface. The structural anisotropy and the polar characteristics of the seed result from also the creation of (FECA, X) cluster or solid solution. It is created even when a FECA forms an interface with $R_S{\equiv}X$ species. One fundamental property of R_S species adatom may probably be the arrangement and distribution of electrical charges within it, as well as the arrangement of electronic orbital and spin angular momenta. Electric and magnetic dipole moments arise from the distribution of these charges. The total electric dipole moment of an adatom is then

$$\vec{p} = \int \rho(\vec{r})\vec{r}d^3r, \tag{4.1}$$

where $\rho(\vec{r})$ is the charge density (including both electrons and nuclei) at position $\vec{r}$ within the adatom, and the integral is over the entire R_S species adatom volume. Similarly, the magnetic dipole moment of the R_S species adatom is

$$\vec{\mu} = (2c)^{-2} \int \vec{r} \times \vec{J}(\vec{r})d^3r, \tag{4.2}$$

where $\vec{J}$ is the volume current density and c is the speed of light. The vector product for this equation is related to the charge's angular momentum. The potential energy of a permanent dipole in an external electric field $\vec{E}$ of the nanoparticle seed is

$$U = -\vec{p}.\vec{E}, \tag{4.3}$$

and the force experienced by the polar adatom is

$$\vec{F} = (\vec{p}.\vec{\nabla})\vec{E}. \tag{4.4}$$

A similar equation for the magnetic force is

$$\vec{F} = (\vec{\mu}.\vec{\nabla})\vec{B}, \tag{4.5}$$

where $\vec{B}$ is the magnetic field. Thus, as an R_S species adatom passes by a nanoparticle seed, it is influenced by both the electric and magnetic forces. The field applies a torque to the permanent dipole in an external electric field $\vec{E}$ of the nanoparticle seed, and it may be given by

$$\tau = \vec{p} \times \vec{E} \text{ or } \tau = \vec{\mu} \times \vec{B}. \tag{4.6}$$

This torque tends to align it along the field lines of the nanoparticle seed. Thus, under the combined effect of the surface tension, the electrostatic attraction between the R_L and the R_S species, the electric force, and the magnetic force, the R_S species adatoms are attracted by the nanoparticle seed and eventually adsorb on its surface. To substantiate it, we cite an example. In an experiment [12] for Zn nanowire growth, using zinc vapor atoms as the R_S species and lead as nanoparticle catalyst, but without a carrier gas, no nanowires could be produced. However, under identical growth conditions, long, thin, uniform Zn nanowires could be produced using ammonia as carrier gas. It happened because ammonia has dipole moment. As Zn vapor particle was bound and carried by ammonia, it was attracted by the nanoparticle seed. It would not be attracted by the nanoparticle seed without this seed having both electric and magnetic fields.

4.4.7 Contact Angle

One key feature of FECA (METANO and SUBSANO)-mediated epitaxial growths of one-dimensional (quasi-two-dimensional) growths of X_mY_n nanomaterials is the capillary stability of the FECA at the X_mY_n nanomaterial tip. This stability is necessary for preserving fixed FECA shape and size during growths. It is necessary also for determining the onset of new layers, for dictating the characteristics of new layers, and for governing the growth direction. The contact angle that it forms with the nanomaterial growth facet serves as tool to achieve the stated goal. The equilibrium value of this contact angle can be obtained in terms of the energies and geometries of the surface and the interface [13].

Figure 4.3a–c exemplifies three different scenarios of a molten $R_L \equiv$(MET, X) droplet at the tip of three different nanowires: (a) a cylindrical nanowire, (b) a tapered nanowire, and (c) an edged nanowire, e.g., a nanowire with an external edge. Figure 4.3a, b shows the droplet wetting the nanowire-side facets. The droplets exhibit the vapor–liquid surface energy γ_{LV} and the radius R. Two other surface energies, namely the solid–liquid surface energy γ_{LS} and the vapor–solid surface energy γ_{SV} of the facet(s), also contribute to the nanowire growths. There is also a contact angle ϕ between the projected nanowire and the liquid interface and a tapering angle θ of the tapered nanowire (see Fig. 4.3b). If γ_{SV}, γ_{SL}, and γ_{LV} are the interfacial free energies per unit area of the solid–vapor, solid–liquid, and liquid–vapor interfaces, respectively, then the contact angle ϕ may be given by the classical equation,

$$\cos\phi = \frac{\gamma_{SV} - \gamma_{SL}}{\gamma_{LV}} \tag{4.7}$$

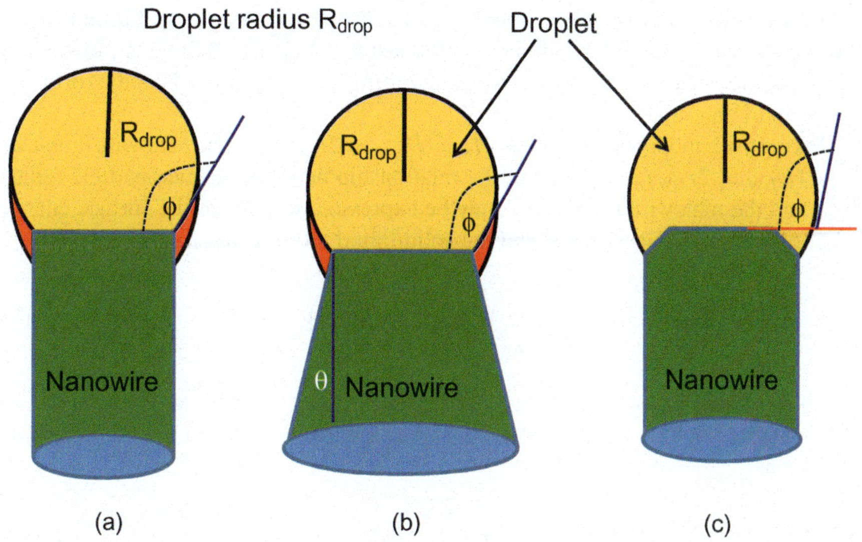

Fig. 4.3 Schematic diagrams of droplets formed at the tips of three different types of nanowire: **a** cylindrical nanowire, **b** tapered nanowire, and **c** nanowire with an external edge

FECA nanoparticle (METANO or SUBSANO), for the VLS growths, becomes FECA droplet. Also, the solid–liquid interface exists only if the FECA nanoparticle is molten. Expressed in terms of the surface energy γ_{nano} of the FECA nanoparticle (METANO or SUBSANO) and the surface energy γ_{SL} of the solid–liquid interface, the contact angle ϕ may be given by Young's well-known equation [14].

$$\gamma_{\text{nano}} \cos\phi = -\gamma_{\text{SL}}. \tag{4.8}$$

For Si nanowire growth with Au droplet, $\gamma_{\text{nano}} = 0.85$ J/m^2. This growth with Sn droplet is however 0.6 J/m^2. We may consider the idealized case in which the liquid droplet is on a horizontal plane. The droplet consequently assumes the shape of a spherical (hemispherical) cap with its base circular and the contact angle ϕ constant around its base. The droplet volume may then be given by

$$V_{\text{drop}} = \frac{\pi D_{\text{drop}}^3}{24}\left[\frac{2 - 3\cos\phi + \cos^3\phi}{\sin^3\phi}\right], \tag{4.9}$$

where D_{drop} is the droplet diameter: $D_{\text{drop}} = 2r_{\text{drop}}$. The system is in equilibrium if the Gibbs free energy G of this system remains constant at its minimum value (dG = 0). The Gibbs energy as a function of r_{drop} is minimized to find a solution. The solution is

$$G_0 = \frac{2\pi r_{\text{drop}}^2 \gamma_{\text{LV}}}{(1 + \text{Cos}\phi)} + \pi r_{\text{drop}}^2 \gamma_{\text{SL}}. \tag{4.10}$$

This result is equivalent to assuming that the droplet is in equilibrium under the horizontal forces that equally act on the droplet [15]. The Nebol'sin–Shchetinin criterion [15] is often used to explain the equilibrium contact angle of catalyst droplets at the tip of nanowires.

Taking the nanoparticle diameter D_{nano} ($D_{\text{nano}} = 2r_{\text{nano}}$) and the nanowire diameter D_{NW} ($D_{\text{NW}} = 2r_{\text{NW}}$) into account, the relationship between the nanoparticle radius r_{nano} and the nanowire radius r_{NW} may be expressed in terms of the surface energy γ_{nano} of the metal nanoparticle and the solid/liquid interface energy γ_{SL} as [14]

$$\left(\frac{r_{\text{NW}}}{r_{\text{nano}}}\right)^2 = 1 - \left(\frac{\gamma_{\text{SL}}}{\gamma_{\text{nano}}}\right)^2. \tag{4.11}$$

Equation (4.11) shows that r_{NW} approaches r_{nano} if the surface energy γ_{nano} is much larger than the surface energy γ_{SL}. Equations (4.8) and (4.11) yield the contact angle ϕ as

$$\left(\frac{r_{\text{NW}}}{r_{\text{nano}}}\right) = 1 - (1 - \cos\phi)^2 = 1 - \cos^2(\phi) \tag{4.12}$$

Equation (4.12) expresses ϕ in terms of the radius of the nanoparticle (droplet) and the radius of the nanowire. Al-Taay et al. [16] employed Sn film to produce Si nanowires by the VLS mechanism. They noted that both the Sn droplet radius r_{nano} and the Si nanowire radius r_{NW} depended on the Sn catalyst film thickness; they obtained the experimental data for r_{NW} and r_{nano} from experiments. Using these data and the data for γ_{nano}, they derived from (4.11) the values of γ_{SL} for different thicknesses of the Sn film. Using these data, we produced various plots shown in Figs. 4.4 and 4.5, respectively. Figure 4.4 indicates that, almost for all Sn film thicknesses, the nanowire radius is smaller than the catalyst droplet radius. Also, diameter of both the nanoparticle and the nanowire increases with increase in Si film thickness. And this finding is not identical, but almost similar to that depicted in Figs. 3.4 and 3.5 of Chap. 3. Figure 4.5 indicates that the solid–liquid surface energy γ_{SL} decreases almost linearly with decrease in the contact angle ϕ. We believe that the solid–liquid surface energy γ_{SL} is a measure of droplet stability. In general, the lower the value of γ_{SL}, the higher is the stability of the droplet. This means nanowire growth by the VLS mechanism must be conducted only with those metal droplets that yield the lowest possible contact angle ϕ.

4.5 The Nanoparticle Core and Shell Structures for Synthesis

The nanoparticle for the growth of some nanomaterials, such as nanowires and nanodots, has uniformly identical bulk characteristics on the top surface. The

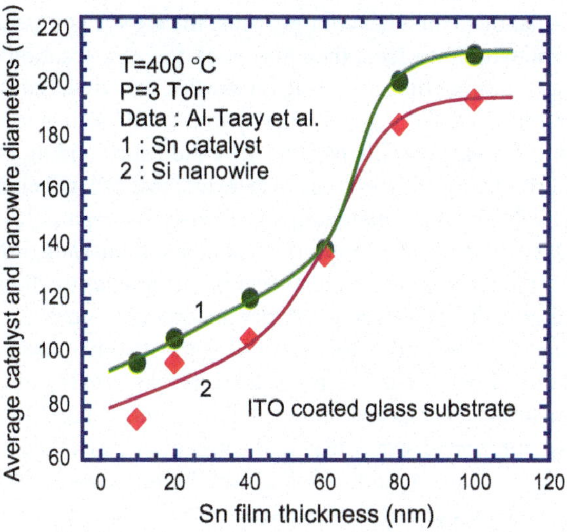

Fig. 4.4 Dependence of the average catalyst diameter and the average Si nanowire diameter on the Sn catalyst film thickness. The plots are obtained with the experimental data by Al-Taay et al. [16]

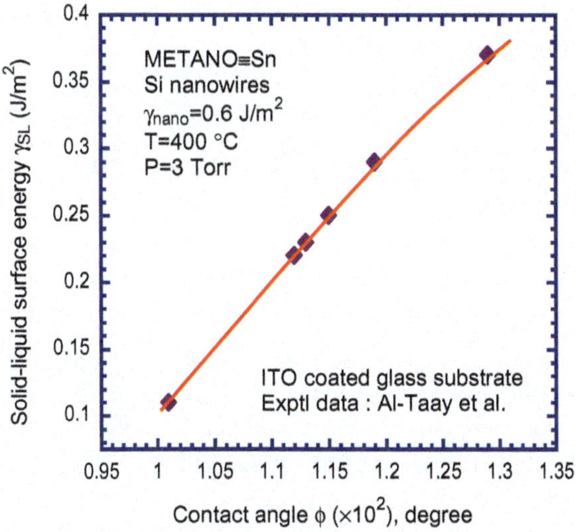

Fig. 4.5 Variation of the solid–liquid surface energy γ_{SL} with the contact angle ϕ. The plot is obtained with the experimental data by Al-Taay et al. [16]

nanoparticle surface for the growth of some other nanomaterials, such as nanotubes and nanorings, has uniformly identical bulk characteristics in the core and distinctly different shell characteristics at the periphery of the nanoparticle. In nanoparticles, for example, for nanowire and nanodot growths, the R_S species are adsorbed everywhere on the nanoparticle surface; they may stick to every location on this surface. For nanotube and nanoring (quantum rings) growths, they undergo surface diffusion from the bulk to the periphery of this nanoparticle. Most, if not all, of them are thus segregated to the nanoparticle periphery. A shell surrounding the core is hence created. A hill surrounding the shell may also be created. We will elaborate later that, in order to effectively mediate nanomaterial growth, the nanoparticle (both METANO and SUBSANO) surface must be appropriately and judiciously disturbed, disordered, amorphized (semi-amorphized), and coarsened. This applies to all nanoparticles for all nanomaterial growths. However, the entire nanoparticle surface, for example, for nanowire and nanodot (quantum dot) growths must also be porous. This porosity may be attributed, at least in part, to its surface amorphicity. In contrast, the core of the nanoparticle surface, for example, for nanotube and nanoring growths must be non-porous. This non-porosity may lead to the absence of adsorbed R_S species in the core and the lack of surface modification caused by the adsorbed R_S species in the bulk (core). The shell of the nanoparticle surface for the same nanotube and nanoring growths must however have adsorbed R_S species and hence be porous. This implies that the Knudsen diffusion of the R_S species for nanowire and nanodot growth takes place through the entire surface of the nanoparticle. But the Knudsen diffusion of the R_S species for nanotube and nanoring growth takes place through the entire shell of the nanoparticle. As a result, the grown nanowire and nanodot are solid, but the grown nanotube and nanoring are hollow.

4.6 Nanoparticle Periphery with Shell and Hill for Synthesis

Surface diffusion of the R_S species, which is the diffusion from the core to the peripheral surface of a nanoparticle, depends on at least three factors:

1. The availability of the R_S species in the core of the nanoparticle surface.
2. The difference in surface energy of the nanoparticle surface and of the R_S species released on the nanoparticle core. The larger the difference between them, the higher is the motive force for the surface diffusion of the R_S species to the peripheral surface.
3. Surface morphology including surface roughness of the nanoparticle surface.

Depending on the three factors stated above, the shell can be wide or narrow. The shell can be surrounded by a hill [17]. If the width of the hill is large due to large accumulation of the R_S species, the width of the shell is small. If the width of the hill is small due to small accumulation of the R_S species, the width of the shell is large. If the width of the hill is large due to large accumulation of the R_S species, the width of the shell can be so small that it accommodates the formation of only one carbon ring (fringe) and hence the formation of SWCNT. If no hill is formed around a wide shell due to small number of R_S species released in the core, this shell can be wide enough to accommodate the formation of multiple concentric carbon rings (fringes). A MWCNT can thus be formed.

4.7 Smaller Nanoparticles Generally Yield SWCNTs

Decrease in the size of catalyst nanoparticle to less than about 5 nm leads this nanoparticle to have increased relative fraction of low coordinated atoms. The immediate consequence of this is

1. Increased curvature of the nanoparticle surface;
2. Increased surface energy of the nanoparticle surface;
3. Increased tendency of this nanoparticle to transfer (attract) low surface energy R_S species (e.g., $R_S{\equiv}C$ of surface energy of only 17 meV) to the peripheral surface;
4. Increased surface reconstruction;
5. Increased tendency of the peripheral surface to be overly saturated with R_S species precipitation; and
6. Increased tendency of this peripheral surface to be partially converted into hill during the pre-nucleation stage of growth.
7. These R_S species atoms (e.g., $R_S{\equiv}C$), exhibiting high concentration, are actually low coordination atoms contributing to the generation of active surface sites.

Fujita et al. [18] found Au nanoparticle, unlike bulk gold, to possess remarkable catalytic activity. A high-angle annular dark-field study, together with scanning transmission electron microscopy analysis, demonstrated a complex arrangement of low- and high-index surfaces and a high density of atomic steps in the internal surfaces of Au nanoparticle. The surface steps had a significant number of kinks. These surfaces also had a high concentration of low-coordination atoms, which were predicted to be one of the important origins of the high catalytic activity of small gold nanoparticles [19–21]. This is attributed to stronger interactions of the low-coordination atoms with the precursor molecules landing on the shell (hill) during the pro-nucleation stage of growth. The local electronic structure of these molecules modified by the low coordination atoms required significantly reduced reaction barriers, for example, in Au nanoparticles relative to those in the close-packed gold surfaces [22]. The catalytic activity of the Au nanoparticles depended as well on the concentration of active surface sites [23].

References

1. A. Moisala, A.G. Nasibulin, E.I. Kauppinen, The role of metal nanoparticles in the catalytic production of single-walled carbon nanotubes—a review. J. Phys. Condens. Matter **15**, S3011–S3035 (2003)
2. S. Hofmann, G. Csanyi, A.C. Ferrari, M.C. Payne, J. Robertson, Surface diffusion: the low activation energy path for nanotube growth. Phys. Rev. Lett. **036101**, 95 (2005)

3. S.N. Mohammad, Systematic investigation of the growth mechanisms for the synthesis of the conventional, doped, and bamboo-shaped nanotubes, primarily the carbon nanotubes. Carbon **75**, 133–148 (2014)
4. S.N. Mohammad, Some possible rules governing the syntheses and characteristics of nanotubes, particularly carbon nanotubes. Carbon **71**, 34–46 (2014)
5. C.J. Smithell, *Smithells Metals Reference Book*, ed. by E.A. Brandes, G. Brook (Butterworth-Heinemann, Oxford, 1992)
6. A.D. Le Claire, *Landolt-Börnstein—Numerical Data and Functional Relationships in Science and Technology*, vol. 26, ed. by H. Mehrer (Springer, Berlin, 1992), p. 471
7. C. Ducati, I. Alexandrou, M. Chhowalla, G.A.J. Amaratunga, J. Robertson, Temperature selective growth of carbon nanotubes by chemical vapor deposition. J. Appl. Phys. **92**, 3299 (2002)
8. O.V. Yazyev, A. Pasquarello, Effect of metal elements in catalytic growth of carbon nanotubes. Phys. Rev. Lett. **100**, 156102 (2008)
9. J.F. Mojica, L.L. Levenson, Bulk-to-surface precipitation and surface diffusion of carbon on polycrystalline nickel. Surf. Sci. **59**(2), 447–460 (1976)
10. R. Heimburger, Solution growth of microcrystalline silicon on amorphous substrates. Ph.D. thesis, Humboldt-Universität, Berlin (1981)
11. C.B. Maliakkal, E.K. Mårtensson, M.U. Tornberg, D. Jacobsson, A.R. Persson, J. Johansson, L.R. Wallenberg, K.A. Dick, Independent control of nucleation and layer growth in nanowires. ACS Nano **14**, 3868–3875 (2020)
12. X. Wen, Y. Fang, S. Yang, Synthesis of ultrathin zinc nanowires and nanotubes by vapor transport. Angew. Chem. Int. Ed. **44**, 3562 (2005)
13. W.D. Kaplan, D. Chatain, P. Wynblatt, W.C. Carter, A review of wetting versus adsorption, complexions, and related phenomena: the rosetta stone of wetting. J. Mater. Sci. **48**, 5681–5717 (2013)
14. V. Schmidt, J.V. Wittemann, S. Senz, U. Gosele, Silicon nanowires: a review on aspects of their growth and their electrical properties. Adv. Mater. **21**, 2681–2702 (2009)
15. V.A. Nebol'sin, A.A. Shchetinin, Role of surface energy in the vapor-liquid-solid growth of silicon. Inorg. Mater. **39**, 899–903 (2003)
16. H.F. Al-Taay, M.A. Mahdi, D. Parlevliet, P. Jennings, Controlling the diameter of silicon nanowires grown using a tin catalyst. Mater. Sci. Semicond. Process. **16**, 15–22 (2013)
17. S. Ganji, Hill model for the base growths and tip growths of doped and undoped carbon nanotubes. J. Nanosci. Nanotechnol. **18**, 7623–7640 (2018)
18. T. Fujita, P. Guan, K. McKenna, X. Lang, A. Hirata, L. Zhang, T. Tokunaga, S. Arai, Y. Yamamoto, N. Tanaka, Y. Ishikawa, N. Asao, Y. Yamamoto, J. Erlebacher, M. Chen, Atomic origins of the high catalytic activity of nanoporous gold. Nat. Mater. **11**, 775–780 (2012)
19. B. Hvolbæk, T.V.W. Janssens, B.S. Clausen, H.F. Claus, H. Christensen, J.K. Nørskov, Catalytic activity of Au nanoparticles. Nano Today **2**, 14–18 (2007)
20. C. Lemire, R. Meyer, S. Shaikhutdinov, H.-J. Freund, Do quantum size effects control CO adsorption on gold nanoparticles? Angew. Chem. Int. Edn. **43**, 118–121 (2003)
21. N. Lopez, T.V.W. Janssens, B.S. Clausen, Y. Xu, M. Mavrikakis, T. Bliggard, J.K. Nørskov, On the origin of the catalytic activity of gold nanoparticles for low-temperature CO oxidation. J. Catal. **223**, 232–235 (2004)
22. L.M. Molina, B. Hammer, Theoretical study of CO oxidation on Au nanoparticles supported by MgO(100). Phys. Rev. B **69**, 155424 (2004)
23. M. Boudart, Turnover rates in heterogeneous catalysis. Chem. Rev. **95**, 661–666 (1995)

Chapter 5
The VLS Mechanism

Abstract Vapor–liquid–solid (VLS) growth mechanism for the growths of one-dimensional and quasi-two-dimensional nanomaterials has been reviewed in some details. The historical background of the VLS mechanism has first been presented. The most important requirements of the VLS mechanism to be effective for growths have then been laid down. Eutectic phase has also been described. Various elements of this phase such as the formation of droplets, incubation time, thermodynamic and kinetic conditions for incubation, supersaturation, nucleation and growth, binary phases and diagrams of binary phases, and the critical elements pertaining to the formation of droplets have then been elaborated. FECA metal selection and the possible means to perform this selection have been narrated. Growth dynamics and the temperature dependency of this growth have been discussed. While examining the failures of the VLS mechanism for nanomaterial growths, important observations pertaining to silicon nanowire growth rate as function of growth parameters, InAs nanowire growth rate as function of growth parameters, and the lack of atomic-scale control over growths by the VLS mechanism have been depicted. While examining the failure of the VLS mechanism in mediating the carbon nanotube growths, location, shape, and size of the catalyst particle during these growths, crystallographic relationship during growths, nanoparticle conditions during growths, and the presumed stages of growths have been chronicled. Finally, various criteria for the VLS growths have been documented.

5.1 Historical Background

During the past years, much of the research pertaining to the VLS growth of one-dimensional X_mY_n nanostructures has been directed to the empirical effect of varying the processing parameters, such as temperature, pressure, and time. As shown in Fig. 5.1, this growth requires the $R_S{\equiv}X$ and $R_S{\equiv}Y$ vapor species released from precursor to pass through a eutectic alloy liquid droplet, to undergo supersaturation and nucleation, and finally to crystallize into solid X_mY_n nanostructures. This alloy droplet adsorbs $R_S{\equiv}X$ and $R_S{\equiv}Y$ vapors, which upon supersaturation yields

S. N. Mohammad, *Synthesis of Nanomaterials*, Springer Series in Materials Science 307, https://doi.org/10.1007/978-3-030-57585-4_5

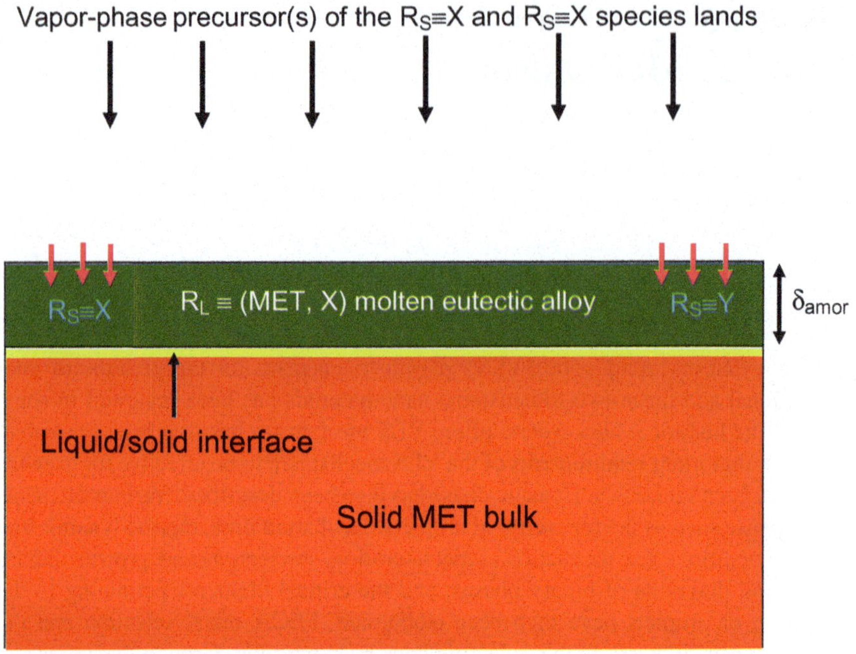

Fig. 5.1 Schematic diagram showing the VLS mechanism for nanomaterial growth. R_L≡(MET, X) molten eutectic alloy of thickness δ_{amor} is formed on the top surface of solid MET bulk. The liquid/solid (L/S) interface is created in between the solid bulk and the nanomaterial (e.g., nanowire) formed on it. R_S≡X and R_S≡Y species, which are generated from their precursor(s) landing on the MET surface, undergo bulk diffusion through the R_L≡(MET, X) molten eutectic alloy to the L/S interface for supersaturation, nucleation, and growth of nanomaterials

solid X_mY_n nanostructures (nanomaterials and nanocrystals). An interface called liquid/solid (L/S) interface is hence created between the liquid droplet and the solid X_mY_n nanomaterial formed underneath it. The alloy liquid droplet is therefore the preferred tool for nanomaterial growths. The precursors of the R_S≡X and R_S≡Y vapor species must crack on the droplet surface at a preferred growth temperature in order to release the R_S≡X and R_S≡Y vapor species for growth.

5.2 Important Requirements

1. The most important requirement of the VLS mechanism to be effective for growth is that a suitable metal, plausibly in the absence of impurity materials(s), must form a eutectic liquid droplet, and that the crystalline nanomaterial (e.g., nanowire) should be grown at an appropriate deposition temperature. In order for this to happen, the distribution coefficient of the metal element in the eutectic alloy must be less than unity at the said deposition temperature.

2. The liquid droplet must experience the minimal equilibrium vapor pressure.
3. Together with impurity, if present in the chamber, the FECA metal used for growth must not react with the nanomaterial source species (e.g., $R_S{\equiv}X$ and $R_S{\equiv}Y$) during growth in the chamber.
4. The interfacial energy dictates wetting angle. It plays a key role for growth. It must be optimal. Some important parameters, such as diameter, of the growing nanowire are greatly influenced by the wetting characteristics of the droplet. In general, the smaller the wetting angle, the larger is the growth area and hence the larger is the nanowire diameter.
5. One of the constituents (viz., metal) of compound semiconductors can serve as the metal element of the droplet for growth.
6. For controlled unidirectional growth, the solid–liquid interface should be defined in a judicious crystallographic order. For this, for example, single-crystal substrate with desired crystal orientation may preferably be chosen.

5.3 The Eutectic Phase

5.3.1 Formation of Droplet

Discovered in 1964, the VLS mechanism [1] is widely employed for the vapor deposition of nanomaterials, particularly nanowires. It follows some rules [2]. Also, for enabling deposition, it makes use of METANOs derived from FECA metals such as transition metals and noble metals [3–6]. The central feature of the said deposition is the eutectic phase of the $R_L{\equiv}$(MET, X, Y) alloy composed of MET, $R_S{\equiv}X$ and $R_S{\equiv}Y$ species formed during the pre-nucleation stage of growth at the eutectic temperature T_E. The $R_S{\equiv}Y$ species is generally volatile and may hence be absent in the R_L species. In fact, most often the R_L species is the $R_L{\equiv}$(MET, X) alloy species. Being eutectic, it is molten and hence a droplet at the eutectic temperature T_E. It serves as the medium for the diffusion of the $R_S{\equiv}X$ and $R_S{\equiv}Y$ vapor species through nanoparticle surface during the pro-nucleation stage of growth for nanomaterial synthesis (see Fig. 5.1). There is an optimal value of SECINI, called SECINI0 (see Sect. 3.3.3), which at $T = T_E$ corresponds to the most molten state of the $R_L{\equiv}$(MET, X) eutectic alloy suitable for the uninterrupted, smoothest diffusion of the $R_S{\equiv}X$ and $R_S{\equiv}Y$ vapor species through the R_L species droplet. The eutectic droplet possesses the highest stability and the minimum-energy configuration at the lowest possible eutectic temperature T_E on the liquidus line of the binary phase diagram. Supersaturation is the difference in chemical potential (hereafter referred to as $\Delta\mu$) between the $R_S{\equiv}X$ species in the vapor and in the solid macrostates; it is the driving force for growths by the VLS mechanism.

Ressel et al. [7] studied the wetting of silicon surfaces by $R_L{\equiv}$(Au, Si) liquid droplets. They observed a constant contact angle of the droplet with the Si surface at temperatures between the eutectic temperature (e.g., $T_E = 363$ °C) and the temperature $T = 650$ °C. The shape of the liquid droplet changed from circular at the

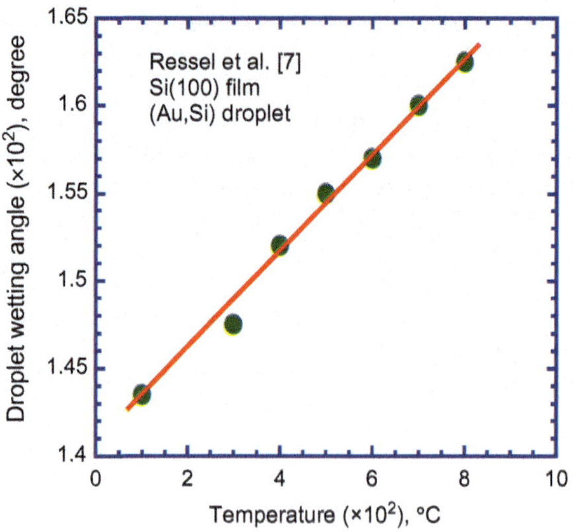

Fig. 5.2 Variation of the wetting angle of R_L≡(Au, Si) droplets formed on Si(1000) surface. The plot was made with experimental data by Ressel et al. [7]

eutectic temperature to hexagonal at higher temperatures. The variation of contact angle as function of temperature of Si(100) surface is depicted in Fig. 5.2. This variation was essentially linear for temperatures between 100 and 800 °C. Detz et al. [8] studied gallium droplet nucleation on silicon(100) substrates with and without the presence of the native oxide. For this study, Ga was deposited under ultrahigh vacuum conditions at temperatures between 580 and 630 °C. It was observed that the droplet volume increased with increase in Ga thickness (see Fig. 5.3), but decreased with annealing time at a temperature of 630 °C (see Fig. 5.4). This implied that the higher the availability of Ga, the higher is the droplet volume. It also implied that Ga atoms are re-evaporated from or diffused into the substrate due to annealing.

5.3.2 *Incubation Time*

Incubation time is the time required for nanowire growth to begin. Unfortunately, the VLS growth of nanowires occurs with a characteristic, temperature-dependent time delay implying that the incubation time for VLS growth is relatively large. It may be on the order of several seconds to several minutes. It is perhaps the inherent property of the VLS growth mechanism, particularly for growths carried out at low temperatures. The variation of incubation time as function of temperature inherently depends on the activation energy of diffusion of the R_S (e.g., R_S≡X, R_S≡Y) species through the droplet. It is hence an important element of the VLS growth kinetics as apparent from an experiment by Kalache et al. [9]. In this experiment, the variations of

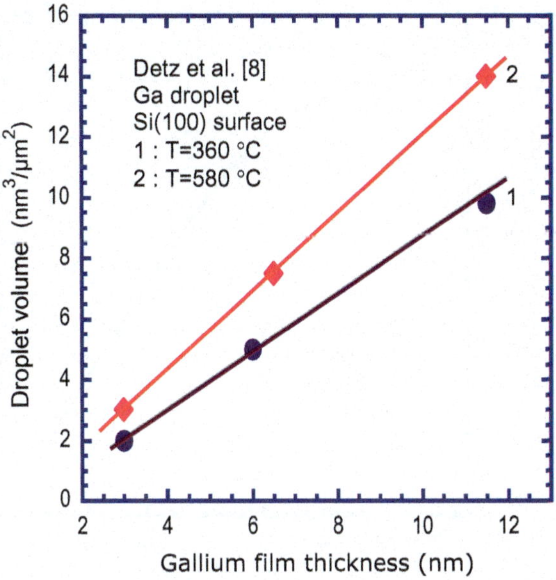

Fig. 5.3 Variation of Ga droplet volume with Ga film thickness. These droplet volumes were produced from Ga film on Si(100) surface by MBE technique at $T = 360$ °C and $T = 580$ °C, respectively. The plots were made with experimental data by Detz et al. [8]

incubation time with growth temperature for the Au- and Cu-mediated VLS growths of Si nanowires were carried out. Figure 5.5 shows the incubation time varying inversely with growth temperature. It suggests that the growth temperature must be higher in order to circumvent excessive delay in nanowire growth.

5.3.3 Thermodynamic and Kinetic Conditions for Incubation Time

We reiterate that the incubation time is the time required for the growth to begin. It is at least a fraction of the time required by the $R_S \equiv X$ species to form the most stable $R_L \equiv (MET, X)$ eutectic alloy droplet. This $R_L \equiv (MET, X)$ eutectic alloy is most stable only under those thermodynamic and kinetic conditions that lead to the activation of the MET element. This means the VLS process is most successful only if the adsorption of the $R_S \equiv X$ species by the METANO surface is so effective that it creates $R_L \equiv (MET, X)$ eutectic alloy at the temperature $T = T_E$. A rapid activation of MET element for the said adsorption is very crucial for growth. We believe that the most important element underlying the creation of the eutectic phase for VLS growth is the SECINI≡SECINI0. SECINI may not be SECINI0 for the non-eutectic phase. The sensitivity of the VLS process to activation and growth is the attribute

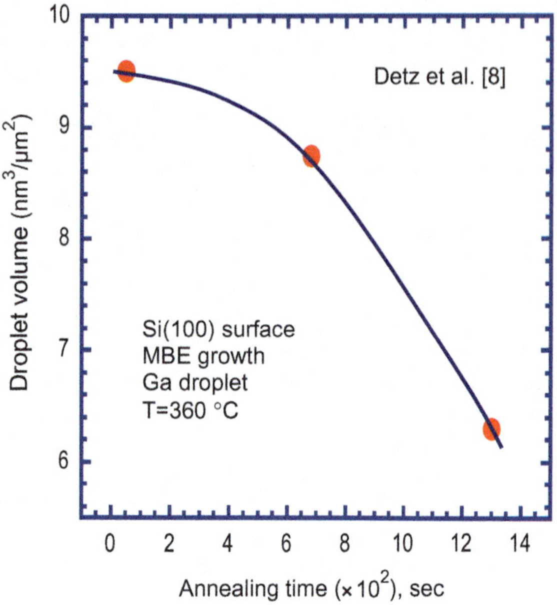

Fig. 5.4 Variation of Ga droplet volume with annealing time. This droplet was produced on Si(100) surface from Ga film by MBE technique. The plot was made with experimental data by Detz et al. [8]

of SECINIO. The formation of a molten phase and the SECINIO of this phase are determined by the growth temperature, chamber pressure, contaminant(s) present in the growth chamber, ambient (e.g., Ar, H_2, N_2, etc.) for growth, and also importantly the structure and properties of METANO and X. It is determined as well by the characteristics of the substrate and support on which the METANO may be formed. The eutectic or non-eutectic composition of the R_L≡(MET, X) alloy also dictates the incubation time. The R_L≡(MET, X) eutectic phase may not be molten if it does not, for example, have SECIINI equal to or close to SECINIO. This phase may result if the alloy formation follows the liquidus line of the phase diagram. It may happen even at a temperature T higher or lower than the eutectic temperature T_E if, for example, the chamber pressure, precursor flow rate, and ambient gases are optimal not at the lowest possible eutectic temperature T_E, but at a temperature $T < T_E$ or $T > T_E$.

5.3.4 Binary Phase Diagrams

For the sake of clarity, the phase diagrams [10], for example, of the binary R_L≡(Au, Si) alloy, R_L≡(Ti, Si) alloy, and R_L≡(Pb, C) alloy are shown in Fig. 5.6a–c. They have distinctly different features. And these are only three of several different features of the R_L≡(MET, X) binary phase diagrams. The phase at the bottom of the V-shaped

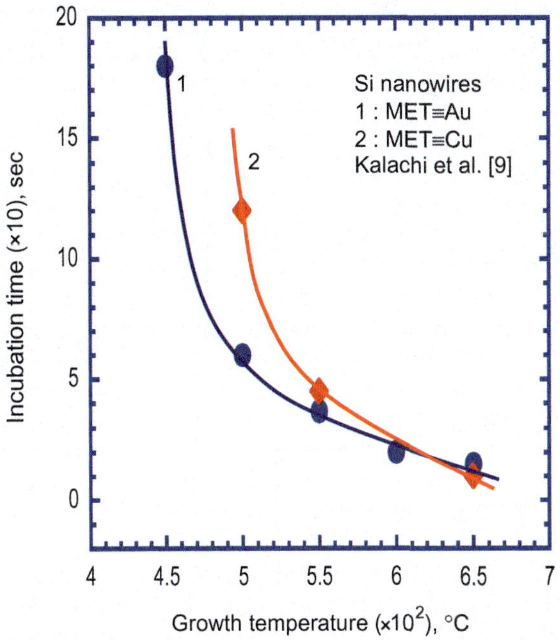

Fig. 5.5 Variation of incubation time with growth temperature for silicon nanowire growths. The growths were made on Au and Cu catalysts, respectively. The plots were made with the experimental data by Kalachi et al. [9]

region of the R_L≡(Au, Si) phase diagram corresponds to the lowest-temperature liquid state. The actual composition of the liquid state on the liquidus line depends on the amount of Si needed for the formation of R_L≡(Au, Si) eutectic alloy droplets. The composition of these R_L≡(Au, Si) alloy droplets is determined by the position (e.g., position A in Fig. 5.6a) on the liquidus line. This means the R_L≡(Au, Si) alloy droplet remains eutectic if Si atoms released from its precursor (e.g., SiH_4) are supplied to the R_L≡(Au, Si) droplet in such a way that the silicon content of it is identical to that needed by the liquidus line at a temperature increased or decreased from its equilibrium value of $T = T_E$. It may otherwise correspond to a location inside or outside of the V-shaped region of the phase diagram. The R_L≡(Au, Si) eutectic droplet may thus be formed at any temperature, for instance, between 363 and 700 °C. The R_L≡(Au, Si) alloy is not, otherwise, eutectic needed for nanomaterial growth by the VLS mechanism. The R_L≡(Al, Si) alloy phase diagram and the R_L≡(Ag, Si) alloy phase diagram resemble the R_L≡(Au, Si) alloy phase diagram. However, the Si content in the R_L≡(Al, Si) alloy is only 12 atomic % at the eutectic temperature of 577 °C and the Si content in the R_L≡(Ag, Si) alloy is only 11 atomic % at the eutectic temperature of 836 °C. These eutectic alloys with lower Si content occur at a temperature much higher than that of the R_L≡(Au, Si) alloy. This means Si is more abundant for the low-temperature VLS growth of Si nanowires with R_L≡(Au, Si)

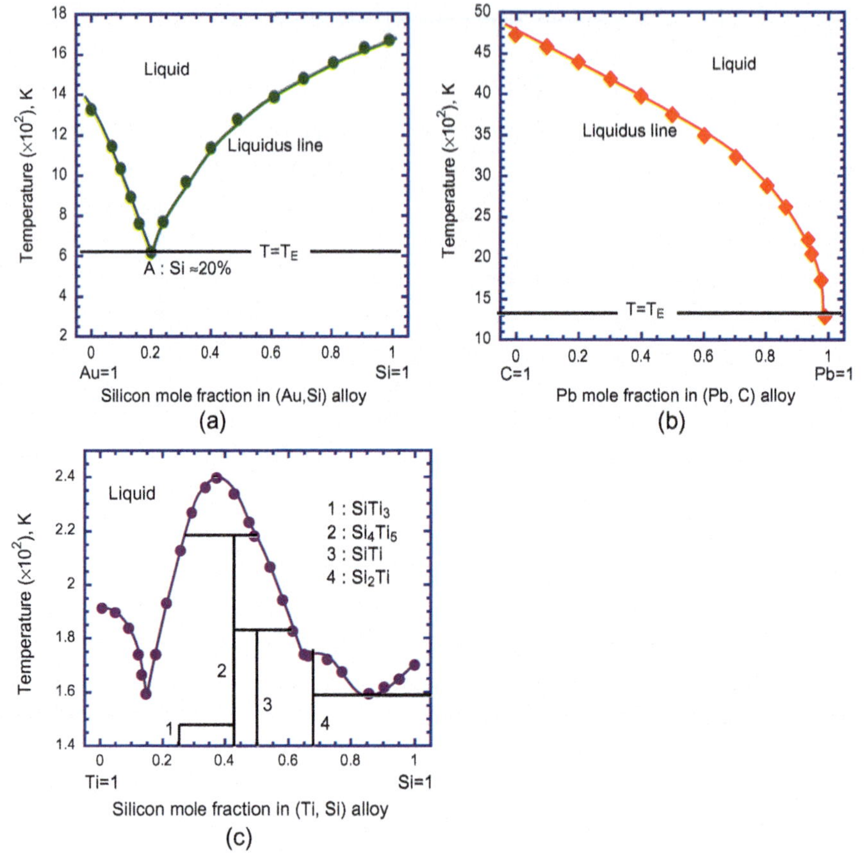

Fig. 5.6 Binary phase diagrams of **a** $R_L \equiv$(Au, Si), **b** $R_L \equiv$(Pb, C), and **c** $R_L \equiv$(Ti, Si) alloys. These diagrams are created with data by Massalski [10]

alloy and Si is less abundant for the high-temperature VLS growth of Si nanowires with the $R_L \equiv$(Al, Si) alloy and $R_L \equiv$(Ag, Si) alloy.

The binary phase diagram of Fig. 5.6b is different from that of Fig. 5.6a. But it resembles the binary phase diagram, for example, of the (Zn, Si) alloy, (Ni, C) alloy, and (Cu, C) alloy. They are dominated by a single eutectic point for very low content of the MET [e.g., MET≡Zn, Ni, or Cu, and X≡C, Si, or Ge]. These METs have proven to be effective FECA metals for nanomaterial growths despite very low solubility of $R_S \equiv$X in them. These growths may not though be the VLS growths. The binary phase diagram of Fig. 5.6c is again different from those of Fig. 5.6a, b. This binary phase diagram resembles many other (MET, X) binary phase diagrams and is characterized by the formation of intermediate non-eutectic alloys (solid solutions) before the formation of eutectic alloy phase. For example, (Ti, Si) binary phase diagram indicates that MET≡Ti will first transform into Ti_5Si_3, then into Ti_5Si_4, next into TiSi, and finally into $TiSi_2$. Such transformations are common for MET-mediated CNT growths. Kamins et al. [11] grew Si nanowires at 640–670 °C; the R_L≡(Ti, Si) alloy species mediating this growth exhibited less than 1 atomic % Si.

5.3.5 *Important Requirements for the Formation of Droplets*

We emphasize that FECA metal, R_S≡X, incubation time, annealing time, annealing temperature, and the ambient(s) for annealing are important elements for the formation of the R_L≡(MET, X) molten alloy. These parameters must yield SECINI≡SECINI0 within a reasonable incubation time period in order to be most conducive to the realization of the R_L≡(MET, X) molten phase. Else, the R_L≡(MET, X) alloy thus formed may be non-eutectic, generally metastable, and even solid. It may be R_L≡(MET, X) solid solution or cluster. Due to the lack of long-range lattice order, this solid solution is disturbed, disordered, and hence amorphous (semi-amorphous and amorphous-like). It has grains, grain boundaries, voids, vacancies, dislocations, and nanopores. This means R_L≡(MET, X) species may easily be decomposed (dissociated) into MET species and R_S≡X species under the influence, for example, of nanoparticle vibration, non-uniform temperature and pressure, and even the injection of some ionic species onto the nanoparticle surface.

We cite several examples. Iacopi et al. [12] employed SiH_4 precursor and MET≡Al for the PECVD growth of Si nanowires. The carrier gas Ar and the precursor SiH_4 were in 1:1 ratio. The eutectic temperature for the R_L≡(Al, Si) eutectic alloy droplet is T_E = 577 °C, and the Si mole fraction for this alloy is ~12.6 atomic %. Two temperatures: T = 500 °C and T = 600 °C were employed for the growth. The temperature T = 500 °C was lower than the eutectic temperature T_E of the R_L≡(Al, Si) eutectic alloy droplet. Si nanowires by the VLS mechanism could not, therefore, be achieved. Instead, MET≡Al was buried under a Si layer, about 100 nm thick. The growth temperature T = 600 °C was, on the other hand, higher than the eutectic temperature T_E for the R_L≡(Al, Si) alloy droplet. And yet no growth of Si nanowires by the VLS mechanism could be achieved. Although a few short nanowires could be produced, the MET≡Al was largely buried under an amorphous Si layer. And it happened for two different reasons. First, the incubation time was too long to yield the R_L≡(Al, Si) eutectic alloy droplet. Second, the Si mole fraction in the R_L≡(Al, Si) alloy was lower than 14.50 atomic %, which was needed for the alloy to be eutectic and molten at T_E = 600 °C. Hence, the R_S≡Si species deposited on the Al nanoparticle surface generated an amorphous coating. This amorphous coating poisoned MET≡Al inhibiting the formation of eutectic phase of the R_L≡(Al, Si) alloy. SECINI was also different from SECINI0, and the R_L≡(MET, X) alloy was actually non-eutectic and plausibly a solid solution. Very often large incubation time and/or the presence of contaminant(s) in the R_L≡(MET, X) alloy could have deterred smooth transformation of this alloy to the eutectic phase. This was apparent when Iacopi et al. [12] repeated their experiment using H_2:SiH_4:Ar gas mixture in the 2:1:1 ratio. H_2 gas of such a gas mixture reacted with SiH_4 yielding Si-H radicals strong enough to etch amorphous silicon coating formed on the nanoparticle surface. The immediate result of this was enhanced silicon flow rate and the formation of the R_L≡(Al, Si) molten alloy of Si atomic % of ~14.50. High density of silicon nanowires was consequently formed at a temperature $T = 600$ °C.

5.4 FECA Metal Selection

5.4.1 Possible Selection Criteria

The selection of a FECA metal appropriate for the processing of the R_S (R_S≡X and R_S≡Y) species into quasi-one-dimensional nanostructures is challenging. This is generally done by taking the equilibrium phase diagram into account. And this phase diagram can be a pseudo-binary phase diagram between FECA metal and R_S≡X. It should preferably be a pseudo-binary phase diagram between FECA metal, R_S≡X, and R_S≡Y, if possible. It must however be recognized that the alloying behavior of FECA metal with at least the R_S≡X species is not straightforward. Nevertheless, FECA metals are essential for nanomaterial growth by the VLS mechanism, and only those metals which satisfy the following requirements can catalyze this growth. First, the FECA metal, together with the R_S≡X species, must form a molten or semimolten R_L≡(MET, X) alloy at a suitable temperature T. This molten alloy should preferably be eutectic at the lowest possible eutectic temperature T_E. Second, assuming that R_S≡Y species is volatile, the solubility of the R_S≡X species in the R_L≡(MET, X) eutectic alloy formed during the pre-nucleation stage of growth should be much higher than that in both the solid FECA metal and the R_L≡(MET, X) non-eutectic alloy. For example, the solubility χ_S of R_S≡X should be such that $\chi_S = (C_s/C_l) < 1$, where C_s is the solubility limit in the solid FECA metal and C_l is the solubility limit in the liquid R_L≡(MET, X) liquid. This is necessary for the R_S≡X and R_S≡X species to easily diffuse through the liquid alloy for growth during the pro-nucleation stage of growth. Third, the vapor pressure P of the FECA metal over the eutectic and molten R_L≡(MET, X) alloy must be small, if not insignificant. This ensures that the FECA metal MET of the eutectic R_L≡(MET, X) alloy does not evaporate eventually disappearing during nanomaterial growth. Fourth, the eutectic R_L≡(MET, X) alloy must be inert to chemical reactions. It suffers otherwise from the degradation of catalytic action. Fifth, the eutectic R_L≡(MET, X) alloy must not have any transformed intermediate phase during growth, as it can cause degradation of its catalytic action [1]. All these are generally, but not always, guaranteed in an environment of constant and uniform temperature and pressure in the growth chamber.

5.4.2 *Illustrative Examples*

An experiment by Nguyen et al. [13] provides an excellent example of the tacit requirements of a metal to serve as FECA metal MET for the VLS growth. Dedicated to examining the suitability of various metals as FECA metals for the VLS growths of SnO_2 nanowires at $T \geq T_E$, this experiment made use of MET≡Ta, W, Ir, Pt, Au, and Al as catalysts. It was found that Ta could hardly produce SnO_2 nanowires. It rather produced three-dimensional well-faceted polygonal and mountainous structures. MET≡Ir, however, produced a few nanowires (~90 nm in diameter and 0.7 μm in length) sporadically spread on the substrate surface. Additionally, this surface was coated with pyramidal nanoparticles (~100 nm). MET≡Pt failed to produce any nanowire at all. Instead, it produced irregular-shaped nanoparticle agglomerates on the substrate surface. In contrast, MET≡Au and MET≡Al produced high-density SnO_2 nanowires. While the nanowire diameter d_{NW} of these nanowires was large with Au, it was small with Al. These observations suggest that the R_L≡(Au, Sn) and R_L≡(Al, Sn) alloys satisfied, but the R_L≡(Ta, Sn), R_L≡(W, Sn), R_L≡(Ir, Sn), and R_L≡(Pt, Sn) alloys did not satisfy the aforementioned requirement(s) at the growth temperature. These requirements include the realization of SECINI = SECINI0 and a reasonably contaminant-free environment for the SnO_2 nanowire growth by the VLS mechanism. The experiment by Nguyen et al. [13] is probably a good example of the selectivity of FECA metal MET and R_S≡X for VLS growths. It demonstrates that only certain metal catalysts are good enough to produce eutectic R_L≡(MET, X) alloy at a certain temperature T with a certain R_S≡X species for the VLS growths. It confirms that the selection process for metal for a certain X_mY_n nanomaterial growth must take into consideration the requirements spelled out above. It must ensure that SECINI for the R_L≡(MET, X) eutectic alloy for this nanomaterial growth is equal or close to SECINI0.

5.5 Growth Dynamics

Most nanowire growths by the VLS mechanism appear to take place under near-equilibrium conditions. Under these conditions, the growth is plausibly driven by thermodynamic processes. Also, the preferred growth mode and probably the growth direction tend to minimize the total free energy. Nevertheless, kinetic effects govern various stages of the nanowire growths. Obviously, they play crucial role in determining the overall nanocrystal characteristics and also growth dynamics. Note that the total free energy of nanocrystal (e.g., nanowire) is the sum total of its bulk energy, the energy at its interface with FECA metal nanoparticle (e.g., METANO), the energy at its interface with vapor (vacuum), and the liquid/solid interface energy. Among the component energies, the interface energies depend on the interface direction, implying that the VLS mechanism has the advantage of yielding nanocrystal in a preferred growth direction. Let us cite an example. Micrometer-scale Si whiskers grown by the VLS mechanism showed a dominant preference along the <1111> direction, and it was determined by the formation of a single lowest-free-energy solid/liquid interface parallel to a single (111) plane [14]. Even though it is recognized that the fundamentals of the nanocrystal growth process are nested in the thermodynamics of this growth, the nanocrystal quality and characteristics are determined primarily by the growth kinetics. They are reliant as well on growth parameters. Depending on these parameters, the FECA metal-mediated precipitation at the liquid/solid interface leads to axial elongation, and the FECA metal-independent direct vapor deposition on the nanocrystal sidewall surface leads to radial thickening and eventual tapering of the nanocrystal (e.g., nanowire) length. There may also be deposition on the substrate (nanoparticle) surface at the nanocrystal root (base).

5.6 Temperature Dependency

5.6.1 Inconsistency in Growth Characteristics

The VLS growth of nanowires is heavily dependent on temperature [15–29]. This dependency is though sometimes puzzling. We cite an example with the temperature-dependent GaAs nanowire growth rate by Tchernycheva et al. [24] and by Soci et al. [29]. Tchernycheva et al. found that the GaAs nanowire growth rate increases with temperature, reaches a peak, and then decreases with further increase in temperature T. Soci et al., however, found that nanowire growth rate increases monotonically with increase in temperature. Soci et al. observed that GaAs nanowire growth rate increases also with increase in time and the trimethylgallium molar flow. We believe that the observation by Tchernycheva et al. is probably due to temperature-dependent change in the molten (semimolten) condition of the droplet. At low temperature, the R_L≡(MET, X) alloy is not molten enough and eutectic enough to create droplet for smooth diffusion of the R_S species and hence for high enough nanowire growth. This diffusion rate is low, and the corresponding nanowire growth rate is also low. As temperature increases, the R_L≡(MET, X) alloy becomes increasingly softer reaching an optimal stage at which the R_L≡(MET, X) alloy is molten enough to create droplet for the highest possible diffusion of the R_S species and hence the highest possible nanowire growth rate. An increase in temperature beyond this limit leads the droplet to be increasingly non-eutectic and unstable and hence increasingly susceptible to a lower rate of nanowire growth.

5.6.2 Several VLS Growth Rates Compared

Growth temperatures and related parameters for the Au-catalyzed VLS growth of some Si nanowires [15–23] and GaAs nanowires [24–29] are listed in Tables 5.1 and

Table 5.1 Au-catalyzed VLS growth characteristics of Si nanowires (NWs)

No.	Growth temp (°C)	Growth technique	Source	Pressure (Torr)	Comments	References
1	365–495	CVD	SiH_4	735	SiH_4 flow rate up to 2000 sccm. Growth rate 0.2 nm/s at 365 °C and 180 nm/s at 495 °C	Kikkawa [15]
2	380	PECVD		0.3–1.8	SiO_2 used as catalyst diffusion barrier Si NWs with Si core and oxide sheath	Hofmann [16]
3	400	LPCVD	SiH_4	0.1–0.3	Undiluted SiH_4 together with N_2 was the silicon source; Au particle size 10 nm	Albuschies [17]
4	520	MBE	SiH_4	5×10^{-8}	SiH_4 flow rate at 40 sccm	Westwater [18]
5	850	CVD	$SiCl_4$	760	Growth rates of intrinsic- and B-doped Si NWs were 91 and 106 nm/s, respectively	Chung [19]
6	500	CVD	SiH_4	–	Average NW diameter 200 ± 54 nm close to the pore size of membranes	Lew [20]
7	550	CVD	SiH_4	0.15	SiH_4 flow rate 50 sccm. NWs grown on Si substrate were doped n-type	Gunawan [21]
8	600	UHVEM	SiH_4	5×10^{-5}	$R_L \equiv$(Au, Si) droplets were formed	Hannon [22]
9	680	CVD	SiH_4	95	SiH_4 flow rate 15 sccm. Au-coated wafer annealed in H_2 at 400–900 °C	Sharma [23]

The eutectic temperature of the $R_L \equiv$(Au, Si) alloy is 363 °C; UHVEM is abbreviation for ultra-high vacuum electron microscopy

5.2, respectively. These growths were performed by several different techniques. Recall that the eutectic temperature $T_E = 363$ °C for the (Au, Si) eutectic alloy (e.g., $Au_{1-z}Si_z$, $z = 0.18$), but 339.4 °C for the (Au, Ga) eutectic alloy (e.g., $Au_{1-z}Si_z$, $z = 0.31$). This means Si nanowire growth should be most conveniently performed at $T \approx 363$ °C, but the GaAs nanowire growth should be most conveniently performed at $T \approx 339.4$ °C, both by the Au-catalyzed VLS mechanism. However, various entries of Tables 5.1 and 5.2 indicate that the growth temperatures for Si nanowires are 363 °C $\leq T \leq$ 680 °C and the growth temperatures for GaAs nanowires are 320 °C $\leq T \leq$ 580 °C. The temperature domains for both the growths are quite wide.

Table 5.2 Au-catalyzed VLS growth characteristics of GaAs nanowires (NWs)

No	Growth temp (°C)	Growth technique	V/III ratio	Comments	References
1	320–620	MBE	2	Droplets after annealing formed a thin deposit of Au on GaAs buffer, 100 nm thick, prior to growth	Tchernycheva [24]
2	450	MOCVD	70	Trimethylgallium and trimethylaluminum sources on Si substrate; NWs free from stacking faults	Huang [25]
3	600	MBE	200	Grown on GaAs substrate using As_4 and Ga, NWs were free from stacking faults	Shtrikman [26]
4	580	MBE	9	Grown on Si substrate, NWs had a mixed crystal structure of the wurtzite and the cubic zinc blende	Ihn [27]
5	420–520	MOCVD	36–90	Trimethylgallium and AsH_3 produced NWs on GaAs substrate	Fortuna [28]
6	400	MOCVD	6.5	Trimethylgallium and AsH_3 produced NWs of minimum diameter 2.75 nm on GaAs substrate	Soci [29]

The eutectic temperature of R_L≡(Au, Ga) alloy is 339.4–348.9 °C

Also, the growth temperature 320 °C is lower than the eutectic temperature T_E = 361 °C of the R_L≡(Au, Ga) alloy (e.g., $Au_{1-z}Si_z$, $z = 0.31$). They represent deviation from the ideal VLS growths. There are several possible reasons for this. First, the molten R_L≡(MET, X) alloy with different X compositions may occur at different temperatures on the liquidus line of the binary phase diagram [10]. This is manifested by an experiment by Harmand et al. [30], who varied growth temperature, but kept all other parameters unaltered. Second, there may be different SECINIO under the influence of different combined effects of parameters such as chamber pressure (low partial pressure of reactants), growth temperature, growth technique, precursor flow rate, and contaminants, during growth. The formation of the R_L≡(MET, X) eutectic alloy takes place only for a certain optimal combination of chamber pressure (partial pressure of reactants), growth technique, precursor flow rate, and contaminant(s) yielding SECINI≡SECINI0. A very high or very low chamber pressure and precursor flow rate, in the presence or absence of contaminants, may, for example, lead the liquid (quasiliquid) binary phase and the SECINI≡SECINI0 to correspond to a temperature higher or lower than the eutectic temperature T_E stated above. So, it may cause the growth to take place at a temperature $T > T_E$ or $T < T_E$ instead of $T = T_E$.

5.6.3 *VLS Growth at Eutectic Temperature*

Kikkawa et al. [15] could grow Si nanowires at $T = T_E = 363$ °C, simply because the formation of the R_L≡(Au, Si) eutectic phase (e.g., $Au_{1-z}Si_z$, $z = 0.18$ with SECINI≡SECINI0) at $T = T_E = 363$ °C was not interfered by any of the said parameters. However, Sharma et al. [23] could grow Si nanowires at 680 °C simply because the formation of the eutectic R_L≡(Au, Si) alloy (e.g., $Au_{1-z}Si_z$, $z = 0.18$) was not realized at $T = T_E = 363$ °C, and the SECINI0 for it probably with Si mole fraction higher than $z = 0.18$ was realized only at a growth temperature $T \approx 680$ °C. Assuming that it was indeed the case, as the R_L≡(Au, Si) alloy was molten mediating growth by the VLS mechanism at $T \approx 680$ °C. Otherwise, it was solid mediating growth not by the VLS mechanism, but by the VSS mechanism. There could be another, rather remote, possible scenario. The R_L≡(Au, Si) alloy was too unstable at $T \approx 680$ °C to mediate nanowire growth. The nanowire growth took place when the temperature was eventually reduced to 363 °C and the R_L≡(Au, Si) species became stable and molten at $T = T_E = 363$ °C. Thus, considering that the $Au_{1-z}Si_z$ alloy was molten at $T \approx 680$ °C with increased Si content in the alloy (based on liquidus line), the definition of a binary liquid state in terms of SECINI0 may be envisioned. It may then better explain nanomaterial growths by the VLS mechanism, and it does it at $T = T_E$, $T > T_E$ and even $T < T_E$ under the combined influence of parameters used for these growths.

5.7 Failures of the VLS Mechanism to Mediate Some Nanowire Growths

5.7.1 *Lack of Atomic-Scale Control Over Growth*

Needless to say, during the past years, the VLS mechanism has been widely used to synthesize nanowires. It could not though provide atomic-scale control over the nanowire composition. A liquid droplet acts as a reservoir of material while mediating growth and also a change in the nanowire composition, if any [31, 32]. Due to fluctuations at high temperature and high pressure during growth, the structure, composition, morphology, and surface energy of the droplet frequently change during this growth. Recent in situ microscopic and spectroscopic studies demonstrate that the growth front, as a result, undergoes cyclical reshaping during dissolution and crystallization. The droplet surface changes also due to competition between adsorption and desorption of the passivating species. They all influence deviation from the most stable condition of the liquid droplet. The liquid droplet cannot therefore be expected to preserve atomic-level control over nanowire composition. Note that the absorption, reaction, and diffusion processes of the source species through the surface of METANOs (e.g., FECA metal nanoparticles) are also highly complicated. They depend largely on the experimental conditions and the source materials used for growth [33]. These experimental conditions and materials system contribute to deviations from the atomic-level control of nanowire growths. The deviations lead, for instance, to nanowire growths of Ge [34], Si [35], GaAs [36], and ZnO [37] nanowires at temperatures below their eutectic points. Wu and Yang [38] demonstrated the Au-catalyzed VLS growth of Ge nanowires at a temperature of 800–900 °C. Figure 5.7 shows that these nanowires had spherical (hemispherical) droplets composed of R_L≡(Au, Ge) alloy. The temperature for this growth was though far higher than the (Au, Ge) eutectic temperature of 361 °C. In situ TEM images recorded during nanowire growth showed (1) Au nanoclusters in solid state at 500 °C; (2) the beginning of the R_L≡(Au, Ge) alloying at 800 °C; (3) liquid (Au, Ge) alloy droplet

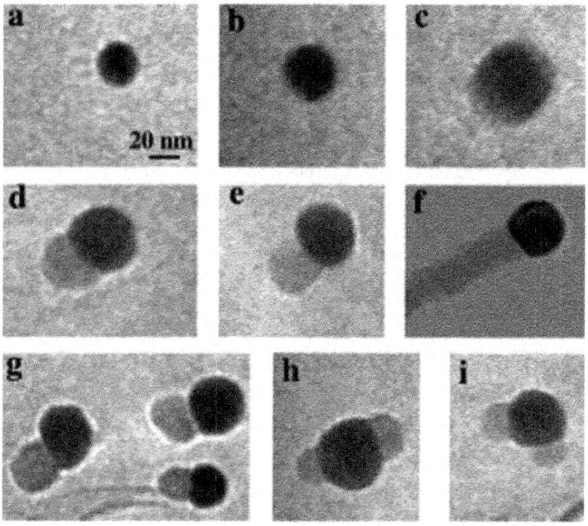

Fig. 5.7 In situ TEM images of Au-catalyzed Ge nanowire growth. **a** Au solid nanoclusters at 500 °C; **b** initiated alloying at 800 °C at which Au is essentially in the solid state; **c** liquid Au/Ge alloy formed due to alloying; **d** the Ge nanocrystal nucleation on the alloy surface; **e** elongated Ge nanocrystal with further condensation of Ge leading to the formation of Ge nanowires; **f** several other nucleations of Ge nanocrystals; and **g** two nucleations on one single alloy droplet. Reproduced from Wu and Yang [38] with copyright permission from the publisher

formed at 800–900 °C; and (4) the nucleation of Ge nanocrystal on the $R_L \equiv$(Au, Ge) alloy surface. Ge nanowires of diameters 20.6 ± 3.2, 24.6 ± 4.0, 29.3 ± 4.5, and 60.7 ± 6.2 nm were grown on the $R_L \equiv$(Au, Ge) clusters of sizes 15.3 ± 2.4, 20.1 ± 3.1, 25.6 ± 4.1, and 52.4 ± 5.3 nm, respectively. Note that these diameters are too large to undergo size-dependent melting point depression [39–41]. VLS growth of Ge nanowires at 800–900 °C could take place only if the Ge mole fraction of the $R_L \equiv$(Au, Ge) alloy, based on (Au/Ge) binary phase diagram [10], could be 0.42–0.48. And it would be possible only under high-pressure GeH_4 precursor. Nevertheless, the fundamentals of the governing physicochemical processes, particularly, the atomic-scale thermodynamic and kinetic phenomena, must be explored and understood in some details.

5.7.2 *Silicon Nanowire Growth Rate as Function of Various Growth Parameters*

Pinion [32] performed CVD growth of Au-catalyzed NWs on Si wafers by using 2 sccm of SiH_4 and 200 sccm of H_2 gases. The furnace temperature was ramped to 450 °C for growth for 5–60 min under a total reactor pressure of 40.0 Torr. The furnace temperature was then lowered to 420 °C. Temperature-dependent growth rates of nanowires, 30–200 nm in diameter and produced at a SiH_4 partial pressure of $P_{SiH4} = 0.4$ Torr, are shown in Fig. 5.8a. There were no statistically significant differences in growth rates for temperature $T \leq 390$ °C, and it was true for nanowires of all different diameters. However, the nanowire growth rates became almost exponential at temperature $T \geq 390$ °C. It was apparent more for larger diameter nanowires than for smaller diameter nanowires. It was probably a reflection of the diffusion coefficient varying exponentially with temperature. Variation of growth rates with nanowire diameter is shown in Fig. 5.8b. This figure shows that there is no dependence of growth rates on diameter at temperature $T < 380$ °C. The situation changed for diameters larger than about 20 nm. The growth rates were always higher at higher temperature. And the growth rate for a certain temperature became essentially diameter independent for diameters larger than about 50 nm. The variation of growth with SiH_4 partial pressure is shown in Fig. 5.8c. This variation for $T = 420$ °C indicates that the growth rate varies nonlinearly with the partial pressure P_{SiH4} for $P_{SiH4} < 0.2$ Torr, but linearly with the partial pressure P_{SiH4} for $P_{SiH4} > 0.2$ Torr. The growth rate always increases with increase in partial pressure P_{SiH4}, which is true for nanowires of all diameters. The growth rate as function of the partial pressure P_{SiH4} was though always higher for large diameter nanowire. Increase in the partial pressure P_{SiH4} led to increase in Si content available for nanowire growth. If the temperature is relatively high (e.g., 420 °C), the diffusivity D_{rs} of Si through droplet may be high enough to cause increase in nanowire growth rate. This is evident from Fig. 5.8c. However, if the temperature is relatively low (e.g., 390 °C), the diffusivity D_{rs} of Si adatoms through droplet may be low. And hence increase in nanowire

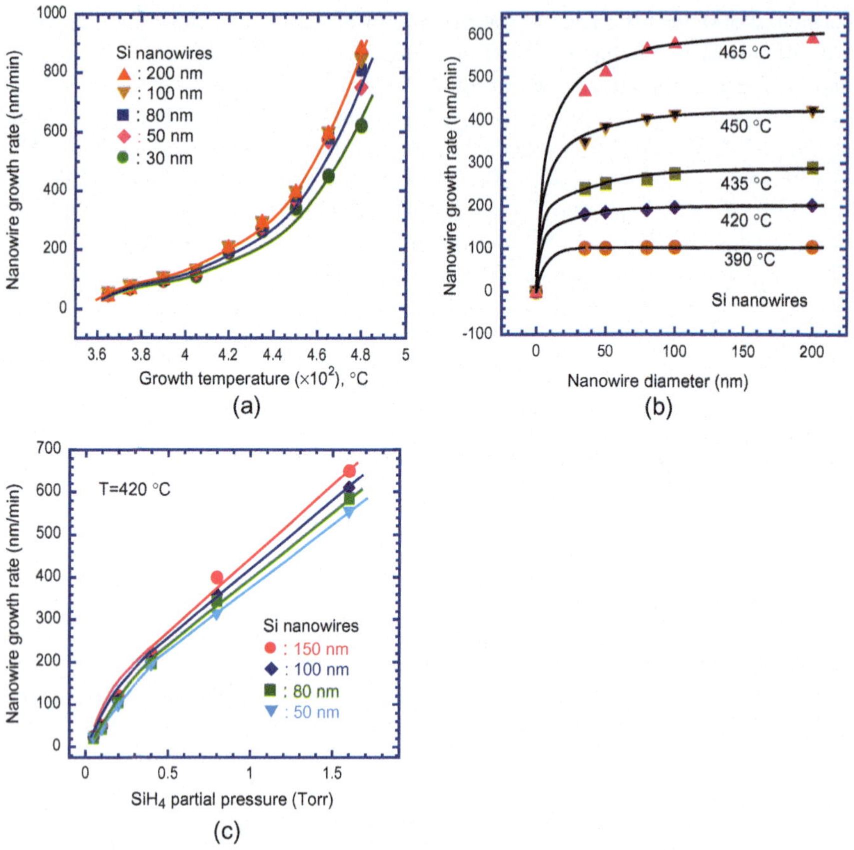

Fig. 5.8 Variation of Au-catalyzed Si nanowire growth rate **a** with temperature for five different nanowire diameters, **b** with nanowire diameters for five different temperatures, and **c** with SiH_4 partial pressure for four different nanowire diameters. The figures are drawn with experimental data by Pinion [32]

growth rate with increase in the partial pressure P_{SiH4} may be insignificant. This is why Pinion [32] found nanowire growth rate versus nanowire diameter plots for 390 °C independent of diameter.

5.7.3 InAs Nanowire Growth Rate as Function of Various Growth Parameters

The variations of InAs nanowire growth rates with temperature T for two different V/III ratios, as observed by Dayeh et al. [42], are shown in Fig. 5.9a. Both the curves thus obtained indicate that the nanowire growth rate increases with temperature,

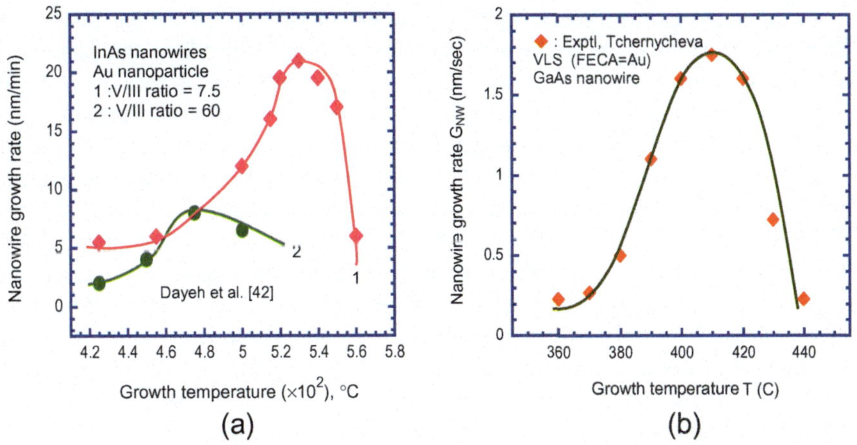

Fig. 5.9 Variation of Au-catalyzed. **a** InAs and **b** GaAs nanowire growth rates with temperature. Figure 5.9a is for two different V/III ratios. Figure 5.9a is drawn with experimental data by Dayeh et al. [42], but Fig. 5.9b is drawn with experimental data by Tchernycheva et al. [24]

reaches a peak, and then decreases with further increase in temperature. Again, increase in temperature led to increase in diffusivity of $R_S \equiv X = In$ and $R_S \equiv Y = As$ through the droplet and hence increase in InAs nanowire growth rate. This continued until a certain optimal temperature at which the droplet became too unstable to permit diffusion of the R_S species through it. The nanowire growth rate then decreased gradually with increase in temperature. The instability might have resulted from migration of contaminants into the droplet and/or change in droplet composition due to change in temperature. Note that this did happen also for GaAs nanowires, rather than for Si nanowires (see Fig. 5.9b) grown by Tchernycheva [24]. The temperature was just high enough to cause instability of the droplet for this growth.

5.7.4 Defects in Nanowires by the VLS Mechanism

Nanowires produced by the VLS mechanism suffer from high densities of planar defects. As noted by Joyce et al. [43], these are twin boundaries, stacking faults, and alternations at the zinc blende–wurtzite phase. Such structural defects deteriorate the nanowire's optical and transport properties. An in situ TEM by Oh et al. [44] suggests that the VLS growth proceeds by a sequence of periodic nucleations at the interface of FECA metal and nanowire. Planar defects and phase alternations result from these successive nucleations. Note that the phase depends on the relative nucleation barriers, which are near equal and vary spontaneously under reaction conditions typical of the VLS mechanism. Nevertheless, spontaneous fluctuations in nucleation barriers give rise to planar defects and phase alternations.

An interesting deviation of VLS growth was observed by Petersen et al. [45]. Using 5- and 30-nm diameter Au nanoparticles, they tried to produce ZnO nanowires at 975 °C. The melting point of a 5 nm Au nanoparticle is depressed to about 750–850 °C. The melting point of a 30 nm Au nanoparticle though remains intact to 1064 °C. Petersen et al. confirmed that 5 nm nanoparticles liquefy, but 30 nm nanoparticles remain solid at the growth temperature of 975 °C. Since no eutectic exists in the (Au, Ti) or (Au, Mo) alloy at the growth temperature, but eutectic exists in the $R_L \equiv$(Au, Zn) alloy at 683 °C [46], the melting of the 5 nm nanoparticles at 975 °C was attributed to nanometer-scale size effects. ZnO nanowires were found to grow only on samples containing 5 nm Au nanoparticles, but not on 30 nm Au nanoparticles. And this growth was mediated by molten Au and not molten $R_L \equiv$(Au, Zn) alloy. Had this alloy been involved in growth, the Zn composition in it, according to binary phase diagram, would be marginally low, about 0.025 atomic %. So, this experiment suggested that molten nanoparticles mediating nanowire growth do not need to be $R_L \equiv$(MET, X) alloys. They can though be molten FECA metal, instead.

5.7.5 *Impact of MET Contamination in Nanowire*

The use of metals such as Au is a matter of concerns as they may cause catalyst metal contamination during the VLS growth. Among various metals, Au is known to produce deep-level traps in silicon. These traps potentially reduce minority carrier lifetime and increase generation–recombination rates. They are therefore detrimental for the performance of nanowire devices. Employing high-angle annular dark-field scanning transmission electron microscopy, Allen et al. [47] mapped out the Au locations in VLS-grown Si nanowires. They found that Au was indeed present in nanowires, and that it had a concentration exceeding the limit of bulk solubility. There was though no concentration gradient along the nanowire length or width. This implied that Au incorporation during growth was more dominant than post-growth Au diffusion. Electron beam-induced current microscopy by Allen et al. revealed that the minority carrier diffusion length in Si nanowires was limited primarily by the surface properties and not by the bulk Au impurities. These were reconfirmed by advances in atom probe topography by Eichfeld et al. [48]. These researchers investigated metal incorporation in Al-catalyzed Si nanowires and observed that the Al concentration in Si exceeded the bulk solubility limit. The finding of Al clusters in Si nanowires suggested that Al is not always electrically active in Si nanowires. The unusually high Al composition in these nanowires could only be attributed to the presence of non-equilibrium defects. An investigation by Chen et al. [49] demonstrated that an increase in VLS growth rate accompanies an increase in the In and Sn catalyst impurity concentration in Si nanowires. Alarmingly, this concentration can reach a level of two orders of magnitude higher than that in their equilibrium solubility. Note that incorporation of catalyst atoms into nanowires is higher if the catalyst atoms are in liquid solution rather than in solid solution.

5.8 Failures of the VLS Mechanism to Mediate Carbon Nanotube Growths

5.8.1 *Location, Shape, and Size of Catalyst Particle During Growth*

Baker et al. [50–53] made considerable effort to understand the mechanism of carbon nanofiber growth. They [52] noted that during CNT growth, the catalyst particle exhibited a pear-like morphology similar to a molten metal droplet [46]. They also noted that FECA metal elements were located almost at the tip of the carbon filament. One side of this particle was chemically bonded to the solid carbon, but the other side of it was exposed to the gas atmosphere. If CNTs were grown by the VLS mechanism, metal carbide should have been formed to mediate this growth. Iron carbide (Fe_3C) was in fact observed by Phillipe [54] by TEM and XRD after the growth of CNTs. Nevertheless, Fe_3C was not believed to catalyze the CNT growths. Instead, Fe_3C was envisioned either to be a deactivated catalyst or formed during cooling after growth.

5.8.2 *Crystallographic Relationship During Growth*

Oberlin et al. [55] and Audier et al. [56] performed HRTEM and SAED analyses to examine the crystallographic relationship between graphene sheets and catalyst metal nanoparticle. They found that the [100] axis of the catalyst nanoparticle coincided with the carbon nanotube axis for metals of body-centered cubic (bcc) structure. The metal/gas interface was also always on the (100) surface. The [110] axis was however parallel to the nanotube axis for metals with a face-centered cubic (fcc) structure. The metal/gas interface was on the (111) plane, as well. These experiments thus demonstrated that there existed a crystallographic relation between carbon walls and the metal nanoparticle. And it seemed to support the surface diffusion, rather than the bulk diffusion as the possible mechanism for growth. Interestingly, it could be true for CNT growth.

5.8.3 *Presumed Stages of Growth*

Baker et al. argued that the carbon filament growth was indeed carried out by the VLS mechanism and that this growth was performed at four successive stages: First, the absorption and decomposition of the carbon-containing gas precursor took place on the METANO surface. Second, the dissolution of carbon atoms in the bulk of METANO led to the formation of a metastable liquid carbide alloy. Third, the diffusion of carbon species was carried out through the nanoparticle surface. And fourth, the precipitation and nucleation of the carbon species resulted in the growth of solid carbon filaments on the nanoparticle surface. The most appealing support for the VLS growth of carbon fibers was thought to be the good agreement of the measured activation energy for the growth of carbon nanofibers with the activation energy for bulk carbon diffusion through the corresponding metal (e.g., Fe, Co, Ni, etc.); it was essentially the growth-rate-limiting process [50, 51]. However, Baker et al. compared the calculated activation energies with those required for bulk carbon diffusion through solid-state metals and not the liquid-state metals. This means the said comparison did not corroborate with the basic tenet of the VLS mechanism. Also, the decomposition of hydrocarbons on the metal surface is generally exothermic. The precipitation of solid carbon is, on the other hand, endothermic. To reconcile this difference, Baker et al. proposed that the driving force for precipitation in catalyst nanoparticle [50] was actually the temperature gradient. This hypothesis does not apply to the growth initiated by the endothermic decomposition of the carbon precursor, for example, alkanes [57–59].

5.8.4 Possible Mechanism for Growth

Tibbetts [60] proposed that the diffusion of carbon through metal nanoparticle is caused by gradient in chemical potential. This is valid only if the metal particle is molten, but not valid if the metal particle is solid during growth. We go back to the hypothesis of carbon precipitation by temperature gradient. This hypothesis may not be valid for small-sized nanoparticles suitable for SWCNT growth. Note that these nanoparticles have high thermal conductivity, and any temperature gradient of these nanoparticles would cause excessively large heat flow [61–63]. Molecular dynamics simulations [61–63] showed that the carbon concentration gradient within the catalytic nanoparticle is necessary for CNT growth. The temperature gradient is not necessary for this growth. Thermodynamic calculations by Klinke et al. [64] confirmed that CNT growth is driven by the carbon concentration gradient in the FECA metal nanoparticle rather than temperature gradient.

5.9 Criteria for Nanomaterial Growths by the VLS Mechanism

We believe that the failure of the VLS mechanism in mediating the growth of some nanowires and most of carbon nanotubes, as described above, has emanated from our failure to understand and hence to enforce the correct growth criteria. A number of these growth criteria were put forth in our previous investigation [2]. They were based on the assumption that the nanowire growths, in particular, by the VLS mechanism were carried out by a technique such as CVD and that the growth chamber has high-temperature reaction zone and low-temperature deposition zone. The functionalized substrate coated with METANO is placed in the deposition zone. The temperature T_{H} of the high-temperature reaction zone is sufficiently high to ensure that the $\mathrm{R_S}{\equiv}\mathrm{X}$ and $\mathrm{R_S}{\equiv}\mathrm{Y}$ vapor species are released from their precursor(s) at this temperature. The $\mathrm{R_S}{\equiv}\mathrm{X}$ and $\mathrm{R_S}{\equiv}\mathrm{Y}$ vapor species are then flown into the deposition zone by one or more suitable carrier gas. The $\mathrm{R_S}{\equiv}\mathrm{X}$ and $\mathrm{R_S}{\equiv}\mathrm{Y}$ vapor species flow over METANO surface inside the furnace at a temperature T_{F} and pressure P_{L}. If not affected by oxygen or some other species, they produce (MET, X) droplet in the deposition zone. Some of the growth criteria are described in the following.

5.9.1 Criterion 1

The formation, composition, overgrowth, and decay of droplets on the METANO nanoparticle surface are crucial for the nanowires grown by the VLS mechanism. These droplets possess surface tension and electronegativity, and both of them dictate the one-dimensional nature of nanowires thus grown. For the growth of these nanowires by the VLS mechanism, the temperature T_{F} and the pressure P_{L} of the chamber must be such that they both facilitate reaction of the METANO nanoparticles with $\mathrm{R_S}{\equiv}\mathrm{X}$ to yield (MET, X) eutectic alloy droplets. The eutectic phase for these droplets should preferably occur at the lowest possible temperature T_{E} and with the highest possible atomic percentage of X in the (MET, X) eutectic alloy of the $\mathrm{R_L}$ species. The choice of temperature T_{F} and pressure P_{L} must take into consideration the size-dependent mesoscopic effect of the droplet formed on the METANO surface. Influenced by the surface energy of the METANO nanoparticle, and depending on the dimension of the METANO nanoparticle, this effect may lead to melting point depression. The melting point T_{L} of the METANO and hence of the $\mathrm{R_L}$ species may consequently be far lower than the melting point T_{M} of the corresponding bulk. If the melting point depression of the METANO at the growth temperature T and the pressure P_{L} is T_{dep}, then, due to size-dependent melting point depression, the real melting temperature of the METANO would be:

$$T_{\mathrm{L}} = T_{\mathrm{M}} - T_{\mathrm{dep}}. \tag{5.1}$$

We can call it T_{L}. Two situations would arise: First, the growth temperature T may be $T \geq T_{\mathrm{L}}$, and hence the METANO would be molten creating droplet without the formation of $\mathrm{R_L}{\equiv}$(MET, X) eutectic alloy. It would mediate $\mathrm{X}_m\mathrm{Y}_n$ nanomaterial growth. Second, the temperature T_{L} may be low enough to cause the melting of the (MET, X) alloy even before this alloy reaches the bulk eutectic temperature T_{E}. Note that this alloy would be a cluster or solid solution, in which X species would remain submerged in molten MET. The eutectic temperature T_{E} of the $\mathrm{R_L}{\equiv}$(MET, X) alloy would not however change due to change in the size-dependent melting point depression of the MET particle. Depending on $\mathrm{R_S}{\equiv}\mathrm{X}$, the depressed melting point of

Table 5.3 List of the values of some MET parameters and of the eutectic alloy of binary compounds from these METs

MET			$R_S{\equiv}X$ species	$R_L{\equiv}$(MET, X) eutectic alloy		$T_M - T_E$ (°C)
Name	T_M (°C)	T_L (°C)		T_E (°C)	X mole fraction	
Au	1064	664	Si	363	0.190	701
Au	1064	664	Ge	361	0.280	703
Au	1064	664	Ga	349	–	725
Al	660	260	Si	577	0.125	83
Al	660	260	Ge	419	0.510	241
Ag	962	562	Si	826	0.100	136
Ag	962	562	Ge	639	0.253	323

The depression in temperature T_{dep} has been assumed to be T_{dep} = 400 °C

the MET particle may though approach the eutectic temperature T_E of the $R_L{\equiv}$(MET, X) alloy. We present Table 5.3 with a list of materials characteristics of some binary alloys. It can be seen that even a reasonably small size-dependent depression of some MET melting temperatures, the resultant melting temperatures of these METs can be identical to the eutectic temperatures T_E of their $R_L{\equiv}$(MET, X) alloys. The $R_L{\equiv}$(MET, X) alloy can be molten at this temperature. Equation (5.1) would however be invalid if the METANO dimension (and hence the droplet dimension) is large enough not to undergo size-dependent melting point depression. In that case, the melting of the (MET, X) droplet would take place at the eutectic temperature T_E.

Note that the melting point, for example, of Au nanoparticles, about 2 nm in diameter, can be lower by over 400 °C than the melting point 1064 °C of Au bulk material. Considering T_{dep} = 400 °C, the values of T_L are also listed in Table 5.3. It may thus be noted that the R_L species can be molten or semimolten at the temperature T_L, if the temperature T_L is lower than the eutectic temperature T_E of the $R_L{\equiv}$(MET, X) alloy. Even when the $R_L{\equiv}$(MET, X) droplet is molten (semimolten) at T_L due to mesoscopic effect [39–41], the substrate may remain solid. The semimolten R_L species can consequently allow (1) the adsorption of the vapor species (e.g., $R_S{\equiv}X$ and $R_S{\equiv}Y$) onto the droplet surface and (2) an easy diffusion of these species from the droplet surface to the liquid/solid interface. While diffusing through the droplet in the deposition zone, the $R_S{\equiv}X$ and $R_S{\equiv}Y$ species may react to form X_mY_n molecules. These molecules can then undergo supersaturation at the L/S interface and subsequently nucleation and crystallization to yield X_mY_n nanomaterial.

5.9.2 *Criterion 2*

A droplet can be most effective in mediating nanowire and possibly nanotube growths if it follows the conditions:

1. The alloyed composition of the droplet mediating X_mY_n nanomaterial growths is ideally R_L≡(MET, X) alloy, and that it is free from contamination during growth.
2. The charge distribution Q_L and the electronegativity ζ_L of the R_L species [e.g., the R_L≡(MET, X) alloy droplet species] are much different from the charge distribution Q_S and the electronegativity ζ_S of the R_S (e.g., R_S≡X and R_S≡Y) species.
3. The MET of the (MET, X) droplet is such that the difference $(\zeta_L - \zeta_S)$ creates a largest possible electric field E_r.
4. The same MET should also be such that the droplet possesses large surface energy γ_L, and that this surface energy creates surface tension preventing it from breakdown under adverse conditions.
5. The electric field E_L and the surface energy γ_L of the droplet (e.g., the R_L species) may indeed be quite high. This is essential for the droplet to be stable enough to attract the precursor(s) of the R_S species, for the precursor(s) of the R_S species to land on the droplet surface, and for the precursor(s) to decompose releasing the R_S species on the droplet surface.
6. The molten (MET, X) alloy droplet of the R_L species may have optimal R_S≡X mole fraction at the temperature T_L, if not at the temperature T_E. This is to insure that the mole fraction of R_S≡X in the R_L≡(MET, X) alloy of the droplet is appropriate; it is not very low. It is not also so large that the difference $(\zeta_L - \zeta_S)$ is too small for the R_S≡X and R_S≡Y species to land on the droplet surface. If the mole fraction of R_S≡X in the R_L≡(MET, X) alloy of the droplet is very large and the difference $(\zeta_L - \zeta_S)$ is very small, the R_S≡X and R_S≡Y species may not land or be released on the droplet surface.
7. Optimal, but not very low mole fraction of X in the R_L≡(MET, X) alloy of the droplet may be essential for smooth diffusion (migration) of the R_S species through the droplet to the L/S interface. The solid of the L/S interface is the METANO (or the substrate) on which the droplet is formed. It is however the substrate (or support) just underneath the METANO if this METANO is fully consumed in generating the R_L≡(MET, X) droplet. The solid of the L/S interface is otherwise the nanowire (nanotube) tip if the growth takes place on this tip.

5.9.3 Criterion 3

Droplet overgrowth with time particularly during nanowire growth may be detrimental for this nanowire growth. This overgrowth takes place for at least two possible reasons: first, change in the R_L≡(MET, X) droplet shape and hence in the droplet contact angle leading to the variation of the contact area at the L/S interface; second, continuous oversupply of the R_S species to the droplet; and third, the lack of control of relative flow rates of R_S≡X and R_S≡Y vapor species in the deposition zone of the growth chamber. If, for example, there occurs excessive flow of R_S≡X vapor species, as compared to R_S≡Y vapor species, some of these R_S≡X species may be integrated into the droplet. The immediate result of this may be the change in composition of the

R_L≡(MET, X) alloy of the droplet or just the modification of the droplet surface with the R_S species existing as the wetted non-alloyed component of this surface. Obviously, the droplet size and even droplet shape are consequently changed, which leads to the formation of cone-shaped nanowires. Note that R_S≡X, R_S≡Y species and/or X_mY_n molecules formed from them diffuse through the alloyed and non-alloyed segments of the droplet to the L/S interface to produce tapered nanowires. Note also that the peripheral region of such nanowire results from the diffusion (migration) of the R_S species through the non-alloyed component of the droplet. This region may exist in entire nanowire or in certain segment of the nanowire length.

5.9.4 Criterion 4

Nanowire growth is severely affected by droplet decay and/or sudden (or gradual) disintegration of droplet during growth. We cite a number of possible reasons for this. First, there can be lateral temperature gradient and/or sudden change in temperature leading to accidental and random breakup of the R_L≡(MET, X) droplets [10]. Second, there can be oversupply of the R_S≡X and/or R_S≡Y species during growth. Such an oversupply may impede diffusion of R_S≡X, R_S≡Y through the droplet, the formation of the X_mY_n molecules inside the droplet, and the diffusion of X_mY_n molecules through the droplet to the L/S interface. There can also be interplay of surface energies of the nanowire and/or of the liquid droplet. There may consequently be droplet oscillation and droplet deformation giving rise to multiple nanowire nucleations [14] during growth. Third, there can be carrier gases and contamination species incident on the droplet surface. Some or all of them may induce viscous and/or inertial hydrodynamic stress χ_L in the droplet. The hydrodynamic stress χ_L tends to negate the droplet surface tension σ_L. While the hydrodynamic stress χ_L tends to deform the droplet from spherical to the ellipsoidal shape reducing its surface energy, the surface tension σ_L tends to preserve the spherical shape of the droplet and hence its surface energy. This means there can be droplet breakup, and even droplet disintegration if $\chi_L > \gamma_L$ due to random (unsystematic) flow of the vapor species during nanowire growth. Even if $\gamma_L > \chi_L$, there may be sporadic and hard landing (incident) of gas particle on the droplet surface. And such a landing may cause MET and/or X particles to be knocked out from the droplet surface. It may particularly happen if the droplet is in highly molten condition at $T \gg T_L$ and $P \ll P_L$. The particles knocked out from the droplet may escape or may stay in the chamber as contaminants. The droplet breakup at high temperatures may indeed take place as apparent from the breakup of the water droplet by high-velocity flow of gas particles [65]. Fourth, there may be tribo-electrification, ion collection, thermionic emission, frictional charging, etc., during nanowire growth. The droplets may be positively or negatively charged by one or more of them. Magnussen et al. [66] placed Au nanoparticles in tube furnace flowing high-purity nitrogen at temperature $T > 600$ °C. These nanoparticles were electrically charged by the nitrogen species. Hwang et al. [67] grew Si nanowires by CVD making use of SiH_2Cl_2 precursor. They found charged silicon clusters formed during the CVD growth of Si nanowires. All these might have happened due to unstable breaking up of droplets in pieces under the influence of surface tension forces [68, 69] on the droplet surfaces.

To reiterate, an electrically charged droplet becomes unstable if its surface is affected more by the repulsive electrostatic forces than by the attractive surface tension forces. The droplet breaks apart [62, 63] under the influence of the dominating electrostatic forces. However, the droplet remains under equilibrated balance if the charge distribution is uniform at the surface of the droplet, and follows the equation known as the Rayleigh limit [62]:

$$\left(\frac{\varepsilon_0 E_{\mathrm{L}}^2}{2}\right) = \frac{2\gamma_{\mathrm{L}}}{r_{\min}}. \tag{5.2}$$

Note that γ_{L} is the surface energy of the droplet, $r_{\min}$ is the minimum radius of the droplet, and E_{L} is the electrostatic field on the droplet surface. The charge Q_{L} resulting in surface tension may be given by

$$Q_{\mathrm{L}} = 8\pi\left(\varepsilon_0 \gamma_{\mathrm{L}} r_{\min}^3\right)^{1/3}. \tag{5.3}$$

The dependence of the electrostatic field E_{L} on the minimum droplet radius $r_{\min}$ is depicted in Fig. 5.10a. This figure indicates that the electrostatic field E_{L} on the surface of small droplets is very large. There occurs however almost an exponential decrease in this field with increase in droplet radius. For a certain radius $r_{\min}$, the electrostatic field increases with increase in MET mole fraction and hence decrease in X mole fraction in (MET, X) droplets. Also, for the same radius $r_{\min}$, a $\mathrm{R_L}$≡(MET, X) droplet with MET mole fraction $\eta_{\mathrm{MET}} > 0$ exhibits much higher surface electric

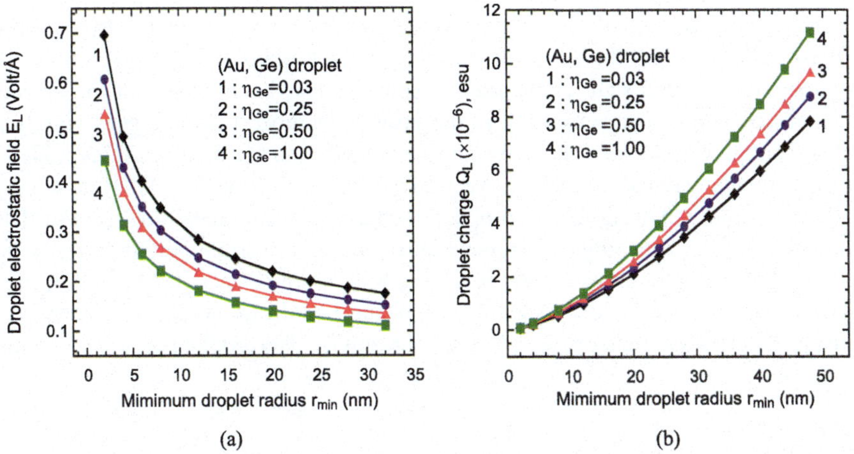

Fig. 5.10 Variations of **a** the droplet electrostatic field E_{L} and **b** the droplet charge Q_{L} with minimum droplet radius $r_{\min}$ for different Ge mole fractions η_{Ge} of the (Au, Ge) droplet

field than an X droplet (e.g., a droplet exhibiting MET mole fraction $\eta_{MET} = 0$ and X mole fraction $\eta_X = 1$). As stated earlier, this electric field degrades the droplet stability. The variation of the Rayleigh limiting charge Q_L with the minimum droplet radius r_{min} is shown in Fig. 5.10b. Recall that such a droplet undergoes breakup if it exhibits a charge higher than the Rayleigh limiting charge Q_L. Interestingly, this charge is smaller for smaller dimension of the $R_L \equiv$(MET, X) droplets. This means a $R_L \equiv$(MET, X) droplet of smaller diameter can suffer from instability even if there occurs relatively small buildup of electrostatic charge on its surface. X droplets and MET droplets are largely free from this problem.

5.9.5 Criterion 5

Appropriate temperature T and pressure P are critical for the droplet stability. There may be higher melting and lower stability of the (MET, X) droplet of the R_L species if the temperature T is higher than the temperature T_L and the pressure P is lower than the pressure P_L. In general, at higher temperature T and lower pressure P, atoms constituting the droplet surface are more loosely bound than the atoms constituting the droplet bulk. The atoms constituting the droplet surface have consequently a higher tendency to free from the droplet surface if the droplet is affected by hydrodynamic stress or by impulse from incident charge particles. This is true particularly at higher temperature T and lower pressure P. And this suggests that, at an appropriate pressure (e.g., 1–10 Torr), the temperature T should preferably be close to the temperature T_L of the MET droplet. It should be close to the temperature T_E of the $R_L \equiv$(MET, X) eutectic alloy droplet. In essence, the temperature and pressure should be such that they provide the droplet with relatively higher stability and higher surface tension σ_L. They would allow the droplet to exert higher attractive electric field on the $R_S \equiv$X and/or $R_S \equiv$Y species. There would thus be smooth bulk diffusion of the $R_S \equiv$X, $R_S \equiv$Y, and/or $X_m Y_n$ molecules through the droplet. Even being close to the temperature T_L, the said temperature T should never be so high that the droplet (1) attracts contaminants existing in the growth chamber, (2) undergoes vibration large enough to be deformed, (3) becomes unstable, and (4) becomes susceptible to unwarranted chemical reaction between MET and the R_S species. All of them may affect the ability of the droplet to properly mediate nanowire (nanotube) growth. The droplet size may thus gradually reduce during entire or certain period of nanowire growth. The droplet may even entirely disappear. Even if the droplet exists, it may suffer from change in (1) its chemical makeup, (2) electronic charge distribution inside it, (3) surface tension, and (4) electronegativity. The droplet may consequently yield nanowires affected by bending, twinning, or splitting.

References

1. R.S. Wagner, W.C. Ellis, Vapor-liquid-solid mechanism of single crystal growth. Appl. Phys. Lett. **4**, 89–91 (1964)
2. S.N. Mohammad, Analysis of the vapor-liquid-solid mechanism for nanowire growth and a model for this mechanism. Nano Lett. **8**, 1532–1538 (2008)
3. D. Wang, Y.-L. Chang, Q. Wang, J. Cao, D.B. Farmer, R.G. Gordon, H. Dai, Surface chemistry and electrical properties of germanium nanowires. J. Am. Chem. Soc. **126**, 11602–11611 (2004)
4. Y. Kim, H.J. Joyce, Q. Gao, H.H. Tan, C. Jagadish, M. Paladugu, J. Zou, A.A. Suvoroval, Influence of nanowire density on the shape and optical properties of ternary InGaAs nanowires. Nano Lett. **6**, 599–604 (2006)
5. V. Schmidt, J.V. Wittemann, U. Gösele, Growth, thermodynamics, and electrical properties of silicon nanowires. Chem. Rev. **110**, 361–388 (2010)
6. K.A. Dick, A review of nanowire growth promoted by alloys and non-alloying elements with emphasis on Au-assisted III-V nanowires. Prog. Crystal Growth Character. Mater. **54**, 138–173 (2008)
7. B. Ressel, K.C. Prince, S. Heun, Wetting of Si surfaces by Au-Si liquid alloys. J. Appl. Phys. **93**, 3886–3892 (2003)
8. H. Detz, M. Kriz, D. MacFarland, S. Lancaster, T. Zederbauer, M. Capriotti, A.M. Andrews, W. Schrenk, G. Strasser, Nucleation of Ga droplets on Si and SiO_x surfaces. Nanotechnology **26**, 315601 (2015)
9. B. Kalache, P.R. Cabarrocas, A.F. Morral, Observation of incubation times in the nucleation of silicon nanowires obtained by the vapor–liquid–solid method. Jpn. J. Appl. Phys. **45**(7), L190–L193 (2006)
10. T.B. Massalski (ed.), *Binary Alloy Phase Diagrams*, vol. 3, 2nd edn. (American Society of Metals, Metals Park, OH, 1986)
11. T.I. Kamins, R.S. Williams, Y. Chen, Y.-L. Chang, Y.A. Chang, Chemical vapor deposition of Si nanowires nucleated by $TiSi_2$ islands on Si. Appl. Phys. Lett. **76**, 562 (2000)
12. F. Iacopi, P.M. Vereecken, M. Schaekers, M. Caymax, N. Moelans, B. Blanpain, C. Detavernier, J. D'Haen, F. Griffiths, *Materials Research Society Symposium Proceedings*, vol. 1017 (2007), pp. 1017-DD01-10-EE01-10
13. P. Nguyen, H.T. Ng, M. Meyyappan, Catalyst metal selection for synthesis of inorganic nanowire. Adv. Mater. **17**, 1773–1777 (2005)
14. R.S. Wagner, *Whisker Technology* (Wiley, New York, 1970)
15. J. Kikkawa, Y. Ohno, S. Takeda, Growth rate of silicon nanowires. Appl. Phys. Lett. **86**, 123109 (2005)
16. S. Hofmann, C. Ducati, R.J. Neill, S. Piscanec, A.C. Ferrari, J. Geng, R.E. Dunin-Borkowski, J. Robertson, Gold catalyzed growth of silicon nanowires by plasma enhanced chemical vapor deposition. J. Appl. Phys. **94**, 6005–6012 (2003)
17. J. Albuschies, M. Baus, O. Winkler, B. Hadam, B. Spangenberg, H. Kurtz, High-density silicon nanowire growth from self-assembled Au nanoparticle. Microelectron. Eng. B **83**, 1530–1533 (2006)
18. J. Westwater, D.P. Gossain, S. Tomiya, S. Usui, H. Ruda, Growth of silicon nanowires via gold/silane vapor-liquid-solid reaction. J. Vac. Sci. Technol. B **15**, 554–557 (1997)
19. S.H. Chung, S. Ramadurgam, C. Yang, Effect of dopants on epitaxial growth of silicon nanowires. Nanomater. Nanotechnol. **4**, 1–6 (2014)
20. K.K. Lew, C. Reuther, A.H. Carim, J.M. Redwing, B.R. Martin, Template-directed vapor-liquid-solid growth of silicon nanowires. J. Vac. Sci. Technol. B **20**, 389–392 (2002)
21. O. Gunawan, S. Guha, Characteristics of vapor-liquid-solid grown silicon nanowire solar cells. Solar Energy Mater. Solar Cells **93**, 1388–1393 (2009)
22. J.B. Hannon, S. Kodambaka, F.M. Ross, R.M. Tromp, The influence of the surface migration of gold on the growth of silicon nanowires. Nature **440**, 69–71 (2006)
23. S. Sharma, T.I. Kamins, R.S. Williams, Synthesis of thin silicon nanowires using gold-catalyzed chemical vapor deposition. Appl. Phys. A **80**, 1225–1229 (2005)

24. M. Tchernycheva, J.C. Harmand, G. Patriarche, L. Travers, G. Cirlin, Temperature conditions for GaAs nanowire formation by Au-assisted molecular beam epitaxy. Nanotechnology **17**, 4025 (2006)
25. H. Huang, X. Ren, X. Ye, J. Guo, Q. Wang, Y. Yang, S. Cai, Y. Huang, Growth of stacking-faults-free zinc blende GaAs nanowires on Si substrate by using AlGaAs/GaAs buffer layers. Nano Lett. **10**, 64–68 (2010)
26. H. Shtrikman, R. Popovitz-Biro, A. Kretinin, M. Heiblum, Stacking-faults-free zinc blende GaAs nanowires. Nano Lett. **9**, 215–219 (2009)
27. S.-G. Ihn, J.-I. Song, T.-W. Kim, D.-S. Leem, T. Lee, S.-G. Lee, E.K. Koh, K. Song, Morphology- and orientation-controlled gallium arsenide nanowires on silicon substrates. Nano Lett. **7**, 39–44 (2007)
28. S.A. Fortuna, J. Wen, I.S. Chun, X. Li, Planar, GaAs nanowires on GaAs (100) substrates: self-aligned, nearly twin-defect free, and transfer-printable. Nano Lett. **8**, 4421–4427 (2008)
29. C. Soci, X.-Y. Bao, D.P.R. Aplin, D. Wang, A systematic study on the growth of GaAs nanowires by metal-organic chemical vapor deposition. Nano Lett. **8**, 4275–4282 (2008)
30. J.-C. Harmand, G. Patriarche, N. Péré-Laperne, M.-N. Mérat-Combes, L. Travers, F. Glas, Analysis of vapor-liquid-solid mechanism for Au-assisted GaAs nanowire growth. Appl. Phys. Lett. **87**, 203101–203103 (2005)
31. J.D. Christesen, C.W. Pinion, X. Zhang, J.R. McBride, J.F. Cahoon, Encoding abrupt and uniform dopant profiles in vapor-liquid-solid nanowires by suppressing the reservoir effect of the liquid catalyst. ACS Nano **8**(11), 11790–11798 (2014)
32. C.W. Pinion, Understanding the vapor-liquid-solid and vapor-solid-solid mechanisms of Si nanowire growth to synthetically encode precise nanoscale morphology. Doctoral thesis, University of North Carolina, Chapel Hill (2017)
33. H. Wang, C.N. Fischman, Role of liquid droplet surface diffusion in the vapor-liquid-solid whisker growth mechanism. J. Appl. Phys. **76**, 1557 (1994)
34. H. Adhikari, A.F. Marshall, C.E.D. Chidsey, P.C. McIntyre, Germanium nanowire epitaxy: shape and orientation control. Nano Lett. **6**, 318–323 (2006)
35. Y.W. Wang, V. Schmidt, S. Senz, U. Gösele, Epitaxial growth of silicon nanowires using an aluminum catalyst. Nat. Nanotechnol. **1**, 186–189 (2006)
36. Y. Wang, X. Zhou, Z. Yang, F. Wang, N. Han, Y. Chen, J.C. Ho, GaAs nanowires grown by catalyst epitaxy for high performance photovoltaics. Crystals **8**, 347 (2018)
37. F.J. Sheini, D.S. Joag, M.A. More, J. Singh, O.N. Srivasatva, Low temperature growth of aligned ZnO nanowires and their application as field emission cathodes. Mater. Chem. Phys. **120**, 691–696 (2010)
38. Y. Wu, P. Yang, Direct observation of vapor-liquid-solid nanowire growth. J. Am. Chem. Soc. **123**, 3165–3166 (2001)
39. K. Dick, T. Dhanasekaran, Z. Zhang, D. Meisel, Size-dependent melting of silica-encapsulated gold nanoparticles. J. Am. Chem. Soc. **124**, 2312–2317 (2002)
40. E. Sutter, P. Sutter, Phase diagram of nanoscale alloy particles used for vapor–liquid–solid growth of semiconductor nanowires. Nano Lett. **8**, 411–414 (2008)
41. F. Gao, Z. Gu, Melting temperature of metallic nanoparticles, in *Handbook of Nanoparticles*, ed. by M. Aliofkhazraei (Springer, Cham, 2016), pp. 661–690
42. S.A. Dayeh, E.T. Yu, D. Wang, III–V nanowire growth mechanism: V/III ratio and temperature effects. Nano Lett. **7**(8), 2486–2490 (2007)
43. H.J. Joyce, J. Wong-Leung, Q. Gao, H.H. Tan, C. Jagadish, Phase perfection in zinc blende and wurtzite III-V nanowires using basic growth parameters. Nano Lett. **10**, 908–915 (2010)
44. S.H. Oh, M.F. Chisholm, Y. Kauffmann, W.D. Kaplan, W. Luo, M. Ruhle, C. Scheu, Oscillatory mass transport in vapor-liquid-solid growth of sapphire nanowires. Science **330**, 489–493 (2010)
45. E.W. Petersen, E.M. Likovich, K.J. Russell, V. Narayanamurti, Growth of ZnO nanowires catalyzed by size-dependent melting of Au nanoparticles. Nanotechnology **20**, 405603 (2009)
46. H. Okamoto, T.B. Massalski, The Au-Zn (Gold-Zinc) system. Bull. Alloy Phase Diagram **10**, 59–69 (1989)

47. J.E. Allen, E.R. Hemesath, D.E. Perea, J.L. Lensch-Falk, Z.Y. Li, F. Yin, M.H. Gass, P. Wang, A.L. Bleloch, R.E. Palmer, L.J. Lauhon, High resolution detection of Au catalyst atoms in Si nanowires. Nat. Nanotechnol. **3**(3), 168–173 (2008)
48. C.M. Eichfeld, S.S.A. Gerstl, T. Prosa, Y. Ke, J.M. Redwing, S.E. Mohney, Local electrode atom probe analysis of silicon nanowires grown with an aluminum catalyst. Nanotechnology **23**(21), 215205 (2012)
49. W. Chen, L. Yu, S. Misra, Z. Fan, P. Pareige, G. Patriarche, S. Bouchoule, P.R. Cabarrocas, Incorporation and redistribution of impurities into silicon nanowires during metal-particle-assisted growth. Nat. Commun. **5**, 4134 (2014)
50. R.T.K. Baker, M.A. Barber, P.S. Harris, F.S. Feates, R.J. Waite, Nucleation and growth of carbon deposits from the nickel catalyzed decomposition of acetylene. J. Catal. **26**, 51–62 (1972)
51. R.T.K. Baker, P.S. Harris, R.B. Thomas, R.J. Waite, Formation of filamentous carbon from iron, cobalt and chromium catalyzed decomposition of acetylene. J. Catal. **30**, 86–95 (1973)
52. R.T.K. Baker, P.S. Harris, *Chemistry and Physics of Carbon*, ed. by P.A. Thrower, vol. 14 (Marcel Dekker, New York, 1978)
53. R.T.K. Baker, J.R. Alonzo, J.A. Dumesic, D.J.C. Yates, Effect of the surface state of iron on filamentous carbon formation. J. Catal. **77**, 74–84 (1982)
54. R. Philippe, B. Caussat, A. Falqui, Y. Kihn, P. Kalck, S. Bordre, D. Plee, P. Gaillard, D. Bernard, P. Serp, An original growth mode of MWCNTs on alumina supported iron catalysts. J. Catal. **263**, 345–358 (2009)
55. A. Oberlin, M. Endo, T. Koyama, Filamentous growth of carbon through benzene decomposition. J. Cryst. Growth **32**, 335–349 (1976)
56. M. Audier, A. Oberlin, M. Coulon, Crystallographic orientations of catalytic particles in filamentous carbon; Case of simple conical particles. J. Cryst. Growth **55**, 549–556 (1981)
57. J. Rostrup-Nielsen, D.L. Trimm, Mechanisms of carbon formation on nickel-containing catalysts. J. Catal. **48**, 155–165 (1977)
58. T. Baird, J.R. Fryer, B. Grant, Structure of fibrous carbon. Nature **233**, 329–330 (1971)
59. E.L. Evans, J.M. Thomas, P.A. Thrower, P.L. Walker, Growth of filamentary carbon on metallic surfaces during the pyrolysis of methane and acetone. Carbon **11**, 441–442 (1973)
60. G.G. Tibbetts, Why are carbon filaments tubular? J. Cryst. Growth **66**, 632–638 (1984)
61. F. Ding, A. Rosén, K. Bolton, The role of the catalytic particle temperature gradient for SWNT growth from small particles. Chem. Phys. Lett. **393**, 309–313 (2004)
62. F. Ding, K. Bolton, A. Rosén, Nucleation and growth of single-walled carbon nanotubes: a molecular dynamics study. J. Phys. Chem. B **108**, 17369–17377 (2004)
63. Z. Xu, T. Yan, F. Ding, Atomistic simulation of the growth of defect-free carbon nanotubes. Chem. Sci. **6**, 4704–4711 (2015)
64. C. Klinke, J.-M. Bonard, K. Kern, Thermodynamic calculations on the catalytic growth of multiwall carbon nano-tubes. Phys. Rev. B **71**, 035403 (2005)
65. H.A. Stone, Dynamics drop deformation and breakup in viscous fluids. Annu. Rev. Fluid Mech. **26**, 65–102 (1994)
66. M.H. Magnussen, K. Deppert, J. Maln, J. Bovin, L. Samuelson, Gold nanoparticles: production, reshaping, and thermal charging. J. Nanoparticle Res. **1**, 243–251 (1999)
67. N.M. Hwang, W.S. Cheong, D.Y. Yoon, D.-Y. Kim, Growth of silicon nanowires by chemical vapor deposition: approach by charged cluster model. J. Cryst. Growth **218**, 33–39 (2000)
68. L. Rayleigh, On the equilibrium of liquid conducting masses charged with electricity. Philos. Mag. **14**, 184 (1882)
69. L. Last, Y. Levy, J. Jortner, Beyond the Rayleigh instability limit for multi-charged finite systems: from fission to Coulomb explosion. Proc. Natl. Acad. Sci. U.S.A. **99**, 9107–9112 (2002)

Chapter 6
Vapor–Solid–Solid Growth Mechanism

Abstract The vapor–solid–solid (VSS) growth mechanism for the growths of one-dimensional and quasi-two-dimensional nanomaterials has been reviewed in some details. This mechanism is widely employed, for example, for nanowire, nanotube, nanofiber, nanodot, and graphene growths. The fundamentals of the VSS mechanism have been articulated. Important features of the VSS mechanism have been described. Characteristics of the nanomaterial growths by this mechanism have been examined. Strengths and weaknesses of this mechanism in mediating the growths have been described. Rate-limiting processes during growths by this mechanism have then been assessed and appraised. Temperature dependence of the nanowire growth rates by this mechanism has also been elucidated. A comparison of the nanowire growth rates by the VSS mechanism and the VLS mechanism have been made. Control of tapering during growths by the VSS mechanism has finally been discussed.

6.1 Basics

Vapor–solid–solid (VSS) growth mechanism for nanomaterial growths has genesis in the vapor precursors and crystalline solid–metal catalysts FECAs [1]. It was realized during the growth of amorphous Si whiskers by thermal decomposition of SiH_4 at a temperature of 500–800 °C in an Ar atmosphere. Crystalline FECA metals such as Pb, Bi, Al, Au, Ag, In, and Sn were deposited on Si substrate to perform this growth. While the amorphous Si whiskers grew most efficiently on MET≡Au surface at 650 °C, the same amorphous Si whiskers grew most efficiently on MET≡Ag surface again at 650 °C. If, for the sake of convenience, the R_L species defined in Chap. 3, Sect. 3.3, is taken into consideration, the growth temperature of Si whiskers on MET≡Au could be found higher than the R_L≡(Au, Si) eutectic species temperature of 363 °C. The growth temperature of the same Si whiskers on MET≡Ag was however lower than the R_L≡(Ag, Si) eutectic temperature of 830 °C. All these suggested that R_L≡(Ag, Si) alloy might have been solid for the Si whisker growth. "Vapor," "solid," and "solid" for this growth were reactant vapor, crystalline solid–metal catalyst, and solid whisker product, respectively. So, while the catalyst for the VLS growth is in

S. N. Mohammad, *Synthesis of Nanomaterials*, Springer Series in Materials Science 307, https://doi.org/10.1007/978-3-030-57585-4_6

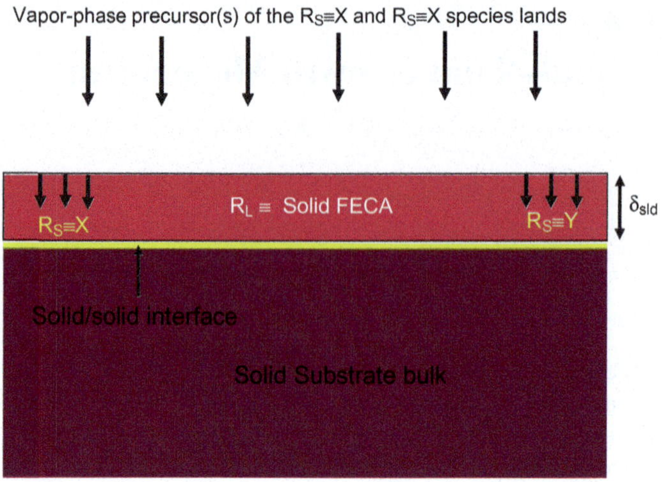

Fig. 6.1 Schematic diagram showing the landing of the precursor (s) of the R_S≡X and R_S≡Y source species on the R_L≡solid surface of FECA nanoparticle and the diffusion of the R_S≡X and R_S≡Y source species through the said FECA surface of thickness δ_{sld}. The FECA nanoparticle is formed on a solid substrate or support. The R_S≡X and R_S≡Y source species are generated on the R_L species surface by thermal decomposition or by some other means. Although solid/solid interface does not really exist, it has been shown only for conceptual reasons

the liquid state, the catalyst for the VSS growth is in the solid state. This is evident from the schematic diagram for the VSS growth on a solid substrate surface depicted in Fig. 6.1.

6.2 Illustrations of Nanowire Growth by the VSS Mechanism

6.2.1 Au-Mediated Low-Temperature ZnO Nanowire Growths

Using a combination of synchrotron X-ray diffraction and high-resolution transmission electron microscopy, Campos et al. [2] studied the Au-mediated low-temperature (T = 350 °C) growth dynamics of ZnO nanowires, 20–100 nm in diameter. The R_L≡(Au, Zn) alloy of FECA nanoparticle catalyzing this growth at this temperature was γ-AuZn. The said growth was believed to be carried out by solid diffusion driven by preferential oxidation of the Zn of the γ-AuZn alloy catalyst. The growth mechanism was hence the VSS mechanism. The highest solubility range of Zn in Au in the β' phase of R_L≡(Au, Zn) alloy is plausibly 38–57 at.%. The highest solubility range of Zn in Au in the $\gamma1$, $\gamma2$, and $\gamma3$ phases of the R_L≡(Au, Zn), however, varies between 63 and 83 at.%. Both the β' phase and the γ-phases of the $R_L \equiv$(Au, Zn) species are solid solutions.

6.2.2 Cu-Mediated Low-Temperature Ge Nanowire Growths

Kang et al. [3] employed MET≡Cu and the GeH_4 precursor for Ge nanowire growth. These nanowires, 7 nm in diameter, could be grown only at a temperature of about 200 °C. It was therefore concluded that the growth took place by the VSS mechanism via the catalytic decomposition of Ge precursors onto the solid-phase Cu catalyst, which is Cu_3Ge. High-resolution transmission electron microscopy images demonstrated that the Ge nanowires were single crystalline preferably along the [110] direction. The growth of these nanowires confirmed that an appropriate combination of solid catalyst and semiconductor is thermodynamically compatible to yield Ge nanowires. Kang et al. [4] synthesized Ge nanowires making use also of $Cu_{1-z}Ni_z$ (z is the Ni mole fraction in CuNi) bulk alloy. They observed that the reaction between GeH_4 precursor and $Cu_{1-z}Ni_z$ bulk alloys could lead selectively to Cu-catalyzed Ge nanowire growth, rather than to Ni-catalyzed Ge nanowire growth. We believe the GeH_4 precursor was decomposed by the $Cu_{1-z}Ni_z$ matrix yielding Ge, and this Ge reacted with $Cu_{1-z}Ni_z$ to yield $R_L \equiv (Cu_3Ge)$ or $R_L \equiv$(Cu, Ge, Ni) solid solution. It was done in a self-organized manner. Subsequent diffusion of Ge through the $R_L \equiv (Cu_3Ge)$ or $R_L \equiv$(Cu, Ge, Ni) solid solution enabled the nucleation and growth of Ge nanowires. We cite another example. The growth rate of ZnO nanowires at a temperature of $T \approx 900$ °C was found to be much higher with the MET≡Cu than with the MET≡Au [5]. It was probably because nanowire growth increases with temperature. Note that, while the eutectic temperature $T_E = 902$ °C for the $R_L \equiv$(Cu, Zn) eutectic alloy [with the atomic % of Cu as 63 and the atomic % of Zn as 36), the eutectic temperature $T_E \approx 683$ °C for the $R_L \equiv$(Au, Zn) eutectic alloy (with the atomic % of Au as 85 and the atomic % of Zn as 15) [6].

6.2.3 Nanowire Growths via Solid FECA Material

Kamins et al. [7] performed CVD growth of Si nanowires at a temperature of 640–670 °C. This growth temperature is about 700 °C lower than the eutectic temperature of 1330°C of the $R_L \equiv$(Ti, Si) eutectic species. Kamins et al. concluded that Si nanowire growth at this temperature was via the VSS mechanism. Unfortunately, the crystallographic quality of Si nanowires thus grown was poorer than that of the Au-catalyzed Si nanowires. Han et al. [8] grew GaAs nanowires on amorphous SiO_2 substrate at 580–620 °C using MET≡Ni. For this, the solid GaAs source was heated to a temperature of 850–950 °C. They argued that the GaAs nanowire growth carried out at $T < 1200$ °C was via the VSS mechanism. Recall that the eutectic temperature of the $R_L \equiv$(Au, Ge) alloy is ~361 °C. Zhu et al. [9] carried out MBE growth of

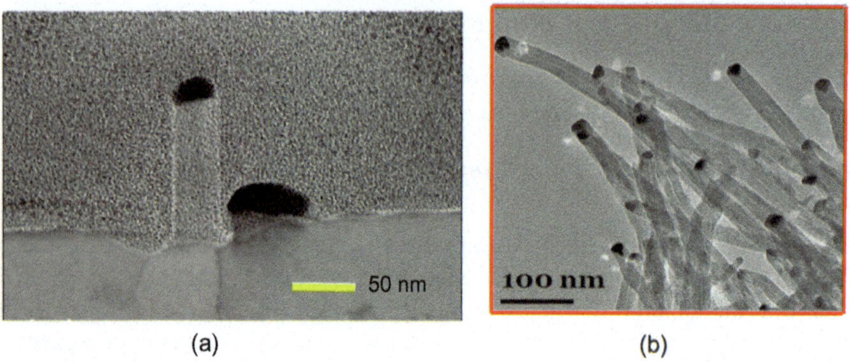

Fig. 6.2 **a** STEM image of the Au-catalyzed Ge nanowire grown at 220 °C on Ge (110) substrate. After Zhu et al. [9]; **b** Ni-catalyzed multiwall carbon nanotubes synthesized at 700 °C by the ECR-PE-CVD process. After Gohier et al. [10]

Au-catalyzed Ge nanowires at 220 °C. This growth at $T < T_E$ was obviously by the VSS mechanism. The STEM image of the nanowire tip shown in Fig. 6.2a indicates that the Au nanoparticle was roughly rounded, and the interface between the Ge nanowire and Au nanoparticle was abrupt. Further, the Au tip was flat rather than spherical or hemispherical suggesting that the Au catalyst was solid and not liquid during growth. Gohier et al. [10] conducted Ni-catalyzed MWCNT growth at 700 °C by the ECR-assisted PECVD process. This temperature was lower than the eutectic temperature 1297 °C of the $R_L \equiv$(Ni, C) eutectic alloy. The MWCNTs were therefore obtained by the VSS mechanism. It is evident from Fig. 6.2b, which depicts these MWCNTs, that they had flat interfaces with the catalyst tips and that the catalyst tips were also flat and not spherical. It implies that the FECA mediating the MWCNT growths was solid and flat. Hibst et al. [11] carried out Pt-catalyzed growth of Si nanowires. MET≡Pt for this growth was deposited by a focused ion or electron beam. Si nanowires thus produced had a polycrystalline silicon tip suggesting that the nanowire growth was via solid, but porous polycrystalline material rather than $PtSi_2$ alloy.

6.2.4 Temperature-Dependent Variation of Nanowire and of MET Catalyzing This Nanowire

The temperature-dependent variation of the diameters of Ge nanowire and of the Au nanoparticle catalyzing the growth of this Ge nanowire is shown in Fig. 6.3. This figure indicates that the Au catalyst nanoparticle diameter depends insignificantly on temperature, but the Ge nanowire diameter depends significantly on temperature. Sidewall growth is a well-known feature of nanowire growth. Also, the diffusion

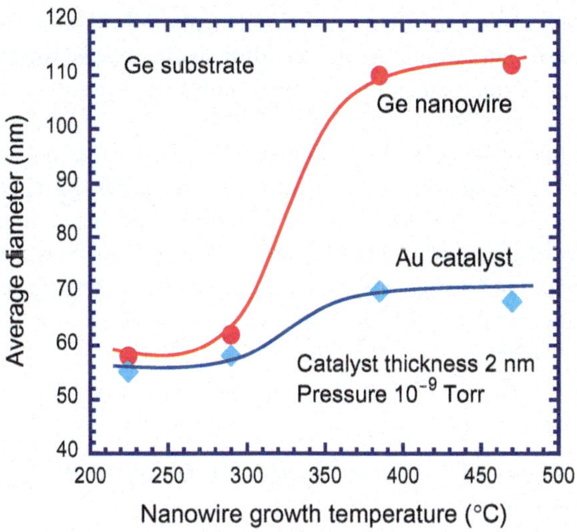

Fig. 6.3 Temperature-dependent variation of the average diameter of Aunanoparticle and of the Ge nanowire grown on this nanoparticle. Au nanoparticle was formed on Ge substrate. The plots were made with the experimental data by Zhu et al. [9]

length of the R_S species (e.g., $R_S{\equiv}Ge$) is temperature-dependent. This diffusion length is therefore smaller at lower temperature. It suggests that, at lower temperature T, smaller number of $R_S{\equiv}Ge$ species is capable of traveling from the impingent location to the nanowire sidewalls just underneath the Au catalyst tip. In contrast, at higher temperature T, the larger number of $R_S{\equiv}Ge$ species is capable of traveling from the impingent location to the nanowire sidewalls just underneath the Au catalyst tip. These $R_S{\equiv}Ge$ species contribute to the thickening of nanowire sidewall. Such a thickening of catalyst nanoparticle does not though take place. All these are evident from Fig. 6.3.

6.3 Illustrations of Carbon Nanofiber Growth by the VSS Mechanism

Yu et al. [12] synthesized CNFs (e.g., carbon nanofibers) at 300 °C by thermal CVD using nickel (II) acetylacetonate as the precursor. CNFs with a diameter of about 50 nm and lengths up to 1 μm were believed to be formed by the VSS mechanism. The FECA metal responsible for the growth was solid. It was confirmed by high-resolution transmission electron microscopy images, which indicated that the FECA metal nanoparticles had lattice spacings of 0.23 nm and 0.43 nm corresponding, respectively, to the {110} and {003} planes of the rhombohedral Ni_3C ($a = 0.458$ nm, $b = 0.458$ nm, $c = 1.299$ nm, and $\beta = 120°$); they were metastable solid solutions.

It was indeed solid solution as confirmed by Nagakura [13], who found the carbon atoms in the Ni_3C lattice possessing one-dimensionally disordered arrangement. Jiao et al. [14] performed quantum chemical molecular dynamic simulations in the framework of self-consistent charge density functional tight-binding method. They also concluded that the Ni_3C lattice becomes relaxed upon thermal annealing at high temperature and is eventually transformed to an amorphous Ni_3C crystal structure. The nickel atoms of Ni_3C were found to be mobile, and hence inner layer carbon atoms could precipitate rapidly out of the surface forming polyyne chains. The SAED pattern confirmed that CNFs were grown from the (001) crystal planes of the Ni_3C nanoparticle. The diameters of the CNFs were, in general, smaller than those of catalytic nanoparticles.

6.4 Illustrations of Carbon Nanotube Growth by the VSS Mechanism

Controversies underlying atomic processes during CNT growths kept the basic growth questions unsettled for a long time. These controversies pertain to the active physical and chemical states of FECA metal during CNT growth. Although the R_L species of the FECA metal nanoparticles in liquid state were found to be necessary for the CNT growths [15], HRTEM, and in particular in situ HRTEM, almost conclusively demonstrated that the VSS mechanism was actually responsible for the CNT growths. Using these tools, Helveg et al. [16] obtained in situ images of MWCNT growing on MET≡Ni. Even minor changes in the FECA metal nanoparticle structure, including dynamic changes in FECA metal morphology, reshaping of FECA metal surface, and also the presence of step edges during the CNT growth could be recorded by HRTEM images. The lattice fringes of the nickel nanoparticle, thus obtained, suggested that Ni nanoparticle mediating MWCNT growth was actually solid during growth, and that surface diffusion of carbon species was the primary transport mechanism during growth. Hofmann et al. [17] shed more light on the chemical and physical states of nickel and iron nanoparticles mediating CNT growths. Using in situ HRTEM and in situ XPS analyses, they concluded that SWCNTs grew also from solid nanoparticles. This means the growth was by the VSS mechanism. Importantly, Hofmann et al. [17] demonstrated that bulk iron carbide was rarely formed and that it played no role in the CNT growth. In fact, iron carbide detected by in situ XPS had very minor contribution to growth. This carbide was actually subsurface carbide with carbon bonded to iron on the surface and dissolved between the first atomic layers. In a subsequent investigation, Hofmann et al. [18] compared the role of Ni, Fe, Pd, and Au FECA metals in CNT growths. These FECA metals differ in their influence on the decomposition of hydrocarbon precursors and also in their ability to dissolve carbon precursor. Viewing the overall CVD growth process as a two-step process, namely (1) the restructuring of the film into catalytically active nanoparticles during pre-treatment and (2) diffusion of carbon species through nanoparticle for growth, they showed again that the FECA metal nanoparticles were in the solid state during growth. Metastable metal carbides, if formed during growth, were intermediates in the transformation of carbon into graphite.

6.5 Strengths of the VSS Mechanism

6.5.1 Superior Crystal-Phase Control

1. Due to the use of solid FECA instead of liquid (molten and semimolten) droplet, the VSS growth mechanism circumvents the problems associated with liquid FECA alloy. It enables atomic-scale control over nanomaterial composition [19]. The crystal-phase control can be realized in nanomaterials (nanowires, nanotubes, etc) due essentially to FECA metal nanoparticle morphologies remaining solid and static during growth by the VSS mechanism [20].
2. The liquid–solid contact angle also remains nearly stable and fixed [21] during growth. This is different from the situation during the VLS growth of nanowires. During this VLS growth, the alloy droplet suffers from morphological fluctuations, and hence the solid–liquid contact angle varies by as much as 20°. A near-constant solid–liquid contact angle during the VSS growth substantially minimizes phase alternations [22]. This is corroborated with the observations that the VSS-grown nanowires are free from planar defects (perpendicular to the nanowire axis) and also from phase alternations [23].
3. The VSS-grown nanomaterials may exhibit single phase and nearly planar-defect-free composition. Lensch-Fal et al. [24] highlighted a number of advantages of the VSS mechanism over the VLS mechanism. Due probably to lower growth temperature, nanowires produced by the VSS mechanism exhibit more uniform diameter distributions, better control of nanowire orientation, increased purity, and more abrupt interfaces for nanowire heterostructures. These are often absent in nanowires produced by the VLS mechanism.
4. With catalyst in the solid state, the VSS mechanism prevents the FECA species from incorporating into nanowires and thus creating deep impurity levels. This mechanism thus helps achieve abrupt interfaces by restraining the so-called reservoir effect.

6.5.2 Abrupt Interface Composition

A wide variety of nanowire applications, such as those in solar cells, tunnel field-effect transistors, and thermoelectric devices can be realized by making use of nanowire heterostructures of abrupt interfaces. Those from Si and Ge can particularly be very useful. They must though have compositionally abrupt interfaces. To be more specific, the abruptness of the interfaces must be tuned to be atomically correct. The job would though be exceedingly difficult if the conventional approach is applied for nanowire growth. One such notable approach is obviously the growth by the VLS mechanism employing, for example, MET≡Au. The primary reason for this would be molten (semimolten) R_L species droplet of the FECA metal nanoparticle appearing as material reservoir during switching between growth materials. The problem can be alleviated by resorting to the growth by the VSS mechanism using, for example, R_L≡($AlAu_2$) alloy, which remains solid during growth even at a reasonably high temperature and yields good growth rate. This temperature must nevertheless be low enough to avoid interdiffusion of Si and Ge during growth. Note that pure Au for the VSS growth of the Si and Ge components of the nanowire would be feasible only at temperatures lower than their eutectic temperature (~360 °C for both Si and Ge), and the growth rate of Si would obviously be low at this temperature.

6.6 Controversy and Weaknesses of the VSS Mechanism

Recall that the VLS mechanism involves the R_L species of a FECA metal in the molten (semimolten) state at a temperature $T = T_E$, but the VSS mechanism involves the R_L species of a FECA metal in the solid state at a temperature $T \langle T_E$ or $T \rangle T_E$. Whether metal-catalyzed X_mY_n nanowire growth can proceed via the VLS or the VSS mechanism depends on the specifics of the respective metal/X binary phase diagram. This binary phase diagram is though often a bulk phase diagram. And hence, the influence of surfaces is not adequately taken into account in this diagram. In the nanoscale, these surfaces have relatively high surface energies. And the most dominant impact of surfaces exhibiting high surface energies on the characteristics of phase diagrams is in shifting the eutectic point to lower temperature. Tanaka and Hara [25] calculated the shift of the eutectic temperature of small spherical alloy droplet of diameter smaller than about 100 nm. For this, they took into consideration the Gibbs free energy and the surface tension of the bulk phase. They observed that the liquid-phase region of phase diagrams was enlarged as the size of the nanoparticle was lowered. Hence, the growth of many X_mY_n nanowires, which is assumed to be done at a temperature lower than the conventional metal/X eutectic temperature, is actually done by the VLS mechanism at the nanometal/X eutectic temperature. Ning et al. [26] synthesized well-shaped, single-crystal GaAs nanowires on cubic NiGa nanoparticle seeds (lattice constant, $a = 0.289$ nm) while the rest were grown with hexagonal Ni_2Ga_3 nanoparticle seeds (lattice constants, $a = 0.405$ nm and $c = 0.489$ nm) at a temperature $T = 600$ °C. This temperature was well below the melting temperature of 1200 °C of both the NiGa and the Ni_2Ga_3 non-eutectic alloys. Recall that Tatsumi et al. [27] also claimed that they synthesized amorphous Si nanowires by a silane CVD process at 650 °C, and that this temperature was well below the eutectic temperature of 830 °C of the R_L≡(Ag, Si) alloy.

Interestingly, Al-mediated Si nanowire growth via the VSS mechanism is related to the peculiarity in the (Al, Si) phase diagram. And this diagram indicates that up to about 1 atom % Si can be dissolved in solid Al at a temperature of 500 °C.

The observation by Tatsumi et al. [27] was later addressed by Weber [28]. A re-evaluation of the Ag/Si phase diagram by him revealed that the solid solubility of Si in Ag was about 0.2 at.% at 650 °C and 0.9 at.% at a temperature close to the eutectic temperature T_E of the R_L≡(Ag, Si) eutectic alloy. This means the VLS growth of Si nanowires with Ag as catalyst was feasible. The conclusion by Wang et al. was also questioned by Wakaser et al. [29].

The observations cited above do not however resolve the puzzle as the downwardly shifted eutectic temperature for many R_L≡(MET, X) alloys may still be higher than the metal-catalyzed growth temperature of some X_mY_n nanowires by the VSS mechanism. It would be true even for very small-sized FECA metal nanoparticles. And hence, the solid–metal-mediated growth of the X_mY_n nanowires would still be carried out by the VSS mechanism. We cite an example. Thombare et al. [30] performed germanium nanowire growth at temperatures 300 °C $\leq T \leq$ 375 °C using a fixed germane partial pressure of 0.75 Torr. They observed no significant nanowire growth at temperature $T <$ 300 °C. The synthesis temperatures (e.g., $T <$ 375 °C) of these nanowires of diameters ~25 nm were lower than the lowest possible eutectic temperature T_E = 762 °C of the R_L≡(Ni, Ge) binary alloy [31] by approximately 400 °C. Such low growth temperatures imply that the growth of nanowires (diameter ~25 nm) can occur only by the VSS mechanism with the R_L species of FECA metal remaining solid during throughout the growth process. Assuming that the nanowire diameter d_{NW} is approximately the same as nanoparticle diameter, the size-dependent melting point depression of such nanoparticle would be marginally low [32].

6.7 Rate-Limiting Steps During the VSS Growths

Thombare et al. [30] suggested four possible rate-limiting steps for the VSS growth of Ge nanowires. These steps may be applicable to the VSS growth of other nanowires, as well. They are:

1. Mass transport of Ge vapor precursor (GeH_4) to the FECA nanoparticle surface and the desorption of H_2 byproduct molecules from this FECA nanoparticle surface;
2. Surface reaction, e.g., desorption of hydrogen molecules from the FECA nanoparticle surface, dissociation (decomposition) of germane on the FECA nanoparticle surface, and adsorption of Ge on the FECA nanoparticle surface;
3. Diffusion of the vapor source species through the FECA nanoparticle to the FECA/nanowire interface; and
4. Addition of Ge atoms to the FECA/nanowire interface.

Thombare et al. [30] plotted the average nanowire length as function of growth temperature for nanowires grown at a fixed germane partial pressure of 0.75 Torr and a fixed growth duration. The nanowire length was found to increase with increase in growth temperature from 300 to 345 °C. Increase in growth temperature beyond

$T = 345$ °C had however marginal effect on the nanowire growth rate. Variation of Ge nanowire length with nanowire diameter obtained at $T = 330$ °C and germane partial pressure of 0.75 Torr showed a very weak dependence of nanowire length on the FECA nanoparticle diameter. The Arrhenius plot obtained by them demonstrated an activation energy, and hence rate-limiting growth mechanism different at $T <$ 340 °C than at $T >$ 340 °C. This activation energy was very small: $\Delta E_a = 0.067$ eV at $T >$ 340 °C. But it (e.g., the activation energy) was substantially larger: $\Delta E_a =$ 0.78 eV at $T <$ 340 °C. It implied that the rate-limiting process is strongly dependent on temperature at $T <$ 340 °C. The extremely small activation energy for growth at $T >$ 340 °C is, on the other hand, inconsistent with the thermally activated process such as solid-state diffusion or a heterogeneous chemical reaction. Rate-limiting processes involving dissociative adsorption of germane, diffusion of Ge through the FECA nanoparticle, or incorporation of Ge into the growth front should depend on temperature so strongly that it becomes inconsistent with the low activation energy. Recall that the activation energy for growth at $T <$ 340 °C is 0.78 eV, which indicates that the rate-limiting process at $T <$ 340 °C is relatively strongly dependent on temperature. However, no evidence of such a trend could be derived by Thombare et al. [30] from their plot of nanowire length versus nanowire diameter [see Fig. 3(d) by these authors]. This suggests that diffusion across the FECA nanoparticle was not probably rate limiting at $T <$ 340 °C. The variation of nanowire length with GeH_4 precursor partial pressure was obtained for nanowires grown at 330 °C, which is in the lower-temperature regime. This variation showed increase in the Ge nanowire growth rate with increasing GeH_4 partial pressure. This happened because a higher rate of GeH_4 reaction at the FECA nanoparticle surface led to an increase in the nanowire growth rate. Alternatively, it happened because increase in chemical potential of Ge in the FECA nanoparticle led to increase in growth rate. Two plausible rate-limiting steps for this could be surface reaction at the FECA nanoparticle surface and incorporation of Ge from the FECA nanoparticle to the Ge nanowire at the solid/solid interface.

6.8 Comparison of VSS Growth Rates with the VLS Growth Rates

Growth rates of various nanowires [33–53] produced by several different techniques employing VLS and VSS mechanisms can shed light on the deeper insight of the VSS mechanism. Most of them are compared in Table 6.1. Entries 1–7 of Table 6.1 are for the VSS growth of Si, Ge, GaAs, InP, InAs, and GaP nanowires. The growth rates for these nanowires are about 0.3, 2.8×10^{-2}, 1.3×10^{-2}, 7.0, 18.0, 7.0, and 13.0 nm/s, respectively. A close look of Table 6.1 would indicate that some of the nanowire growth rates are marginally small, while others are remarkably large. If the R_S species diffuse through solid FECA nanoparticle before supersaturation at the liquid/solid (L/S) interface, the nanowire growth rate by the VSS mechanism must be

Table 6.1 Comparison of growth rates of various nanowires (NWs) produced by employing VLS, VSS, SCG, OAG, and VS mechanisms

No	NW	Mechanism (FECA)	G_{RH} (nm/sec)	T (°C)	T_E (°C)	Technique	Substrate	Comment	References
1	Ge	VSS (Au)	1.3×10^{-2}	320	361	CVD		$Au_{0.29}Ge_{0.71}$-mediated growth	Kodambaka et al. [37]
2	Si	VSS (Al)	0.30	430–490	577	CVD	Si	P-type nanowires	Wang et al. [35]
3	Si	VSS (Au)	2.8×10^{-2}	350	363	CVD	Si	Anodic Al_2O_3 template used	Kuo et al. [36]
4	GaAs	VSS (Au)	7.00	475	450	MOVPE	–	As solubilityin Au < 1%	Dick et al. [38]
5	InP	VSS (Au)	18.00	420		MOVPE	–	Growth at 380–440 °C	Dick et al. [38]
6	InAs	VSS (Au)	7.00	480		MOVPE	–	Growth at 400–500 °C	Dick et al. [38]
7	GaP	VSS (Au)	13.00	475	450	MOVPE	–	No stable Au_2P_3 found	Dick et al. [38]
8	InP	VSS (Cu)	0.11	350	676	MOVPE	InP(111)B	Round nanoparticle	Hillerich et al. [39]
9	InP	VSS (Cu)	0.25	350	676	MOVPE	InP(111)B	Faceted nanoparticle	Hillerich et al. [39]
10	InAs	VLS (Au)	23.50	430		MOVPE	InAs	V/III ratio = 7.5	Dayeh et al. [40]
11	InAs	VLS (Au)	7.20	485		MOVPE	–	V/III ratio = 7.5	Dick et al. [41]
12	Si	VLS(Au)	10.00	415	363	CVD	Si	Activation energy 230 kJ/mol	Kikkawa et al. [34]
13	Si	VLS (Au)	11.50	500	363	MOVPE	–	Activation energy 22 kcal/mol	Lew and Redwing [42]
14	Si	VLS (Au)	133.33	1000	363	CVD	Si	Activation energy 130 kJ/mol	Jeong et al. [43]
15	Si	VLS (Pt)	416.66	1000	363	CVD	Si	Activation energy 80 kJ/mol	Jeong et al. [43]

(continued)

Table 6.1 (continued)

No	NW	Mechanism (FECA)	G_{RH} (nm/sec)	T (°C)	T_E (°C)	Technique	Substrate	Comment	References
16	GaN	VLS (Ga)	36.11			HVPE	Sapphire	GaCl precursor used	Avit et al. [44]
17	ZnO	VLS (Au)	5.5–36.0	870–940		CVD	Silica	AuZn alloy formed	Simon et al. [45]
18	InP	SCG	19.15	380	–	MOCVD	–	V/III ratio = 190	Novotny and Wu [46]
19	GaN	SCG	8.00	850–900	–	CVD	GaN	Chamber pressure ~ 7 torr	–
20	InAs	OAG	17.00	580	–	MOVPE	–	Assisted by SiO_z ($1 \leq z < 2$)	Dick et al. [38]
21	Cu_2S	OAG	2.8×10^{-3}	27	–	–	Cu	Had Cu_2O shell	Wang et al. [47]
22	Si	OAG	133.33	920–950	–	CVD	Al_2O_3	Growth rate > 10 mg/hour	Zhang et al. [48]
23	GaAs	VS	0.37	620	–	MBE	Si	Ga and As beams used	Ambrosini et al. [33]
24	Si	VS	33.33	490	–	LPCVD	Si	SiH_4 used as precursor	Kim et al. [49]
25	Ge	VS	16.66	520	–	LPCVD	Si	GeH_4 used as precursor	Kim et al. [49]
26	$W_{18}O_{49}$	VS	0.0166	700	–	Heating	Mo	Thermal evaporation used	Zhang et al. [50]
27	GaN	VS	0.20	780	–	MBE	Si	V/III ratio was 5	Chèze et al. [51]
28	GaN	VS	0.66	1050	–	MOCVD	SiC	Harsee et al	Hersee et al. [52]
29	InAsSb	VS	0.17	510	–	MOVPE	Si	V/III ratio = 50	Du et al. [53]
30	InAsSb	VLS	1.00	510		MOVPE	Si	V/III ration = 5	Du et al. [53]

CBE is chemical beam epitaxy, G_{RH} is the highest growth rate, and Subst is the abbreviation of substrate

substantially lower than that by the VLS mechanism. While growing Ge nanowires, Kodambaka et al. [37] found the VSS growth rate about 10–100 times lower than the VLS growth rate at the same temperature and pressure. And this was attributed to weaker surface catalytic reactivity and lower diffusivity through the solid catalyst. This was attributed also to lower temperatures (typically a temperature lower than the eutectic temperature) needed for the R_L species alloy of the FECA nanoparticle to be solid. The results by Kodambaka et al. are consistent with the Si nanowire growth rates measured by Wang et al. [35] and Kuo et al. [36], both listed in Table 6.1. This is however inconsistent with the GaAs, InAs, InP, and GaP nanowire growth rates measured by Dick et al. [38]and also listed in Table 6.1. Entries 8–15 of this table are for the VLS growth of InAs nanowires [40, 41], Si nanowires [34, 42, 43], GaN nanowires [44], and ZnO nanowires [45]. The growth rates for these nanowires vary between 7.2 and 446.6 nm/s, respectively. And it increases with increase in temperature [34, 43]. Table 6.1 indicates that the VSS growth rates of GaAs, InAs, InP, and GaP nanowire are, in general, comparable to the VLS growth rates of InAs, GaN, Si, and ZnO nanowires, which is not realistically possible. And this points to the lack of understanding of the most crucial elements of the VSS mechanism.

6.9 Temperature-Dependent Growths

6.9.1 Temperature-Dependent Growth Rate of Nanowires

Temperature-dependent nanowire growth rates shed light on the characteristics of VSS mechanism. Dick et al. [38] measured such temperature-dependent growth rates of the VSS nanowires and found that the temperature-dependent growth rate of the VSS nanowires is comparable with the temperature-dependent growth rate of the VLS nanowires [42]. In steady state, the rate of adatom incorporation at the liquid–solid interface is equal to the rate of release of the R_S species from their precursor and the adsorption of these R_S species onto the nanoparticle surface. This surface may be in the form of a droplet. The diffusion through the nanoparticle surface is not rate limiting. The event(s) mentioned above requires appropriate temperature ranges over which nanowire growth can take place. Even within this temperature range, low temperatures may not suit the formation of the liquid droplet and also of the desorption of oxide present in the growth chamber. On the other hand, within the same temperature range, high temperatures may cause increase in material solubility in the nanoparticle surface and a decrease in supersaturation at the liquid/solid interface. As a result, there may be decrease in nucleation and hence reduction in nanowire growth rate.

6.9.2 Temperature-Dependent Growth Rate of Carbon Nanotubes

CNT growth rate as function of temperature has been investigated by a number of researchers. They all suggested that this growth rate increased nonlinearly as function of temperature. Lee et al. [54] performed thermal CVD using Fe as catalyst and C_2H_2 as carbon precursor. They found that the CNT growth rate increased exponentially from 1.6 to 28 μm/min due to increase in temperature from 800 to 1100 °C. Similarly, Kim et al. [55] performed pyrolysis of phthalocyanine as catalyst and carbon source to grow CNTs. They found that the CNT growth rate increased exponentially from 0.075 to 3.5 μm/min due to increase in growth temperature from 700 to 1000 °C. Figure 6.4 shows the variation of the CNT growth rate as function of temperature for CNTs grown via pyrolysis of iron, nickel, and cobalt phthalocyanines. These CNTs showed nonlinear increase of the growth rate with temperature for Fe, Co, and Ni phthalocyanine nanoparticles, respectively. The CNT growth rate with Fe phthalocyanine was about 2 times higher than those with nickel and cobalt phthalocyanines. These observations are in line with those on the growth of carbon filaments [56]. These observations appear to support that the catalytic growth of CNTs is a thermally activated process and that the dependence of the growth rate on temperature obeys the Arrhenius equation [57]:

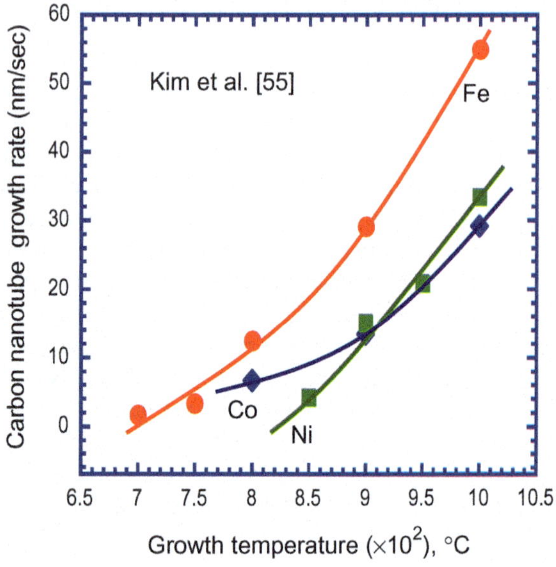

Fig. 6.4 Variation of CNT growth rate as a function of synthesis temperature. The CNTs were grown via pyrolysis of iron, nickel, and cobalt phthalocyanines, respectively. The plots were made with the experimental data by Kim et al. [55]

$$G_R = B_0 \exp\left[-\frac{E_A}{k_B T}\right], \tag{6.1}$$

where G_R is the CNT growth rate, E_A is the activation energy required for this growth rate, k_B is the Boltzmann constant, T is the absolute temperature, and B_0 is a proportionality constant.

Thermal CVD is characterized by an activation energy. It is the result of a combined effect of (1) the decomposition of the precursor molecule(s) on the FECA nanoparticle surface, (1) adsorption of the R_S species on this surface, (3) diffusion of the R_S species through this surface, and (4) nucleation and incorporation of carbon into the growing CNT structure. It can be estimated by taking into consideration the Arrhenius equation (6.1). Taking the natural logarithm of both sides of (6.1), we obtain

$$\ln(G_R) = \ln(B_0) - \frac{E_A}{k_B T}, \tag{6.2}$$

which shows that the natural logarithm of the growth rate is a linear function of the inverse growth temperature. The slope of this function is $-E_A/k_B$. Thus, the variation of ln (G_R) as function of $(1/T)$ yields the activation energy for CNT growth. Figure 6.5 shows the Arrhenius plot by Lee et al. [54] for MWCNT growth by thermal CVD at 800–1100 °C using C_2H_2 precursor and Fe catalyst. The experimental data fit well with the calculated solid line yielding an activation energy of 1.3 eV. It is reasonable considering that the observed activation energies vary between 0.1 and 2.8 eV.

Control of tapering is important for nanowire growth. Following an experiment by Kim et al. [58], it may be achieved by a judicious choice of growth temperature. Note that tapering is caused by diffusing adatoms assimilated onto the nanowire sidewalls. While they can be assimilated more at higher temperature, they can be assimilated less at lower temperature. Nanowires of almost uniform diameters can be realized at an optimally low temperature due essentially to no assimilation of diffusing adatoms onto the sidewalls.

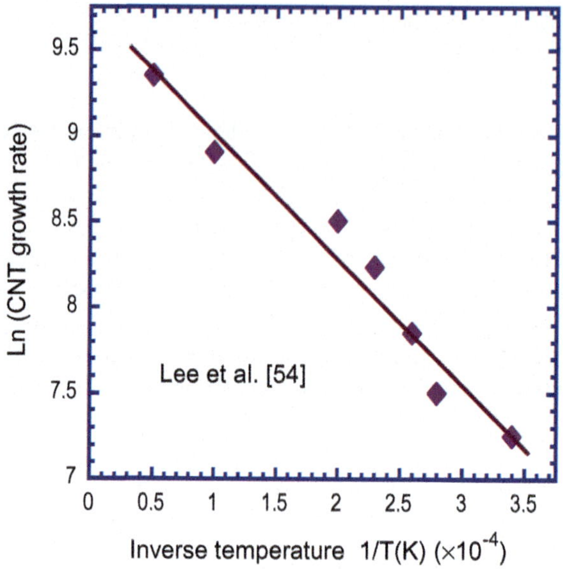

Fig. 6.5 Arrhenius plot for the growth rates of MWCNTs synthesized by thermal CVD at 800–1100 °C. For the growth, C_2H_2 was the carbon [54] source precursor and Fe was the catalyst. The experimental data show the activation energy to be 1.3 eV. The plots were made with the experimental data by Lee et al.

References

1. S. Ganji, Nanowire growths, and mechanisms of these growths, for developing novel nanomaterials. J. Nanosci. Nanotechnol. **19**, 1849–1874 (2019)
2. L.C. Campos, M. Tonezzer, A.S. Ferlauto, V. Grillo, R. Magalhães-Paniago, S., Oliveira, L.O. Ladeira, R.G. Lacerda, Vapor-solid-solid growth mechanism driven by epitaxial match between solid AuZn alloy catalyst particle and ZnO nanowire at low temperature, ,Adv. Mater. **20**, 1499–1504 (2008)
3. K. Kang, D.A. Kim, H.-S. Lee, C.-J. Kim, J.-E. Yang, M.-H. Jo, Low-temperature deterministic growth of Ge nanowires using Cu solid catalysts. Adv. Mater. **20**, 4684–4690 (2008)
4. K. Kang, G.H. Gu, D.A. Kim, C.G. Park, M.-H. Jo, Self-organized growth of Ge nanowires from Ni-Cu bulk alloys. Chem. Mater. **20**, 6577–6579 (2008)
5. S. Eustis, L.H. Robins, B. Nikoobakht, Patterns of ensemble variation of the optical properties of ZnO nanowires grown with copper and gold catalysts. J. Phys. Chem. C **113**, 2277–2285 (2009)
6. A.V. Davydov, National Institute of Standard and Technology, Gaithersburg, MD (private communications)
7. T.I. Kamins, R.S. Williams, Y. Chen, Y.-L. Chang, Y.A. Chang, Chemical vapor deposition of Si nanowires nucleated by $TiSi_2$ islands on Si. Appl. Phys. Lett. **76**, 562 (2000)
8. N. Han, F. Wang, A.T. Hui, J.J Hou, G. Shan, F. Xiu, T.- F. Hung, J.C. Ho, Facile synthesis and growth mecha-nism of Ni-catalyzed GaAs nanowires on non-crystalline substrates, Nanotechnology **22**, 285607 (2011)
9. Z. Zhu, Y. Song, Z. Zhang, H. Sun, Y. Han, Y. Li, L. Zhang, Z. Xue, Z. Di, S. Wang, The vapor-solid-solid growth of Ge nanowires on Ge (110) by molecular beam epitaxy. J. Appl. Phys. **122**, 094304 (2017)

10. A. Gohier T.M. Minea, S. Point, J.-V. Mevellec, J. Jimenez, M.A. Djouadi, A. Granier, Early stages of the carbon nanotube growth by low-pressure CVD and PECVD, Diam. Relat. Mater. **18**, 61–65 (2009)
11. N. Hibst, P. Knittel, J. Biskupek, C. Kranz, B. Mizaikoff, S. Strehle, The mechanisms of platinum-catalyzed silicon nanowire growth. Semicond. Sci. Technol. **31**, 025005 (2016)
12. B. Yu, S. Wang, Q. Zhang, Y. He, H. Huang, J. Zou, Ni_3C-assisted growth of carbon nanofibres at 300 °C by thermal CVD. Nanotechnology **25**, 325602 (2014)
13. S. Nagakura, Study of metallic carbides by electron diffraction, Part II. Crystal structure analysis of nickel carbide, J. Phys. Soc. Jpn. **13**, 1005–1014 (1958)
14. M. Jiao, K. Li, W. Guan, Y. Wang, Z. Wu, A. Page, K. Morokuma, Crystalline Ni_3C as both carbon source and catalyst for graphene nucleation: a QM/MD study. Sci. Rep. **5**, 12091 (2015)
15. A.R. Harutyunyan, T. Tokune, E. Mora, Liquid as a required catalyst phase for carbon single-walled nanotube growth. Appl. Phys. Lett. **87**, 051919 (2005)
16. S. Helveg, C. Lopez-Cartes, J. Sehested, P.L. Hansen, B.S. Clausen, J.R. Rostrup-Nielsen, F. Abild-Pedersen, J.K. Norskov, Nature **427**, 426–429 (2004)
17. S. Hofmann, R. Sharma, C. Ducati, G. Du, C. Mattevi, C. Cepek, M. Cantoro, S. Pisana, A. Parvez, F. Cervantes-Sodi, A.C. Ferrari, R. Dunin-Borkowski, S. Lizzit, L. Petaccia, A. Goldoni, J. Robertson, Nano Lett. **7**, 602–608 (2007)
18. S. Hofmann, R. Blume, C.T. Wirth, M. Cantoro, R. Sharma, C. Ducati, M. Havecker, S. Zafeiratos, P. Schnoerch, A. Oestereich, D. Teschner, M. Albrecht, A. Knop-Gericke, R. Schlogl, J. Robertson, State of transition metal catalysts during carbon nanotube growth. J. Phys. Chem. C **113**, 1648–1656 (2009)
19. C.W. Wen, M.C. Reuter, J. Bruley, J. Tersoff, S. Kodambaka, E.A. Stach, F.M. Ross, Formation of compositionally abrupt axial heterojunctions in silicon-germanium nanowires. Science **326**, 1247–1250 (2009)
20. K.-L. Wu, Y. Chou, C.-C. Su, C.-C. Yang, W.-I. Lee, Y.-C. Chou, Controlling bottom-up rapid growth of single crystalline gallium nitride nanowires on silicon. Sci. Rep. **7**, 17942 (2017)
21. Y.-C. Chou, C.-Y. Wen, M.C. Reuter, D. Su, E.A. Stach, F.M. Ross, Controlling the growth of Si/Ge nanowires and heterojunctions using silver-gold alloy catalysts. ACS Nano **6**, 6407–6415 (2012)
22. F. Wang, W.E. Buhro, Crystal-phase control by solution-solid-solid growth of II–VI quantum wires. Nano Lett. **16**, 889–894 (2016)
23. C.Y. Wen, M.C. Reuter, J. Tersoff, E.A. Stach, F.M. Ross, Structure, growth kinetics, and ledge flow during vapor-solid-solid growth of copper-catalyzed silicon nanowires. Nano Lett. **10**, 514–519 (2010)
24. J.I. Lensch-Falk, E.R. Hemesath, D.E. Perea, L.J. Lauhon, Alternative catalysts for VSS growth of silicon and germanium nanowires. J. Mater. Chem. **19**, 849–857 (2009)
25. T. Tanaka, S. Hara, Thermodynamic evaluation of binary phase diagrams in small particle systems. Zeit. Für Metallkunde. **92**(5), 467–472 (2001)
26. N. Han, A.T. Hui, F.-Y. Wang, J.J. Hou, F. Xiu, T.-F. Hung, J.C. Ho, Crystal phase and growth orientation dependence of GaAs nanowires on Ni_xGa_y seeds via vapor-solid-solid mechanism. Appl. Phys. Lett. **99**, 083114 (2011)
27. Y. Tatsumi, M. Hirata, M. Shigi, Characteristics of whisker growth in amorphous silicon. Jpn. J. Appl. Phys. **18**, 2199–2206 (1979)
28. L. Weber, Equilibrium solid solubility of silicon in silver. Metall. Mater. Trans. A **33A**, 1145–1150 (2002)
29. B.A. Wakaser, M.C. Reuter, M.M. Khayyat, C.-Y. Wen, R. Haight, S. Guha, F.M. Ross, Growth system, structure, and doping of aluminum-seeded epitaxial silicon nanowires. Nano Lett. **9**, 3296–3301 (2009)
30. S.V. Thombare, A.F. Marshall, P. C. , McIntyre, Kinetics of germanium nanowire growth by the vapor-solid-solid mechanism with a Ni-based catalyst. APL Mater. **1**, 061101 (2013)
31. A. Nash, P. Nash, Ge-Ni (germanium-nickel), *Binary Alloy Phase Diagrams*, vol. 2, 2nd edn, ed. by T.B. Massalski (ASM International, Cleveland, OH, 1990)

32. S.N. Mohammad, General hypothesis and shell model for the synthesis of semiconductor nanotubes, including carbon nanotubes. J. Appl. Phys. **108**, 064323 (2010)
33. S. Ambrosini, M. Fanetti, V. Grillo, A. Franciosi, S. Rubini, Vapor-liquid-solid and vapor-solid growth of self-catalyzed GaAs nanowires. AIP Adv. **1**, 042142 (2011)
34. J. Kikkawa, Y. Ohno, S. Takeda, Growth rate of silicon nanowires. Appl. Phys. Lett. **86**, 123109 (2005)
35. Y.W. Wang, V. Schmidt, S. Senz, U. Gösele, Epitaxial growth of silicon nanowires using an aluminum catalyst. Nat. Nanotechnol. **1**, 186–189 (2006)
36. C.Y. Kuo, C. Gau, Vapor-solid-solid growth of crystalline silicon nanowires using anodic aluminum oxide template. Thin Solid Film **519**, 3603–3607 (2011)
37. S. Kodambaka, J. Tersoff, F.M. Ross, Ge nanowire growth below the eutectic temperature. Science **316**, 729–732 (2007)
38. K.A. Dick, K. Deppert, L.S. Karlsson, L.R. Wallenberg, L. Samuelson, W. Seifert, A new understanding of Au-assisted growth of III–V semiconductor nanowires. Adv. Funct. Mater. **15**, 1603–1610 (2005)
39. K. Hillerich, K.A. Dick, M.E. Messing, K. Deppert, J. Johansson, Simultaneous growth mechanisms for Cu-seeded InP nanowires. Nano Res. **5**, 297–306 (2012)
40. S.A. Dayeh, E.T. Wu, D. Wang, III-V nanowire growth mechanism: V/III ratio and temperature effects. Nano Lett. **7**, 2486–2490 (2007)
41. K.A. Dick, K. Deppert, T. Martensson, B. Mandl, L. Samuelson, W. Seifert, Failure of the vapor-liquid-solid mechanism in Au-assisted MOVPE growth of InAs nanowires. Nano Lett. **5**, 761–764 (2005)
42. K.K. Lew, J.M. Redwing, Growth characteristics of silicon nanowires synthesized by vapor-liquid-solid growth in nanoporous alumina templates. J. Cryst. Growth **254**, 14–22 (2003)
43. H. Jeong, T.E. Park, H.K. Seong, M. Kim, U. Kim, H.J. Choi, Growth kinetics of silicon nanowires by platinum assisted vapor-liquid-solid mechanism. Chem. Phys. Lett. **467**, 331–334 (2009)
44. G. Avit, K. Lekhal, Y. André, C. Bougerol, F. Réveret, J. Leymarie, E. Gil, G. Monier, D. Castelluci, A. Trasso-udaine, Ultralong and defect-free GaN nanowires grown by the HVPE process. Nano Lett. **14**, 559–562 (2014)
45. H. Simon, T. Krekeler, G. Schaan, W. Mader, Metal-seeded growth mechanism of ZnO nanowires. Cryst. Growth Design. **13**, 572–580 (2013)
46. C.J. Novotny, P.K.L. Wu, Vertically aligned, catalyst-free InP nanowires grown by metalorganic chemical vapor deposition. Appl. Phys. Lett **87**, 203111 (2005)
47. N. Wang, K.K. Fung, S. Wang, S. Yang, Oxide-assisted nucleation and growth of copper sulphide nanowire arrays. J. Cryst. Growth **233**, 226–232 (2001)
48. Y.F. Zhang, Y.H. Tang, C. Lam, N. Wang, C.S. Lee, I. Bello, S.T. Lee, Bulk-quantity Si nanowires synthesized by SiO sublimation. J. Cryst. Growth **212**, 115–118 (2000)
49. B.-S. Kim, T.-W. Koo, J.-H. Lee, D.S. Kim, Y.C. Jung, S.W. Hwang, B.L. Choi, E.K. Lee, J.M. Kim, D. Whang, Catalyst-free growth of single-crystal silicon and germanium nanowires. Nano Lett. **9**, 864–869 (2009)
50. Z. Zhang, Y. Wang, H. Li, W. Yuan, X. Zhang, C. Sun, Z, Zhang, Atomic-scale observation of vapor−solid nanowire growth via oscillatory mass transport. ACS Nano **10**, 763–769 (2016)
51. C. Chèze, L. Geelhaar, O. Brandt, W.M. Weber, H. Riechert, S. Münch, R. Rothemund, S. Reitzenstein, A. Forchel, T. Kehagias, P. Komninou, G.P. Dimitrakopulos, T. Karakostas, Direct comparison of catalyst-free and catalyst-induced GaN nanowires. Nano Res. **3**, 528–536 (2010)
52. S.D. Hersee, X. Sun, X. Wang, The controlled growth of GaN nanowires. Nano Lett. **6**, 1808–1811 (2006)
53. W. Du, X. Yang, H. Pan, X. Wang, H. Ji, S. Luo, X. Ji, Z. Wang, T. Yang, Two different growth mechanisms for Au-free InAsSb nanowire growth on Si substrate. Cryst. Growth Des. **15**, 2413–2418 (2015)
54. Y.T. Lee, J. Park, Y.S. Choi, H. Ryu, H.J. Lee, Temperature-dependent growth of vertically aligned carbon nanotubes in the range 800–1100 °C. J. Phys. Chem. B **106**, 7614–7618 (2002)

55. N.S. Kim, Y.-T. Lee, J. Park, J.B. Han, Y.S. Choi, S.Y. Choi, J. Choo, G.H. Lee, Vertically aligned carbon nanotubes grown by pyrolysis of iron, cobalt, and nickel phthalocyanines. J. Phys. Chem. B **107**, 9249–9255 (2003)
56. R.T.K. Baker, Catalytic growth of carbon filaments. Carbon **27**, 315–323 (1989)
57. I. Chorkendorff, J.W. Niemantsverdriet, *Concepts of Modern Catalysis and Kinetics*, 2nd edn. (Wiley-VCH, Weinheim, Germany, 2007)
58. J.H. Kim, S.R. Moon, H.S. Yoon, J.H. Jung, Y. Kim, Z.G. Chen, J. Zou, D.Y. Choi, H.J. Joyce, Q. Gao, H.H. Tan, C. Jagadish, Taper-free and vertically oriented Ge nanowires on Ge/Si substrates grown by a two-temperature process. Cryst. Growth Des. **12**, 135–141 (2012)

Chapter 7
Vapor–Solid Growth Mechanism

Abstract The vapor–solid (VS) growth mechanism for the growths of one-dimensional and quasi-two-dimensional nanomaterials has been reviewed in some details. This mechanism is one of the most widely employed mechanisms for nanomaterial growths. The basics of this mechanism have been articulated. Important features of this mechanism have been presented. Characteristic features of nanomaterial growths by this mechanism have then been illustrated and critically examined. Based on these, the strengths and weaknesses of this mechanism in mediating nanomaterial growths have been described. Nanowire growths by this mechanism have particularly been studied. For this, the comparisons of nanowire growth by the VS and the VLS mechanisms have been made. Carbon nanotube growths and semiconductor nanobelt growths by the VS mechanism have also been investigated. The roles of SUBSANOs in growths by the VS mechanism have been assessed. The basics of the water-assisted growths of nanowires and nanotubes by the VS mechanism have been thoroughly explored.

7.1 Basics

The previous two chapters (e.g., Chaps. 5 and 6) describe the FECA-metal -mediated nanomaterial growths. They highlight the concern in using FECA metals for these growths. To alleviate this concern, catalyst-free vapor–solid (VS) growth mechanism illustrated in Fig. 7.1 was proposed. Both the patterned growths and the selective area epitaxial growths are best performed by this epitaxy. InAs nanowires were thus grown, for example, by electron beam lithography and wet chemical etching in patterned regions of Si(111) substrate [1]. A thin SiO_2 film served as a mask for the growth. This mask defined the diameters and the desired locations on the substrate for growth.

A wide variety of nanomaterials, such as nanowires, nanotubes, nanofibers, and nanobelts, are currently grown by the metal-catalyst-free VS mechanism. It is believed that the nanomaterial growths by this mechanism are carried out on SUBSANO nanoparticles (see Chap. 3) of crystalline lattice structure. The general

S. N. Mohammad, *Synthesis of Nanomaterials*, Springer Series in Materials Science 307, https://doi.org/10.1007/978-3-030-57585-4_7

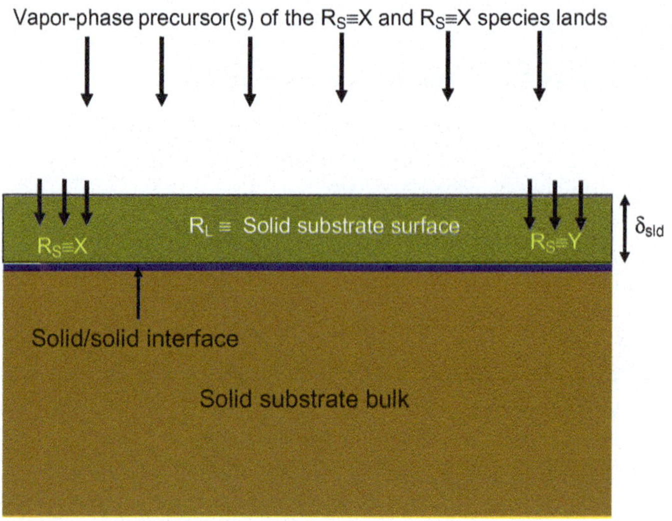

Fig. 7.1 Schematic diagram showing the landing of the precursor(s) of the R_S≡X and R_S≡Y source species on the R_L≡solid substrate surface and the diffusion of the R_S≡X and R_S≡Y source species through the said substrate surface of thickness δ_{sld}. The R_S≡X and R_S≡Y source species are generated from their precursor(s) on the R_L species surface by thermal decomposition or some other means. Solid/solid interface does not really exist. Yet it is shown only for conceptual purpose

terminology of the R_L species defined in Chap. 3, Sect. 3.3 may again be utilized to describe the growths. The pre-nucleation stage and the pro-nucleation stage of growths defined in Chap. 4, Sect. 3 would also be used to describe the growths. An important attribute of the VS mechanism is that it does not necessitate METANO seed [2, 3] for growths; yet the growth is realized, and directly from the vapor-phase source species condensed to the solid-phase nanomaterials. Arguably, crystalline surface of SUBSANO plays key role in this growth. Due to the absence of energetically favored METANO for growth, the temperature for this growth may be higher than those by the VLS and the VSS mechanism.

Recall that the bulk diffusion of source species through eutectic droplet to the liquid/solid interface is key to the VLS mechanism for nanomaterial growths. However, FECA metal (e.g., METANO) is absent for growths by the VS mechanism (see Fig. 7.1). So, the interface(s) with metal is also absent for growth. The growth by the VS mechanism is consequently simpler. This mechanism involves only two interfaces, namely the vapor/SUBSANO interface and the solid/SUBSANO interface. Note that the SUBSANO surface is normally crystalline, and hence the interface between vapor and the crystalline SUBSANO surface is the one which is transformed to an interface between solid nanomaterial surface and the crystalline SUBSANO surface. The basic tenet of the VS mechanism is that vapor condenses into solid to form nucleus at the vapor/SUBSANO interface. And this is governed primarily by deposition temperature, pressure, and SUBSANO surface. And, due to the absence of FECA metal (e.g., METANO), the nanomaterials by the VS mechanism may be free from contamination. Further, based on thermodynamic and kinetic consideration, these nanomaterials by the VS mechanism could actually be the result of self-catalytic growth, oxide-assisted growth, anisotropic growth, defect-induced growth, and/or the growth by Frank's screw dislocation. The exact mechanism of growths has though not yet been determined. Some of these growths may lack compelling thermodynamic and kinetic validation.

7.2 Illustrations of Nanowire Growth by the VS Mechanism

7.2.1 Nanowire Growths on Non-stoichiometric SiO_z and GeO_z Wafer Surfaces

Kim et al. [4] etched hydrogen-terminated silicon wafer with ultrapure water to generate islands of a reactive silicon-rich oxide (SiO_z) layer on the crystalline Si substrate surface. They found that such surface was conducive to Si nanowire growth at a temperature of 520 °C and a pressure of 15 Torr using SiH_4 (10% diluted in H_2) precursor. Anisotropic growth of single-crystal Si nanowires could also be grown at a reduced temperature of 490 °C. Ge nanowires could similarly be grown with a similar two-step growth method making use of a mixture of GeH_4 and H_2 gases. Interestingly, Kim et al. [4] could also grow Si nanowires on the surface of quartz wool, which is actually silica fiber. X-ray diffraction study by Choi et al. [5] indicated that these fibers had the characteristics of non-stoichiometric amorphous silica. The most notable feature of the experiments by Kim et al. [4] was the non-stoichiometric SiO_z and GeO_z ($1<z<2$) wafer surfaces suited to yield Si and Ge nanowires. These surfaces were not crystalline and stoichiometric, but were reactive without the presence of metal catalyst. This was probably why high-density silicon nanowires could be selectively grown [4] on the regions covered with the reactive thin non-stoichiometric SiO_z film, but not on regions covered with stoichiometric SiO_2 film. And this demonstrates that reactive SiO_z layer plays a key role in the nucleation of metal-free nanowires by the VS mechanism.

7.2.2 Nanowire Growths on Activated Amorphous Carbon

Gao et al. [6] synthesized SiC nanowires, 20–100 nm in diameter and 10–100 μm in length, at 1380 °C in a fused quartz tube. A graphite crucible placed in the quart tube was fed with 99.9% pure SiO and activated amorphous carbon powder. This activated amorphous carbon powder was placed on the top of SiO powder. Nanowires thus produced had a thin amorphous layer, 1–3 nm thick, at their tip. It consisted of Si, C, and O implying that the nanoparticle surface that catalyzed the nanowire growth had $R_L \equiv$(Si, C, O) species, and that this species had disordered, amorphous (semi-amorphous) lattice structure. Ho et al. [7] grew vertically aligned ZnO nanowires on sapphire, glass, and Si substrates by direct thermal evaporation of the source materials (e.g., zinc and graphite powders) and without the assistance of FECA metal(s). The substrates were however subjected to mechanical scratching, mechanical polishing, and chemical etching that yielded surface roughness of the substrate surfaces, which became disturbed and disordered due to scratching and etching. Noticeably, nanowires grew abundantly on the scratched areas of the rough and disordered surface, but rarely on the unscratched, smooth, crystalline, and ordered areas of the surface.

7.2.3 Nanowire Growths on Au Cluster Surface

Tang et al. [8] grew InN nanowires at a growth rate of 1.1–2.8 nm/s by placing a mixture of In metal and In_2O_3 powder in a quartz boat inside a quartz tube reactor. Au clusters deposited on a SiO_2-coated Si substrate and placed in the downstream of the reactor were used as nanoparticles. While the temperature of the source species zone of the reactor was 700 °C, the temperature of the substrate-containing zone of the reactor was 525 °C. In vapor diffused into Au clusters creating R_L≡(Au, In) alloy. Continued supply of this vapor together with the nitrogen vapor from ammonia diffused into the (Au, In) alloy and reacted together to form InN molecules. Supersaturation of these molecules yielded InN nanowires at a growth rate of 6.6–16.6 nm/s. The supersaturation generally takes place at the liquid/solid interface, which is absent for the VS growth. So, it is not clear how nucleation of InN nanowires took place. Also, a close look of the (Au, In) binary phase diagram suggests that the (Au, In) compound has the highest eutectic temperature of 486 °C for $Au_{0.45}$, $In_{0.55}$ alloy and that this compound at 525 °C could likely be in the liquid state.

7.2.4 Chemical Vapor Deposition of Tungsten Oxide Nanowires

Zhang et al. [9] made use of environmental TEM to carry out a real-time atomic-scale investigation of the tungsten oxide ($W_{18}O_{49}$) nanowire growth. For this, a pure tungsten filament was first heated to ~700 °C within the TEM column. Pure oxygen was then introduced into the filament area for growth. Metal(s) could be present at the nanowire tip by vapor deposition of metal(s) or by a solid metal substrate for nanowire growth. The metallic source present at the nanowire tip could then drive this nanowire growth. At the end of experiment by Zhang et al. [9], no FECA metal was however detected on the nanowire tips (see Fig. 7.2). Had the metal been originated at the metal substrate, it would be present at the nanowire tip at least via surface diffusion or twin boundary diffusion of metal ions. Yet, tungsten oxide nanowires were found to grow on the tungsten filament. This means that diffusion of metal ions from the metal substrate to nanowire tip along the nanowire sidewall did not take place. It was not hence responsible for nanowire growth. Then what was responsible

Fig. 7.2 TEM images of tungsten oxide ($W_{18}O_{49}$) nanowires grown on the tungsten filament by two-hour thermal oxidation under an oxygen pressure of 0.095 Pa and at a temperature of ~700 °C. Reproduced with copyright permission from the publisher; Ref. Zhang et al. [9]

for growth? To address it, a tungsten lamella was affixed on a molybdenum (Mo) grid, and the experiment was repeated. Tungsten oxide nanowire growth took place on both the tungsten lamella and the Mo grid. The nanowires grew layer by layer on also the top of the Mo grid. This suggested that the tungsten oxide nanowire growth was similar to that by chemical vapor deposition in which chemical vapors from the tungsten lamella were deposited onto the Mo substrate in the presence of oxygen flow. The in situ high-resolution TEM images confirmed the tip growth dynamics of the nanowires grown on the tungsten filament. The growth mechanism was not therefore driven by screw dislocation [10] or by any other means. Yazdi et al. [11] grew AlN nanowires by the VS mechanism. The top of the nanowires was flat suggesting that it was grown not mediated by hemispherical catalyst droplet. This was evident from the TEM image presented in Fig. 7.3. This image clearly showed no metal present at the nanowire tip.

Fig. 7.3 SEM image of the top flat surface of AlN nanowires, a characteristic of the VS growth. These AlN nanowires were grown at 1750 °C on vicinal SiC substrate. The nanowires are hexagonally shaped and perfectly aligned along the [0001] direction. Reproduced with copyright permission from publisher; Ref. Yazdi et al. [11]

7.2.5 Comparison of Nanowires by the VS and the VSS Mechanisms

Chèze et al. [12] found two main differences between GaN nanowires produced at 780 °C by the catalyst-free VS mechanism and the nanowires produced by the Ni-catalyst-induced VSS mechanism. These nanowires were grown by plasma-assisted MBE under identical nitrogen-rich conditions. The catalyst-free nanowires were grown in silicon substrate, but the catalyst-assisted nanowires were grown on sapphire substrate. Majority of the nanowires by the VS mechanism was shorter than the nanowires by the VSS mechanism. Both the catalyst-free and catalyst-induced nanowires were Ga-polar. The nanowires produced by the catalyst-induced VSS mechanism had high density of basal-plane stacking faults; no other extended defects were found in these nanowires. The nanowires produced by the catalyst-free VS mechanism were, on the other hand, largely free from stacking faults. Further, the photoluminescence peaks of the nanowires by the catalyst-free VS mechanism were far more intense and narrower than the photoluminescence peaks of nanowires produced by the catalyst-mediated VSS mechanism. And it was attributed to the Ni contamination and/or the presence of an additional phase. In other words, Ni contamination was the cause of the significantly reduced photoluminescence intensity of the nanowires by the Ni-catalyzed VSS mechanism. Both the structural and optical qualities of catalyst-free VS nanowires were superior to those of catalyst-mediated VSS nanowires. In a subsequent report, Chèze et al. [13] demonstrated that the GaN nanowires by the Ni-catalyzed VSS mechanism grow faster than the GaN nanowires by the self-induced VS mechanism. The VS nanowire growth rates were identical on both Si(111) and Si(001) substrates. This difference in growth rates by the VS and the VSS mechanisms resulted from the higher diffusion coefficient, lower activation barrier, and the enhanced incorporation of the source species in Ni-catalyzed growth. Such an enhancement was absent in nanowires produced by the self-induced VS mechanism. This investigation thus points to the superiority of the nanomaterials produced by the VS mechanism as compared to the nanomaterials produced by the VSS mechanism. An important attribute of the plasma-assisted MBE is that the substrate (catalyst) surface is damaged by plasma bombardment. So, it is conceivable that the silicon substrate surface was not crystalline, but disturbed and disordered for catalyst-free growth. Likewise, the Ni catalyst surface was not crystalline, but disturbed and disordered for catalyst-mediated growths.

7.3 Illustrations of Carbon Nanotube Growths by the VS Mechanism

7.3.1 CNT Growths on Nanosized Diamond

Takagi et al. [14] and later Varanasi et al. [15] performed solid-phase CNT growths by CVD using nanosized diamond. Note that diamond has very strong and nearly isotropic tetrahedral sp^3 bond, but no ability to decompose carbon precursors such as methane and ethanol. Thus, it lacks ability to serve as catalyst for CNT growths. Yet, acting as nanoparticle seed it produced SWCNT at 850 °C from ethanol precursor. This nanoparticle seed did not fuse at the growth temperature, but could nevertheless produce CNTs. This means CNTs were synthesized from diamond nanoparticles in the solid phase, rather than in the liquid phase. And remarkably, this synthesis by diamond nanoparticles could take place on any substrate, such as that from semiconductors (Si, SiC, etc.), insulators (SiO_2, sapphire, etc.), metals (Pt, Au, etc.), and metal oxides (Al-oxide, Mg-oxide, etc.).

7.3.2 Basics of CNT Growths on Nanosized Diamond

Koga et al. [16] noted that the bulk diffusion of carbon in diamond is negligible as compared to that in metals such as Fe. This bulk diffusion of carbon hence contributes marginally to CNT growth. Surface diffusion of carbon though takes place through nanodiamond particles. In the CVD environment, hydrogen termination of the diamond surface is inhibited. The density of atomic hydrogen in this surface is too low to induce any hydrogen termination. In view of this, the diamond surface is transformed to a π-bonded sp^2-type relaxed surface from the tetrahedral sp^3 bonded surface. Further, graphite is not etched from the nanodiamond surface due to depletion of atomic hydrogen. In the CVD environment, a small domain of graphite island is, as a result, formed on the π-bonded relaxed diamond surface. Graphite is a soft, black, layered semimetal with extremely strong trigonal sp^2 planar bonds, but weak interlayer π-bonds [17]. This island has a nanosized curvature. But it contains amorphous (semi-amorphous) carbon materials, which might have been most crucial for CNT growths on nanodiamond by the VS mechanism. And it is a good example of a novel means of CNT growth by the VS mechanism.

7.3.3 MWCNT Growths on Defective Surfaces

MWCNTs could be produced from carbon black by Lin et al. [18]. Being fullerene-like, the carbon black surface played important roles in the CNT growth by the VS mechanism. Previous studies by Hahn et al. [19] demonstrated that the fullerene-like structures could serve as nanoparticle seeds for fabricating SWCNTs. High surface energy sites of the fullerene-like surface probably decomposed the precursor into carbon and hydrogen. In a subsequent investigation, Lin et al. [20] produced also MWCNTs directly on defective graphite (flake graphite and highly oriented pyrolytic graphite) surface at 850 °C by CVD using helium-diluted ethylene. Although methane and ethylene were suitable for growing CNTs on the treated surface, alcohols and cyclohexane were not suitable for CNT growths on this surface. High surface energy sites of the surface were not probably strong enough to decompose these precursors (e.g., alcohols and cyclohexane) to release C species. These precursors could produce only amorphous carbon on this surface. Lin et al. [21] grew MWCNTs and carbon nanostructures as well from the metastable carbon layers self-assembled on the Au thin-film surface, which is actually a bulk material substrate. Huang et al. [22] made use of a clean Si wafer with a 1 μm thick layer of SiO_2, which was scratched by a diamond blade for CNT growths at 900 °C. But no catalyst was formed (deposited) on the substrate surface. SEM observations indicated that CNTs were formed at the edge of the scratched areas. The same experiment performed on a scratched quartz plate yielded similar results. However, no CNTs could be found, on the nanoscopic level, both smooth and crystalline.

7.3.4 SWCNT Growths on SiO_2 Surfaces

Liu et al. [23] presented a FECA-metal-free CVD technique to produce CNTs. For this, a 30-nm-thick SiO_2 film was first sputter deposited onto Si or on 1-μm-thick thermally grown SiO_2 layer on Si, which served as substrate. This substrate was then placed into CVD chamber which had a temperature of 900 °C and which was under an Ar flow of 100 sccm. Maintaining the same temperature of 900 °C, CH_4 and H_2 were also introduced into the chamber for 20 min. Both of them had the flow rate of 500 sccm. These led to the initiation of CNT growths. The thickness of the SiO_2 layer was very important for this. Liu et al. [23] carried out FECA-metal -free CNT growths also by using a clean Si/SiO_2 wafer, which was scratched by sharp tip of another Si/SiO_2 wafer. Both of these wafers had Si covered with a 1-μm-thick thermally grown SiO_2. The second wafer was used to scratch the first wafer just to ensure that the process was completely free from metal. The scratched wafer was then introduced into a CVD chamber, and the growth was initiated at 900 °C with a flow of Ar, CH_4, and H_2. Raman spectra confirmed the growth of SWCNTs. In a subsequent experiment, Liu et al. [24] reported SWCNT growth on silica. In situ TEM studies revealed that this silica served actively as catalyst nanoparticle, and CNTs grew on non-stoichiometric SiO_z nanoparticles, but not on Si nanoparticles, during the entire experiment. No SWCNTs grew on Si nanoparticles even under a reasonably broad range of process parameters such as temperatureand pressure. With similar spirit, Steiner et al. [25] made use of zirconium oxide as catalyst nanoparticles to produce CNTs. They found that the adsorption of the carbon source on the non-stoichiometric ZrO_z nanoparticle surface, subsequent carbon source decomposition on this surface, the carbon species diffusion through this surface, and nanotube nucleation on this surface were essential for CNT growths.

Using over 30 different samples of elemental carbon, including a variety of activated carbons (ACs), carbon blacks, nanostructured carbons, fullerenes, graphite, glassy carbon, and synthesized diamond powders, Muradov [26] studied methane decomposition reaction under atmospheric pressure, at a temperature of 850 °C and with methane flow rate of 5.0 ± 0.2 mL/min. Catalytic activity of these carbon samples for methane decomposition was determined by their origin, structure, and surface area. Muradov found that the catalytic activity of carbon precursors was determined solely by their origin and surface properties.

7.4 Illustrations of Nanobelt Growth by the VS Mechanism

Nanobelts are quasi-one-dimensional nanomaterials exhibiting control over chemical composition, crystallographic structure, and surfaces. These surfaces include top and bottom surfaces, and side surfaces. Nanobelts can be both oxide and non-oxide nanobelts. Among them, oxide nanobelts have been synthesized extensively from both the corresponding bulk semiconductor materials and bulk functional oxides (e.g., ZnO, SnO_2, In_2O_3, Ga_2O_3, CdO, PbO_2, etc). The VS mechanism is believed to be responsible for their growths. The nanobelts, thus obtained, can be single-crystalline and dislocation-free exhibiting atomically flat surfaces. While hollow nanotubes and solid nanowires have cylindrically symmetric cross sections, nanobelts have controlled rectangular cross sections. The morphology of oxide nanobelts can have cations of many different valence states and materials of many different crystallographic structures. The structural characteristics of various oxide nanobelts are quite similar even though bulk oxides from which they are made have varied semiconducting properties.

Pan et al. [27] synthesized many different oxide nanobelts, including ZnO nanobelts. They performed, for example, thermal evaporation of ZnO powders at 1400 °C for 2 h to produce these ZnO nanobelts on the surface of alumina plate. SEM images showed that the lengths of these nanobelts were several tens to several hundred microns, and some of them were even millimeter long. The nanobelts had a uniform width (50–300 nm) along their entire lengths. There was no trace of nanoparticle at the tips of the nanobelts. The thickness and width-to-thickness ratios of them were in the range of 10–30 nm and 5–10, respectively, Single-crystalline SnO_2 nanobelts of rutile structure were similarly synthesized by thermal evaporation of SnO_2 powders at 1350 °C or of SnO powders at 1000 °C. The scanning electron microscopy images and energy-dispersive spectroscopy analyses suggested that the materials evaporated on alumina plate were ultralong SnO_2 nanobelts. These rectangle-like nanobelts had width-to-thickness ratios of 5–10. Based on high-resolution transmission electron microscopy, they were single crystalline and dislocation free. In_2O_3 nanobelts were synthesized by the same method. The evaporation of In_2O_3 powders at 1400 °C yielded In_2O_3 nanobelts. TEM observations showed that most of these nanobelts had uniform width and thickness, both along their lengths. The widths were in the range of 50–150 nm and the lengths in the range of several tens to several hundreds of micrometers. Electron diffraction analysis showed that they were single crystalline. CdO nanobelts were also synthesized by evaporating CdO powders at 1000 °C. The lengths of these nanobelts were less than 100 mm, and the widths typically 100 to 500 nm. They were substantially wider and shorter than those of ZnO, SnO_2, and In_2O_3 nanobelts. The nanobelt-like structures of them were also obtained from other oxide materials, such as Ga_2O_3 and PbO_2. Considering that the only source material used for nanobelt synthesis was pure oxide powders and that the oxide nanobelt composition was similar or identical to that of the starting oxide material, the nanobelt growth appeared to involve oxide vapor evaporation at higher temperature zone, but direct deposition of it on a substrate nanoparticle at lower temperature zone of the growth chamber.

Growth parameters and crystal characteristics of some semiconductor nanobelts [28–38] are presented in Table 7.1. A close look of this table would indicate that most of these nanobelts were grown on FECA metal-free SUBSANO nanoparticles, but some were also grown on METANO nanoparticles. Most of the nanobelts were produced simply by thermal evaporation of the powder of the bulk material, and the temperatures for these evaporations were quite high. While the crystal structures of some nanobelts are again identical to those of the corresponding bulk, the crystal structures of some other nanobelts are different from those of the bulk. For example, the crystal structure of both the Ga_2O_3 bulk and the Ga_2O_3 nanobelt is monoclinic. In contrast, the crystal structure of ZnO bulk is wurtzite. But the crystal structure of ZnO nanobelts, as noted by Zhang et al., is hexagonal. The typical width of nanobelts is between 60 and 300 nm. The nanobelts preferentially grow along the minimum surface energy of the bulk material. This means CdO nanobelt should grow along $\langle 100 \rangle$ direction, because the [100] surfaces possess the lowest surface energy. The CdO nanobelt by Pan et al. however grew along the [001] surfaces.

7.5 Nanomaterial Growth Rates by the VS Mechanism

Table 6.1 of Chap. 6 compares the growth rates of nanowires produced by several different mechanisms, including the VS mechanism. This table indicates that the nanowire growth rates by the VS mechanism by Ambrosini et al. [3], Cheze et al. [12], and Zhang et al. [9] are very low. Ishiyama et al. [39] presented catalyst-free growth of Si nanowires on Si substrate, while Gomes et al. [40] presented the catalyst-free growth of GaN nanowires. Gomes et al. [40] measured variation of GaN nanowire length as function of growth temperature. If we assume that the increase in nanowire length at a certain temperature was due to increase in time, but the growth was at a constant growth rate, then the nanowire growth rate was 0.16 nm/s at a temperature of 480 °C, but 0.22 nm/s at a temperature of 540 °C. This shows that the growth rate of nanowires produced by the VS mechanism is indeed very low. And interestingly, this growth is comparable to the GaN nanowire growth rate by Cheze

Table 7.1 Growth parameters and crystal characteristics of some semiconductor nanobelts

Product	Growth temperature (° C)		Catalyst	Source species	Crystal structure	Growth direction/plane	References
	Nanowire	Nanobelt					
ZnO	–	700	–	$ZnCl_2$, O_2	Hexagonal	$[01\bar{1}0]$	Zhang [28]
Ga_2O_3	–	850	–	Ga, SiO_2 powder	Monoclinic	[100]	Geng [29]
ZnS	–	1200	Au	ZnS powder	–	$[00\bar{1}0]$	Li [30]
In_2O_3	–	1400	–	In_2O_3 powder	C-rare earth	[001]	Kong [31]
CdO	–	1000	–	CdO powder	NaCl	[001]	Wang [32]
PbO_2	–	950	–	PbO_2 powder	Rutile	[010]	Pan [33]
SnO_2	850	1000	Au	Sn powder, O_2 gas	Tetragonally rutile	–	Mondal [34]
Sn_3O_4	–	1200	–	SnO_2 powder	Triclinic	[010]	Berengue [35]
α-Si_3N_4	–	1450	Ni	Si powder, N_2 gas	Hexagonal	[101]	Huang [36]
SiO_2	–	1300	–	SiO powder	Cristobalite	[100]	Zhang [37]
AlN	–	1200	Al	Al powder, N_2 gas	Hexagonal	[0001]	Xie [38]

et al. [12]. If we assume that the increase in InN nanowire growth is due to enhanced thermally activated diffusion, for example, of In adatoms through the R_L species, and that it takes place at higher temperature, then this diffusion through the R_L species increases slowly with increase in temperature. Interestingly, in sharp contrast to the findings made above, the nanowire growth rate by the VS mechanism by Dick et al. [41] is comparable to those by other mechanisms, particularly the VLS mechanism [41]. This points to a fundamental flaw and a serious problem in the basics of growth by the VS mechanism. The nanowire growth rate by the VS mechanism should not be comparable to that by the VLS mechanism.

7.6 The Role of SUBSANO in Nanomaterial Growths

We argued that SUBSANO plays a crucial role in nanomaterial growth by the VS mechanism. An experiment by Gomes et el [40]. on GaN nanowire growth demonstrated that a high-temperature annealing at 750° C not only removes the weakly bound organic contaminants from the Si substrate surface, but also yields reconstructed islands on this surface. These reconstructed islands are the SUBSANOs. They have disordered surface lattice structure. We believe lattice mismatch between GaN and AlN interlayers on Si(111) substrate [42] yielded such a SUBSANO of the R_L≡(Ga, N, Si) species. An important element of the study by Gomes et al. is the low-temperature nucleation step needed to prevent thermal desorption of In from Si(111) substrates and high temperature needed to initiate anisotropic growth, e.g., the axial elongation of nanowire dominating over their radial extension. Gomes et al. [40] showed that InAs nanowires could nucleate on nanoscale islands without forming any In droplets, and observed that the SUBSANO density was momentously reduced due to increase in temperature to 450 ± 10 °C. This confirmed the formation of R_L≡(In, As) species in various islands called SUBSANOs. The density of this species and hence of the SUBSANOs decreased due to desorption of In adatoms from Si surface at increased temperature. The SUBSANOs essentially disappeared from the substrate surface due to further increase in temperature to 480 ± 10 °C. We contend that these happened because the disturbance (disorder) and amorphicity of the SUBSANO surface were dramatically altered by an increase in temperature. The loosely bound surface species became more energetic at higher temperature, and hence migrated to the neighboring sites or were evaporated away from the SUBSANOs. Another important element of the study by Gomes et al. [40] is the increase in nanowire length, but decrease in nanowire diameter with increase in temperature. As shown in Fig. 7.4, nanowire length gradually increased with temperature, reached its highest point at around 520 °C, and then gradually decreased with further increase in temperature. However, nanowire diameter gradually decreased with temperature, reached its lowest point at around 520 °C, and then gradually increased with further increase in temperature. All these happened due to variation in temperature-dependent enhancement of thermally activated diffusion of In adatoms through the R_L species for growth and also a variation in temperature-dependent decrease in the sidewall incorporation of In adatoms. Nanowires grow not only axially but also radially. There occurs, as a result, variation in nanowire diameter

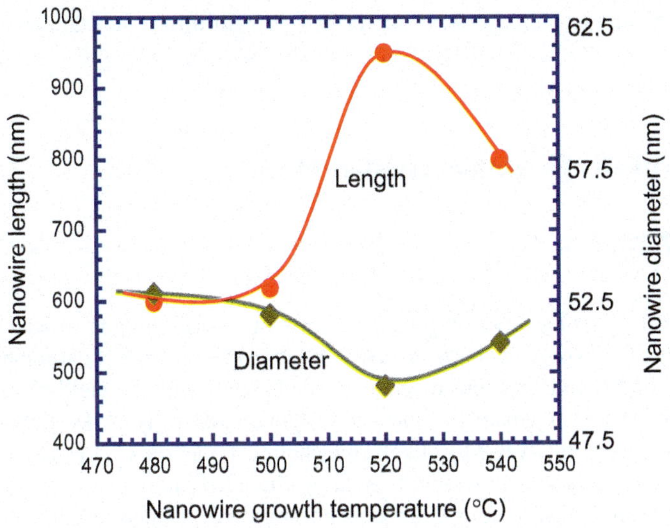

Fig. 7.4 Variation of length and diameter with the growth temperature of InAs nanowires produced by the VS mechanism. Produced with experimental data by Gomes et al. [40]

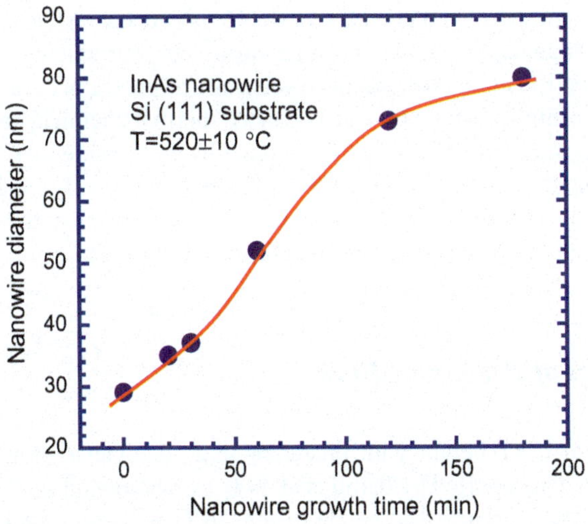

Fig. 7.5 Variation of InAs nanowire diameter with nanowire growth time. The nanowire was grown by the VS mechanism on Si(111) substrate at a temperature of 520 ± 10 °C. Tertiarybutylarsine pressure was 3.00 Torr, and the trimethyindium pressure was 0.20 Torr. Produced with experimental data by Gomes et al. [40]

with growth time. The measured time dependence of the InAs nanowire diameters, as obtained by Gomes et al., is shown in Fig. 7.5. Note that the radial growth of nanowire for InAs nanowires depends on the V/III ratio. Under this scenario, the nanowire diameter increases with increase in growth time, and that this increase is nonlinear.

7.7 Growths by Water-Assisted Means

7.7.1 Basics

It was gratifying [43, 44] that the introduction of a small amount of water vapor in the catalytic CVD growth chamber vastly enhanced the growth of carbon nanotubes. The catalytic lifetime was thus extended, and also the nanotube yield was increased. The investigations suggested that the water vapor etched away carbon precipitate that covered the metal catalyst. Most likely, the chemical reaction $C + H_2O \rightarrow CO + H_2$ was believed to be responsible for the etching which led catalytic activity to continue. The said etching was, of course, dependent on growth temperature. Note that there was a balance between the etching and the carbon precipitation, which was crucial for an effective CNT growth. There was also a need for an optimized water concentration. Recall that Lin et al. [21] produced CNTs on SUBSANOs generated on Au substrate surface. The growth was conducted by the VS mechanism without the addition of water inside the growth chamber. Szabo et al. [45] produced, on the other hand, CNTs on SUBSANOs generated on Al substrate surface by introducing water inside the growth chamber. Liu et al. [46] produced vertically aligned CNTs on a planar glossy carbon-like thin-film substrate by water-assisted thermal chemical vapor deposition. They noted that such growths of CNTs on metallic and nonmetallic thin films were highly advantageous, and hence these growths were subsequently carried out by many other researchers. The investigation by Liu et al. [46] suggests that water-assisted CNT growths could be carried out by also the VS mechanism.

7.7.2 Effect of Precursor Flow

The variation of CNT height with precursor (e.g., acetylene) flow rate in water-assisted growth was reported by Patoli et al. [44]. As shown in Fig. 7.6, the height of the CNT was 200 ± 15 nm. The height of the CNTs increased with increase in the flow rate of the acetylene precursor. The maximum height of CNT thus reached to 1.4 ± 0.002 mm at an acetylene flow rate of 200 sccm. Thereafter, the CNT height decreased with increase in acetylene flow rate. The height of the CNT was 270 ± 20 μm at a flow rate of 500 sccm of this acetylene. This observation was identical to that by Liu et al. [46]. They indicated that the feed stock flow rate played an important role in the superior growth of the CNTs. At a lower acetylene flow rate, the CNTs were neither densely packed nor vertically aligned. The decrease in the CNT height from that observed at the peak was most likely due to decrease in the activity and

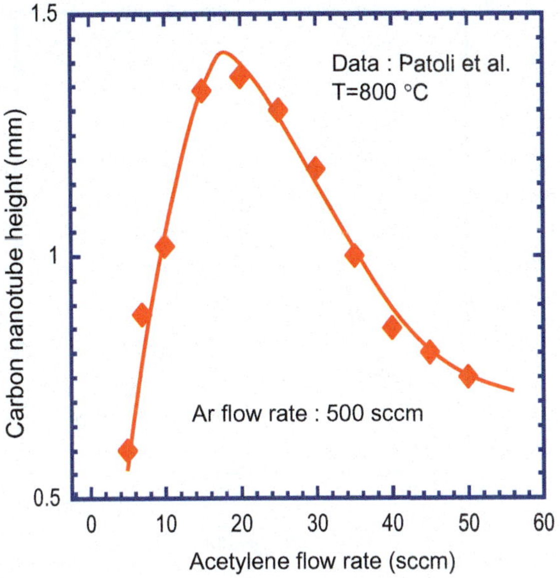

Fig. 7.6 Variation of CNT height with precursor (e.g., acetylene) flow rate in water-assisted CNT growth. The figure was obtained with experimental data by Patoli et al. [44]

lifetime of the SUBSANOs (METANOs) generated from Fe metal sheet. While the onset of oxidation of Fe sheet was predominant at a lower acetylene flow rate, the excessive carbonization hindered the growth of the CNTs at higher acetylene flow rate. A balance between oxidation and carbonation was achieved only under optimal conditions which led to the realization of the largest CNT heights.

7.7.3 Effect of Temperature on CNT Growth

Considering that the oxidation and carbonation of the catalyst surface are the primary elements dictating the CNT height, growth temperature can best monitor catalyst oxidation and control of CNT height. This hypothesis was tested by studying the effect of temperature on the CNT growth. The optimum condition for temperature was therefore set for superior CNT growth. CNT growth was conducted at different temperatures keeping the precursor flow rate constant and at an optimal level. The synthesis was carried out at the optimum condition of the gases. Figure 7.7 shows that there was a variation in the growth rate as function of the gas injection temperature. The highest growth rate was 1.55 μm/s for gas injection at room temperature. However, the growth rate decreased from 1.50 to ~0.83 μm/s for an increase in temperature from 100 to 800 °C. The oxidation of the catalyst was the lowest for gas injection at room temperature, which resulted in high growth rate. However,

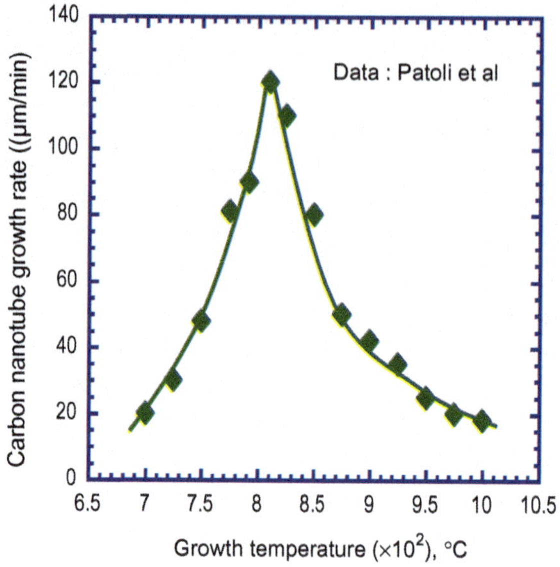

Fig. 7.7 Variation of the CNT growth rate as function of the gas injection temperature. The figure was obtained with experimental data by Patoli et al. [44]

as the temperature was increased, the catalyst oxidation gradually increased, which hindered the growth rate.

References

1. C.-W. Yang, W.-C. Chen, C. Chou, H.-H. Lin, Selective area growth of InAs nanowires from SiO_2/Si(1 1 1) templates direct-written by focused helium ion beam technology. J. Cryst. Growth **484**, 50–63 (2018)
2. Y.S. Won, Y.S., Kim, O. Kryliouk, T.J. Anderson, Growth mechanism of catalyst- and template-free group III-nitride nanorods. J. Cryst. Growth **310**, 3735–3740 (2008)
3. S. Ambrosini, M. Fanetti, V. Grillo, A. Franciosi, S. Rubini, Vapor-liquid-solid and vapor-solid growth of self-catalyzed GaAs nanowires. AIP Adv. **1**, 042142 (2011)
4. B.-S. Kim, T.-W. Koo, J.-H. Lee, D.S. Kim, Y.C. Jung, S.W. Hwang, B.L. Choi, E.K. Lee, J.M. Kim, D. Whang, Catalyst-free growth of single-crystal silicon and germanium nanowires. Nano Lett. **9**, 864–869 (2009)
5. S.-S. Choi, S.G. Lee, S.S. Im, S.H. Kim, Y.L. Joo, Silica nanofibers from electrospinning/sol-gel process. J. Mater. Sci. Lett. **22**, 891–893 (2003)
6. H.H. Gao, Y. Bando, K. Kurushima, T. Sato, SiC nanorods prepared from SiO and activated carbon. J. Mater. Sci. **37**, 2023–2029 (2002)
7. S.-T. Ho, K.-C. Chen, H.-A. Chen, H.-Y. Lin, C.-Y. Cheng, H.-N. Lin, Catalyst-free surface-roughness-assisted growth of large-scale vertically aligned zinc oxide nanowires by thermal evaporation. Chem. Mater. **19**, 4083–4086 (2007)
8. T. Tang, S. Han, W. Jin, X.L. Liu, C. Li, D.-H. Zhang, C.-W. Zhou, B. Chen, J. Han, M. Meyyappan, Synthesis and characterization of single-crystal indium nitride nanowires. J. Mater. Res. **19**, 423–426 (2004)

9. Z. Zhang, Y. Wang, H. Li, W. Yuan, X. Zhang, C. Sun, Z. Zhang, Atomic-scale observation of vapor-solid nanowire growth via oscillatory mass transport. ACS Nano **10**, 763–769 (2016)
10. M.J. Bierman, Y.K.A. Lau, A.V. Kvit, A.L. Schmitt, S. Jin, Dislocation-driven nanowire growth and eshelby twist. Science **320**, 1060–1063 (2008)
11. G.R. Yazdi, P.O.Å. Persson, D. Gogova, R. Fornari, L. Hultman, M. Syväjärvi, R. Yakimova, Aligned AlN nanowires by self-organized vapor-solid growth. Nanotechnology **20**, 495304 (2009)
12. C. Chèze, L. Geelhaar, O. Brandt, W.M. Weber, H. Riechert, S. Münch, R. Rothemund, S. Reitzenstein, A. Forchel, T. Kehagias, P. Komninou, G.P. Dimitrakopulos, T. Karakostas, Direct comparison of catalyst-free and catalyst-induced GaN nanowires. Nano Res. **3**, 528–536 (2010)
13. C. Chèze, L. Geelhaar, B. Jenichen, H. Riechert, Different growth rates for catalyst-induced and self-induced GaN nanowires. Appl. Phys. Lett. **97**, 153105 (2010)
14. D. Takagi, Y. Kobayashi, Y. Homma, Carbon nanotube growth from diamond. J. Am. Chem. Soc. **131**, 6922–6923 (2009)
15. C. Varanasi, J. Petry, L. Brunke, B.T. Yang, W. Lanter, J. Burke, H. Wang, J.S. Bulmer, J. Scofield, P.N. Barnes, Growth of high-quality carbon nanotubes on free-standing diamond substrates. Carbon **48**, 2442–2446 (2010)
16. K.T. Koga, M.J. Walter, E. Nakamura, K. Kobayashi, Carbon self-diffusion in a natural diamond. Phys. Rev. B **72**, 024108 (2005)
17. H.O. Pierson, *Handbook of Carbon, Graphite, Diamond and Fullerenes: Properties, Processing and Applications* (Noyes Publications, Park Ridge, 1993)
18. J.H. Lin, C.-S. Chen, H.-L. Ma, C.-W. Chang, C.-Y. Hsu, H.-W. Chen, Self-assembling of multi-walled carbon nanotubes on a porous carbon surface by catalyst-free chemical vapor deposition. Carbon **46**, 1619–1623 (2008)
19. J. Hahn, S.B. Heo, J.S. Suh, Catalyst-free synthesis of high-purity carbon nanotubes by thermal plasma jet. Carbon **43**, 2638–2641 (2005)
20. J.H. Lin, C.S. Chen, M.H. Rümmeli, A. Bachmatiuk, Z.Y. Zeng, H.L. Ma, B. Büchner, H.W. Chen, Growth of carbon nanotubes catalyzed by defect-rich graphite surfaces. Chem. Mater. **23**, 1637–1639 (2011)
21. J.-H. Lin, C.-S. Chen, M.H. Rummeli, Z.-Y. Zeng, Self-assembly formation of multiwalled carbon nanotubes on gold surfaces. Nanoscale **2**, 2835–2840 (2010)
22. S. Huang, Q. Cai, J. Chen, Y. Qian, L. Zhang, Metal-catalyst-free growth of single-walled carbon nanotubes on substrates. J. Am. Chem. Soc. **131**, 2094–2095 (2009)
23. B. Liu, W. Ren, L. Gao, S. Li, S. Pei, C. Liu, C. Jiang, H.-M. Cheng, Metal-catalyst-free growth of single-walled carbon nanotubes. J. Am. Chem. Soc. **131**, 2082–2083 (2009)
24. B. Liu, D.-M. Tang, C. Sun, C. Liu, W. Ren, F. Li, W.-J. Yu, L.C. Yin, L. Zhang, C. Jiang, H.-M. Cheng, Importance of oxygen in the metal-free catalytic growth of single-walled carbon nanotubes from SiO_x by a vapor-solid-solid mechanism. J. Am. Chem. Soc. **133**, 197–199 (2011)
25. S.A. Steiner, T.F. Baumann, B.C. Bayer, R. Blume, M.A. Worsley, W.J. MoberlyChan, E.L. Shaw, R. Schlogl, A.J. Hart, S. Hofmann, B.L. Wardle, Nanoscale zirconia as a nonmetallic catalyst for graphitization of carbon and growth of single- and multiwall carbon nanotubes. J. Am. Chem. Soc. **131**, 12144–12154 (2009)
26. N. Muradov, Catalysis of methane decomposition over elemental carbon. Catalysis Commun. **2**, 89–94 (2001)
27. Z.W. Pan, Z.R. Dai, Z.L. Wang, Nanobelts of semiconducting oxides. Science **291**, 1947–1949 (2001)
28. J. Zhang, W. Yu, L. Zhang, Fabrication of semiconducting ZnO nanobelts using a halide source and their photo-luminescence properties. Phys. Lett. A **299**, 276–281 (2002)
29. B. Geng, L. Zhang, G. Meng, T. Xie, X. Peng, Y. Lin, Large-scale synthesis and photoluminescence of single-crystalline β-Ga_2O_3 nanobelts. J. Cryst. Growth **259**, 291–295 (2003)
30. Q. Li, C. Wang, Fabrication of wurtzite ZnS nanobelts via simple thermal evaporation. Appl. Phys. Lett. **83**, 359–361 (2003)

31. X.Y. Kong, Z.L. Wang, Structures of indium oxide nanobelts. Solid State Commun. **128**, 1–4 (2003)
32. Z.L. Wang, Functional oxide nanobelts: materials, properties and potential applications in nanosystems and biotechnology. Annu. Rev. Phys. Chem. **55**, 159–196 (2004)
33. Z.W. Pan, Z.R. Dai, Z.L. Wang, Lead oxide nanobelts and phase transformation induced by electron beam irradiation. Appl. Phys. Lett. **80**, 309–311 (2002)
34. S.P. Mondal, S.K. Ray, J. Ravichandran, I. Manna, Temperature dependent growth and optical properties of SnO_2 nanowires and nanobelts. Bull. Mater. Sci. Indian Acad. Sci. **33**, 357–364 (2010)
35. O.M. Berengue, R.A. Simon, A.J. Chiquito, C.J. Dalmaschio, E.R. Leite, H.A. Guerreiro, F.E.G. Guimarães, Semiconducting Sn_3O_4 nanobelts: growth and electronic structure. J. Appl. Phys. **107**, 033717 (2010)
36. J. Huang, S. Zhang, Z. Huang, Y. Wen, M. Fang, Y. Liu, Catalyst-assisted synthesis and growth mechanism of ultra-long single crystal a-Si_3N_4 nanobelts with strong violet–blue luminescent properties. CrystEngComm **14**, 7301–7305 (20112)
37. Z.Y. Zhang, X.L. Wu, L.L. Xu, J.C. Shen, G.G. Siu, P.K. Chu, Synthesis, growth mechanism, and light-emission properties of twisted SiO_2 nanobelts and nanosprings. J. Chem. Phys. **129**, 164702 (2008)
38. T. Xie, M. Ye, Z. Jiang, Y. Qin, Y.-C. Wu, G.-W. Meng, L.-D. Zhang, Chloride assisted growth of aluminum nitride nanobelts and their enhanced dielectric responses. Chin. J. Chem. Phys. **21**, 586–590 (2008)
39. T. Ishiyama, S. Nakagawa, T. Wakamatsu, Growth of epitaxial silicon nanowires on a Si substrate by a metal-catalyst-free process. Sci. Report **6**, 30608 (2016)
40. U.P. Gomes, D. Ercolani, N.V. Sibirev, M. Gemmi, V.G. Dubrovskii, F. Beltram, L. Sorba, Catalyst-free growth of InAs nanowires on Si(111) by CBE. Nanotechnology **26**, 415604 (2015)
41. K.A. Dick, K. Deppert, L.S. Karlsson, L.R. Wallenberg, L. Samuelson, W. Seifert, A new understanding of Au-assisted growth of III-V semiconductor nanowires. Adv. Funct. Mater. **15**, 1603–1610 (2005)
42. V. Consonni, M. Knelangen, L. Geelhaar, A. Trampert, H. Riechert, Nucleation mechanisms of epitaxial GaN nanowires: origin of their self-induced formation and initial radius. Phys. Rev. B **81**, 085310 (2010)
43. K. Hata, D.N. Futaba, K. Mizuno, T. Namai, M. Yumura, S. Iijima, Water-assisted highly efficient synthesis of impurity-free single-walled carbon nanotubes. Science **306**, 1362 (2004)
44. S.P. Patoli, P.S. Alegaonkar, H.-C. Lee, J.-B. Yoo, Optimization of water assisted chemical vapor deposition parameters for super growth of carbon nanotubes. Carbon **46**, 1987–1993 (2008)
45. A. Szabó, E. Kecsenovity, Z. Pápa, T. Gyulavári, K. Németh, E. Horvath, K. Hernadi, Infuence of synthesis parameters on CCVD growth of vertically aligned carbon nanotubes over aluminum substrate. Sci. Report **7**, 9557 (2017)
46. Xianming Liu, K. H. R. Baronian, A. J. Downard, Direct growth of vertically aligned carbon nanotubes on a planar carbon substrate by thermal chemical vapor deposition, Carbon 47, 500–506 (2009)

Chapter 8
Solution and Supercritical Fluid-Based Growth Mechanisms

Abstract The solution-liquid-solid (SoLS) mechanism and the supercritical fluid-liquid-solid (SFLS) mechanism for the growths of one-dimensional and quasi-two-dimensional nanomaterials have been described. Both of these mechanisms are employed for nanowire growths in the solution phase environment. And both of them are metal catalyst-based mechanisms. They are solution -based analogs of the VLS mechanism. The distinctions between them have been detailed. The basics of them have been interpreted. The supercritical fluid conditions of the solution needed for nanomaterials growths have been discussed. Various characteristic features of them have been described. Novelty of the nanomaterials growths by them has been laid down. Strengths and weaknesses of them in mediating nanomaterials growths have been articulated. The importance and feasibility of the solution-solid-solid (SoSS) mechanism and the supercritical fluid-solid-solid (SFSS) mechanism have been demonstrated. Both of them are considered to be solution-based analogs of the VSS mechanism.

8.1 Background

The solution-phase nanowire syntheses are analogous to the VLS nanowire syntheses. The former, unlike the latter, though use the low melting MET catalysts to induce the asymmetric crystallization of nanowires, and at temperatures compatible with the solution chemistry ($T < 400$ °C) of them. Further, the reactions in the solution-phase syntheses are conducted in the presence of some mild coordinating solvents, which are suitable for modulating the growth kinetics and for passivating the product surfaces. The CVD technique for nanomaterials syntheses has been very successful during the past years [1–7]. Following this technique, the nanomaterials synthesis can proceed by also the solution-based approach. During this synthesis, the MET catalyst particle is converted to metallic alloy eutectic droplet. The MET catalyst consequently catalyzes the decomposition of metallo-organic precursors via VLS-type process leading to crystalline nanowire growth. This approach has nevertheless both strength and weakness in the throughput of nanomaterials. Following the syntheses by this technique, the solution-based approach utilizes free floating of the nanoparticle seeds

S. N. Mohammad, *Synthesis of Nanomaterials*, Springer Series in Materials Science 307, https://doi.org/10.1007/978-3-030-57585-4_8

to produce larger quantities of nanomaterials. These syntheses are obviously variants of the VLS syntheses; although conducted in solution or solvent dispersion. The solution-based mechanism can be solution-liquid-solid (SoLS) mechanism [8] or supercritical fluid-liquid-solid (SFLS) mechanism [9], as shown schematically in Fig. 8.1. To repeat, colloidal nanowires of various compositions, structures, and properties produced by these mechanisms have, during the past years, facilitated a great variety of new applications. Indeed the solution-based approach is a promising approach for mass production of metal, semiconductor, and oxide nanomaterials with excellent controls of the shape, size, and composition. It guarantees high reproducibility, and hence has great potential for applications in nanoelectronic and biological systems. To describe all of them, we will make use of the terminologies such as METANO and SUBSANO for nanoparticlesdefined in Chap. 3, the R_S ($R_S \equiv X$ and $R_S \equiv Y$) source species defined in Chap. 1, and the pre-nucleation and pro-nucleation stages

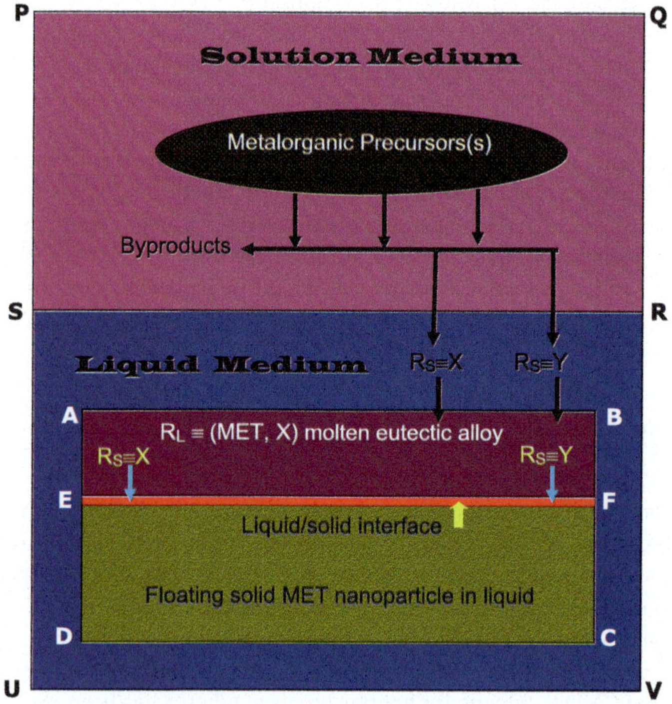

Fig. 8.1 Schematic diagram of the nanomaterials growth by the SoLS and SFLS mechanisms. PQRS is the solution-phase medium containing metalorganic precursors. But RSUV is the liquid phase medium containing ABCD solid MET nanoparticle. This ABCD solid MET nanoparticle floats in the RSUV liquid phase medium. $R_S \equiv X$ and $R_S \equiv Y$ species released from the decomposition of metalorganic precursor(s) migrate from the solution phase medium into the liquid phase medium where they diffuse through the surface of the MET nanoparticle and create $R_L \equiv$(MET, X) eutectic alloy in the ABFE segment of the MET nanoparticle leaving the CDEF segment of the MET nanoparticle still solid

of growths defined in Chap 4, Sect. 3. We will also utilize the R_L species defined again in Chap. 3, Sect. 3.3.

8.2 Various Features of Solution and Supercritical Fluid-Liquid-Solid Mechanisms

8.2.1 Basics

To be specific, we first touch on the solution-based SoLS syntheses, which are conducted by Wang et al. [10] in a solution and generally created with organic solvents. Various steps of this synthesis, as shown in Fig. 8.2, involve the pre-nucleation stage and the pro-nucleation stage of growths. The pre-nucleation stage roughly includes the decomposition (dissociation) of the precursors of $R_S{\equiv}X$ and $R_S{\equiv}Y$ species, the adsorption of $R_S{\equiv}X$ species onto the MET cluster to form

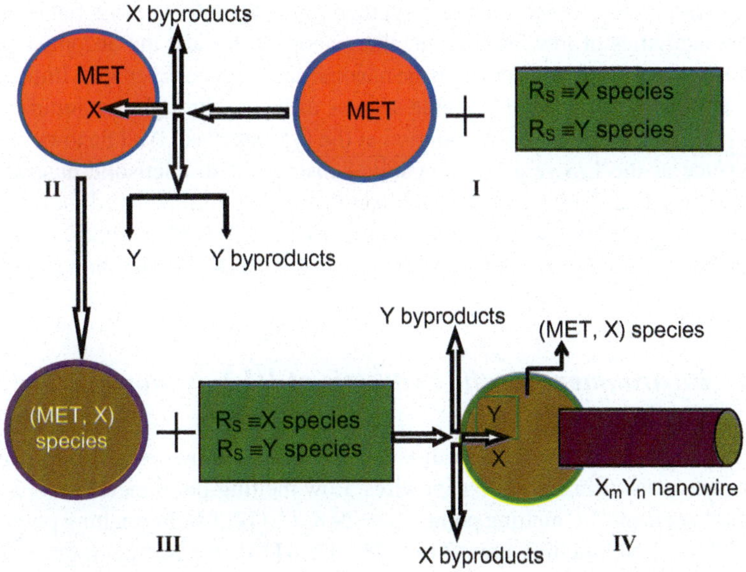

Fig. 8.2 Schematic diagrams of various steps of SFLS growth of nanowires. Step I: Dissociation of the precursors of $R_S{\equiv}X$ and $R_S{\equiv}Y$ species upon landing on the MET nanoparticle surface. Step II: Adsorption of the $R_S{\equiv}X$ into the MET nanoparticle to form $R_L{\equiv}$(MET, X) nanoparticle seed during the pre-nucleation stage of growth, but the desorption of $R_S{\equiv}Y$ species and of the by-products of the precursors of $R_S{\equiv}X$ and $R_S{\equiv}Y$ species during this stage of growth. Step III: Melting of the $R_L{\equiv}$(MET, X) species at an appropriate growth temperature during the pre-nucleation stage of growth and diffusion of $R_S{\equiv}X$ and $R_S{\equiv}Y$ into this molten $R_L{\equiv}$(MET, X) species during the pro-nucleation stage of growth. Step IV: Removal of the by-products of the precursors of X and Y and the nucleation of X_mY_n nanowires from the molten $R_L{\equiv}$(MET, X) species

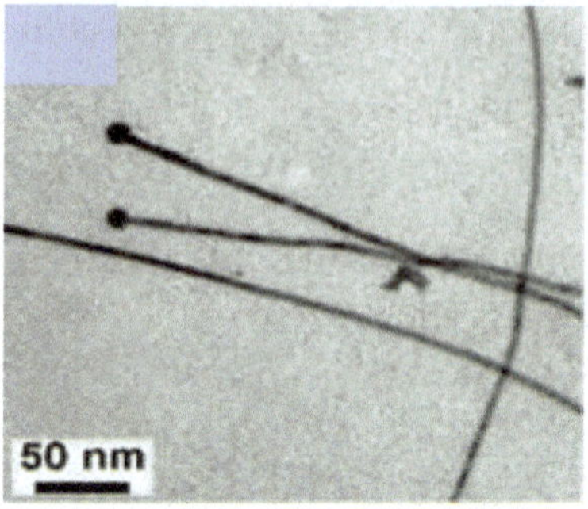

Fig. 8.3 TEM images of InP quantum wires, 5.2 nm thick. Reproduced from Wang et al. [11] with copyright permission from the publisher

R_L=(MET, X) alloy species, and the diffusion of R_S≡X and R_S≡Y species through the molten R_L=(MET, X) alloy species. The R_L=(MET, X) species may be eutectic and hence a droplet. The pro-nucleation stage involves supersaturation of the R_S species at the L/S interface leading to the nucleation and crystallization of nanowires. Typical temperature range for these syntheses is 200–350 °C. This range suffices catalytic activities of low-melting metallic nanoparticles and the decomposition of many metallo-organic precursors on the nanoparticle surface. Even with limited solubility, the R_S (R_S≡X, R_S≡Y) source species diffusing through the liquid metallic catalyst droplet undergo supersaturation at the L/S interface. And nanowire growth takes place at the L/S (e.g., droplet/solid-substrate or droplet/solid-nanowire-tip) interface (see Fig. 8.1). Obviously, the nanowires thus formed are dissolved in the liquid. During the past years, many nanowires have thus been synthesized. They include Bi-catalyzed InAs nanowires grown by Wang et al. [11] and shown in Fig. 8.3.

8.2.2 *Environment for the Synthesis of III-V Nanowires*

Researchers [12–16] showed that di-isopropylbenzene provides good environment for the synthesis of many III-V nanowires. Low melting-point metal nanoparticles, such as those from Ga (melting point $T_M = 29.8$ °C) [12, 13], In (melting point $T_M = 156.6$ °C) [14], Sn (melting point $T_M = 231.9$ °C) [15], and Bi (melting point $T_M = 271.4$ °C) [16] have been found useful for the SoLS growth of nanowires under typical reaction conditions at 200–350 °C. High melting-point metal nanoparticles such as those from Au (melting point $T_M = 1064$ °C) are also useful for SoLS growths, but only if they have depressed melting points due to size-dependent mesoscopic effects [17]. SoLS growth may not therefore be suitable for the Au-catalyzed synthesis of Si and Ge nanowires of diameters larger than about 10 nm. The synthesis of both of these nanowiresrequires much higher synthesis and crystallization temperatures. Hanrath et al. [18] concluded that this problem can be alleviated by pressurizing and heating the solvent above its critical pressure and temperature. Solution-based reactions can then be carried out in supercritical fluid environments by the SFLS mechanism.

8.2.3 Supercritical Fluid Conditions

Recall that fluids maintain supercritical fluid (SCF) conditions if they are at temperature beyond their critical temperature and at pressure beyond their critical pressure. Conventional solvents are good to provide flexible medium for the syntheses of a variety of nanomaterials. Supercritical fluids are good though to provide a unique medium for the syntheses of these nanomaterials. Note that, at temperatures and pressures above the critical point, the vapor-liquid coexistence curve of the generic pressure-temperature-phase diagram disappears; and only a single-phase fluid exists. The density, viscosity, and diffusivity in the single-phase supercritical region are therefore different from those in the biphasic region. They are intermediate of those of liquids and gases. They vary continuously from having "gas-like" behavior to "liquid-like" behavior even due to small changes in temperature, pressure, or both. There is no liquid/gas-phase boundary in a supercritical fluid. This means there is no surface tension in a supercritical fluid. This fluid can be "tuned" to be more liquid-like or more gas-like just by changing its pressure and temperature. The density of the fluid, which relates to solvent strength, is also sensitive to temperature and pressure. It can be tuned easily by varying temperature and pressure, as evident from the n-hexane equilibrium-phase diagram [19, 20]. The high diffusivity and low viscosity of supercritical fluids can particularly be very useful. Savage et al. [21] noted that both of them can provide an environment conducive to transport-limited fast reactions in the liquidic phase.

Common organic supercritical solvents include hexane, cyclohexane and toluene. With them as solvents, supercritical fluids can be ideal intermediate of the medium for the vapor-based syntheses and the medium for the liquid-based synthesis. They can provide a superior environment for the synthesis of semiconductor nanowires. The solvent strengths of these fluids can particularly allow dispersion of size selected nanoparticle seeds and solubilization in high precursor concentrations. They can also provide high synthesis-temperature required for precursor decomposition, for the release of the R_S ($R_S{\equiv}X$, $R_S{\equiv}Y$) source species from their respective precursor, and material recrystallization into nanowire. Scalable and continuous flow operation of the fluid can thus enable the synthesis of technologically significant quantities of nanomaterials. Strikingly, they can create room for integration of supercritical fluid synthesis and the modifications of solution-based surface chemistry.

8.2.4 *First Realization of Supercritical Fluid Condition*

Heath and LeGoues [22] synthesized Ge nanowires, probably for the first time, in supercritical fluid environment. To achieve this environment, they reduced $GeCl_4$ and phenyl-$GeCl_3$ compounds by sodium in hexane heated and pressurized just above their critical points (275 °C and 10 MPa). Although they could produce nanowires, the yield was quite low and the nanowires were riddled with crystallographic defects. The supercritical fluid environment nevertheless provided a growth mechanism with nanoparticle seed crystals. And this mechanism was probably superior to those with CVD-based mechanisms. The fluid-based mechanism had also an added advantage; the products by this mechanism are scalable for the manufacture of technologically significant nanowire quantities. These products can as well be easily compatible to various surface chemistry modifications. Due to the high surface-to-volume ratio of the nanowires, the said modifications are of paramount importance for nanowire processing and utilization.

8.2.5 *Inference*

The SoLS and SFLS growth mechanisms are, no doubt, attractive for nanomaterials syntheses. Recall that nanomaterials possess unique size-dependent properties. The utilization of these properties requires synthesis mechanism suited to specific diameters and monodisperse distributions better than those attainable with the vapor-based mechanisms. Holmes et al. [23] noted that the SFLS mechanism with size selected gold nanocrystals capped with dodecanethiol in supercritical hexane could effectively seed the synthesis of Si nanowire of diameter smaller than 10 nm and length of several micrometers. This mechanism has since been extended to achieve the synthesis of milligram quantities of a wide variety of high-quality single crystal nanowires [24–26].

8.3 Solution- and Supercritical-Fluid-Based Growth Techniques

The supercritical-fluid-based and solution-based growths are performed in a flow reactor [9, 27, 28] (see Fig. 8.2) instead of a closed reaction vessel. Precursors in these growths are introduced in liquid and supercritical media, respectively. For example, gold colloidal particles together with a well-defined amount of Si precursor, e.g., diphenylsilane, are fed into the solution as dispersions for Si nanowire growths. They are transferred next into a high-pressure heated reactor, where the nanowire synthesis via the SoLS or the SFLS growth mechanism takes place. Nanowire diameters in the SoLS and SFLS mechanisms can be determined, as in the VLS mechanism, by the size of the METANO nanoparticles. Crystalline nanowires produced by these mechanisms can have diameters as small as 5 nm and lengths up to several microns [9]. High-resolution transmission electron microscopy of Au-catalyzed silicon nanowires grown via the SFLS mechanism using monophenylsilane as precursor demonstrated that these nanowires could have perfect crystallinity.

8.4 SFLS Nanowire Characteristics

8.4.1 *Nanowire Size Distribution*

The nanowire size distribution is broadened primarily by two factors: (1) METANO nanoparticle agglomeration and (2) kinetics not favorable for the decomposition of precursor, for example, TEG [tetraethylgermane; $(CH_3CH_2)_4Ge$] for Ge nanowire growth. The agglomeration of the nanoparticle seed takes place in the early stages, plausibly during the pre-nucleation stage of the reaction. It is hence dependent on the synthesis approach, which involves batch reaction, injection reaction, or flow reaction. The injected precursor solution contains relatively size-monodisperse and sterically stabilized METANO nanoparticles. Agglomeration of these METANO nanoparticles or the formation of $R_L \equiv$(MET, X) eutectic liquid alloy droplets generally takes place at expectedly high temperatures leading to broad nanowire diameter distributions. A control experiment by Hanrath [18] demonstrated that the MET≡Au nanoparticles during synthesis and even in the absence of germanium precursor underwent conspicuous agglomeration. Yet they were sufficiently stable to yield reasonable nanowire size distributions. With diphenylgermane [DPG; dipropylene glycol, $(C_6H_5)_2H_2Ge$] used as precursor, the standard deviations of these distributions were less than $\pm$ 30%. Size distribution broadening appeared to be caused primarily by the nanowire growth kinetics. The TEG [tetraethylgermane; $(CH_3CH_2)_4Ge$] has yielded larger nanowires with very broad size distributions. The DPG [e.g., dipropylene glycol, $C_6H_5)_2H_2Ge$], on the other hand, produced smaller nanowires with relatively narrow size distributions. These differences were attributed to different decomposition kinetics of the two different precursors.

8.4.2 *Example of Ge Products from Low-Temperature Reactions*

Lu et al. [28] found that Ge products resulting from the low temperature reactions were much lower with TEG than with DPG. This suggests that TEG is kinetically more stable than DPG. The first step in decomposing DPG was presumably the homolytic cleavage of the Ge-H bond, and then the addition of diphenylgermyl radical to an aromatic ring in diphenylgermane. The first step to decompose TEG during the pre-nucleation stage of growth was, on the other hand, slow; it was via a homogeneous unimolecular reaction [29]. Slow precursor decomposition resulted in slow Ge supply to the R_L species, which is the seed of the catalyst nanoparticle, and which was particularly crucial for the early-stage of synthesis of nanowires. Fast decomposition of DPG led, on the other hand, to efficient diffusion of the R_S species through the R_L species and supersaturation of the Au-based R_L nanocrystals with R_S species to initiate nanowire growth during the pro-nucleation stage of growth. The slower decomposition of TEG allowed more time to elapse before the nanocrystals were saturated enough to produce nanowires. Nanoparticle agglomeration taking place during this "lag time," gave rise to larger average nanowire diameters with TEG than with DPG.

8.4.3 SFLS Growth of Si and Ge Nanowires

Heitsch et al. [30] demonstrated silicon nanowire growth in solution atmospheric pressure employing Au and Bi catalysts, respectively. Trisilane (Si_3H_8) was used as silicon precursor. The reaction(s) for growth was carried out in a vessel containing long-chain, low-vapor-pressure hydrocarbon. The temperature for Au-catalyzed nanowire synthesis was higher than the eutectic temperature of the R_L≡(Au, Si) eutectic alloy. The temperature for Bi-catalyzed nanowire synthesis was similarly higher than the eutectic temperature of the R_L≡(Bi, Si) eutectic alloy. High-resolution transmission electron microscopy images of Au-catalyzed Ge nanowires formed at 370 °C and 10 MPa using DPG had cubic diamond crystal structure. Au-catalyzed Ge nanowires formed at 400 °C and 15 MPa using TEG also had crystalline structure. Au nanoparticles, injected into supercritical hexane at 450 °C with diphenylsilane, produced crystalline Si nanowires with a diameter close to that of the Au nanoparticles. These growth temperatures were higher than the R_L≡(Au, Ge) eutectic temperature of 361 °C. and the R_L≡(Au, Si) eutectic temperature of 363 °C. Similar is the case with Si and Ge nanowires produced by the VLS mechanism. Hence, it was evident that a solution-liquid-solid mechanism was indeed analogous to the vapor-phase VLS mechanism for nanowire growths. Using the SFLS mechanism, Heitsch et al. [30] could produce micrometer long, crystalline Si nanowires, about 25 nm in diameter. The superiority of the growth was evident from thin nanowire products of good crystalline quality obtained in large quantities. Compared to other nanowire syntheses, the yield was excellent. The only problem with the synthesis was though that a controlled, in-place, epitaxial growth could hardly be realized. Holmes et al. [23] produced Si nanowires by making use of monodispersed alkanethiol-capped gold nanocrystals, 2.5 nm in diameter in supercritical hexane. The crystal structure of the nanowires by the SFLS mechanism could be controlled by changing the reaction pressure. The growth direction of the Si nanowires was preferentially <110> at a pressure of 270 bar, but <100> at a pressures of 200 bar. The nanowires by the SFLS mechanism were unfortunately entangled in a mesh.

8.5 Strengths

8.5.1 Growth of Highly Promising Ge Nanowires

Lower reaction temperatures and solution access are the hallmarks of the SoLS nanowire growths. They are important for supporting nanowire-assembly engineering via the use of patterned substrates, the patterned deposition of catalyst nanparticles, and soft templating. Like Si, Ge is a Group-IV semiconductor material. It has same diamond cubic crystal structure as Si. It has certain properties such as higher charge carrier mobility and a larger Bohr exciton radius [24.3 nm), which are superior to those of Si. For this reason, interest in Ge nanowires has flourished during the past years, and the goal was to obtain Ge-based nanowire electronic devices with functionality superior to those of Si-based nanowire electronic devices. Recent interests in Ge nanowires have led to studies of lithium-ion batteries, field-effect transistors, memory devices, photovoltaics, and nanoelectromechanical systems (NEMS). Ge nanowires have also been employed to investigate doping, transport modulation, and band offset efficiency in core–shell nanowires. A key element of Ge nanowires for new technology development is low-temperature growths by the solution-based mechanisms. This is evident from Table 8.1, which lists low growth temperature and related parameters for Ge nanowire growths by a variety of FECA metal catalysts [31–37].

Table 8.1 Reported growth temperatures for the metal-assisted SFLS and SoLS growth of Ge nanowires

MET	Mechanism	Growth Temp (°C)	MET melting Temp (°C)	(MET, Ge) eutectic temp (°C)	Fluid	References
Ga	SoLS	170	29.8	29	–	Pertl [31]
In	SoLS	180	156.6	156	Toluene	Meshgi [32]
Sn	SoLS	270	231.9	231	Oleylamine	Bodnarchuk [33]
Pb	SFLS	250	327.5	327	–	Seifner [34, 35]
Ni	SFLS	390	1455.0	762	Oleylamine	Barth [36]
Bi	SFLS	300	271.4	271	Trioctylphosphine	Lu [37]

8.5.2 High Crystallinity of Nanowires

Semiconductor nanowires grown by SFLS and SoLS mechanisms can have crystallinity almost identical to those grown by the VLS mechanism. This may happen despite the fact that these nanowires by SFLS and SoLS mechanisms are grown at temperatures much lower than those by the VLS mechanism. Solutions, particularly, the supercritical fluids provide chemical flexibility and synthetic tunability needed for the realization of highly controlled solution-based and template-grown nanomaterials. These solvents are unique in the sense that they offer a number of advantages over conventional solvents for the synthesis of nanomaterials. These advantages include low surface tension, low viscosity, and high diffusivity. These fluids are unique in the formation of ordered arrays of nanowires within mesoporous phases due to the high penetrating power of them into the mesopores. In addition, the techniques for supercritical fluid-based growths can be easily applied for the growths of a range of many semiconductor and metallic nanowires. A notable advantage of the solution-based growth is that nanowire synthesis can be performed in a continuous process instead of a batch process. Another advantage of this growth is the variability in the choice of the precursor for growth. It offers extra freedom for growth optimization. This advantage allows Si nanowire growth, for example, both with monophenylsilane (C_6H_8Si), and diphenylsilane ($C_{12}H_{12}Si$) as precursor. The use of monophenylsilane, unlike diphenylsilane, provides higher product yield together with a smaller amount of carbonaceous by-product.

8.5.3 Other Advantages

The strengths of the solution-based growths are reflected also from large qualities of the nanowires produced by the solution-based mechanisms. This is done at lower costs than those by typically more complex high-vacuum systems. These growths can produce narrow size and shape distributions of many different kinds of nanomaterials exhibiting crystalline structures with coated surfaces and important optical and electronic properties. The Ge nanowires obtained by SFLS mechanism by Lu et al. [28] employing supercritical cyclohexane or hexane and gold nanoparticle R_L species seeds exhibited almost defect-free crystalline cores. HRTEM images indicated that, under optimum growth conditions, less than 5% of the nanowires had traces of defects. Crystalline interface between the tip alloy and the nanowire was quite abrupt, which is important for future applications, particularly, the contact designs of devices. The SoLS and SFLS mechanisms can also be combined with a templated method to guide the catalyst-mediated growth of nanowires within anodic alumina membranes [38]. The templated growth has some notable advantages over conventional growth. Templated growth can, for example, lead to pre-aligning of nanowires during the growth stage. Also, solution-based and supercritical fluid-based growths, unlike vapor-based growths, offer promising possibility of larger-scale nanowire syntheses.

These growths involve enhanced solubilities of reactants and products minimizing transport restrictions on reaction rates. Indeed they endure significant chemical flexibility and synthetic tenability. Table 8.2 lists metal-assisted high-temperature growth of Si and Ge nanowires [15, 23, 39–46] by the SFLS mechanism.

Table 8.2 List of metal-assisted high-temperature growths of Si and/or Ge nanowires by the solution -phase synthesis mechanisms

Metal catalyst	FECA form	Nanowire formed	Growth temp (°C)	Fluid medium	References
Sn	Discrete nanoparticle seed	Si	450	SCF	Lu [15]
Au	Discrete nanoparticle seed	Si	500	HBS (hexane)	Holmes [23]
Ni	Discrete nanoparticle seed	Ge	450	SCF (toluene)	Tuan [39]
Co	Discrete nanoparticle seed	Si, Ge		SCF (toluene)	Tuan [40]
Al, Ag	Nanoparticle formed from bulk metal foils	Si	450–550	SCF	Yuan [41]
Pb	Bulk Pb foil	Si		SCF	Yuan [41]
In	Seeds formed in situ from evaporated metal layer	Si		HBS vapor	Gearney [42]
Bi	Discrete nanoparticle seeds	Si	420	HBS	Heitsch [43]
Cu	In situ Cu_3Ge seeds formed from bulk Cu foil	Ge	425	HBS vapor	Gearney [44]
Mn	Discrete nanoparticle seed	Si, Ge	450	SCF	Tuan [40] Gearney [5]
Sn	Derived from $Sn(HMDS)_2$	Si	450	SCF (toluene)	Bogart [46]

High boiling point organic solvent is abbreviated as HBS; supercritical fluid is abbreviated as SCF; and bis(bis(trimethylsilyl)amino)tin is abbreviated as $[Sn(HMDS]_2$

8.6 Weaknesses

8.6.1 Challenges

Nevertheless, challenges remain for the solution-based synthesis and for the high yields of products by this synthesis. There are also challenges to producing CNTs, in particular SWCNTs, by the solution-based mechanisms. Note that significant amounts of oligomeric by-products are formed during solution-based synthesis, and these by-products can severely limit the reaction yield. These challenges must be met before nanowires such as Si and Ge nanowires can be produced in high-volume commercial quantities. There are other challenges, as well. No doubt, colloidal nanowires can be synthesized at low cost. However, the large-scale assembly of solution-based nanowires into integrated, functional devices is highly challenging.

8.6.2 Low-Temperature Growths

Although solution-based nanowire syntheses are believed to be variants of the vapor-based VLS syntheses, Si nanowire growth in supercritical fluid environment and catalyzed, for example, by Cu, Ni, and Co nanoparticles by Tuan et al. [39, 47] could be found to grow at temperatures well below the eutectic melting point T_E of the corresponding (MET, Si) eutectic alloy. Ge nanowires obtained in this environment could grow at reaction temperatures as low as 300 °C. In contrast, Au-catalyzed GaAs nanowires by Davidson et al. [25], 25 nm in diameter and tens of micrometer in length, were found to grow in supercritical hexane at 500 °C. GaAs nanowire growth thus took place at ~150 °C above the eutectic temperature $T_E = 339.4$ °C to 348.9 °C of the (Au, Ga) pseudo-binary alloy [48]. Si and GaAs nanowire growths cited above were different from the Ge nanowire growths by the same mechanism. The HRTEM image of Ge nanowires formed at 370 °C and 10 MPa using DPG exhibited cubic diamond crystal structure with the [110] growth direction. This was a VLS-type growth at temperature close to the (Au, Ge) eutectic temperature of 361 °C. Hanrath and Korgel [24] found that an 8 nm diameter Ge nanowire grown at a temperature of 400 °C and a pressure of 5 MPa using TEG had a crystalline $Au_{0.72}Ge_{0.28}$alloy at the nanowire tip. The AuGe crystal can be molten at 400 °C if it has a Ge mole fraction of at least ~ 29 atomic %. Based on EDS and crystallographic characterization, the Ge mole fraction in the crystal was however 28 at%. This means it was in the metastable β hcp $Au_{0.72}Ge_{0.28}$ phase, and not the eutectic phase, formed during the rapid quenching of the reaction after the elapsed synthesis time.

8.6.3 Solution-Phase Mechanisms Inferior to Vapor-Phase Mechanisms

The solution-phase mechanisms may be inferior to vapor-phase mechanisms at least in two avenues [49] of some nanowire growths. First, nanowire nucleation and growth by the SFLS and SoLS mechanisms take place at a much higher rate than by the VLS and VSS mechanisms. These are on the order of micrometers per minute by the SFLS and SoLS mechanisms, but nanometers per minute by the VLS mechanism. Most of the SFLS and SoLS reactions are over within a few minutes. These are too fast to make quantitative analysis of growth kinetics and to characterize the growth mechanisms. Second, SFLS and SoLS syntheses are "single-pot" syntheses. They make use of rapid injection of precursors and the release of the $R_S \equiv X$ and $R_S \equiv Y$ on the low-melting-point FECA metal nanoparticle (for example, In, Sn, and Bi) in an environment of hot solvent, precursor, and surfactant (ligand) mixture (typically at 200–300 °C). Worryingly, in most cases, reactant concentrations are hardly constant or controllable during the very rapid process of reaction. Further, the inherent limitations of flask-based chemistry resulting from very rapid growth and changes in composition along the nanowires length are largely compromised in growth by the SFLS and SoLS mechanisms. In contrast, the sequential introduction of reactants and modulation of their concentration is allowed in growths by the VLS mechanism. This eases the direct fabrication of axial heterostructures.

8.6.4 Illustrative Demonstrations of SFLS and SoLS Mechanisms

While trying to demonstrate the inferiority of the solution-based mechanisms to the vapor-based mechanisms, Laocharoensuk et al. [49] made some important observations. They studied the variations of nanowire length and nanowire diameter with nanowire growth time, which are depicted in Fig. 8.4. They noted that the nanowire diameter slowly increases, attains its largest value and then decreases with increase in growth time. In contrast, the nanowire length slowly decreases, attains its smallest value and then increases with increase in growth time. They also studied the variation of nanowire growth rate with nanowire diameter for CdSe and ZnSe nanowires. As shown in Fig. 8.5, almost always the growth rate increased with diameter, reached the peak and then slowly decreased with diameter. It was more pronounced for higher temperature growth of CdSe nanowires.

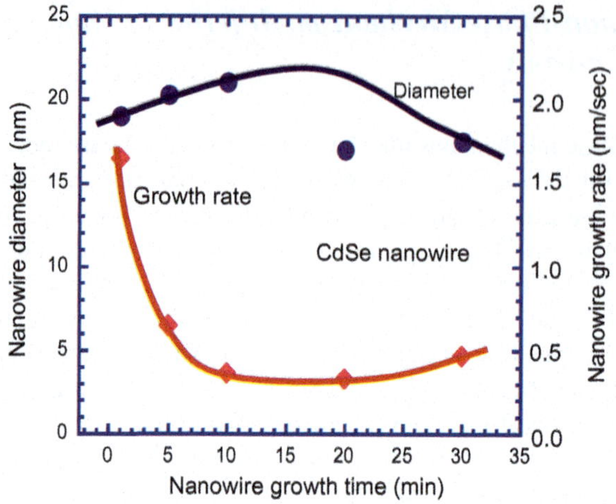

Fig. 8.4 Variations of the diameter and growth rate of CdSe nanowire with growth time. CdSe nanowires were synthesized from 5 nm Bi layers at 300 C using a constant precursor flow rate of 0.02 ml/minute. The plots were made with experimental data by Laocharoensuk et al. [49]

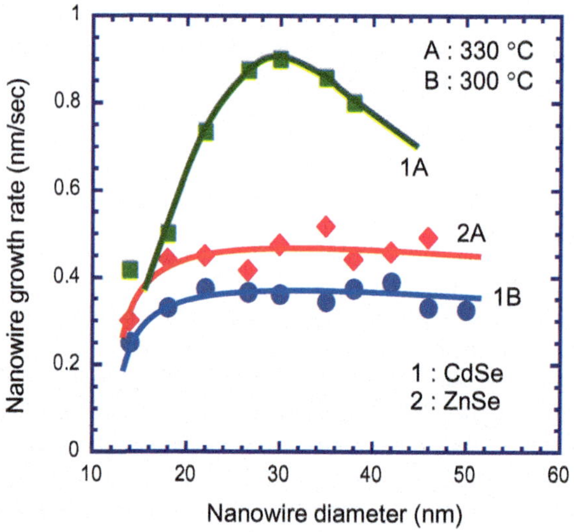

Fig. 8.5 Variation of nanowire growth rate with nanowire diameter of CdSe and ZnSe nanowires. The nanowires were grown at $T = 300$ and 330 °C. The plots were obtained with experimental data by Laocharoensuk et al. [49]

8.7 SoSS (SFSS) Mechanism

8.7.1 *Background*

Wang et al. [4] produced Ge nanowires using germane precursor and CVD technique at temperatures as low as 275 °C. This could have happened by one of the two possible reasons: First, the Au/Ge eutectic temperature was significantly lowered for the formation of $R_L \equiv$(Au, Ge) eutectic alloy droplets as suggested by Wu and Yang [50]. It could otherwise have happened by nanometer-sized solid particle. And this situation is identical to the one observed by Kamins et al. [51, 52], who performed Ti nucleated Si nanowire growth at temperatures up to 500 °C below the eutectic temperature T_E. Both of them highlight the option of the R_S ($R_S \equiv X$, $R_S \equiv Y$) source species exhibiting sufficiently high diffusion rates through a solid nucleation nanoparticle to permit nanowire growth. The observation of Cu, Ni, and Co nanoparticles catalyzing Si nanowire growths at temperatures well below the eutectic melting point T_E of the corresponding (MET, Si) eutectic alloy [40, 47] may hence be a good example of the solution-based analog of the VSS mechanism, a mechanism in which solid metal nanoparticle catalyzes the nanowire growths.

8.7.2 *Specifics of the SoSS and the SFSS Mechanisms*

The solution-based growth mechanism called the solution-solid-solid (SoSS) mechanism or supercritical fluid-solid-solid (SFSS) mechanism is a liquid-based variant of the vapor-phase VSS mechanism. It is an interesting growth mechanism suitable for the synthesis of semiconductor nanowires, nanorods, and related heterostructures [53–60] in the solution phase environment. The SoSS mechanism differs from the conventional SoLS mechanism in that solid, nanoporous, (microporous, mesoporous) nanoparticles, namely those from chalcogenide Ag_2S, Cu_2S, Ag_2Se, etc., rather than those from low-melting metals, are used as catalysts for SoSS growths. Wang et al. [61] used artificial atoms as uniform building blocks to construct ordered mesoporous, and probably quasiliquid (quasisolid) Ag, Ag_2S, and Ag_2Se materials. Kristl et al. [62] examined the sonochemical synthesis routes for the synthesis of chalcogenide materials.

8.7.3 *Interesting Feature of Chalcogenite Nanoparticles*

An interesting feature of the said chalcogenite nanoparticles is that they themselves can simultaneously serve as catalysts and also as reactants, of course, under some optimized condition. The immediate result of this is the growth of binary, ternary, or even quaternary nanowires, with component elements from both the R_S source species and the catalyst nanoparticle. These nanowires could grow at temperatures lower than 250 °C, which can be much lower than the melting points of the said nanoparticles. We cite some example. Zhu and Xu [59] found that, due to fast ion mobility of the Ag cation, the R_L species of the solid Ag_2S nanoparticles catalyzes the growth of ZnS and CdS nanorods (nanowires) and ZnS-CdS heterosegmented nanorods. At the end of the growth, the R_L species material(s) was found to be attached to the nanowire tips.

8.7.4 Nanowire Quality Dependent on Precursor

Tuan et al. [63] performed Ni-seeded SFSS synthesis of Si nanowires using monophenylsilane (MPS) as a precursor in toluene at a chamber pressure of 23.4 MPa and in the presence of MET≡Ni nanocrystals. They found reaction temperature to significantly influence the quality of the nanowires. The best growth temperature at the said pressure was 460 °C, which produced straight crystalline nanowires longer than 10 μm. The nanowires grown predominantly in the ⟨110⟩ growth direction had Si diamond cubic crystal structure and some dislocation defects. Also, TEM images indicated that the nanowire surface had a significant amount of roughness. The diameter fluctuated by only ~1 nm along the entire length of the nanowire and had essentially no sidewall growth.

The Si nanowires were produced also by using octylsilane and trisilane as precursors. Although they were crystalline and relatively long, they had crystal quality not as good as those with MPS. The nanowires with both octylsilane and trisilane had significant amounts of amorphous sidewall deposition. The sidewall deposition from trisilane was expected, as it underwent rapid dehydrogenation. Sidewall growth could be eliminated to some extent by using higher Ni/Si ratio. The best Si nanowires were obtained from trisilane by using nearly two orders of magnitude larger Ni/Si ratio than with MPS. One problem with very high Ni/Si ratio was that the Si supply to the metal seeds could be starved, and that it would generate crystallographic defects. Sidewall-deposited Si nanowire from octylsilane was amorphous. But unlike the one with trisilane, it suffered from carbon contamination.

References

1. C.-C. Chen, C.-C. Yeh, C.-H. Chen, M.-H. Yu, H.-L. Liu, J.-J. Wu, K.-H. Chen, L.-C. Chen, J.-Y. Peng, Y.-F. Chen, Catalytic growth and characterization of gallium nitride nanowires. J. Am. Chem. Soc. **123**, 2791–2798 (2001)
2. Y. M. Manawi, Ihsanullah, A. Samara, T. Al-Ansari, M. A. Atieh, A Review of carbon nanomaterials' synthesis via the chemical vapor deposition (CVD) method, Materials 11, 822 (2018)

3. S. Mathur, H. Shen, V. Sivakov, U. Werner, Germanium nanowires and core–shell nanostructures by chemical vapor deposition of [Ge(C_5H_5)$_2$]. Chem. Mater. **16**, 2449–2456 (2004)
4. D. Wang, H Dai, Low-temperature synthesis of single-crystal germanium nanowires by chemical vapor deposition. Angew. Chem., Int. Edn. **41**, 4783–4786 (2002)
5. S. Ganji, Nanowire growths, and mechanisms of these growths for developing novel manomaterials. J. Nanosci Nanotechnol. **19**(4), 1849–1874 (2019)
6. M. Royo, M. De Luca, R. Rurali, I. Zardo, A review on III–V core–multishell nanowires: growth, properties, and applications. J. Phys. D Appl. Phys. **50**, 143001 (2017)
7. M.H. Huang, Y. Wu, H. Feick, N. Tran, E. Weber, P. Yang, Catalytic growth of zinc oxide nanowires by vapor transport. Adv. Mater. **13**, 113–116 (2001)
8. T.J. Trentler, K.M., Hickman, S.C. Goel, A.M. Viano, P.C. Gibbons, W.E. Buhro, Solution-liquid-solid growth of crystalline III-V semiconductors: an analogy to vapor-liquid-solid growth. *Science, 270*, 1791–1794 (1995)
9. H. Hanrath, B.A. Korgel, Supercritical fluid-liquid-solid (SFLS) synthesis of Si and Ge nanowires seeded by colloidal metal nanocrystals. Adv. Mater. **15**, 437–440 (2003)
10. F. Wang, A. Dong, J. Sun, R. Tang, H. Yu, W.E. Buhro, Solution-liquid-solid growth of semiconductor nanowires. Inorg. Chem. **45**, 7511–7521 (2006)
11. F. Wang, A. Dong, W.E. Buhro, Solution–liquid–solid synthesis, properties, and applications of one-dimensional colloidal semiconductor nanorods and nanowires. Chem. Rev. **116**, 10888–10933 (2016)
12. A. Dong, H. Yu, F. Wang, W.E. Buhro, Colloidal GaAs quantum wires: solution-liquid-solid synthesis and quantum-confinement studies. J. Am. Chem. Soc. **130**, 5954–5961 (2008)
13. J. Sun, C. Liu, P. Yang, Surfactant-free, large-scale, solution-liquid-solid growth of gallium phosphide nanowires and their use for visible-light-driven hydrogen production from water reduction. J. Am. Chem. Soc. **133**, 19306–19309 (2011)
14. H. Yu, W.E. Buhro, Solution–liquid–solid growth of soluble GaAs nanowires. Adv. Mater. **15**, 416–419 (2003)
15. X. Lu, C.M. Hessel, Y. Yu, T.D. Bogart, B.A. Korgel, Colloidal luminescent silicon nanorods. Nano Lett. **13**, 3101–3105 (2013)
16. Y. Yu, J.B. Li, R.A. Loomis, P.C. Gibbons, L.W. Wang, W.E. Buhro, Cadmium selenide quantum wires and the transition from 3D to 2D confinement. J. Am. Chem. Soc. **125**, 16168–16169 (2003)
17. S. Kan, T. Mokari, E. Rothenberg, U. Banin, Synthesis and size-dependent properties of zinc-blende semiconductor quantum rods. Nat. Mater. **2**, 155–158 (2003)
18. T. Hanrath, Germanium nanowires: synthesis, characterization, and utilization, *Doctoral Thesis* (University of Texas, Austin, 2004)
19. E.W. Lemmon, M.O. McLinden, D.G. Friend, *NIST Chemistry WebBook, NIST Standard Reference Database Number 69* 2004 (2003), https://webbook.nist.gov
20. R. Span, *Multiparameter Equations of State* (Springer, Berlin, 2000), p. 367
21. P.E. Savage, S. Goplan, T.I. Mizan, C.J. Martino, E.E. Brock, Reactions at supercritical conditions: applications and fundamentals. AIChE J. **41**, 1723–1778 (1995)
22. J.R. Heath, F.K. LeGoues, A liquid solution synthesis. Chem. Phys. Lett. **208**, 263–266 (1993)
23. J.D. Holmes, K.P. Johnston, R.C. Doty, B.A. Korgel, Control of thickness and orientation of solution-grown silicon nanowires. Science **287**, 1471–1473 (2000)
24. T. Hanrath, B.A. Korgel, Nucleation and growth of germanium nanowires seeded by organic monolayer-coated gold nanocrystals. J. Am. Chem. Soc. **124**, 1424–1429 (2002)
25. F.M. Davidson III., A.D. Schricker, R.J. Wiacek, B.A. Korgel, Supercritical fluid-liquid-solid synthesis of gallium arsenide nanowires seeded by alkanethiol-stabilized gold nanocrystals. Adv. Mater. **16**, 646–649 (2004)
26. F.M. Davidson III., R. Wiacek, B.A. Korgel, Supercritical fluid-liquid-solid synthesis of gallium phosphide nanowires. Chem. Mater. **17**, 230–233 (2005)
27. X.M. Lu, T. Hanrath, K.P. Johnston, B.A. Korgel, Growth of single crystal nanowires in supercritical silicon solution from tethered gold particles on a silicon substrate. Nano Lett. **3**, 93–99 (2003)

28. X.M. Lu, M. De la Mata, J. Arbiol, B.A. Korgel, Colloidal silicon-germanium nanorod heterostructures. Chem. Mater. **29**(22), 9786–9792 (2017)
29. R.L. Geddes, E. Mack, The thermal decomposition of gaseous germanium tetraethyl. J. Am. Chem. Soc. **52**, 4372–4380 (1930)
30. A.T. Heitsch, D.D. Fanfair, H.Y. Tuan, B.A. Korgel, Solution-liquid-solid (SLS) growth of silicon nanowires. J. Am. Chem. Soc. **130**, 5436–5437 (2008)
31. P. Pertl, M.S. Seifner, C. Herzig, A. Limbeck, M. Sistani, A. Lugstein, S. Barth, Solution-based low-temperature synthesis of germanium nanorods and nanowires. Monatshefte Für Chemie - Chemical Monthly **149**, 1315–1320 (2018)
32. M.A. Meshgi, S. Biswas, D. McNulty, C. O'Dwyer, G.A. Verni, J. O'Connell, F. Davitt, I. Letofsky-Papst, P. Poelt, J.D. Holmes, C. Marschner, Rapid, low temperature synthesis of germanium nanowires from oligosilylgermane precursors. Chem. Mater. **29**, 4351–4360 (2017)
33. M.I. Bodnarchuk, K.V. Kravchyk, F. Krumeich, S. Wang, M.V. Kovalenko, Colloidal tin–germanium nanorods and their Li-ion storage properties. ACS Nano **8**, 2360–2368 (2014)
34. M.S. Seifner, P. Pertl, J. Bernardi, S. Biswas, J.D. Holmes, S. Barth, Lead-supported germanium nanowire growth. Mater. Lett. **73**, 248–251 (2016)
35. M.S. Seifner, M. Sistani, F. Porrati, G. Di Prima, P. Pertl, M. Huth, A. Lugstein, S. Barth, Direct synthesis of hyperdoped germanium nanowires. ACS Nano **12**, 1236–1241 (2018)
36. S. Barth, M.M. Koleśnik-Gray, K.P. Donegan, V. Krstić, J.D. Holmes, Diameter-controlled solid-phase seeding of germanium nanowires: Structural charactesrization and electrical transport properties. Chem. Mater. **23**, 3335–3340 (2011)
37. X. M. Lu, D.D. Fanfair, K.P. Johnston, B.A. Korgel, High yield solution-liquid-solid synthesis of germanium nanowires. J. Am. Chem. Soc. **127**, 15718–15719 (2005)
38. N. Petkov, P. Birjukovs, R. Phelan, M. Morris, D. Erts, J. Holmes, Growth of ordered arrangements of one-dimensional germanium nanostructures with controllable crystallinities. Chem. Mater. **20**, 1902–1908 (2008)
39. H.-Y. Tuan, D.C. Lee, T. Hanrath, B.A. Korgel, Germanium nanowire synthesis: an example of solid-phase seeded growth with nickel nanocrystals. Chem. Mater. **17**, 5705–5711 (2005)
40. H.-Y. Tuan, D.C. Lee, B.A. Korgel, Nanocrystal-mediated crystallization of silicon and germanium nanowires in organic solvents: the role of catalysis and solid-phase seeding. Angew. Chem. **45**, 5184–5187 (2006)
41. F.-W. Yuan, H.-J. Yang, H.-Y. Tuan, Seeded silicon nanowire growth catalyzed by commercially available bulk metals: broad selection of metal catalysts, superior field emission performance, and versatile nanowire/metal architectures. J. Mater. Chem. **21**, 13793–13800 (2011)
42. H. Gearney, T. Kennedy, C. Dickinson, E. Mullane, A. Singh, F. Laffir, K.M. Ryan, High density growth of indium seeded silicon nanowires in the vapor phase of a high boiling point solvent. Chem. Mater. **24**, 2204–2210 (2012)
43. A.T. Heitsch, V.A. Akhavan, B.A. Korgel, Rapid SFLS synthesis of Si nanowires using trisilane with In situ alkyl-amine passivation. Chem. Mater. **23**, 2697–2699 (2011)
44. H. Gearney, C. Dickinson, C.A. Barrett, K.M. Ryan, High density germanium nanowire growth directly from copper foil by self-induced solid seeding. Chem. Mater. **23**, 4838–4843 (2011)
45. H. Geaney, E. Mullane, K.M. Ryan, Solution phase synthesis of silicon and germanium nanowires. J. Mater. Chem. C **1**, 4996–5007 (2013)
46. T.D. Bogart, X. Lu, B.A. Korgel, Precision synthesis of silicon nanowires with crystalline core and amorphous shell. Dalton Trans. **42**, 12675 (2013)
47. H.-Y. Tuan, A. Ghezelbash, B.A. Korgel, Silicon nanowires and silica nanotubes seeded by copper nanoparticles in an organic solvent. Chem. Mater. **20**, 2306–2313 (2008)
48. T.B. Massalski (ed.), *Binary Alloy Phase Diagrams*, vol. 3, 2nd edn (American Society of Metals, Metals Park, OH, 1986)
49. R. Laocharoensuk, K. Palaniappan, N.A. Smith, R.M. Dickerson, D.J. Werder, J.K. Baldwin, J.A. Hollingsworth, Flow-based solution–liquid–solid nanowire synthesis. Nat. Nanotechnol. **8**, 660–666 (2013)
50. Y. Wu, P. Yang, Melting and welding of semiconductor nanowires in nanotubes. Adv. Mater **13**, 520–523 (2001)

51. T.I. Kamins, R.S. Williams, Y. Chen, Y.L. Chang, Y.A. Chang, Chemical vapor deposition of Si nanowires nucleated by $TiSi_2$ islands on Si. Appl. Phys. Lett. **76**, 562 (2000)
52. T.I. Kamins, R.S. Williams, Ti-catalyzed Si nanowires by chemical vapor deposition: Microscopy and growth mechanisms. J. Appl. Phys. **89**, 1008 (2001)
53. A.K. Guria, S. Sarkar, B.K. Patra, N. Pradhan, Efficient superionic conductor catalyst for solid in solution-solid-solid growth of heteronanowires. Jss. Phys. Chem. Lett. **5**, 732–736 (2014)
54. Y.J. Zhang, S.L. Shen, Q.B. Wang, Controllable growth of Ag_2S-CdS heteronanostructures. CrystEngComm **16**, 9501–9505 (2014)
55. J.L. Wang, C.M. Yang, Z.P. Huang, M.G. Humphrey, D. Jia, T.T. You, K.M. Chen, Q. Yang, C. Zhang, Seed-catalyzed heteroepitaxial growth and nonlinear optical properties of zinc selenide nanowires. J. Mater. Chem. **22**, 10009–10014 (2012)
56. Q. Li, C. Zou, L.L. Zhai, L.J. Zhang, Y. Yang, X. Chen, S.M. Huang, Synthesis of wurtzite $CuInS_2$ nanowires by Ag_2S catalyzed growth. CrystEngComm **15**, 1806–1813 (2013)
57. C. Zou, M. Li, L.J. Zhang, Y. Yang, Q. Li, X. Chen, X.J. Xu, S.M. Huang, Ag_2S-catalyzed growth of quaternary $AgInZn_7S_9$ semiconductor nanowires in solution. CrystEngComm **13**, 3515–3520 (2011)
58. S.T. Connor, C.M. Hsu, B.D. Weil, S. Aloni, Y. Cui, Phase transformation of biphasic Cu_2S-$CuInS_2$ to monophasic $CuInS_2$ nanorods. J. Am. Chem. Soc. **131**, 4962–4966 (2009)
59. W. Han, L. Yi, N. Zhao, A. Tang, M. Gao, Z. Tang, Synthesis and shape-tailoring of copper sulfide/indium sulfide-based nanocrystals. J. Am. Chem. Soc. **130**, 13152–13161 (2008)
60. G. Zhu, Z. Xu, Controllable growth of semiconductor heterostructures mediated by bifunctional Ag_2S nanocrystals as catalyst or source-host. J. Am. Chem. Soc. **133**, 148–157 (2011)
61. D. Wang, T. Xie, Q. Peng, Y. Li, Ag, Ag_2S, and Ag_2Se nanocrystals: Synthesis, assembly, and construction of mesoporous structures. Am. Chem. Soc. **130**, 4016–4022 (2008)
62. M. Kristl, S. Gyergyek, J. Kristl, Synthesis and characterization of nanosized silver chalcogenides under ultrasonic irradiation. Mater. Express **5**, 359–366 (2015)
63. H.-Y. Tuan, D.C. Lee, T. Hanrath, B.A. Korgel, Catalytic solid-phase seeding of silicon nanowires by nickel nanocrystals in organic solvents. Nano Lett. **5**, 681–684 (2005)

Chapter 9
Solid–Liquid–Solid Growth Mechanism

Abstract The solid–liquid–solid (SLS) mechanism for the growths of one-dimensional and quasi-two-dimensional nanomaterials has been reviewed. This mechanism is employed for nanowire growths in the solid crystalline environment. It is the solid-phase analog of the VLS mechanism. It is hence metal catalyst-based mechanism. The basics of this SLS mechanism have been described. The general hypothesis of the growths by the SLS mechanism has been detailed. Characteristic features of it have been depicted. It has been noted to be primarily employed for nanowire growths. So nanowire growths by it (e.g., the SLS mechanism) have been illustrated and examined. Strengths and weaknesses of it in mediating nanowire growths have been elucidated. The importance and feasibility of it have been demonstrated. The nanowires by the SLS mechanism are often amorphous. And, the reasons of this have been detailed.

9.1 Various Features of the Synthesis

9.1.1 Basics

The solid–liquid–solid (SLS) mechanism [1–7] is also an important mechanism for nanomaterials growths. During the past years, it has been employed for the growth of many different X_mY_n nanowires. The most important feature of this mechanism is that no vapor-phase or liquid-phase R_S ($R_S \equiv X$ and $R_S \equiv Y$) source species are supplied from outside for the SLS growth. Instead, X_mY_n substrate serves as the source of the $R_S \equiv X$ and $R_S \equiv Y$ species for this growth. To be specific, a FECA metal (e.g., MET) film, generally 40–50 nm in thickness, is used to catalyze the growth. This thin MET film is deposited on the X_mY_n substrate during the pre-nucleation stage of growth (see Fig. 9.1). As described in previous reports [8–12], X_mY_n nanowires are grown on this substrate under high vacuum condition, and at a temperature, suppose, $T = T_{sls}$. This temperature is set in such a way that it is high enough to satisfy two essential conditions: (1) the fragmentation of FECA metal film into nanoparticles during heating to a temperature $T = T_{sls}$. (2) The composition

S. N. Mohammad, *Synthesis of Nanomaterials*, Springer Series in Materials Science 307, https://doi.org/10.1007/978-3-030-57585-4_9

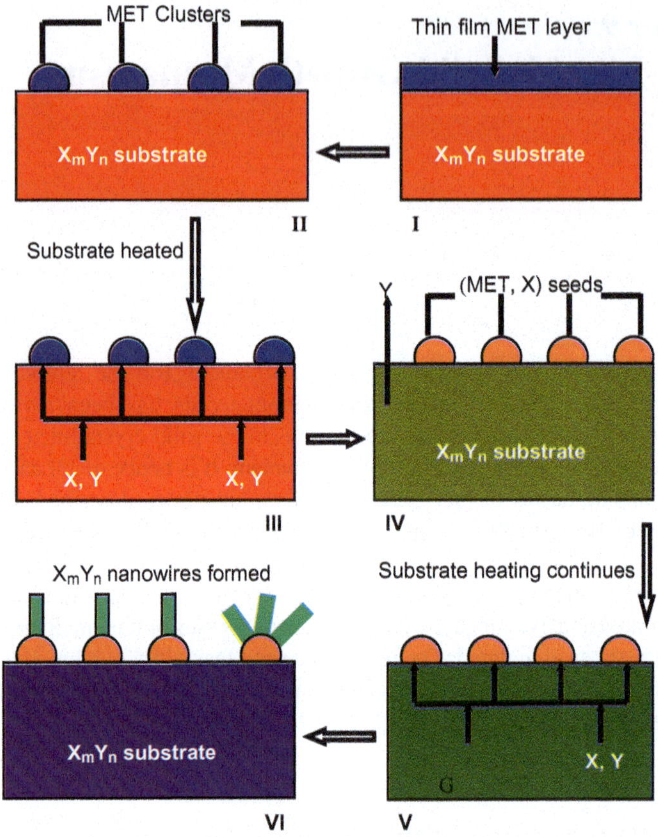

Fig. 9.1 Schematic diagrams of various steps of solid–liquid–solid (SLS) growth of nanowires. Step I: Formation of thin-film MET layer on the X_mY_n substrate. Step II: Formation of MET clusters on the X_mY_n substrate surface upon annealing. Step III: Release of $R_S{\equiv}X$ and $R_S{\equiv}Y$ from the X_mY_n substrate upon high-temperature heating of this substrate. Step IV: Desorption of $R_S{\equiv}Y$ and the formation of $R_L{\equiv}$(MET, X) species droplets upon diffusion of $R_S{\equiv}X$ into the MET clusters. Step V: Continuation of substrate heating and diffusion of $R_S{\equiv}X$ and $R_S{\equiv}Y$ into the molten (semimolten) $R_L{\equiv}$(MET, X) species. Step VI: Nucleation of nanowires via the molten (semimolten) $R_L{\equiv}$(MET, X) species droplets

of the R_L species of the nanoparticles becomes $R_L{\equiv}$(MET, X, Y) or $R_L{\equiv}$(MET, X) alloy due to diffusion of X and Y, or just of X onto the FECA metal nanoparticle surface, if Y species is volatile. The R_L species may also contain contaminant present in the growth chamber. If eutectic, this $R_L{\equiv}$(MET, X, Y) or $R_L{\equiv}$(MET, X) alloy species is droplet. Otherwise, it is a solid solution. A high density of FECA metal nanoparticles may be created on the X_mY_n substrate surface and at a temperature $T > T_E$. We emphasize that there must be a large motive force F_{mot} and a concentration gradient C_{rs} of the $R_S{\equiv}X$ and $R_S{\equiv}Y$ species created on the substrate surface at a temperature $T = T_{sls}$. These are very crucial, and to achieve them, the $R_S{\equiv}X$ and

$R_S{\equiv}Y$ species may be generated, for example, by successive heating and cooling of substrate, for example, at a rate of ~5 °C/min during high-temperature annealing. Generally, $T_{sls} \gg T_E$ for the $R_L{\equiv}$(MET, X) alloy species exhibiting low eutectic temperature. But T_{sls} is comparable with T_E for the $R_L{\equiv}$(MET, X) alloy species exhibiting high eutectic temperature. We cite an example; $T_E \approx 361$ °C and $T_{sls} \approx 1000$ °C for Si nanowire growth by MET≡Au, but $T_E \approx 993$ °C and $T_{sls} \approx 950$ °C for Si nanowire growth by MET≡Ni. Note that $R_L{\equiv}$(MET, X) alloy species created on nanoparticle surface are due to (1) migration of the $R_S{\equiv}X$ species from inside the substrate to the substrate surface, and (2) subsequent diffusion of the $R_S{\equiv}X$ species through the FECA metal nanoparticle to the FECA metal surface.

9.1.2 General Hypothesis for Growths

General hypothesis of X_mY_n nanowire growth by the SLS mechanism [13] elegantly describes the dependence of this nanowire growth on a number of factors: First, the temperature-dependent diffusion of $R_S{\equiv}X$ and $R_S{\equiv}Y$ species onto the nanoparticle surface is comparable. This surface has a composition $R_L{\equiv}$(MET, X) alloy created during the pre-nucleation stage (see Chap. 4, Sect. 3) prior to the pro-nucleation stage of growth at $T = T_{sls}$. This allows $R_S{\equiv}X$ and $R_S{\equiv}Y$ species to be in close proximity and to react to form X_mY_n molecules: $m\text{X} + \text{Y}n \rightarrow X_mY_n$ inside the R_L species alloy. Second, the time-sensitive diffusion of $R_S{\equiv}X$ and $R_S{\equiv}Y$ species onto the nanoparticle surface of composition $R_L{\equiv}$(MET, X) alloy at $T = T_{sls}$ requires optimal concentration gradients and motive forces of both the $R_S{\equiv}X$ and the $R_S{\equiv}Y$ species. They both enable the $R_S{\equiv}X$ and $R_S{\equiv}Y$ species to move up to the growth front from inside the substrate to the substrate surface (see Fig. 9.2) even against gravity. These are important also for the supersaturation of X_mY_n molecules at the liquid/solid interface. This liquid/solid interface is at the tips of the nanoparticle for the SLS growth, but at the bottom of the nanoparticle for the VLS growth. Third, with the eutectic temperature T_E relatively low, the growth temperature T_{sls} is not comparable to T_E; also, the X mole fraction in the $R_L{\equiv}$(MET, X) alloy may be lower than that required for it to be molten. This growth temperature T_{sls} must be such that it causes the R_L species of the nanoparticle surface to be (1) molten (semimolten) due to eutectic effect at a temperature $T_{sls} \approx T_E$ or (2) disturbed, disordered, coarsened, and porous exhibiting grains, grain boundaries, pits, hillocks, semi-melting, etc., at a temperature $T_{sls} \neq T_E$. The X_mY_n molecules, upon diffusion through the R_L species, are eventually supersaturated at the liquid/solid interface. X_mY_n nanowires formed from X_mY_n molecules eject subsequently out of the nanoparticle surface. If summarized, there are thus five distinct stages of the SLS growth:

1. the formation of the $R_L{\equiv}$(MET, X) species on nanoparticle surface;
2. the diffusion of the $R_S{\equiv}X$ and $R_S{\equiv}Y$ species into the R_L species of nanoparticle;
3. the reaction of the $R_S{\equiv}X$ and $R_S{\equiv}Y$ species leading to the formation of X_mY_n molecules in the R_L species of the nanoparticle;

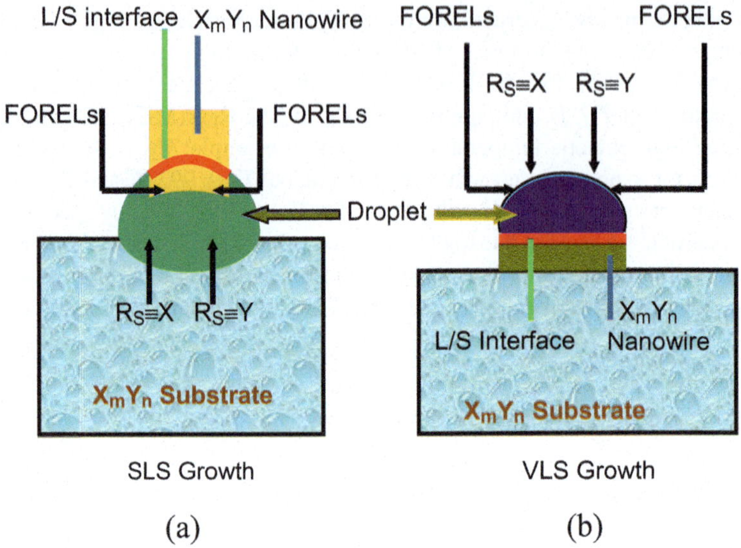

Fig. 9.2 Schematic comparison of the SLS and the VLS growth of nanowires. Foreign elements (contaminants including oxygen) have been called FORELs, and the liquid/solid (L/S) interface is represented by thick red line between the droplet and the nanowire. The L/S interface is at the tip of the droplet in the SLS process, but at the base of the droplet in the VLS process. Nanowire grows at the tip of the droplet in the SLS process, but at the base of the droplet in the VLS process

4. the supersaturation of the X_mY_n molecules at the liquid/solid interface; and
5. the nucleation crystallization, and growth of X_mY_n nanowires on the nanoparticle surface.

9.1.3 *Why Nanowire Growths Do Not Take Place at* $T < T_{sls}$

It has been elucidated in the general hypothesis for nanowire growths [13] that there are a number of possible reasons for nanowire growth not taking place at $T < T_{\text{sls}}$. These are stated in the following:

1. R_L≡(MET, X) alloy needed for growth may not be formed at $T < T_{\text{sls}}$.
2. The concentration gradient and motive force needed for the R_S≡X and R_S≡Y species to diffuse into the nanoparticle surface of composition R_L≡(MET, X) alloy from inside the substrate, to form X_mY_n molecules, and to reach supersaturation may be insufficient at $T < T_{\text{sls}}$.
3. The composition of the R_L≡(MET, X) alloy of the nanoparticle may not be porous enough for the large number of X_mY_n molecules needed to produce X_mY_n nanowires at $T < T_{\text{sls}}$.
4. The atomic % of X in the R_L≡(MET, X) alloy species at $T < T_{\text{sls}}$ may not be sufficient to generate liquid-like (quasiliquid-like, quasisolid-like) structural

feature, and to create dipole moment and charge asymmetry large enough for the $R_S \equiv X$ and $R_S \equiv Y$ species to be attracted to the nanoparticle of R_L species composition $R_L \equiv (MET, X)$ or $R_L \equiv (MET, X, \vartheta)$ or $R_L \equiv (MET, X, O, \vartheta)$.

9.2 Why SLS Nanowires Are Often Amorphous or Have Amorphous Shell?

The $R_L \equiv (MET, X)$ alloy in both the SLS and the VLS mechanisms possesses dipole moment and electric field. However, the $R_L \equiv (MET, X)$ alloy in the VLS mechanism attracts $R_S \equiv X$ and $R_S \equiv Y$ species from the vapor accumulated above the substrate surface. In contrast, the $R_L \equiv (MET, X)$ alloy in the SLS mechanism attracts $R_S \equiv X$ and $R_S \equiv Y$ species from the solid existing inside the substrate. All of them though attract contaminant ϑ (e.g., O, N, H, dopants, foreign elements, oxides, etc.) from the vapor above the substrate surface. The growth chamber often contains oxygen derived from species, such as CO_2 and H_2O. It contains also residual atmospheric oxygen. As the temperature is increased during the SLS growth, large numbers of $R_S \equiv X$ and $R_S \equiv Y$ species diffuse through the $R_L \equiv (MET, X)$ species under the influence of concentration gradient and motive force. Contaminants also enter the $R_L \equiv (MET, X)$ species. The oxygen incorporation into the chamber generally causes the SLS nanowires to be amorphous. An important difference between the SLS mechanism and the OAG mechanism is that, unlike the growth kinetics of the OAG mechanism, the growth kinetics of the SLS mechanism does not always generate driving force needed for phase separation and segregation. And hence, with some exceptions, there occurs no formation of structures containing crystalline core and amorphous shell. It happens particularly if the oxygen content is too large to cause amorphization even of the core material. One of these exceptions is the Si nanowire growth by Lee et al. [2]. These researchers found that Si nanowires grown in their laboratory by the SLS mechanism at 1000 °C had crystalline core of an average diameter of 10 nm and a thick amorphous shell of thickness of 20–30 nm. They also found that nanowire diameter increased with increase in temperature. The variation of nanowire diameter with temperature T_{sls} as observed by them is displayed in Fig. 9.3. This variation is near exponential in nature. It suggests that the concentration gradient and the motive force prompting the R_S species to diffuse into the R_L species increase nearly exponentially with increase in temperature. Another of these exceptions is the Fe-catalyzed Si nanowire growth by Nie et al. [6]. They produced core–shell nanowires having ultrathin Si cores of diameter of 5–8 nm inside much thicker amorphous SiO_z shells. As shown in Fig. 9.4, the catalyst nanoparticles at the tip of nanowires had FeSi core, (Fe, Si, O) inner shell, and SiO_z outer shell. The synthesis temperature, 1100 °C, was lower than the (Fe, Si) eutectic temperature of 1207 °C. The $Fe_{0.67}Si_{0.33}$ alloy formed at the growth temperature is different from the $Fe_{0.64}Si_{0.36}$ eutectic alloy formed at 1207 °C. The observation by Nie et al. confirms that the SLS growth process is largely a non-eutectic process mediated by (MET, X) non-eutectic alloy (solid solution).

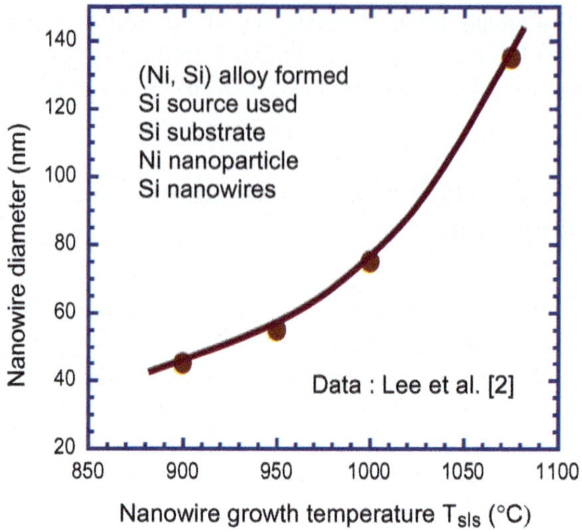

Fig. 9.3 Variation of the Si nanowire diameter as a function of the growth temperature of Si nanowires produced by the SLS mechanism. The figure was made with experimental data by Lee et al. [2]

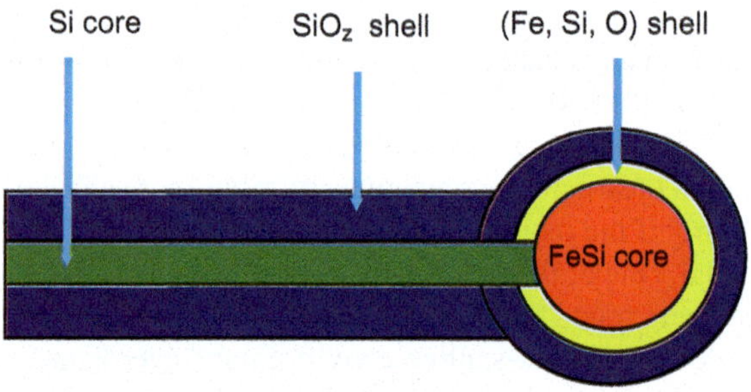

Fig. 9.4 Illustration of the schematic structure of the core–shell silicon nanowire exhibiting SiO_z shell. This nanowire was grown under the influence of Fe catalyst. The (FeSi) core at the nanowire tip was covered with (Fe, Si, O) shell. Illustration was drawn by following the sketch by Nie et al. [6]

9.3 Illustrations of the Catalyst and Temperature for the SLS Growth

Growth parameters of various Si nanowires produced by the SLS (solid–liquid–solid) mechanism have been listed in Table 9.1. A close look of Table 9.1 would indicate that SLS growth of Si nanowires mediated by $R_L \equiv$(Pt, Si) alloy took place at $T = T_{sls} = 1100$ °C. Following binary phase diagram [14], the eutectic composition of the $R_L \equiv$(Pt, Si) alloy at the eutectic temperature $T_E = 979$ °C is $Pt_{0.7}Si_{0.3}$. The

Table 9.1 Growth parameters of various Si nanowires (NWs) produced by the SLS (solid–liquid–solid) mechanism

No.	MET, T_M (°C)	Si in (MET, Si) T_E (°C)	T_{sls} (°C); Carrier gas	ζ_{approx}	Comments	References
1	Au, 1064.2	(Si: 19 at.%) 363	1000; Ar, H or N, H	$Au_{0.51}Si_{0.49}$	Au Si at NW tips; a-Si NWs grown; t_{film} = 20–50 nm; no NW growth at $T < 1000$ °C	[1]
2	Ni, 1453.0	(Si: 59 at.%) 993	900; Ar	$Ni_{0.65}Si_{0.35}$	Amorphous or polycrystalline Si wafer, NiSi at NW tips; No NW growth at $T < 900$ °C. t_{film} = 40 nm	[2]
3	Ni, 1453.0	(Si: 59 at.%) 993	950; Ar, H	Solid Ni_zSi_{1-z} (z = 0.33, 0.5, 0.6, 0.67, 0.72, 0.77, 0.88)	a-Si NWs; NW growth rate 30 nm/s; ambient pressure 750 Torr; $NiSi_2$droplet; t_{film} = 40 nm thick Ni	[3]
4	In, 156.6	–	400	$In_{0.999}Si_{0.001}$	a-Si wafer; droplet at the NW tip	[4]
5	Pt, 1772	(Si: 29 at.%) 979	1100; Ar, H	Liquid Pt_2Si, solid PtSi	Pt_3Si droplet corresponding to ~850 °C; no droplet at the NW tips; (Pt, Si) seeds only on the substrate; t_{film} = 5 nm Pt	[5]
6	Fe, 1535	(Si: 74 at.%) 1207	1200; Formic gas	Solid Fe_zSi_{1-z} (z = 0.3, 0.5, 0.67, 0.84)	Fe_2Si droplet (at 1150 °C in oxygen); no NW growth at $T < 1000$ °C. t_{film} = 1 nm	[6]
7	Ni, 1453	(Si: 59 at.%) 993	950; Ar, H	–	No Ni/Si droplet at NW tips; no NW growth at $T < 900$ °C; growth rate higher with (Ar, H) than with Ar	[7]

T is the growth temperature, T_E is the (MET, Si) eutectic temperature, T_M is the MET melting temperature, T_{min} is the minimum growth temperature, d_{NW} is the nanowire diameter, and ζ_{approx} is the approximate droplet composition at the minimum growth temperature T_{sls} based on MET/X binary-phase diagram. Also, t_{film} is the MET film thickness liquidus line of this diagram [14] suggests that the molten alloy composition would be $Pt_{0.81}Si_{0.19}$, $Pt_{0.58}Si_{0.42}$, $Pt_{0.41}Si_{0.59}$, or $Pt_{0.29}Si_{0.71}$. However, the observed composition [5] of the $R_L \equiv$(Pt, Si) alloy that catalyzed the SLS growth was $Pt_{0.75}Si_{0.25}$. This composition [14] corresponds to the temperature $T \approx 900$ °C. This means that although the furnace temperature for the SLS growth was $T_{sls} = 1100$ °C, the $R_L \equiv$(Pt, Si) alloy composition mediating nanowire growth was not stable until the temperature was lowered down to 900 °C, which is lower than even the eutectic temperature of 979 °C. Note that the composition $Pt_{0.75}Si_{0.25}$ of the $R_L \equiv$(Pt, Si) alloy at the growth temperature of 900 °C had Si atomic % lower than the Si atomic % of the eutectic alloy $Pt_{0.7}Si_{0.3}$ at the eutectic temperature T_E. This Si atomic % was lower than that of the $Pt_{0.67}Si_{0.33}$ composition at $T = T_{sls}$.

Si-SiO_z core–shell nanowires produced at 1000 °C by Nie et al. [6] on Fe-coated Si substrate had catalyst particles at the nanowire tips. Based on high-field TEM contrast, these particles consisted of a dark core and a relatively light shell. They had SiO_z shell formed at the outer peripheral region of the nanoparticle. The EDS micrograph in the scanning TEM mode identified the catalyst nanoparticle to compose of Fe and Si with a small amount of O. The shell contained, on the other hand, a significant amount of O. The catalyst nanoparticle was hexagonally structured with lattice parameters of $a = 0.405$ nm and $c = 0.509$ nm. The surface of this nanoparticle was obviously disturbed (disordered) due to the presence of oxygen; the R_L species of this surface was plausibly $R_L \equiv$(Fe, Si, O) solid solution or cluster. Due to high oxygen content, the R_L species of the shell was though $R_L \equiv$(Fe, Si, O) amorphous alloy. The Si source for Si nanowire growth was Si atoms diffused out of the silicon wafer through the $R_L \equiv$(Fe, Si, O) species solid solution (cluster) of the catalyst nanoparticle. No Fe was found in the nanowires. This means the primary role of this Fe was just to generate a solid solution for mediating nanowire growth. The nanowire growth took place at a temperature of 1000 °C, which is lower than the (Fe, Si) eutectic temperature of 1207 °C. The catalyst particle was consequently in the solid solution state. Nie et al. [6] observed that silicon nanowires grown at 1200 °C had thicker amorphous SiO_z shell than nanowires grown at 1100 °C. This was expected as growth at the higher growth temperature of 1200 °C accompanied higher oxygenation and/or higher oxidation rate than the growths at 1000 and 1100 °C.

Unlike Nie et al. [6], Wong et al. [15] grew silicon nanowires from silicon substrates coated with a thin layer of gold. These nanowires grew at the substrate temperature of 1100 °C and in the presence of nitrogen ambient. The nanowire growth could be achieved only at temperatures beyond 900 °C. Also, the nanowires had larger diameter (~200 nm) at temperatures higher than about 1100 °C. The $R_L \equiv$(Au, Si) species resulting from the diffusion of Si in Au could be eutectic at the eutectic temperature of 363 °C. It did not happen during increase in temperature during growth. And hence, the temperature had to be increased to 900 °C to achieve nanowire growth. Alloyed islands were obviously formed at the growth temperature. The nanoparticle size became larger at higher temperature. Thicker nanowires were consequently produced at higher temperature.

Poulose et al. [1] and Lee et al. [2] observed multiple nanowires emerging from one single nanoparticle, and that the nanowire diameter d_{NW} increased with the increase in temperature. These happened despite no apparent change in nanoparticle dimension with temperature. These could happen only if nanoparticle had multiple domains of R_L≡(MET, X) species, and each of these domains separated by grain boundary might have X mole fraction different from that of other domains. Interestingly, Wang et al. [5] observed such multiple domains generated in some of the nanoparticles. They called them "buds" and found nanowires emerging from each of these buds of a nanoparticle. Poulose et al. [1] fabricated nanowires at 1000 °C $\leq T \leq$ 1200 °C, which is way above $T_E = 363$ °C for the R_L≡(Au, Si) eutectic alloy. For these experiments, temperature was increased keeping the Au layer thickness fixed. As a result, number density of nanowires increased with increase in temperature. The reason for this is again simple: R_S≡X and possibly R_S≡Y diffusions in R_L≡(MET, X) alloy during the pro-nucleation stage of growth are slower at lower temperature, but faster at higher temperature. Also, new R_L≡(MET, X) alloys were created on nanoparticle surface at higher temperature yielding new nanowires. All the R_L≡(MET, X) alloys did not participate in nanowire growth at lower temperature. Some of these alloys remained inactive at lower temperature. They became active at elevated temperatures. The nanoparticle size might also have increased at elevated temperatures due to Ostwald effect. And more R_S≡X and R_S≡Y species diffused into R_L≡(MET, X) alloys at these temperatures. So they yielded higher number density of nanowires.

9.4 Illustrations of Oxide and Nitride Nanowires Growth by the SLS Mechanism

9.4.1 SiO_z Nanowire Growths

Many different oxide and nitride nanowires have been produced by the SLS mechanism. Wang et al. [5] grew SiO_z nanowires using Si source directly from the solid Si(100) wafer substrate. Using Ar and H_2 as the activation agents and Pt as catalyst, the furnace temperature was gradually increased from 800 to 1400 °C to insure that the reactions proceed under equilibrium condition. A gradual increase in temperature led first to the formation of Pt_3Si catalyst dot and next to the formation of an amorphous SiO_z oxide layer on the substrate surface. Oxygen for this oxide layer appeared to emerge from a small amount of oxygen unintentionally mixed with the commercially available Ar. Also, almost always growth chamber contains contaminants such as oxygen. This oxygen had partial pressure high enough to oxidize Si from Si substrate. As temperature was increased further, Si and oxygen from this oxide layer, or the SiO_z molecules themselves began to diffuse onto the catalyst dots. Upon supersaturation at an appropriate temperature (~1100 °C), they yielded SiO_z nanowires.

Fig. 9.5 SEM image of In_2O_3 nanowires, 10–50 nm in diameter and several hundred nanometers to micrometers in length, grown by the SLS mechanism after annealing at 550 °C. It was reproduced from a report by Chen et al. [16] with copyright permission from the publisher

9.4.2 In_2O_3 Nanowire Growths

In_2O_3 has a direct band gap of around 3.6 eV. It is promising for many different kinds of applications, such as window heaters, solar cells, field-effect transistors, field-emission devices, and sensors. Chen et al. [16] used solid (Cu, In) compound in N_2 atmosphere to synthesize In_2O_3 nanowires (see Fig. 9.5), 10–50 nm in diameter and several hundred micrometer in length. The synthesis was carried out at a relatively low temperature of 550 °C. Cu acted as the catalyst and In as the solid source for the growth. Note that Cu and Ni are adjacent to one another in the periodic system of elements. They are completely miscible in both the liquid state and solid state. They form alloys over the whole concentration range and in a face-centered cubic lattice. $R_L \equiv$(Cu, In) liquid droplets might have been initially formed in part of (Cu, In) compound. Liquid/Solid interface between solid substrate and liquid $R_L \equiv$(Cu, In) compound might have then been formed. Nanowire nucleation took place at this interface.

9.4.3 Si_3N_4 Nanowire Growths

Liu et al. [17] grew α-Si_3N_4 nanowires, 80–150 nm in diameter, on clean Si substrate at 1150 °C without using flowing gases (N_2, CH_4, Ar, NH_3, etc.). The clean Si substrate was dipped into 0.01 m nickel nitrate $Ni(NO_3)_2$ aqueous solution and dried at room temperature. The solidified $Ni(NO_3)_2$ served as the source of N atoms for α-Si_3N_4 nanowires. These nanowires were sheathed with amorphous silicon oxide. They had uniform diameter. The EDS micrograph recorded from the bottom of nanowire showed marks of dotted rings. This means catalytic bases existed at the bottom of nanowires. EDS analysis suggested that these catalytic bases were composed mainly of Si, Ni, and N elements implying that the R_L species catalyzing the growth was R_L≡(Si, Ni, N) solid solution. Liu et al. [17] concluded that the growth of these nanowires was carried out by the SLS mechanism. For this growth, Si atoms diffused out of the Si substrate, but the N_2 vapor resulted from the solidified $Ni(NO_3)_2$. They both diffused into the R_L≡(Si, Ni, N) alloy cluster. While diffusing through this alloy cluster, they reacted together yielding Si_3N_4 molecules. Upon supersaturation, they generated Si_3N_4 nanowires. Lee et al. [18] made some important observations on the nucleation and growth of Si nanowires by the SLS mechanism. They noted the formation of islands composed of R_L≡(Au, Si). These islands had disordered lattice and thermally etched pyramidal pits formed during high-temperature annealing. They noted that the nucleation of nanowires was uniquely facilitated by these pits. Obviously, these pits were integral elements of the R_L species catalyzing the growth with annealing time as an agent to dictate the nanowire diameter. This is evident [19] from Fig. 9.6, which suggests that nanowire diameter increases almost linearly with increase in the duration of thermal annealing. Like Yu et al. [3] and Xing et al. [7], Tan et al. [20] produced amorphous Si nanowires. The nanowires by all of them were produced at 950 °C using MET≡Ni. An interesting feature of Tan's nanowires is that they had growth rates (~30 nm/s) comparable to those by the VLS mechanism.

Yu and Cabarrocas [4] investigated the SLS growth and dynamic behavior of Si nanowires by indium droplets. The growth at 500 °C transformed the hydrogenated amorphous silicon (a-Si:H) into crystalline Si nanowires. The In catalyst was deposited directly on the top of indium tin oxide (ITO)-coated glass substrates. Melting point of In is 156.6 °C. So, In solid was transformed into In droplet at the growth temperature of 500 °C. With Si adsorbed onto In at $T = T_{sla} = 500$ °C, porous R_L≡(In, Si) cluster (solid solution), rather than R_L≡(In, Si) eutectic alloy, was

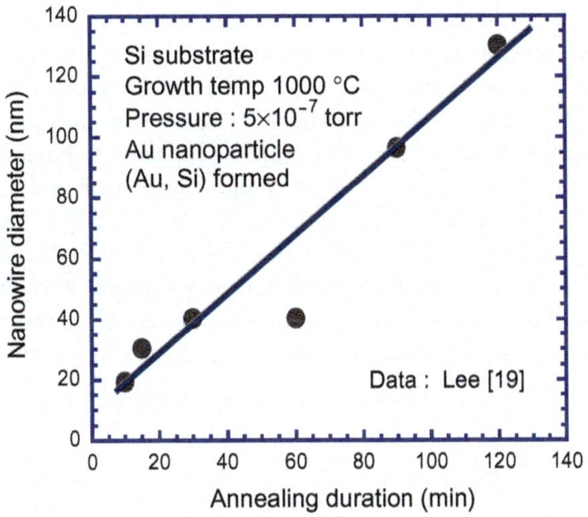

Fig. 9.6 Dependence of Si nanowire diameters on annealing duration. The plot was made employing experimental data by Lee [19]

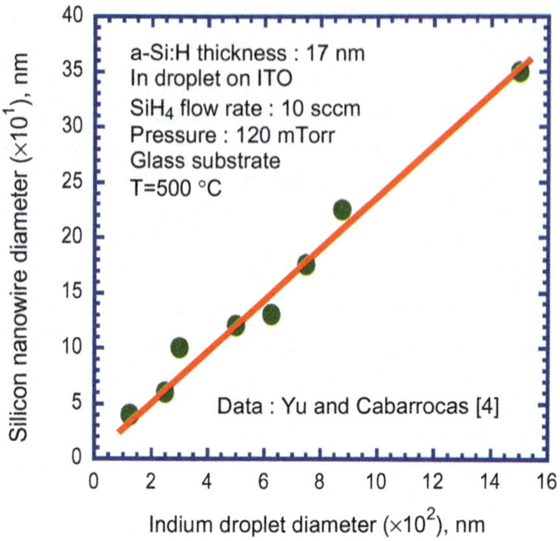

Fig. 9.7 Variation of the silicon nanowire diameter with the diameter of the (In, Si) cluster responsible for the SLS nanowire growths. The figure was produced with experimental data by Yu and Cabarrocas [4]

formed on the substrate surface. And this cluster served as nanoparticle to mediate Si nanowire growth. The variation of Si nanowire diameter with the variation of the diameter of R_L≡(In, Si) cluster, as obtained by Yu and Cabarrocas [4], is shown in Fig. 9.7. One can note that the diameter of the cluster is much larger than that of the Si nanowire grown on it implying that only a segment of this cluster was porous and the diffusion of the R_S≡Si species through this segment led to the nanowire growth. Increase in Si nanowire diameter with increase in the diameter of (In, Si) cluster was linear because the size of the said porous segment increased linearly with increase in the size of the R_L≡(In, Si) cluster. The thickness of the a-Si:H layer and the growth temperature (450–500 °C) appeared to have marginal impact on the Si nanowire diameter. The variation of nanowire growth rate with the nanowire diameter, as obtained by Yu and Cabarrocas [4], is shown in Fig. 9.8. This variation indicates that the growth rate decreases almost linearly with increase in nanowire diameter. And, it happened because the number density and/or the dimension of the nanopores did not proportionately increase with increase in nanowire diameter. The growth rate, as a result, decreased with increase in nanowire diameter.

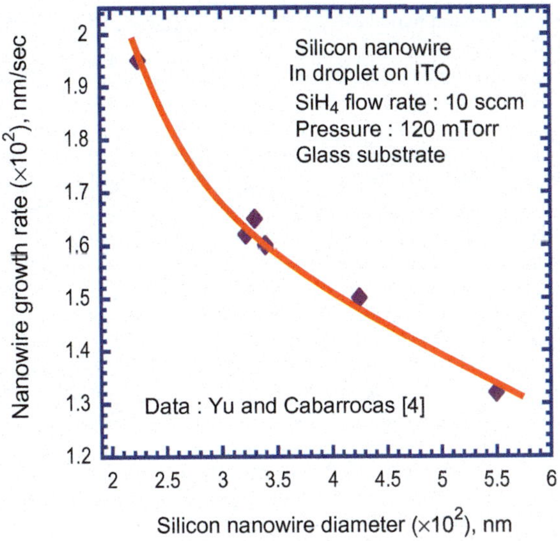

Fig. 9.8 Variation of silicon nanowire growth rate with the diameter of the silicon nanowire produced by the SLS mechanism. The figure was produced with the experimental data by Yu and Cabarrocas [4]

References

1. M. Poulose, O.K. Varghese, C.A. Grimes, Synthesis of gold-silica composite nanowires through solid-liquid-solid phase growth. J. Nanosci. Nanotechnol. **3**, 341–346 (2003)
2. E.K. Lee, B.L. Choi, Y.D. Park, Y. Kuk, S.Y. Kwon, H.J. Kim, Device fabrication with solid-liquid-solid grown silicon nanowires. Nanotechnology **19**, 185701 (2008)
3. D.P. Yu, Y.J. Xing, Q.L. Hang, H.F. Yan, J. Xu, Z.H. Xi, S.Q. Feng, Controlled growth of oriented amorphous silicon nanowires via a solid-liquid-solid (SLS) mechanism. Phys. E **9**, 305–309 (2001)
4. L. Yu, P.R. Cabarrocas, Initial nucleation and growth of in-plane solid-liquid-solid silicon nanowires catalyzed by indium. Phys. Rev. B **80**, 085313 (2009)
5. C.-Y. Wang, L.-H. Chan, D.-Q. Xiao, T.-C. Lin, H.C. Shih, Mechanism of solid-liquid-solid on the silicon oxide nanowire growth. J. Vac. Sci. Technol. B **24**, 613 (2006)
6. T.-X. Nie, Z.-G. Chen, Y.-Q. Wu, J.-L. Wang, J.-Z. Zhang, Y.-L. Fan, X.-J. Yang, Z.-M. Jiang, J. Zou, Metallic and ionic Fe induced growth of Si–SiO_x core-shell nanowires. J. Phys. Chem. C **114**, 15370–15376 (2010)
7. Y.-J. Xing, Z.-H. Xi, D.P. Yu, Q.-L. Hang, H.-F. Yan, S.-Q. Feng, Z.-Q. Xue, Growth of silicon nanowires by heating Si substrate. Chin. Phys. Lett. **19**, 240 (2002)
8. K. Watanabe, N. Koguchi, Y. Gotoh, Fabrication of GaAs quantum dots by modified droplet epitaxy. Jpn. J. Appl. Phys. Lett. **39**, L79–L81 (2000)
9. S. Huang, Z. Niu, Z. Fang, H. Ni, Z. Gong, J. Xia, Complex quantum ring structures formed by droplet epitaxy. Appl. Phys. Lett. **89**, 031921 (2006)
10. C.Z. Tong, S.F. Yoon, Investigation of the fabrication mechanism of self-assembled GaAs quantum rings grown by droplet epitaxy. Nanotechnology **19**, 365604 (2008)
11. M.A. Carignano, A. Mohammad, S. Kais, Temperature dependent electron binding in $(H_2O)_8$. J. Phys. Chem. A **113**, 10886–10890 (2009)
12. M. Yazawa, M. Koguchi, A. Muto, M. Ozawa, K. Hiruma, Effect of one monolayer of surface gold atoms on the epitaxial growth of InAs nanowhiskers. Appl. Phys. Lett. **61**, 2051 (1992)

13. S.N. Mohammad, General hypothesis for nanowire synthesis. II: Universality. J. Appl. Phys. **110**, 054312 (2011)
14. T.B. Massalski (ed.), *Binary Alloy Phase Diagrams*, 2nd edn., vol. 3 (American Society of Metals, Metals Park, OH, 1986)
15. Y.Y. Wong, M. Yahaya, M.M. Salleh, B.Y. Majlis, Controlled growth of silicon nanowires synthesized via solid–liquid–solid mechanism. Sci. Technol. Adv. Mater. **6**, 330–334 (2005)
16. G. Chen, L. Wang, X. Sheng, H. Liu, X. Pi, Y. Zhang, D. Li, D. Yang, Growth of In_2O_3 nanowires catalyzed by Cu via a solid–liquid–solid mechanism. Nanoscale Res. Lett. **5**, 898–903 (2010)
17. H. Liu, Z. Huang, J. Huang, M. Fang, Y.-G. Liu, X. Wu, X. Hu, S. Zhang, Novel, low-cost solid-liquid-solid process for the synthesis of α-Si_3N_4 nanowires at lower temperatures and their luminescence properties. Sci. Rep. **5**, 17250 (2015)
18. J.H. Lee, M.A. Carpenter, R.E. Geer, Novel growth mode of solid–liquid–solid (SLS) silica nanowires. J. Mater. Res. **26**, 2232–2239 (2011)
19. J.H. Lee, Templated Si-based nanowires via solid-liquid-solid (SLS) and vapor-liquid-solid (VLS) growth, in *Novel Growth Mode, Synthesis, Morphology Control, Characteristics, and Electrical Transport* (InTech Europe University Campus STeP Ri Slavka Krautzeka 83/A, 51000, Rijeka, Croatia)
20. H.F. Yan, Y.J. Xing, Q.L. Hang, D.P. Yu, Y.P. Wang, J. Xu, Z.H. Xi, S.Q. Feng, Growth of amorphous silicon nanowires via a solid–liquid–solid mechanism. Chem. Phys. Lett. **323**, 224–228 (2000)

Chapter 10
Oxide-Assisted Growth Mechanism

Abstract The oxide-assisted growth (OAG) mechanism for the growths of one-dimensional and quasi-two-dimensional nanomaterials has been described. The basics and the experimental base of the OAG mechanism have been presented. The general hypothesis of the growths by the OAG mechanism has been examined. High-temperature reaction, formation of clusters, and their impacts in the growths by the OAG mechanism have been investigated. Characteristic features of it have been elaborated. These features suggest that it is primarily employed for nanowire growths. Nanowire growths by it have therefore been illustrated and reviewed. The roles of metals and the roles of sulfur in the growths by the OAG mechanism have been accounted for. The formation of oxide sheaths in nanowires by the OAG mechanism has been demonstrated. Strengths and weaknesses of it in mediating nanowire growths have been stated. The importance and feasibility of it have been documented.

10.1 Introduction

10.1.1 Basics

The synthesis mechanism, in which oxides, rather than metals play critical role in inducing the nucleation and growth of X_mY_n nanomaterials, is the oxide-assisted growth (OAG) mechanism. The mechanism was found to produce large quantities of silicon nanowires having preferential growth direction, uniform size, and large growth rate even without the use of a FECA metal. It was discovered [1–3] during the VLS growth of Si nanowires using pure Si powder target. Si is easily oxidized in air, and hence, traces of Si oxide can be present in pure Si target. It was hence assumed that Si oxide was, at least partly, responsible for Si nanowire growth. And it appeared justified when Si nanowire growth was found to be greatly enhanced by thermal evaporation of SiO_2-containing Si powder target. This target was obviously solid source material. Nanowires were produced by using SiO_2 together with Si powder, and these nanowires were pure silicon in composition. On the other hand, nanowires were not produced by using only Si powder. This means Si powder itself lacked

S. N. Mohammad, *Synthesis of Nanomaterials*, Springer Series in Materials Science 307, https://doi.org/10.1007/978-3-030-57585-4_10

a self-catalytic effect and that Si-oxide powder served as assistant in Si nanowire growths. The temperature gradient and the vacuum condition were probably the two important parameters influencing the formation of nanowires. It was done by means of evaporation in vacuum (or in the presence of some inert gas) and at elevated temperatures. Note that solid source materials may not sublime in normal atmosphere; they are not transformed into vapor unless under high vacuum conditions. A solid source material is directly vaporized in normal atmosphere only if it reacts with some other suitable material(s) in this atmosphere. For example, ZnO powder does not sublime in normal atmosphere even at 1000 °C unless it reacts with carbon powder.

10.1.2 High-Temperature Reaction, Formation of Clusters, and Their Impacts

The yield from a target composed of 10 wt% Si and 90 wt% SiO_2 was higher than that from pure Si target [1]. There was yield even from a target composed of 1 wt% Si and 99 wt% SiO_2. On the other hand, there was no yield from a pure SiO_2 target. These observations led to the following presumptions: (1) Si and SiO_2 reacted at high temperature yielding highly reactive silicon oxide species (e.g., SiO_z, $1 \leq z < 2$); (2) several different species of Si oxides were generated during growth at high temperature; and (3) oxygen atoms were desorbed from Si oxides at high temperature. With these assumptions as centerpiece, the influences of oxygen and oxides on the growth of various nanowires were systematically studied [1–4]. These nanowires were formed in the low-temperature zone of the growth chamber due to deposition of the source materials directly from vapor and importantly in the absence of a metal catalyst. The source materials were placed in the high-temperature zone of the growth chamber. Recall that Si does not have self-catalytic effect implying that Si nanowire growths were mediated by Si-oxide. The experiments thus highlighted one important observation: oxides do play a crucial role in nanowire growths, and that these growths are probably the vapor–solid growths without the formation of an intermediate metal liquidic state. There may be a cluster of oxides, such as Si_nO_m (n, m = 1–8) for Si nanowires growths. These oxides [5–12] may both be high-temperature oxides (HXOs) and low-temperature oxides (LXOs), as shown in Table 10.1. They may both be needed for oxide-assisted nanowire growths [5–12]. High-temperature oxides are the ones stable primarily at high temperatures. Low-temperature oxides are, on the other hand, the ones stable primarily at low temperatures. It was suggested [1–4] that Si_nO_m clusters have affinity for binding to Si vapor ($R_S{\equiv}X = Si$) species and hence $R_S{\equiv}X = Si$ species adhere to the cluster facilitating nanowire nucleation. Such adherence serves as driving force for nanowire growth. And this was probably the basis of the oxide-assisted growth mechanism (see Fig. 10.1) for nanowire growth. Characteristic features of growths of some nanowires [5–15] are listed in Table 10.2. These features include HXOs (see column 7, Table 10.2) and the reactions for

Table 10.1 $R_S{\equiv}X$, $R_S{\equiv}Y$, low-temperature oxides (LXOs) and high-temperature oxides (HXOs) for the oxide-assisted growths of some representative nanowires

Notation	Nanowire (NW)							
X_mY_n NW	GaP	GaN	Ge_3N_4	GaAs	InP	Si	Ge	InAs
NW diameter, nm	22	20–120	20–200	10–120	200	–	–	–
$R_S{\equiv}X$	Ga	Ga	Ge	Ga	In	Si	Ge	In
$R_S{\equiv}Y$	P	N	N	As	P	–	–	As
LXO	Ga_2O_3	Ga_2O_3	GeO_2	Ga_2O_3	In_2O_3	SiO_2	GeO_2	In_2O_3
HXO	Ga_2O	Ga_2O	GeO	Ga_2O	In_2O	SiO	SiO	In_2O
References	Shi et al. [5]	Shi et al. [6]	Xie et al. [7]	Shi et al. [8]	Tang et al. [9]	Zhang et al. [10]	Zhang et al. [11]	Park et al. [12]

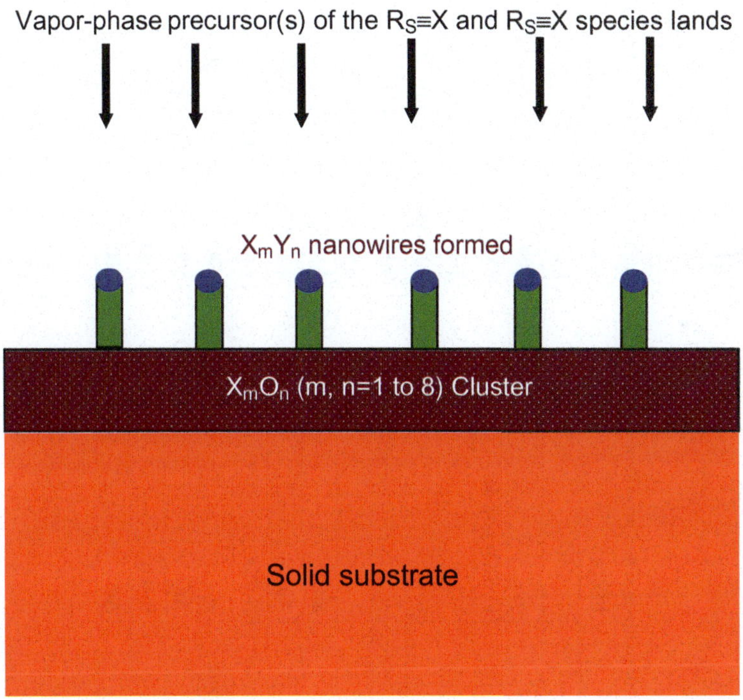

Fig. 10.1 Schematic diagram showing the landing of the precursor(s) of the $R_S{\equiv}X$ and $R_S{\equiv}Y$ source species on the oxide cluster, and the binding of the $R_S{\equiv}X$ and $R_S{\equiv}$ source vapor species to the oxide cluster leading to nanowire growth. The oxide cluster is formed on a solid substrate (or support formed on the substrate)

Table 10.2 Characteristics of oxide-assisted growth of nanowires (NWs) at a temperature T; TMI is trimethyl indium

NW	Substrate	FECA	Tip	T (°C)	Source	Reactive oxide (HXO)	References
GaP	Si	–	Ga_xO	690	GaN, powder Ga_2O_3	Ga_2O	[5]
GaN	Si	–	–	800	GaN, powder Ga_2O_3	Ga_2O	[6]
Ge_3N_4	Al_2O_3	C, SiO_2 powder	–	650	NH_3, Ge powder		[7]
GaAs	Si	–	Ga_xO	730	GaAs, powder Ga_2O_3	Ga_2O	[8]
InP	Al_2O_3	–	In, P, O	400–500	InP, In_2O_3	In_2O	[9]
Si	–	–	Si, O	900	Si, SiO_2	SiO	[10]
Ge	–	–		630	Ge, GeO_2	GeO	[11]
InAs	InAs	SiO	–	520–620	TMI, AsH_3	In_2O	[12]
InAs	InAs	Au	Au, In	400–505	TMI, AsH_3	–	[12]
InAs	InAs	Au/SiO_x	–	480–580	TMI, AsH_3	In_2O	[12]
GaN	Al_2O_3 membrane	–	–	800	Ga_2O, NH_3	Ga_2O	[13]
InN	Si	Au	Au, In_2O_3	500	In, NH_3	–	[14]
InN	Si	In_2O_3	–	500	In, NH_3	In_2O	[14]
Si	Si	Au	Au	900	SiO	SiO	[15]

Table 10.3 Chemical reactions leading to the formation of peripheral sheath during oxide-assisted growth of nanowires (NWs) at a temperature T; TMI is trimethyl indium

NW	Reaction during oxide-assisted growth	Oxide sheath	References
GaP	$3Ga_2O + 4P \rightarrow 4GaP + Ga_2O_3$	Ga_2O_3	[5]
GaN	$Ga_2O_3 + 4Ga \rightarrow 3Ga_2O$ $3Ga_2O + 4N \rightarrow 4GaN + Ga_2O_3$	Ga_2O_3	[6]
Ge_3N_4	$SiO_2 + Ge \rightarrow SiO + GeO$ $3GeO + 4NH_3 \rightarrow Ge_3N_4 + 3H_2O + 3H_2$	–	[7]
GaAs	$3Ga_2O + 4As \rightarrow 4GaAs + Ga_2O_3$	Ga_2O_3	[8]
InP	$3In_2O + 4P \rightarrow 4InP + In_2O_3$	In_2O_3	[9]
Si	$2SiO \rightarrow Si + SiO_2$ $2SiO_2 \rightarrow 2SiO + O_2$	SiO_2	[10]
Ge	$2GeO \rightarrow Ge + GeO_2$	GeO_2	[11]
InAs	$SiO + 2InAs \rightarrow Si + 2As + In_2O$ $3In_2O + 4As \rightarrow 4InAs + In_2O_3$	In_2O_3	[12]
InAs	–	–	[12]
InAs	$3In_2O + 4As \rightarrow 4InAs + In_2O_3$	In_2O_3	[12]
GaN	$Ga_2O + 2NH_3 \rightarrow 2GaN + H_2O + 2H_2$	–	[13]
InN	$2In + 2NH_3 \rightarrow 2InN + 3H_2$ $4In + 3O_2 \rightarrow 2In_2O_3$	In_2O_3	[14]
InN	$2In + 2NH_3 \rightarrow 2InN + 3H_2$ $4In + 3O_2 \rightarrow 2In_2O_3$	In_2O_3	[14]
Si	$2SiO \rightarrow Si + SiO_2$	SiO_2	[15]

the decomposition of HXOs (see column 2, Table 10.3) on the substrate surface. This decomposition was followed by precipitation, which we believe resulted in the formation of nanoparticles. We also believe that the growth of nanowires (e.g., Si, Ge, GaAs, GaN, GaP, InN, and InP nanowires, etc.) took place on these nanoparticles. The nanowires thus produced were covered with a shell of LXO [1]. We cite an example. Ge nanowires by Zhang et al. [2] were found to be covered with Ge-oxide sheath, as shown in Fig. 10.2.

10.1.3 Experimental Demonstrations Of The OAG Mechanism

A simple experiment [1, 16] might have shed light on the OAG mechanism. In this experiment, pure SiO powder was sealed. Otherwise, Si and SiO_2 were mixed (1:1) yielding SiO_z ($1 \leq z < 2$) vapor in an evacuated (vacuum < 10 Torr) quartz tube. This was followed by placing the quartz tube inside a furnace preheated to 1250–1300 °C. There was no ambient gas introduced into the chamber. Also, one end of the quartz tube was kept outside the furnace. It was to ensure that a temperature gradient

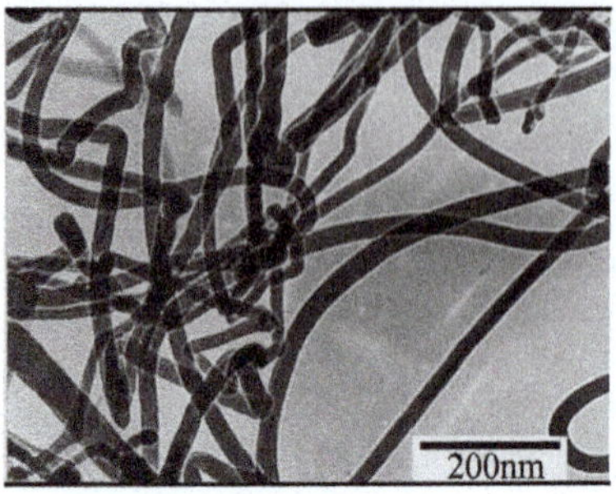

Fig. 10.2 Ge nanowires grown by the oxide-assisted growth mechanism. Reproduced from Zhang et al. [11] with copyright permission from the publisher

between the high-temperature location of source material and the low-temperature location of the nanowire growth is duly created. Following a thermal annealing for 20–30 min, a high yield of sponge-like Si nanowires was observed to grow in the low-temperature zone of the quart tube. The temperature in this zone was about 800–1000 °C.

In an analogous experiment by Zhang et al. [11], high-purity Si nanowires were grown in bulk quantity from SiO in a high-temperature tube furnace. Thermal sublimation of SiO powders yielded SiO vapors. When transported to low-temperature zone, they were deposited on the inner wall of the tube at about 930 °C. Interestingly, as the vapor pressure was gradually increased, nanowire yield was increased, but at the cost of SiO. As shown in Fig. 10.3, nanowire yield was the lowest when the

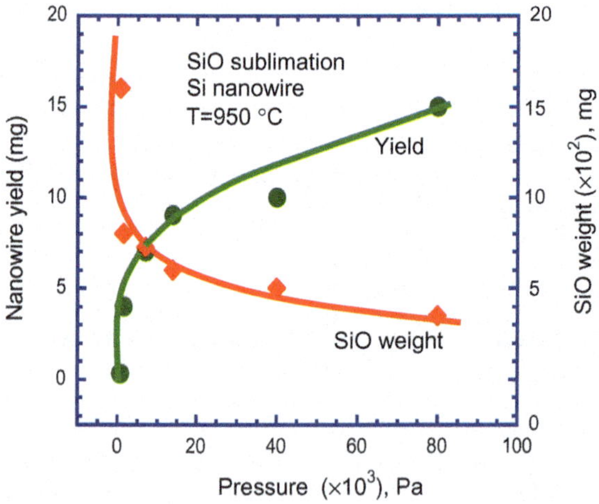

Fig. 10.3 Pressure dependence of the yield of Si nanowires and the weight loss of SiO during Si nanowire growth. The sublimation temperature was kept at 1300 °C during this growth. Plots are made with experimental data by Zhang et al. [2]

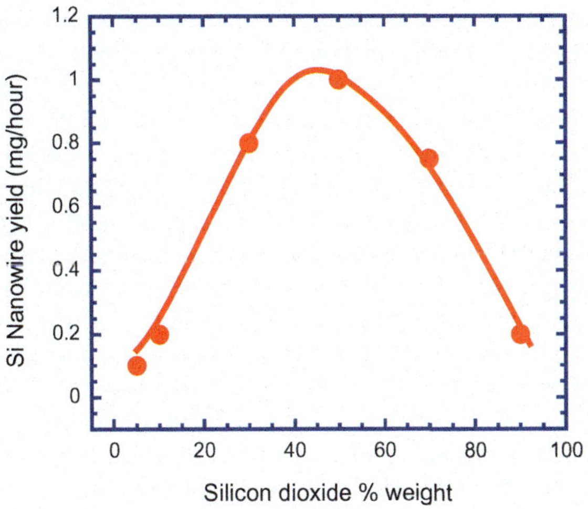

Fig. 10.4 Variation of Si nanowire yield with the percentage of SiO_2 in the target for silicon nanowire growth. The plots were made with experimental data reported by Lee et al. [1]

SiO weight was the highest. But the nanowire yield was the highest, when the SiO weight was the lowest. This suggests that Si nanowire grew at the expense of SiO. The dependence of Si nanowire yield on SiO_2 in the target during oxide-assisted growth [1] is shown in Fig. 10.4. One can see that nanowire yield gradually increased with increase in SiO_2 in the target until this yield reached its highest limit. Beyond this limit, the nanowire yield gradually decreased with increase in SiO_2 in the target. This happened probably because the excessive SiO_2 on the substrate surface contaminated the substrate surface creating an amorphous layer on the substrate surface and hence reducing the nanowire nucleation.

10.2 Advantages and Disadvantages of the OAG Mechanism

10.2.1 Advantages

The OAG mechanism has some important advantages. First, the X_mY_n nanowires grown by this mechanism are free from contamination of metal catalyst. Second, nanowire doping by this mechanism can be easily achieved as the experimental setup for OAG is very similar to that for the laser ablation growths. Recall that the laser ablation is suitable for depositing solid dopant materials during growth. Third, nanowires, grown by the OAG mechanism, possess uniform diameter, on the order of about 20 nm, even when the nanowire length is several hundred micrometers. These nanowires can even be as thin as ~1 nm in diameter [17, 18], which is hard to achieve by the VLS mechanism. Fourth, the growth direction of Si nanowires is primarily ⟨112⟩ and ⟨110⟩. This growth direction of Si nanowires by the VLS mechanism is however ⟨111⟩. Wu et al. [19] attributed this growth direction to the formation of lowest energy liquid/solid (111) interface. They argued that the surface energetics drive the faceting of the growth. The surface energetics are more dominant for the OAG process than for the VLS process It is hence reasonable that the most stable small-diameter Si nanowires by the OAG mechanism had the ⟨112⟩ growth direction.

While studying nanowire tapering, Hawley et al. [20] noted that the control and prevention of radial growth of Ge nanowires could be achieved by combining the OAG mechanism with the traditional CVD technique. Nanowires produced by the OAG mechanism were wrapped in an amorphous oxide shell, which prevented them from radial growth and tapering. As a result, the oxide-assisted mechanism yielded nanowires with a constant, uniform diameter. According to Zhang et al. [3], the OAG mechanism can produce several different morphologies such as chains, rods, wires, ribbons, and coaxial structures. However, oxide-assisted growth of carbon nanotubes and graphene has been quite rare. We came across just one report on oxide-assisted growth of carbon nanotubes [21] and another report on the oxide-assisted growth of graphene [22]. Importantly, highly poisonous precursor gases such as silane (SiH_4) or tetrachlorosilane ($SiCl_4$) can be avoided for example, for Si nanowire growths by the OAG mechanism.

10.2.2 Disadvantages

The main weaknesses of the OAG mechanism are the high reaction temperatures; for example, it is 850–1200 °C for Si nanowire growth. The formation of peripheral oxide shell (sheath) is also an artifact. This shell must be removed before any electrical or optical characterization of the nanowire is carried out. Additionally, X_mY_n nanowire growth, for example, on Si substrate would be difficult as the substrate during growth would rapidly oxidize. Pham et al. [23] fabricated Si nanowires via thermal evaporation. The Si nanowires thus produced had Si core coved with SiO_z shell. Photoluminescence study of these nanowires at room temperature showed intense emission centered at 650 nm. Three emission bands with peaks roughly at around 455, 510, and 690 nm however appeared at low temperature (10 K). While the red band was attributed to quantum confinement effect in the thin crystalline Si nanowire, the blue and green band resulted from thermal ionization of charged carriers trapped in defect centers located at the Si–SiO_z interface or in the shell. Intensities decreased rapidly as a result of thermal ionization of charged carriers trapped in defect centers located in the interface or shell region.

10.3 Analysis of the Observed Oxide-Assisted Growths

The descriptions of X_mY_n nanowire growth made above (see Sect. 10.1.1) provide no clear view of how the nanoparticles are created and how these nanoparticles serve as nucleation sites for nanowire growth. The nucleation of Si nanowires has however been observed to occur on the substrate due to the decomposition or precipitation of Si-rich oxide. Also, the nucleation of Si nanowire during the initial growth stage was found to follow the piling up of nanoparticles on the matrix surface, and it happened soon after the start of the decomposition of SiO. The oxide-assisted growth of Si nanowires was believed to be dictated by a number of factors:

1. Defect, disorder, and amorphicity;
2. The melting point depression at the nanowire tips;
3. SiO_2 shell formed around the silicon nanowire;
4. Suppression of the lateral (radial) growth of nanowires;
5. Electric charging of nanowire tips and its effect on growth; and
6. Reactivities of Si suboxides.

Experiments indicated that the nanowire facets had high density of stacking faults and microtwins in the Si nanowire core adjacent to the nanowire tip. These defects and microtwins were assumed to give rise to atomic steps and kinks that facilitated the deposition of Si atoms.

Zhang et al. [3] stated (see Fig. 7 by these authors) that metal-assisted X_mY_n nanowire growth via the VLS process at 1100–1200 °C was driven by supersaturation of Si in the metal-Si melt; but the X_mY_n nanowire growth by the OAG mechanism at 850–1050 °C was assisted by the semiliquid SiO_z. This contradicts the assertion by Kang et al. [24] that metal nanoparticle, if present during growth, remains solid at a temperature $T < T_E$. This contradicts also the in situ atomic-scale, video-rate environmental TEM, and in situ time-resolved XPS (X-ray photoelectron spectroscopy) observation by Hoffman et al. [25], which confirmed that nanoparticle remains solid during CNT growth. The melting point of SiO_2 is about 1600 °C. The melting point of SiO_z ($1 \leq z < 2$), which depends on the mole fraction z, may not be much different from this. Yet SiO_z was found [16] to be molten at 850–1050 °C. Size-dependent depression of melting temperature T_M (see Sect. 3.3.1 of [16]) could be one possible reason for this. However, our investigation [26] suggests that the size-dependent depression of the melting temperature T_M is essentially insignificant for nanoparticle size larger than about 10 nm. If the size of SiO_z nanoparticle is the same as the size of nanowire, which may be reasonable, then the SiO_z size as evident from entries of Table 10.1 was as large as 200 nm. It is beyond the limit of size-dependent melting point depression. Also, we could not find any justification [16] of why nanowires would be produced directly from vapor without supersaturation. How semiliquid SiO_z assisted the nanowire nucleation was not articulated by Zhang et al. [3].

10.4 Formation of the Oxide Sheath

Investigation of the OAG mechanism for the growth of X_mY_n nanowires carried out in our laboratory [27] may have revealed some basic physics and chemistry underlying this mechanism. This investigation suggests that R_L species (defined in Chap. 3, Sect. 3.3) is formed on the nanoparticle surface. One or the only component of this R_L species formed during the pre-nucleation stage of growth (see Chap. 4, Sect. 3) could be HXO. This R_L species could also be R_L≡(X, Y, O, HXO). While still a component of the R_L species of nanoparticle, this high-temperature oxide HXO could decompose yielding the low-temperature oxide LXO and releasing R_S≡X: HXO → X + LXO. It was due to phase transition during growth. It appears to be more probable than the dissociation HXO → X + O_2 (for example, $2In_2O \rightarrow 4In + O_2$ for InP nanowire growths). This may be confirmed based on the reaction $2SiO \rightarrow Si + SiO_2$, for example, for Si nanowire growths, as observed by Lee et al. [28]. Also, during the pro-nucleation stage of growth (see Chap. 4, Sect. 3), the subsequently released species: X, Y, O, and HXO diffuse through the R_L species; they are in close proximity to the nanoparticle surface. As a result, they undergo a series of reactions among themselves yielding, for example, Y + HXO → X_mY_n + LXO. This is evident from the reaction: $3Ga_2O + 4As \rightarrow 4GaAs + Ga_2O_3$ for GaAs nanowire growths [8]. There also occurs the reaction: $mX + nY \rightarrow X_mY_n$. Obviously, most, if not all, of them, take place on the surface of the R_L species or during diffusion through the R_L species. Two main products resulting from the reactions are X_mY_n and LXO. Of them the low-temperature oxide, viz., LXO (for example, SiO_2 for Si nanowire growth) cannot leave the surface of the R_L species; they are just bound by the surface tension of the R_L species. Further, they are wetted by the R_L species and hence no longer in pure vapor phase. So, X_mY_n and LXO remain in close proximity as they diffuse through the R_L species to the liquid/solid (L/S) interface for supersaturation and growth. X_mY_n remains surrounded with LXO sheath when nanomaterials are formed.

There is another reason for this to happen. The formation of the core and of the shell of core–shell heterostructured nanomaterials depend on the surface energy of the materials forming the core and the shell [29]. The surface energy of a heterostructure nanomaterial is higher at the surface than in the core. For the sake of stability, the material with lower surface energy therefore forms the shell and the material with the higher surface energy forms the core. Oxide of a metal or semiconductor generally has surface energy lower than the corresponding metal or semiconductor [29]. For example, the surface energy of Si is 1140 mJ/m^2. But the surface energy of SiO_2 is 320 mJ/m^2. In view of this, metal (semiconductor) nanowires grown by the OAG mechanism has oxide sheath made of the oxide of the same metal (semiconductor) as of the core.

10.5 Role of Metal in Oxide-Assisted Growth

Nanowire growth by the OAG mechanism is credited to be a metal-free growth mechanism. Nevertheless, Kolb et al. [30] grew a large quantity of crystalline silicon nanowires using the evaporation and disproportionation of silicon monoxide in the presence of MET≡Au. These nanowires were sheathed with an amorphous oxide shell. The SiO and the gold, both on silicon substrate, were heated from room temperature to about 1125 °C and 950 °C, respectively, for growth. It was believed that the gold dots on silicon substrate served as catalytic agents for initiating the nanowire growth. We believe that, during the pre-nucleation stage of growth, the Au dots (nanoparticles) became molten yielding R_L≡(Au, Si) alloy dots at the growth temperature. During the pro-nucleation stage of growth, some of the remaining SiO and Si were vaporized. They underwent diffusion through the R_L species. During this diffusion, SiO underwent phase separation into Si and SiO_2 on the surface or inside of molten R_L≡(Au, Si) alloy droplets. While Si formed the core, SiO_2 formed the sheath of nanowires thus grown. In order for the R_L≡(Au, Si) alloy to be molten at the growth temperature of 950 °C, the Si content in this alloy should be 46 at.%. And this would happen only under Si-rich condition. The growth was probably conducted under such a Si-rich condition that the rapid diffusion of only high-density Si through the R_L≡(Au, Si) molten alloy could lead to the growth of very long and very thick Si nanowires. Figure 10.5 suggests that, under the prevailing growth conditions, the

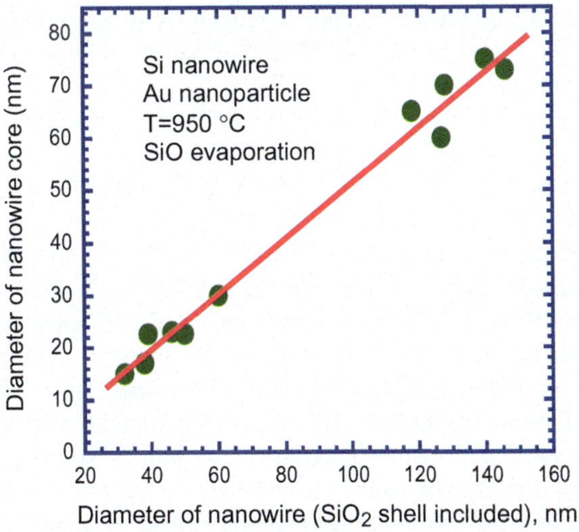

Fig. 10.5 Dependence of the silicon nanowire core on the nanowire diameter (e.g., the diameter of the nanowire covered with SiO_2 shell). The linear relation between the two suggests that the growth of the silicon core and the oxide shell occurred through the same process. The figure is drawn with experimental data by Kolb et al. [30]

diameter of nanowire core increased almost linearly with increase in the sheathed nanowire diameter (nanowire core and oxide shell included).

10.6 Role of Sulfur in Oxide-Assisted Growth

Our descriptions in the previous sections of this chapter may suggest that the mechanism of oxide-assisted growth is not yet very clear. Ishiyana et al. [31] studied the oxide-assisted growth of Si nanowires to shed some light on this mechanism. The well-aligned Si nanowires were thus grown on a (111)-oriented Si substrate by chemical vapor deposition. Just thermal treatment conducted at approximately 1200 °C was enough for growth. This thermal treatment was obviously done in the presence of oxygen. For this growth, no metal catalysts were employed. However, silicon sulfide was used as a Si source. The Si nanowires had needle-like shapes. They also had a core–shell structure with a single-crystalline Si core grown along the same direction as the Si substrate. The core was covered with a SiO_2 sheath. It was believed [31] that the silicon sulfide served as the liquid phase facilitating the nucleation and nanowire growth. If the temperature was increased to 1300 °C, oxygen-rich nanowires possessing crystalline Si at the tip and lumpy silicon oxide on the body were observed. This implies that the epitaxial growth of Si nanowires was disturbed. Energy dispersive X-ray (EDX) fluorescence analysis of the nanowire tip confirmed that the nanowires were silicon containing a small amount of oxygen. To be more specific, the elemental composition analysis suggested 80 at.% Si and 20 at.% oxygen at the nanowire tip, and 90 at.% Si and 10 at.% oxygen in the central portion of the nanowire.

Previously, sulfur and catalytic Au were used for the growth of Si nanowires by a VLS method [32]. For this growth by Kohno and Takeda [32], sulfur produced islands of Si sulfide by etching the Si substrate. Also, instead of SiH_4, silicon sulfide vapor was used as the vapor-phase source. The crystallinity of the nanowires was not though good and some of the nanowires had bends and kinks. Nevertheless, we argue that sulfur and oxygen created islands of SUBSANOs (see Chap. 3) on the substrate surface. These SUBSANOs had composition of R_L≡(Si, S, O) or R_L≡(Si, Au, S, O) species on the substrate surface. The R_L species might have some other contaminant(s) present in the growth chamber. A comparison of the growths by Ishiyama et al. [31] and by Kohno and Takeda [32] suggests that sulfur served the same purpose as sulfur and Au together did. Silicon sulfide (SiS_2) has a melting point of 1090 °C. Then, a growth by Ishiyama et al. [31] at 1200 °C means silicon sulfide or R_L≡(Si, S) solid solution was molten (semimolten) at the growth temperature and that this R_L≡(Si, S) played a critical role in the nucleation and growth of nanowires. So sulfur was found in the nanowires. This means sulfur primarily modified the substrate surface for growth, which was carried out by the same mechanism as the OAG.

The studies by Ishiyama et al. [31] may have the same spirit as that of Li et al. [33], who conducted thermal evaporation of Si/Fe, Si disk, or Si/SiO_2 powder mixture in quartz tube at 1200 °C under Ar atmosphere to produce Si nanowires. They argued that the growth by the OAG mechanism was not done just by the thermal evaporation of Si/SiO_2. To justify it, they examined commercial quartz tube used for the growth. They found that this quart tube was not made of 100% SiO_2, but contained certain amount of metallic elements, such as Al and Cu. This means those metals actually acted as catalysts for growth. At the least they modified the Si substrate surface yielding SUBSANOs of the $R_L \equiv$(Si, Al, Cu) species composition on the substrate surface.

References

1. S.T. Lee, N. Wang, Y.F. Zhang, Y.H. Tang, Oxide-assisted semiconductor nanowire growth. MRS Bull. 36–42 (1999)
2. Y.F. Zhang, Y.H. Tang, C. Lam, N. Wang, C.S. Lee, I. Bello, S.T. Lee, Bulk-quantity Si nanowires synthesized by SiO sublimation. J. Cryst. Growth **212**, 115–118 (2000)
3. R.Q. Zhang, Y. Lifshitz, S.-T. Lee, Oxide-assisted growth of semiconducting nanowires. Adv. Mater. **15**, 635–640 (2003)
4. R.Q. Zhang, *Growth Mechanisms and Novel Properties of Silicon Nanostructures from Quantum Mechanical Calculations* (Springer, Berlin, Heidelberg, 2014)
5. W.S. Shi, Y.F. Zheng, N. Wang, C.S. Lee, S.T. Lee, Synthesis and microstructure of gallium phosphide nanowires. J. Vac. Sci. Technol. B **19**, 1115–1118 (2001)
6. W.S. Shi, Y.F. Zheng, N. Wang, C.S. Lee, S.T. Lee, Microstructures of gallium nitride nanowires synthesized by oxide-assisted method. Chem. Phys. Lett. **345**, 377–380 (2001)
7. T. Xie, Z. Jiang, G. Wu, X. Fang, G. Li, L. Zhang, Characterization and growth mechanism of germanium nitride nanowires prepared by an oxide-assisted method. J. Cryst. Growth **283**, 286–290 (2005)
8. W.S. Shi, Y.F. Zheng, N. Wang, C.S. Lee, S.T. Lee, Oxide-assisted growth and optical characterization of gallium-arsenide nanowires. Appl. Phys. Lett. **78**, 3304–3306 (2001)
9. C. Tang, Y. Bando, Z. Liu, D. Goldberg, Synthesis and structure of InP nanowires and nanotubes. Chem. Phys. Lett. **376**, 676–678 (2003)
10. R.Q. Zhang, T.S. Chu, H.-F. Cheung, N. Wang, S.T. Lee, Mechanism of oxide-assisted nucleation and growth of silicon nanostructure. Mater. Sci. Eng. C **16**, 31–35 (2001)
11. Y.F. Zhang, Y.H. Tang, N. Wang, C.S. Lee, S.T. Lee, Germanium nanowires sheathed with an oxide layer. Phys. Rev. B **61**, 4518–4521 (2000)
12. H.D. Park, S.M. Prokes, M.E. Twigg, R.C. Cammarata, A.-C. Gaillot, Si-assisted growth of InAs nanowires. Appl. Phys. Lett. **89**, 223125 (2006)
13. G.S. Cheng, L.D. Zhang, Y. Zhu, G.T. Fei, L. Li, C.M. Mo, Y.Q. Mao, Large-scale synthesis of single crystalline gallium nitride nanowires. Appl. Phys. Lett. **75**, 2455–2457 (1999)
14. Z.H. Lan, W.M. Wang, C.L. Sun, S.C. Shi, C.W. Hsu, T.T. Chen, K.H. Chen, C.C. Chen, Y.F. Chen, L.C. Chen, Growth mechanism, structure, and IR photoluminescence studies of indium nitride nanorods. J. Cryst. Growth **269**, 87–94 (2004)
15. S.T. Lee, R.Q. Zhang, Y. Lifshitz, Oxide-assisted growth of silicon and related nanowires: growth mechanism, structure and properties, in *The Chemistry of Nanomaterials: Synthesis, Properties and Applications*, ed. by C.N.R. Rao, A. Muller, A.K. Cheetham (Wiley-VCH Verlag GmbH, Weinheim, Germany, 2004)
16. N. Wang, Y. Cai, R.Q. Zhang, Growth of nanowires. Mater. Sci. Eng. R **60**, 1–51 (2008)
17. D.D.D. Ma, C.S. Lee, F.C.K. Au, S.Y. Tong, S.T. Lee, Small-diameter silicon nanowire surfaces. Science **299**, 1874–1877 (2003)
18. C.P. Li, X.H. Sun, N.B. Wong, C.S. Lee, B.K. Teo, Ultrafine and uniform silicon nanowires grown with zeolites. Chem. Phys. Lett. **365**, 22–26 (2002)

19. Y. Wu, Y. Cui, L. Huynh, C.J. Barrelet, D.C. Bell, C.M. Lieber, Controlled growth and structures of molecular-scale silicon nanowires. Nano Lett. **4**, 433–436 (2004)
20. C.J. Hawley, T. McGuckin, J.E. Spanier, Selective epitaxial growth on germanium nanowires via hybrid oxide-stabilized vapor–liquid–solid growth. Cryst. Growth Des. **13**, 491–496 (2013)
21. Y.H. Tang, Y.F. Zheng, C.S. Lee, N. Wang, S.T. Lee, T.K. Sham, Carbon monoxide-assisted growth of carbon nanotubes. Chem. Phys. Lett. **342**, 259–264 (2001)
22. Y. Wang, Y. Cheng, Y. Wang, S. Zhang, X.-W. Zhang, S.-G. Yin, M. Wang, Y. Xia, Q.-Y. Li, P. Zhao, H.-G. Wang, Oxide-assisted growth of scalable single-crystalline graphene with seamlessly stitched millimeter-sized domains on commercial copper foils. RSC Adv. **8**, 8800–8804 (2018)
23. V.T. Pham, V.N. Le, A.T. Chu, T.T. Pham, N.K. Tran, H.D. Pham, T.H. Pham, Silicon nanowires prepared by thermal evaporation and their photoluminescence properties measured at low temperatures. Adv. Nat. Sci. Nanosci. Nanotechnol. **2**, 035004 (2011)
24. K. Kang, G.H. Gu, D.A. Kim, C.C. Park, M.H. Jo, Self-organized growth of Ge nanowires from Ni-Cu bulk alloys. Chem. Mater. **20**, 6577–6579 (2008)
25. S. Hoffman, R. Sharma, C. Ducati, G. Du, C. Mattevi, C. Cepek, M. Cantoro, S. Pisana, A. Parvez, F. Cervantes-Sodi, A.C. Ferrari, R. Dunin-Borkowski, S. Lizzit, L. Petaccia, A. Goldoni, J. Robertson, In-situ observations of catalyst dynamics during surface-bound carbon nanotube nucleation. Nano Lett. **7**, 602–608 (2007)
26. S.N. Mohammad, General hypothesis and shell model for the synthesis of semiconductor nanotubes, including carbon nanotubes. J. Appl. Phys. **108**, 064323 (2010)
27. S.N. Mohammad, Investigation of the oxide-assisted growth mechanism for nanowire growth and a model for this mechanism. J. Vac. Sci. Technol. B **26**, 1993 (2008)
28. S.T. Lee, N. Wang, C.S. Lee, Semiconductor nanowires: synthesis, structure and properties. Mater. Sci. Eng., A **286**, 16–23 (2000)
29. S. N. Mohammad, General hypothesis for nanowire synthesis. II: Universality, J. Appl. Phys. 110, 054312 (2011)
30. F.M. Kolb, H. Hofmeister, R. Scholz, M. Zacharias, U. Gosele, D.D. Ma, S.-T. Lee, Analysis of silicon nanowires grown by combining SiO evaporation with the VLS mechanism. J. Electrochem. Soc. **151**, G472–G475 (2004)
31. T. Ishiyama, S. Nakagawa, T. Wakamatsu, Growth of epitaxial silicon nanowires on a Si substrate by a metal-catalyst-free process. Sci. Rep. **6**, 30608 (2016)
32. H. Kohno, S. Takeda, Silicon nanoneedles grown by a simple thermal treatment using metal-sulfur catalysts. Jpn. J. Appl. Phys. **41**, 577–578 (2002)
33. F.J. Li, S. Zhang, J.-W. Lee, Rethinking of the silicon nanowire growth mechanism during thermal evaporation of Si-containing powders. Thin Solid Films **558**, 75–85 (2014)

Chapter 11
Self-catalytic Growth (SCG) Mechanism

Abstract The self-catalytic growth (SCG) mechanism for the growths of one-dimensional and quasi-two-dimensional nanomaterials has been reviewed. This is a metal-catalyst-free mechanism. The basics of it have been presented. Nanowire growths by this mechanism by various catalyst-free means have been investigated. Strengths and weaknesses of it in mediating nanowire growths have been stated. The ability of this mechanism to mediate both nanowire and nanotube growths has been highlighted. The ability of it also to produce highly pure defect-free nanowires and even ternary and quaternary nanowires by this mechanism has been commended. Understanding self-catalytic nanowire growth taking into consideration factors such as edges, tips, hillocks and roughness of the nanoparticle surface and also grains, and grain boundaries formed prior to growths has been elucidated. The differences between the selective area epitaxy and the self-catalytic growth have been described. The novelty of the SCG mechanism has been discerned. The role of oxide in the growths by the SCG mechanism has been examined. Finally, the superiority of the SCG mechanism to other mechanism has been elaborated.

11.1 Introduction

Needless to say, FECA metal elements and alloy nanoparticles, such as those from Au, Ni, or AuNi, have been successfully used [1] to grow nanomaterials. However, some nanomaterials, such as III–V nanowires produced on these metal nanoparticles (e.g., METANOs), do not possess thermal stability. To be more specific, InAs nanowires produced on them show sign of catalytic decomposition [2] at a temperature of ~500 °C. Ni nanoparticles react with semiconducting InAs nanowires at 220–300 °C yielding highly conducting $Ni_z(InAs)_{1-z}$ nanowires [3]. Metal diffusion into the semiconducting nanowire can change the nanowire properties. It can, for example, form deep-level traps in nanowires. Indeed, diffusion of Au in ZnSe [4], InAs [5], and Si [6] nanowires has been recorded and the presence of Au in Si nanowires has been found to affect the performance of Si nanowire-based photovoltaic solar cells [7]. These catastrophic effects of FECA metals on nanowires limit

S. N. Mohammad, *Synthesis of Nanomaterials*, Springer Series in Materials Science 307, https://doi.org/10.1007/978-3-030-57585-4_11

the processing temperature of these nanowires. Additionally, they compromise the compatibility of nanowire devices with the state-of-the-art current Si-based technologies. To overcome these problems, nanowires may preferably be produced in metal-free environments, for example, by the self-catalytic growth (SCG) mechanism. This mechanism is desirable, for example, for FECA-metal free growths of III−V nanowires, and for their applications and integrations into Si-based technologies. To describe the growths of them, we will make use of the general terminologies such as METANO and SUBSANO for nanoparticles, as defined in Chap. 3. We may also utilize the general terminology of the R_L species defined again in Chap. 3, Sect. 3.3 and the pre-nucleation and pro-nucleation stages of growths defined in Chap 4, Sect. 3).

11.2 Approaches to Obtain Met-Free Nanoparticle Seeds

During the past years, different approaches within the framework of the SCG mechanism had been tried to obtain MET-free nanowires. For the sake of convenience, these nanowires may, in general, be denoted by X_mY_n nanowire made from X and Y component elements. They may be produced on nanoparticles, such as those from

1. X_mY_n thin film converted subsequently to X_mY_n nanoparticles,
2. X thin film converted subsequently to X_mY_n nanoparticle, or
3. X-film converted subsequently to X nanoparticles,
4. X_mY_n thin film converted subsequently to X nanoparticles. It would happen due to desorption of Y from the X_mY_n thin film or from the X_mY_n nanoparticles formed from this film.

Discovered plausibly for the first time by He et al. [8–15], the SCG mechanism does not make use of MET catalyst for growth. Unlike the VLS [16, 17] mechanism (see Chap. 5) and the VSS [18] mechanism (see Chap. 7), it mediates the growth [8–15] without the assistance of MET catalyst (e.g., METANO). He et al. argued that the key element of the SCG mechanism is molten (semimolten) X nanoparticle surfaces intrinsically formed (see Fig. 11.1a) on a X_mY_n matrix (substrate) to promote the X_mY_n nanowire growth. The X_mY_n matrix may otherwise be formed on a non-X_mY_n substrate. X-rich islands of this matrix may be the X nanoparticle seeds, which are formed on X_mY_n matrix, for example, by thermal annealing of the X_mY_n matrix. X nanoparticle seeds may as well be formed by thermal annealing of a very thin X thin film layer (1–5 nm in thickness) deposited on the X_mY_n matrix. Thin X-film layer is fractured upon thermal annealing at a suitable temperature creating X nanoparticles. These nanoparticles have R_L species. They are composed of disturbed (disordered, amorphous, semi-amorphous) and coarsened lattice structure (see Fig. 11.1b). The amorphicity is though controlled, as described in the Appendix. In general, the smaller the size of the nanoparticles, the larger is the coarsening of the R_L species [19]. The R_L species exhibit pits, nanopores, and surface melting (semi-melting). To be more specific, the amorphous morphology of the nanoparticle

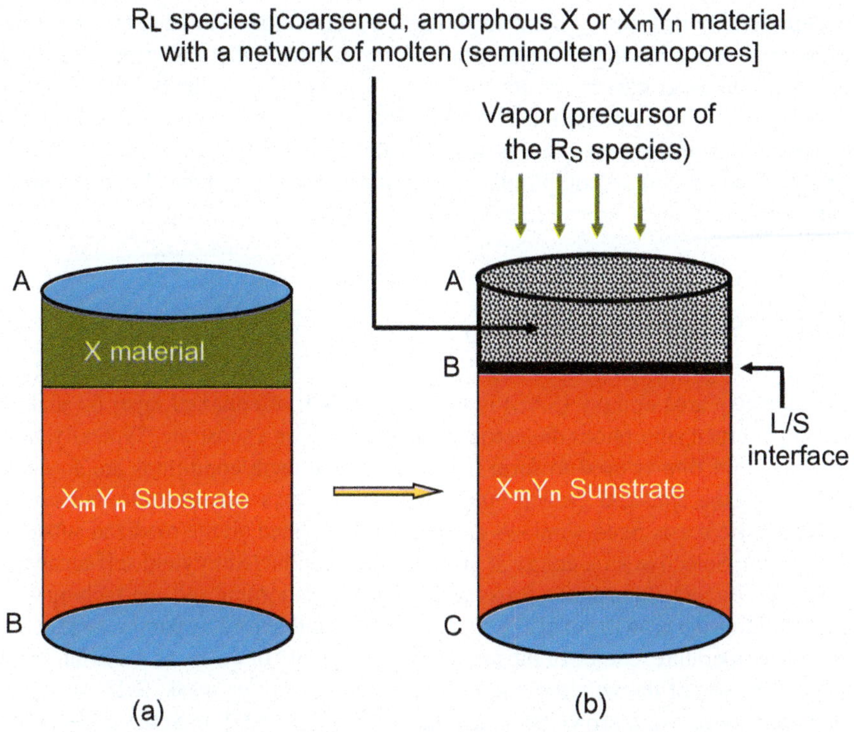

Fig. 11.1 **a** Solid crystalline X nanoparticle intrinsically formed on X_mY_n matrix (substrate) not suitable for self-catalytic growth; **b** solid X (X_mY_n) nanoparticle intrinsically formed on X_mY_n matrix (substrate), but coarsened, amorphous (semi-amorphous), and molten suitable for self-catalytic growth

is the result of surface disorder, surface coarsening, and the generation of pits and hillocks [8–15]. Nanoparticles possess large surface energy [20, 21]. The surface atoms of these nanoparticles under the influence of such surface energy, and also due to vibration and fluctuation of the surface, are relatively loose. The nanoparticle dimensions are also small. Hence, they may undergo melting (semi-melting) due to the size-dependent mesoscopic effects [22–26]. The X seeds act, as a result, as droplets or as solid solution (cluster) exhibiting nanopores. They may appear as solid and yet act as droplets. They produce nanowires of superior quality.

11.3 Examples of Met-Free Nanowire Growths

The $R_S \equiv X$ and $\equiv Y$ vapor species diffuse through the molten (semimolten) nanopores to the liquid/solid (L/S) interface. They are supersaturated at this interface yielding nanowires. Recall that the nanowires, and the matrix on which nanowires are formed, may preferably be from the same material. If so, the matrix/nanowire interface may experience marginal lattice mismatch and mismatch of thermal expansion coefficients. The SCG nanowires are thus free from interface defects. They are free also from metal contamination. The VLS mechanism is not very suitable for metal nanowire growths because the growing X metal nanowires would react with the FECA metal tending to compromise the pristine crystal structure of the nanowires thus grown.

11.3.1 Selective Area Epitaxy

The selective area epitaxy (SAE) is the local growth of epitaxial layer by making use of patterned amorphous dielectric mask deposited on a substrate. Typically, SiO_2 or Si_3N_4 thin film is used to form the mask. The growth conditions are so determined that the epitaxial growth takes place on the exposed substrate, but not on the dielectric mask. Seed windows are selectively opened on the oxidized substrate for growth. Selective area epitaxial growths were conducted to yield self-catalyzed nanowires on SiO_2/Si and SiO_2/InAs substrates. Pre-patterning and wet-etching were used to define the growth sites [27]. The growth was influenced by parameters such as growth temperature, choice of the mask material, quality of the mask material, shape and orientation of the window, and etchants used to etch the mask. Self-assembled GaAs nanowires were grown by molecular beam epitaxy [28] on Si facets generated by the cleavage of Si(100) substrates. But vertically aligned InP nanowires were grown on silicon (111) substrate by MOVPE using indium droplets [29]. An interesting feature of this growth by Gao et al. [29] is that the interfacial strain induced by the 8.1% lattice mismatch between InP and Si appeared to be fully accommodated by elastic deformation. Among others, mechanically exfoliated and CVD-grown graphene [30, 31] were employed to perform van der Waals epitaxial growth of InAs nanowires. Ledges, kinks, and defects were considered to be preferred nucleation sites for these growths. They all however go through complex additional pre-growth stages, such as pre-deposition and patterning of SiO_2 layer, and transfer of graphene, which are challenging for large-scale production.

11.3.2 CVD Growth of III–V Nitride Nanowires

Plausibly for the first time, He et al. [8–15] grew GaN nanowires and nanotubes (see Fig. 11.2a, b) in CVD chamber through direct reaction of gallium vapor with ammonia. Depending on temperature and ammonia flow rate, the diameters of the products ranged between 26 and 100 nm; the lengths of some of them were up to 500 μm. About 3 g of pure metal Ga ~99.999% pure was put in a small boron nitride boat and placed at the bottom of a liner made of boron nitride or quartz. Ammonia ~99.999% pure was introduced into the quartz tube through a mass-flow controller

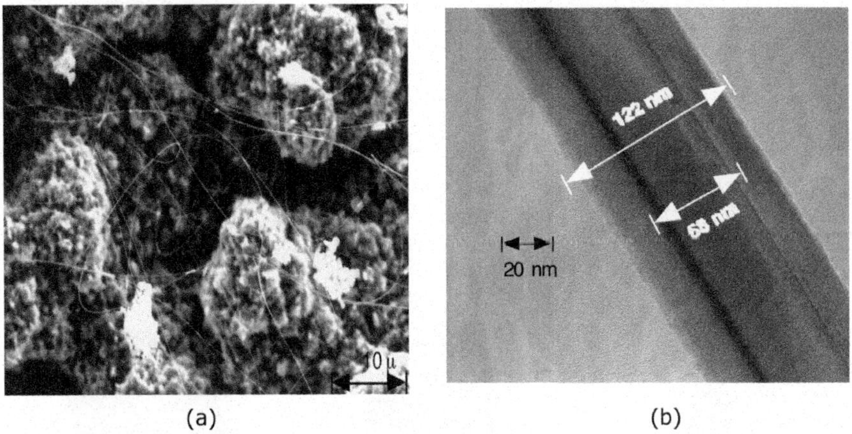

Fig. 11.2 **a** Magnified SEM image of GaN nanowires grown by CVD for 4 h at a temperature of 900 °C and under a NH_3 flow rate of 80 sccm and a pressure of 15 Torr; **b** high-resolution TEM image of a GaN nanotube grown under the same conditions. Reproduced from He et al. [8] with copyright permission from the publisher

at the rate of 50–100 sccm. During growth, the pressure was kept at 15 Torr. The temperature of the oven was 850–900 °C for 3–4 h. As shown by SEM images, amorphous GaN matrix with pits and hillocks was initially formed during heating at the pre-nucleation stage of growth; it served as thermal annealing. Many pits and hillocks together with small crystal platelets were formed. At a temperature of about 850 °C, ammonia underwent stepwise decomposition into NH_2, NH, and N on the surfaces of boats and liners. Liquid Ga formed at this temperature vaporized and deposited also on the surfaces of the same boat and liner. During the initial stage of growth, thus they created dark-gray GaN matrix. During the next stage, very thin hexagon-like GaN crystal platelets were however formed on the matrix. These platelets had polycrystalline hillocks accompanied by pits. The edges and corners of these thin hexagonal GaN crystal platelets were sharp; they were rich in defects. To reiterate, they had porous amorphous (semi-amorphous and amorphous-like) structures accompanying sharp edges, corners, and pits. They also underwent size-dependent melting point depression [22–26]. As shown schematically in Fig. 11.3a–c, they functioned as nanoparticle seeds. Nanowires grew from them. Nanotubes grew from the flat side of the platelets. The nanowire diameters were approximately identical to the platelet dimension at the location where nanowires grew.

11.3.3 Vapor Deposition of InAs Nanowires

Almost two decades later, Zi et al. [1] performed self-catalyzed InAs nanowire growth in a simple vapor deposition chamber exhibiting a multizone tube furnace. InAs

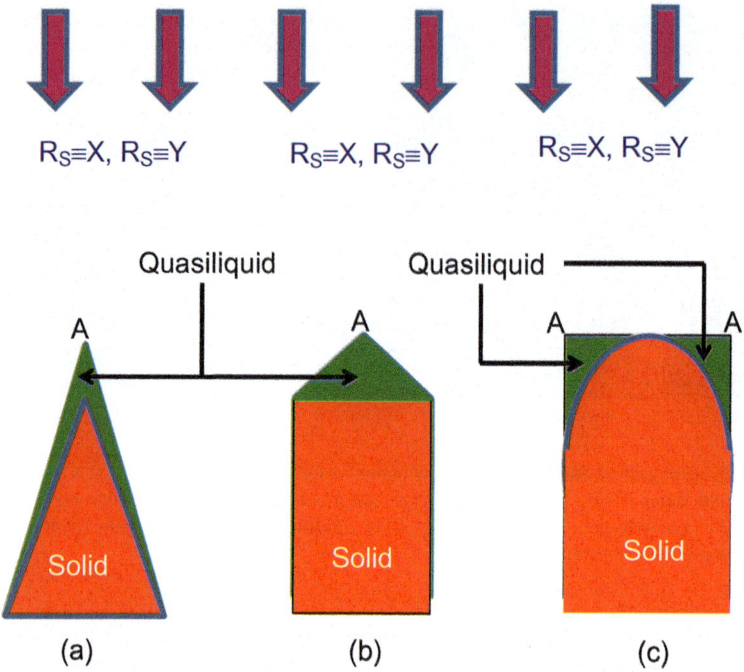

Fig. 11.3 Schematic diagrams showing quasiliquid (quasisolid) state of the sharp edges of **a** tips, **b** hillocks, and **c** corners of nanoparticle surface due to semi-melting by size-dependent mesoscopic effect. While the bulks of the edges and hillocks shown in red are crystalline solid, the sharp edges of them shown in green are molten (semimolten). The surface energy primarily of the sharp edges is very high, and under the influence of this surface energy, the sharp edges (e.g., highly pointed areas) are defective (amorphous and semi-amorphous)

powder (99.9999% pure from Alfa Aesar) placed in the upstream zone of the furnace served as the solid precursor. Growth was carried out on different III–V substrates such as those made of InAs, GaSb, and GaAs thin films. These substrates cleaned by sonication in acetone, isopropyl alcohol, and ethanol were placed in the downstream zone of the chamber. This chamber was evacuated to a base pressure of a few mTorr prior to growth. Both the upstream zone containing precursor and the downstream zone containing substrate were heated to a temperature of 820–850 °C. The substrate temperature of 490–500 °C and the substrate pressure of 2.2 Torr were maintained by the carrier gas H_2. Such increase in temperature served initially as thermal annealing leading to the formation of islands of disturbed (disordered, amorphous, semi-amorphous) and coarsened surface lattice structure. These islands were actually nanoparticles on which nanowires grew. Zi et al. [1] observed that the surface of the GaSb(111)A substrate was smooth before it was loaded inside the furnace, but had pits after thermal annealing of it at high temperature.

11.3.4 Patterned Growth of InAs Nanowires

Mandle et al. [27] made successful attempts to perform patterned nanowire growths by the SCG mechanism. For these growths, ultrathin silicon oxide layers, thick SiO_2 layers with pitch holes, or lithographically patterned layers were employed. InAs nanowires were thus grown at different temperatures on InP(111)B surfaces, which were coated with 1.3 nm SiO_z ($1 \leq z < 2$) prior to growth. This SiO_z layer appeared to serve as catalyst for nanowire growth. The nanowires grew spontaneously on the InP(111)B surface. They were standing vertically on the surface and appeared to emerge as homogeneously thick. The growth rate of these nanowires as function of temperature is shown in Fig. 11.4. One can see that the growth rate increases with temperature and attains a peak at about 600 °C. It then gradually decreases with further increase in temperature. In a subsequent investigation, Mandl et al. [32] grew InAs nanowires on an InAs(111)B substrate coated with ~1.3-nm-thick SiO_z layer for 8 min. The density of these nanowires was found to depend on temperature. This dependence shown in Fig. 11.5 appears to be linear plausibly owing to an activation energy-based process.

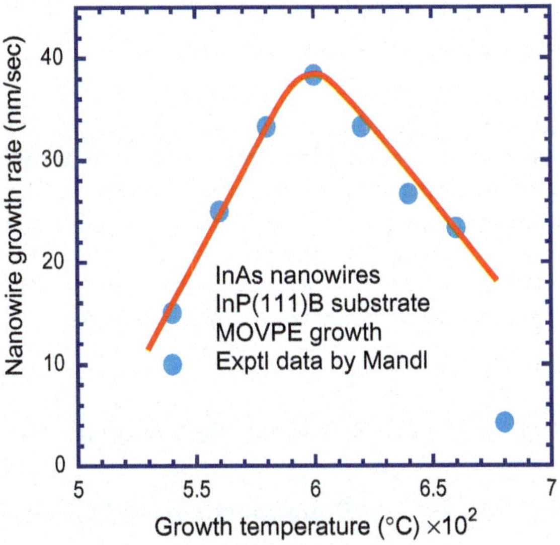

Fig. 11.4 Variation of nanowire growth rate with growth temperature for InAs nanowires. These nanowires were grown on InP(111B) substrate by the MOVPE technique. The plot was obtained with experimental data by Mandl et al. [27]

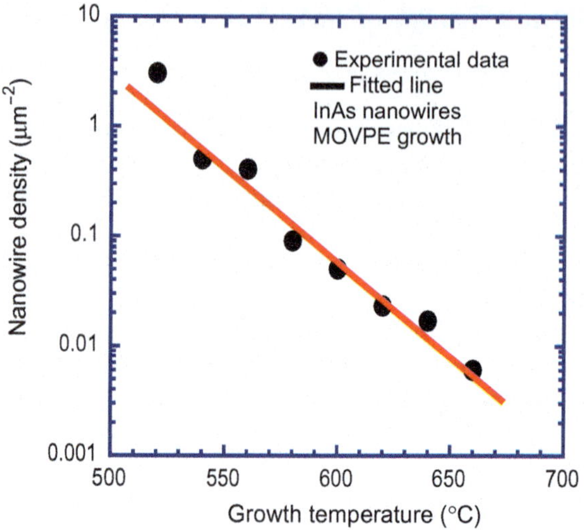

Fig. 11.5 Variation of InAs nanowire density with the growth temperature. The plot is obtained with experimental data by Mandl et al. [32]

11.3.5 Boron Nitride Nanotube Growths

Boron nitride nanotubes were synthesized in large quantity by the SCG mechanism [33] using a continuous CO_2 laser ablation reactor. HRTEM analyses indicated that the nanotubes were organized in "ropes" and that the majority of the nanotubes were zigzag. TEELs analyses suggested that they were composed of pure boron covered with BN fullerene-like "cages". They resulted from "root base" with the nanotubes emerging from boron nanoparticles dispersed on the substrate. Note that selective area epitaxy [34] is a well-known form of nanowire synthesis.

11.4 Understanding the Self-catalytic Nanowire Growth

11.4.1 Tips, Hillocks, and Roughness Crucial for Growths

Understanding the self-catalytic nanowire growth is important, for example, for the control of nanowire position, diameter, and density. As stated earlier, Zi et al. [1] suggested that thermally created pits on III–V substrates had some correlation with the self-catalyzed growth. In fact, it was found in some prior studies [8–15] that edges and corners of platelets and the tips of hillocks are crucial for the self-catalyzed growth. The surface roughness assists the formation of Ga droplets, for example, for GaAs nanowire growth. It acts also as the growth catalyst or template for nanobowls, which host the formation of nanowires after emerging above the sample surface.

The roughness of cleaved {110} Si surface was exploited for the formation of these nanowires. Note that {110} Si surfaces are not easily obtained cleavage planes; the cleavage led to the generation of steps and facets on the surface. Roughness was found also to induce the growth of GaAs nanoneedles by Moewe et al. [35] and of GaN nanowires by Koester et al. [36].

11.4.2 Grains and Grain Boundaries Formed Prior to Growths

Prior to carrying out self-catalyzed nanowire growth, Cuscunà et al. [37] modified the Si substrate surface. Both SEM and AFM analyses carried out by them revealed that the grains, 50–100 nm in dimension, were, as a result, formed on the substrate surface. This surface was flat but coarsened exhibiting hillocks of heights of 10–15 nm on it. This means the substrate surface was also amorphous (amorphous-like). The average roughness of the surface was hence about 3 nm, which was much larger than the average roughness of less than 1 nm of the flat regions. Si nanowires could be grown on the coarsened region of the surface at a temperature of 320–430 °C and a pressure of 1 Torr. Thinner nanowires could also be grown on the surface if this surface had smaller roughness. Interestingly, the idea of Cuscunà et al. [37] to generate hillocks on Si substrate for Si nanowire growth is not different from that of He et al. [8–15] to generate hillocks on GaN substrate for GaN nanowire growth. And this very idea motivated Jabeen et al. [28] to cleave Si wafers prior to the MBE growth of Si nanowires at 580–650 °C. Gao et al. [29] doped Si substrate to achieve the same goal. They fed the MOVPE chamber with trimethylindium before loading the substrate into this growth chamber. Both the cleaved and doped substrate surface had grains, disorders, and surface roughness. Cross-sectional TEM images of as grown sample by Zi et al. [1] confirmed that pits of grains and grain boundaries are actually the preferred sites for nanowire growth. The finding enabled Zi et al. to develop pre-growth annealing strategy suitable for yielding nanowires of high density by creating high density of grains and pits in those grains.

11.5 Role of Oxide in Nanowire Growth

Cuscunà et al. [37] suggested that oxide does not play any role in Si nanowire growth. This observation is in line with that of Hofmann et al. for CNT growth [19]. While investigating the nanoparticle surface most suitable for nanowire growth, Cuscunà et al. made use of flat Si surfaces of different orientations. These surfaces had native oxide on the top of them. Cuscunà et al. found no nanowire growth successful on these surfaces. Attempts were then made to expose the crystalline Si surfaces to RIE process and thus to create rough surface. Yet this surface was not successful in producing nanowires. Different etchant compositions were tried, but the results were nevertheless poor; the density of nanowires thus obtained was marginally low. A more controlled experiment by laser annealing of amorphous Si layers was tried as well. SEM and AFM images of these layers showed that circular regions surrounded by protrusions were relatively flat. As determined by AFM images, the surface roughness of the surface was primarily due to the amorphous ridges at the grain boundaries. This surface did not produce nanowires of any significant density. All these suggest that coarsening of the entire surface exhibited pits, hillocks, sharp edges, and amorphicity. They were not the results of oxidation of the nanoparticle surface. Also, just oxidation of the surface is not good enough for growth.

11.6 Various Stages of Nanowire Growth

We suggested [7–15, 38, 39] that the self-catalytic growth involves at least four different stages:

1. Formation of matrices;
2. Formation of SUBSANOs on matrices;
3. Formation of R_L species comprising cluster, solid solution, and/or droplets in the SUBSANOs;
4. Supersaturation, nucleation, crystallization, and nanowire growths on the SUBSANOs.

For example, for the X_mY_n (X≡Ga, In, Al, etc., and Y ≡N) nanowire growth with NH_3, without a substrate, there occurs direct reaction between NH_3 and the metal X (e.g., R_S≡X≡Ga, In, or Al). The growth temperature is slowly increased to ensure that the low-melting-point metal melts. The R_S≡X metal melts into liquid, some of which is converted to gas. The vapor of the R_S≡X metal diffuses onto the surface of the chamber yielding matrices, and grains on the surface of the matrices. These grains on matrices may be ESNOs serving as nanoparticles. Almost always the growth chamber contains oxygen. Some of the R_S≡X may hence be oxygenated. The lattice structures of some of the grains of matrices may therefore have the R_L species as R_L≡(X, O, ϑ) or R_L≡(X, Y, O, ϑ) solid solutions due to the presence of oxygen and ϑ contaminant(s). The R_L species of the nanoparticle may have sharp edges, pits, hillocks, and platelets. They serve as nucleation sites and produce nanomaterial. The growth parameters must however be judiciously chosen to ensure that nanomaterial grows. The most important parameters, for example, for CVD growths, are:

1. The source-substrate proximity between the source species and the nanoparticles;
2. Sharpness of the edges, pits, hillocks, and platelets of the R_L species of nanoparticles formed on the substrate;
3. Composition, shape, and size of nanoparticles;
4. The chamber pressure, the precursor flow rate, and the growth temperature; and

5. Carrier gases of optimal flow rate introduced at a suitable temperature into the chamber.

Notably almost all the nanomaterials (whiskers, nanowires, nanotubes, etc.) may grow on nanoparticles at the edges and corners of the substrate (see Fig. 11.3a–c). The judicious choice of temperature and pressure of the growth chamber may be critical for nanomaterial growths at these edges and corners.

11.7 Novelty of the SCG Mechanism

An investigation [11] in our laboratory highlights the novelty of the SCG mechanism for nanomaterial growths. This investigation demonstrated that a variety of nanowires produced by the SCG mechanism can be long (up to 1 mm), uniform, defect-free, well controlled, and single crystalline. These nanowires had properties dependent on their shape, size, and geometry, and that these shapes, sizes, and geometries are functions of some specific growth conditions. The nanowires were groups II–VI and III–V binaries, ternaries, quaternaries, and most importantly heterostructures produced under suitable conditions. Nanotubes could also be produced by the same mechanism. However, some conditions suited the growth of nanowires, some other conditions suited the growth of whiskers, some other conditions suited the growth of nanotubes, and some other conditions suited the growth of a mixture of nanowires and whiskers. The products had simultaneous control over dimensions, morphology (or shape), and monodispersity (or uniformity) under these conditions. An important feature of the mechanism was its ability to successfully synthesize single-crystal ternary and quaternary alloys such as InGaAs and InGaN ternary nanowires and quaternary InGaAsN nanowires. Interestingly, these quaternary nanowires are most suitable for optical devices of wavelengths spanning from infrared to ultraviolet. These nanowires grown in our laboratory are shown in Fig. 11.6a, b.

To confirm the quality of SCG nanomaterials, we performed PL measurements of GaN whisker [11] (see Fig. 11.7). For the sake of comparison, the PL micrograph of a high-quality GaN film grown by MOCVD is also shown in the figure. Both the said micrographs were obtained at 6 K by a visible–ultraviolet spectrometer CdHe 325 nm laser beam focused on the GaN whisker ($2 \times 200\ \mu m^2$). The PL intensity of the whisker was generated with a power level of 2.1 MW. It was very strong; it was stronger than that of GaN bulk. As evident from Fig. 11.7, the peaks of the two spectra were at 3.48 eV and 3.42–3.44 eV, respectively. While the peak for the whisker was at 3.480 eV, the peak for the bulk GaN film was at 3.465 eV. They resulted from band-edge emission and had remarkable similarity. Nevertheless, the spectrum for the whisker was slightly broader than that for the bulk. The broadband spectrum at ~3.42–3.44 eV resulted plausibly from deep-level nitrogen vacancies. The FWHM for the GaN bulk was 4.1 meV, but for GaN whisker was 9.2 meV. This highlights the superiority and novelty of the self-catalytic growth mechanism to produce nanomaterials almost free from defects.

(a)

(b)

Fig. 11.6 SEM images of **a** InGaAs and **b** InGaAsN whiskers and nanowires. These nanowires and whiskers were grown by reaction of GaAs with In at 620 °C for 3 h, while nitrogen was used as carrier gas. For InGaAsN nanowire growth, the reaction gas was ammonia. The diameter ranged from 150 nm to 1 μm, and the length was up to 40 μm. Reproduced from He and Mohammad [11] with copyright permission from the publisher

11.8 Differences Between Selective Area Epitaxy and Self-catalytic Growth

The primary element of the selective area epitaxy (SAE) [34] is the mask layer that hinders growth everywhere except in specific openings in the mask. The thin SiO_z film formed on the substrate surface may appear as such a mask layer. However, the one-dimensional growth by the selective area epitaxy takes place on the very slowly growing low-surface-energy side facets. No nanoparticle is used for growth

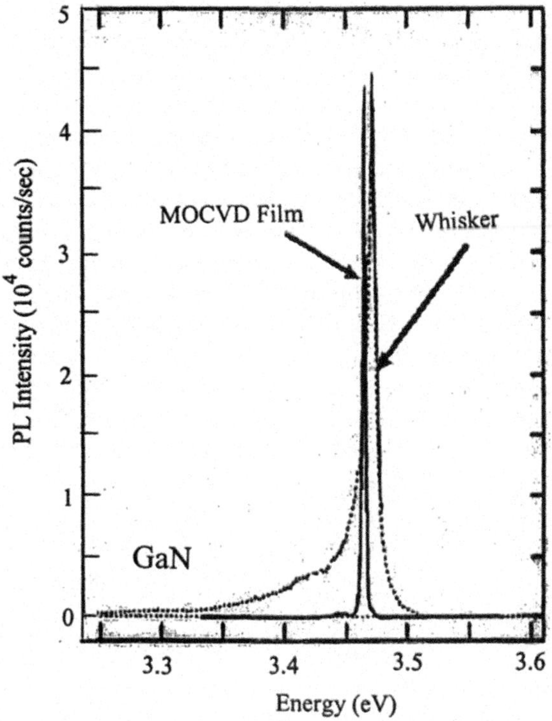

Fig. 11.7 Comparison of microphotoluminescence spectra of GaN whisker and GaN thin film obtained at a temperature of 6 K. Reproduced from He and Mohammad [11] with copyright permission of the publisher

by this mechanism. Unlike self-catalytic growth, the SAE growth is inconsistent with decrease in density of nucleated structures on the surface due to increase in temperature. Note that decrease in density of nucleated structures with increase in temperature would mean decrease in the number of openings of the mask layer. A stable mask is essential for stable nanowire growth by the selected area epitaxy. In contrast, a change in the density of nucleated structures with change in temperature is normal in growth by the SCG mechanism. This was confirmed by Mandl et al. [27, 32] who studied the temperature stability of the SiO_z layer by directly annealing it on the InAs(111)B substrate.

11.9 Nanoparticles Crucial for Nanowire Growths

11.9.1 Porous SiO_z Film Formed on Substrate

Koblmüller et al. [40] sputter-deposited approximately 25–45 nm-thick amorphous layer on the epi-ready side of the Si(111) substrates prior to InAs nanowire growth. After HCl etch, the final thickness of the SiO_z film was reduced to about 2.5–6.0 nm.

For III–V nanowire growths, SiO_z layers of comparable thicknesses were also previously employed on InP, Si(111), and GaAs substrates. The InAs nanowires produced by Koblmüller et al. were predominantly vertical and in a direction related to the {111} family of orientations. Being in the same direction, these nanowires had direct relationship with the underlying Si(111) substrate.

To investigate the effect of SiO_z on the InAs nanowire growths, Park et al. [41] made use of InAs(111)B and Si(111) substrates. About 1.3-nm-thick SiO_z film was deposited on both of them. While the growth direction of the nanowires on the InAs(111)B substrate was along the ⟨111⟩ direction, the growth direction of them on the Si(111) substrate was random. This suggested again a direct relationship between the nanowire and the underlying substrate. Recall that SiO_z undergoes a phase separation reaction (e.g., $2SiO \rightarrow Si + SiO_2$) at a temperature $T > 400$ °C yielding nanometer-sized Si clusters [42, 43]. It would therefore be reasonable to assume that similar phase separation would take place at the growth temperature of 580 °C. Notably, Zhang [44] also reported Si clusters formed in silicon suboxide, and that these clusters were energetically very reactive; SiO_2 is however less reactive but far more stable than them. Considering the formation of Si clusters, one may expect that these clusters would actually be the Si-based ESNO. We argue that Si of the Si clusters react (interact) with the R_S≡In species yielding R_L≡(Si, In) species, which is a cluster or solid solution. This cluster (solid solution) has nanopores; nanowire growth at 580 °C took place due to diffusion of In through the nanopores of this solid solution (cluster).

Kim et al. [45] performed Si nanowire growth on the SiO_z surface at 520 °C and 15 Torr. Kim et al. [45] concluded that the precursors were adsorbed on the reactive thin oxide surfaces and decomposed to form nanocrystalline nanoparticle seeds, e.g., R_L species. Low-magnification TEM image clearly identified these nanoparticle seeds on Si substrate. These nanoparticle seeds formed on Si substrate were actually clusters having molten (semimolten) nanopores. Kim et al. noted that, for the formation of nuclei, reactive surface sites were required for convenient deposition of the Si species and thus for the formation of nuclei. Interestingly, it was corroborated with high-density silicon nanowires growing selectively on the regions covered by the reactive thin SiO_z film, but no nanowires growing in the area covered with stable SiO_2 (Fig. 11.2c). This observation confirmed that the thin reactive SiO_z layer on the substrate surface had high-energy sites essential for the decomposition of precursors.

11.9.2 Buffer Layer Formed on Substrate

Bertness et al. [46] performed self-catalytic growth of GaN nanowires on Si(111) substrates by using AlN buffer layers. These nanowires by MBE route grew spontaneously under high N-to-Ga ratios at a temperature of about 810–830 °C. Field-emission scanning electron microscopy (FESEM) revealed the nucleation conditions. It demonstrated that a GaN formed a "matrix layer" grown along the [0 0 0 1] direction perpendicular to the substrate surface. This layer contained small, dense hexagonal pits needed for nanowire nucleation. It served as nanoparticle. FESEM, together with AFM, identified pit facets as $\{10\bar{1}2\}$ planes. The AlN buffer layer was 40–50 nm thick. The use of this buffer layer was found to be essential for the formation of nanowires and matrix layers under the prevailing growth conditions. GaN nanowire growth was driven by differences in growth rates among crystallographic planes under N-rich conditions. Pattinson et al. [47] observed that variation in ammonia concentration leads to considerable morphological changes of the catalyst surface and that catalyst restructuring is necessary for the epitaxial growth of carbon nanotubes. Formation of mobile species at the growth temperature and under elevated oxygen pressure accelerated sintering of the catalyst particles yielding porous surface.

11.9.3 Both X and X_mY_n Nanoparticles Can Lead to X_mY_n Nanowire Growths

It was suggested by He et al. [8–15] and also stated in Sect. 11.2 that both X nanoparticles and X_mY_n nanoparticles can lead to X_mY_n nanowire growths by the SCG mechanism. While growing GaAs nanowires at 580–650 °C on cleaved Si(111) substrate by the SCG mechanism, Jabeen et al. [28] studied the effect of Ga pre-deposition on growths. To be more specific, they conducted growths with or without the pre-deposition of Ga during the pre-nucleation stage of growths. They observed that GaAs nanowires were produced in both cases, and remarkably under identical growth conditions. However, the nanowires produced with Ga pre-deposition had Ga droplet at their tips. They had a wurtzite lattice structure in regions underneath the Ga droplet. The GaAs nanowires produced without Ga pre-deposition ended, in contrast, with pyramid-shaped GaAs structure (morphology). These nanowires had mainly zinc blende crystal lattice structure except in only a small region. Nanowires, if any, had wurtzite crystal lattice in these regions. The Ga-ended nanowires of wurtzite phase were longer than the other ones of zinc blende phase; the former ones were thinner on the average. Jabeen et al. [28] suggested that two different growth mechanisms were responsible for two different kinds of nanowires produced by them.

The growth of III–V nanowires exhibiting wurtzite phase has been found to have some relationship with the nanowire diameter. It is believed to involve a large surface contribution to the formation energy of the nanowire, which makes the wurtzite phase more favorable [48]. For example, the wurtzite phase was found more favorable for InP nanowires if these nanowires had diameter smaller than about 10 nm [48]. The value of 10 nm is one order of magnitude smaller than the typical diameter of wurtzite nanowires by Jabeen et al. Glas et al. [49] proposed a model that suggests that the formation of the wurtzite phase in VLS-grown GaAs nanowires was favored for certain ranges of the interface energies at which the nucleation of the nanowire was believed to take place at the vapor–liquid–solid triple-phase line. Staudinger et al. [50] reported on µm-sized InP layers epitaxially deposited inside hollow SiO_2 cavities formed on the top of an InP(001) substrate. They demonstrated that the crystal structure of the nanowires could be controlled by confining and guiding the growth along specific crystalline directions. It could, for example, be zinc blende for diffusion along the ⟨100⟩ direction, but wurtzite for diffusion along the ⟨110⟩ direction.

We argue that the cleaved Si(100) substrate surface was porous and coarsened at the growth temperature of 580–650 °C. However, Ga deposited during the pre-nucleation stage of growth was adsorbed onto the cleaved Si(100) substrate surface enhancing the surface roughness of the R_L≡(Ga, Si) solid solution thus created on this surface. This R_L species became significantly porous. The melting point of Ga is 29.76 °C. While a constituent of the R_L≡(Ga, Si) solid solution, the Ga atoms were molten. Hence, the R_L≡(Ga, Si) solid solution was molten (semimolten) at the growth temperature of 580–650 °C. Although still a solid solution, and not a eutectic alloy, the R_L≡(Ga, Si) species appeared as droplet. The nanopores of it became straight, smooth, narrow, but unidirectional. The R_S species could smoothly diffuse through these nanopores to the liquid/solid interface. The diffusion of the R_S species through the cleaved Si(111) surface also took place, but it was not largely smooth and straight. This surface is the surface of relative weakness, although due to splitting it could have a few thin straight nanopores. It could have acted as the R_L species. As a result, the nanowires produced by the R_L≡(Ga, Si) species were thin and long, and the crystal lattice of this nanowire was wurtzite. Except in a few cases, the nanowires produced by the R_L≡Si species were not however thin and long, and the crystal lattice of this nanowire was zinc blende. The nanowires produced in a few exceptional cases were though long and uniform and had wurtzite crystal lattice structure.

11.9.4 Superiority of SEG Mechanism to Other Mechanisms

The superiority of the SCG mechanism to other mechanisms may be best illustrated by just considering the growths of X_mY_n nanowires. Figure 11.8a shows the X_mY_n nanowire growth by the VLS mechanism. For this growth, the R_L≡(MET, X) species or more specifically the R_L species in droplet is created on a substrate. It is unstable. Contaminants can easily migrate into it. Due to lattice mismatch, interface states are generated at the substrate/R_L species interface. And these surface states together with some component element(s) of the R_L species migrate deep into the nanowire; they remain also at the droplet/nanowire interface until the end of the growth. Traces of MET may thus remain inside the nanowire. Figure 11.8b shows the X_mY_n nanowire growth by the oxide-assisted growth mechanism. For this growth, the R_L species in the form of oxygen-induced ESNO [e.g., R_L≡(oxygen-induced ESNO)] is created on an oxide matrix formed on the substrate. The nanowire thus produced is free from the droplet/nanowire lattice mismatch, but suffers from the traces of oxide-related elements, particularly oxygen, migrating deep inside the nanowire. For nanowire growth by the SCG mechanism, the R_L species is actually the disordered X_mY_n material created on the X_mY_n substrate. X-material is deposited onto the X_mY_n film for

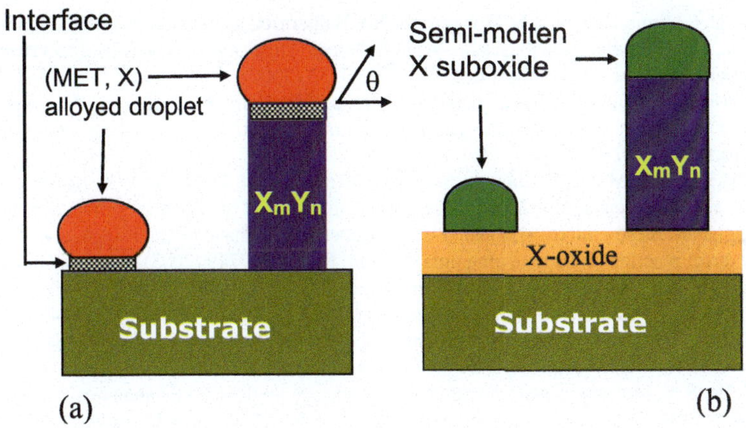

Fig. 11.8 Comparative diagrams of the X_mY_n nanowires grown by **a** the VLS mechanism and **b** the OAG mechanism

X_mY_n nanomaterial growth. Kendrick et al. [51] grew GaN nanowires employing the self-catalytic growth mechanism. Ga-based R_L≡ESNO formed on the GaN matrix mediated the growth; it served as nanoparticle seed. GaN nanowires grown on GaN matrix had Ga at the tips. EDX analysis indicated that the tip had solely Ga atoms. HRTEM showed that the Ga-based R_L≡ESNO composition was amorphous. So, the nanoparticle (exhibiting the composition of the R_L species) had grain boundaries, semi-grain-boundaries, and/or grain-boundary-like regions and hence nanopores, which became semimolten due to size-dependent mesoscopic effects. They served as diffusion paths for the R_S species for nanowire growth.

References

1. Y. Zi, S. Suslov, C. Yang, Understanding self-catalyzed epitaxial growth of III–V nanowires toward controlled synthesis. Nano Lett. **17**, 1167–1173 (2017)
2. S.A. Dayeh, E.T. Yu, D. Wang, High electron mobility InAs nanowire field-effect transistors. Small **3**, 1683–1687 (2007)
3. Y.-L. Chueh, A.C. Ford, J.C. Ho, Z.A. Jacobson, Z. Fan, C.-Y. Chen, L.-J. Chou, A. Javey, Formation and characterization of Ni_xInAs/InAs nanowire heterostructures by solid source reaction. Nano Lett. **8**, 4528–4533 (2008)
4. E. Carlino, F. Martelli, S. Rubini, A. Franciosi, Catalyst incorporation in ZnSe nanowires. Philos. Mag. Lett. **86**, 261 (2006)
5. D.E. Perea, J.E. Allen, S.J. May, B.W. Wessels, D.N. Seidman, L.J. Lauhon, Three-dimensional nanoscale composition mapping of semiconductor nanowires. Nano Lett. **6**, 181–185 (2006)
6. S.H. Oh, K. Van Benthem, S.I. Molina, A.Y. Borisevich, W. Luo, P. Werner, N.D. Zakharov, D. Kumar, S.T. Pantelides, S. Pennycook, Point defect configurations of supersaturated Au atoms inside Si nanowires. Nano Lett. **8**, 1016–1019 (2008)
7. L. Tsakalakos, J. Balch, J. Fronheiser, B.A. Korevaar, O. Sulima, J. Rand, Silicon nanowire solar cells. Appl. Phys. Lett. **91**, 233117 (2007)

8. M. He, I. Minus, P. Zhou, S.N. Mohammad, J.B. Halpern, R. Jacobs, W.L. Sarney, L. Salamanca-Riba, R.D. Vispute, Growth of large-scale GaN nanowires and nanotubes by direct reaction of Ga with ammonia. Appl. Phys. Lett. **77**, 3731–3733 (2000)
9. M. He, P. Zhou, S.N. Mohammad, G.L. Harris, J.B. Halpern, R. Jacobs, W.L. Sarney, L. Salamanca-Riba, Growth of GaN nanowires by direct reaction of Ga with ammonia. J. Cryst. Growth **231**, 357 (2001)
10. A.M.S El Ahl, M. He, P. Zhou, L. Salamanca-Riba, F. Felt, H. Shaw, A. K. Sharma, M. Jah, D. Lakins, T. Steiner, S.N. Mohammad, A systematic study of effects of growth conditions on the (nano-, meso-, micro) size and (1-, 2-, 3-dimensional) shape of GaN single crystals grown by direct reaction of Ga with ammonia. J. Appl. Phys. **94**, 7749 (2003)
11. M. He, S.N. Mohammad, Novel chemical vapor deposition technique for the synthesis of high-quality single-crystal nanowires and nanotubes. J. Chem. Phys. **124**, 064714 (2006)
12. M. He, M.E.E. Fahmi, S.N. Mohammad, InAs nanowires and whiskers grown by reaction of indium with GaAs. Appl. Phys. Lett. **82**, 3749 (2003)
13. M. He, S.N. Mohammad, Structural characteristics of single-crystal nanowires grown by self-catalytic chemical vapor deposition method. J. Vac. Sci. Technol. B **25**, 1909 (2007)
14. M. He, S.N. Mohammad, Novelty of self-catalytic nanowire growth: a case study with InN nanowires. J. Vac. Sci. Technol. B **25**, 940 (2007)
15. M. He, A. Motayed, S.N. Mohammad, Phase separations of single-crystal nanowires grown by self-catalytic chemical vapor deposition method. J. Chem. Phys. **126**, 064704 (2007)
16. R.S. Wagner, W.C. Ellis, Vapor-liquid-solid mechanism of single crystal growth. Appl. Phys. Lett. **4**, 89 (1964)
17. S.N. Mohammad, Analysis of the vapor-liquid-solid mechanism for nanowire growth and a model for this mechanism. Nano Lett. **8**, 1532–1538 (2008)
18. Y. Tatsumi, M. Hirata, M. Shigi, Characteristics of whisker growth in amorphous silicon. Jpn. J. Appl. Phys. **18**, 2199–2206 (1979)
19. S. Hofmann, R. Blume, C.T. Wirth, M. Cantoro, R. Sharma, C. Ducati, M. Havecker, S. Zafeiratos, P. Schnoerch, A. Oestereich, D. Teschner, M. Albrecht, A. Knop-Gericke, R. Schlogl, J. Robertson, State of transition metal catalysts during carbon nanotube growth. J. Phys. Chem. C **113**, 1648–1656 (2009)
20. S.N. Mohammad, Systematic investigation of the growth mechanisms for the synthesis of the conventional, doped, and bamboo-shaped nanotubes, primarily the carbon nanotubes. Carbon **75**, 133–148 (2014)
21. S.N. Mohammad, Some possible rules governing the syntheses and characteristics of nanotubes, particularly carbon nanotubes. Carbon **71**, 34–46 (2014)
22. K. Dick, T. Dhanasekaran, Z. Zhang, D. Meisel, Size-dependent melting of silica-encapsulated gold nano-particles. J. Am. Chem. Soc. **124**, 2312–2317 (2002)
23. Ph. Buffat, J.P. Borel, Size effect on the melting temperature of gold particles. Phys. Rev. A **13**, 2287–2298 (1976)
24. E. Sutter, P. Sutter, Phase diagram of nanoscale alloy particles used for vapor−liquid−solid growth of semiconductor nanowires. Nano Lett. **8**, 411–414 (2008)
25. K.K. Nanda, Size-dependent melting of nanoparticles: hundred years of thermodynamic model. Pramana **72**, 617–628 (2009)
26. F. Gao, Z. Gu, Melting temperature of metallic nanoparticles, in *Handbook of Nanoparticles*, ed. by M. Aliofkhazraei (Springer, Cham, 2016), pp. 661–690
27. B. Mandl, J. Stangl, T. Mårtensson, A. Mikkelsen, J. Eriksson, L.S. Karlsson, G. Bauer, L. Samuelson, W. Seifert, Au-free epitaxial growth of InAs nanowires. Nano Lett. **6**, 1817–1821 (2006)
28. F. Jabeen, V. Grillo, S. Rubini, F. Martelli, Self-catalyzed growth of GaAs nanowires on cleaved Si by molecular beam epitaxy. Nanotechnology **19**, 275711 (2008)
29. L. Gao, R.L. Woo, B. Liang, M. Pozuelo, S. Prikhodko, M. Jackson, N. Goel, M.K. Hudait, D.L. Huffaker, M.S. Goorsky, S. Kodambaka, R.F. Hicks, Self-catalyzed epitaxial growth of vertical indium phosphide nanowires on silicon. Nano Lett. **9**, 2223 (2009)

30. M. Batzill, The surface science of graphene: metal interfaces, CVD synthesis, nanoribbons, chemical modifications, and defects. Surf. Sci. Rep. **67**, 83–115 (2012)
31. Y. Zhang, L. Zhang, C. Zhou, Review of chemical vapor deposition of graphene and related applications. Acc. Chem. Res. **46**, 329–2339 (2013)
32. B. Mandl, J. Stangl, E. Hilner, A.A. Zakharov, K. Hillerich, A.W. Dey, L. Samuelson, G. Bauer, K. Deppert, A. Mikkelsen, Growth mechanism of self-catalyzed group III-V nanowires. Nano Lett. **10**, 4443–4449 (2010)
33. R. Lee, J. Gavillet, M. Chapelle, J.Cochon, D. Pigache, J. Thibault, A. Loiseau, Catalyst-free synthesis of single-wall boron nitride nanotubes via laser ablation. MRS Proc. **633**, A15.3 (2000)
34. P. Mohan, J. Motohisa, T. Fukui, Controlled growth of highly uniform, axial/radial direction-defined, individually addressable InP nanowire arrays. Nanotechnology **16**, 2903–2907 (2005)
35. M. Moewe, L.C. Chuang, S. Crankshaw, C. Chase, C. Chang-Hasnain, Atomically sharp catalyst-free wurtzite GaAs/AlGaAs nanoneedles grown on silicon. Appl. Phys. Lett. **93**, 023116 (2008)
36. R. Koester, J.S. Hwang, C. Durand, S. Dang Die, J. Eymery, Self-assembled growth of catalyst-free GaN wires by metal-organic vapor phase epitaxy. Nanotechnology **21**, 015602 (2010)
37. M. Cuscunà, A. Convertino, L. Mariucci, G. Fortunato, L. Felisari, G. Nicotra, C. Spinella, A. Pecora, F. Martelli, Low-temperature, self-catalyzed growth of Si nanowires. Nanotechnology **21**, 255601 (2010)
38. S.N. Mohammad, Self-catalysis: a contamination-free, substrate-free growth mechanism for single-crystal nanowire and nanotube growth by chemical vapor deposition. J. Chem. Phys. **125**, 094705 (2006)
39. S.N. Mohammad, Self-catalytic solution for single-crystal nanowire and nanotube growth. J. Chem. Phys. **127**, 244702 (2007)
40. G. Koblmüller, S. Hertenberger, K. Vizbaras, M. Bichler, F. Bao, J.-P. Zhang, G. Abstreiter, Self-induced growth of vertical free-standing InAs nanowires on Si(111) by molecular beam epitaxy. Nanotechnology **21**, 365602 (2010)
41. H.D. Park, S.M. Prokes, M.E. Twigg, R.C. Cammarata, A.-C. Gaillot, Si-assisted growth of InAs nanowires. Appl. Phys. Lett. **89**, 223125 (2006)
42. H. Rinnert, M. Vergnat, G. Matchal, A. Burneau, Strong visible photoluminescence in amorphous SiO_x and SiO_x: H thin films prepared by thermal evaporation of SiO powder. J. Lumin. **80**, 445–448 (1999)
43. D. Nesheva, C. Raptis, A. Perakis, I. Bineva, Z. Aneva, Z. Levi, S. Alexandrova, H. Hofmeister, Raman scattering and photoluminescence from Si nanoparticles in annealed SiO_x thin films. J Appl. Phys. **29**, 4678 (2002)
44. R.Q. Zhang, *Growth Mechanisms and Novel Properties of Silicon Nanostructures from Quantum Mechanical Calculations* (Springer, Berlin, Heidelberg, 2014)
45. B.-S. Kim, T.-W. Koo, J.-H. Lee, D.S. Kim, Y.C. Jung, S.W. Hwang, B.L. Choi, E.K. Lee, J.M. Kim, D. Whang, Catalyst-free growth of single-crystal silicon and germanium nanowires. Nano Lett. **9**, 864–869 (2009)
46. K.A. Bertness, A. Roshko, L.M. Mansfield, T.E. Harvey, N.A. Sanford, Nucleation conditions for catalyst-free GaN nanowires. J. Cryst. Growth **300**, 94–99 (2007)
47. S.W. Pattinson, V. Ranganathan, H.K. Murakami, K.K.K. Koziol, A.H. Windle, Nitrogen-induced catalyst restructuring for epitaxial growth of multiwalled carbon nanotubes. ACS Nano **6**, 7723–7730 (2012)
48. T. Akiyama, K. Nakayama, T. Ito, Structural stability and electronic structures of InP nanowires: role of surface dangling bonds on nanowire facets. Phys. Rev. B **73**, 235308 (2006)
49. F. Glas, J.-C. Harmand, G. Patriarche, Why does wurtzite form in nanowires of III-V zinc blende semiconductors? Phys. Rev. Lett. **99**, 146101 (2007)
50. P. Staudinger, S. Mauthe, K.E. Moseland, H. Schmid, Concurrent zinc-blende and wurtzite film formation by selection of confined growth planes. Nano Lett. **18**(12), 7856–7862 (2018)
51. C.E. Kendrick, R. Tilley, M. Kobayashi, R.J. Reeves, S.M. Durbin, Electroluminescence from single 3D GaN nanowire grown by self-catalytic molecular beam epitaxy. Mater. Res. Soc. Symp. Proc. **955**, 0955–107–52 (2007)

Chapter 12
VQS Mechanism for Nanomaterials Syntheses

Abstract The fundamentals, applicability, and novelty of the vapor-quasiliquid-solid (vapor-quasisolid-solid) growth mechanism called the VQS growth mechanism have been described. First, the quasiliquid (quasisolid) medium has been defined. The structure and morphology of the catalyst nanoparticles most suitable for growths by the VQS mechanism have then been investigated. The importance of the nanoparticle surfaces influenced by various parameters has been discussed. Based on all of them and also their surface characteristics, they have been differentiated into NP1 and NP2 nanoparticles. And depending on whether the nanoparticle is a metal nanoparticle (e.g., METANO) or a substrate nanoparticle (SUBSANO), the VQS mechanism has been the metal-assisted VQS mechanism and the metal-free VQS mechanism. The significance of surface amorphicity of the nanoparticle surface, of surface coarsening of the nanoparticle surface, of surface looseness and surface porosity of the nanoparticle surface, of high-energy sites (HETs) of the nanoparticle surface, and of the surface melting (semi-melting) of the nanoparticle surface for the best-favored growth by the VQS mechanism have been articulated. The relevance of phase transition and phase transformation to the growth by the VQS mechanism and the experimental evidences of them have been comprehended. Further, the role of catalyst support and dipole moment in FECA nanoparticle surface resulting from surface functionalization has been found crucial. Also, the need of streamlining of the growth species for the most useful and uninterrupted interaction and diffusion of the source species through nanoparticle, and the amphoteric characteristics of catalyst support for modifying the nanoparticle surface properties have been highlighted, Nanomaterial's growths by low-meltingpoint metals employing the VQS mechanism have been manifested.

12.1 Forwarding Note

Some research, carried out in the past suggested that the growths of one-dimensional and quasi two-dimensional nanostructures, such as high-aspect-ratio whiskers, have genesis in crystal defects, particularly, screw dislocations parallel to the growth axis

S. N. Mohammad, *Synthesis of Nanomaterials*, Springer Series in Materials Science 307, https://doi.org/10.1007/978-3-030-57585-4_12

[1, 2]. This research mandated that the intersection of the screw dislocation axis and the top whisker surface gives rise to perpetual steps at the tip of the whisker surface. These perpetual steps provide the most favorable sites for the adsorption and attachment (adhesion) of the R_S ($R_S \equiv X$, $R_S \equiv Y$) species to the growing whisker. These species have been defined in Chap. 1. Taking the said adsorption and attachment into account, the growth is faster along the axis of the screw dislocations than along the sides peripheral or perpendicular to the screw dislocations. Wagner and Ellis [3] however ruled out the Frank screw dislocations as tools to produce one-dimensional nanomaterials. They argued that impurities and/or FECAs are instead necessary for one-dimensional nanomaterials growths. Some of our concepts for these growths may overlap with those of Wagner and Ellis. They would be carried out by one of the routes put forth in Chap. 3 and would have pre-nucleation stage and pro-nucleation stage elucidated in Chap. 4. In order to describe them and to deal with related topics, generic terminologies such as FECANO, METANO, and SUBSANO for nanoparticles described in Chap. 3, and the pre-nucleation and pro-nucleation stages of growths defined in Chap. 3, would be utilized. SECINI, SECINI0, MET, and the R_L species defined in Chap. 3 would also be exploited.

12.2 The Concept of Quasiliquid (Quasisolid) Nanoparticle Surface

12.2.1 Quasiliquid (Quasisolid) Medium Defined

Quasiliquid medium is defined as a medium which is neither fully solid nor fully liquid, but rather partly liquid and partly solid. It may, for example, be 99% liquid and 1% solid, or 1% liquid and 99% solid. This medium may also be called a quasisolid medium as it lies between a solid and a liquid. It is similar to solid in some respect, as it can support its own weight and hold its shape. But at the same time, it shares some properties of liquid as it is semimolten and hence has the ability to flow under pressure. The words quasisolid, quasiliquid, semi-solid, and semi-liquid may be used interchangeably for the same medium. For nanomaterials synthesis, it is an essential of nanoparticle. It may both be METANO and SUBSANO with surface characteristics of interests for nanomaterials growths. It has a depth of δ_{amor} from the top. It is quasiliquid (quasisolid) up to the depth of δ_{amor} in the sense that, at the microscopic scale, it lacks long-range lattice order and molten (semimolten). Depending on the degree of its quasiliquid (quasisolid) state, it may otherwise have disturbed, disordered, amorphous, and coarsened surface lattice structure up to this depth. The disturbance, disorder, amorphicity, and coarsening are not arbitrary. They are judiciously controlled, guided by the effective amorphicity $\alpha_{amoreff}$, related to each other, and even dependent on each other, as stated earlier in Chap. 3, and described in some details in the Appendix.

12.2.2 Structure and Morphology of Nanoparticle

The most suitable lattice structure and surface morphology of a FECA nanoparticle may result from surface treatment and surface functionalization. This structure may be a crystalline solid, a cluster, an ionic species, a non-stoichiometric species, a eutectic alloy, a non-eutectic alloy, a non-eutectic solid solution, or a non-eutectic liquid solution. It may be quasisolid (quasiliquid) for a number of reasons:

1. It has the size-dependent melting point depression [4–6], which depends on nanoparticle size smaller than about 10 nm.
2. It has the mixing of the metal MET, $R_S \equiv X$ and/or $R_S \equiv Y$ species together with contaminant(s), if any, in such a way that, at the growth temperature T, it yields a eutectic alloy, a non-eutectic alloy, cluster, mixture, or solid solution of melting temperature lower than that of the metal MET.
3. It has the mixing of one or more metals, $R_S \equiv X$ and/or $R_S \equiv Y$ species together with contaminant(s), if any, in such a way that, at the growth temperature T, the end product has two or more phases. There can be a molten (semimolten) shell surrounding a solid (semi-solid) core. And both of them can be non-eutectic alloy, cluster, mixture or solid solution, plausibly of different compositions. It would happen if particularly the metal(s) has low melting temperature.

Certain metal(s), if mixed with $R_S \equiv X$, may be semimolten at a temperature T lower than the MET/X eutectic temperature T_E, of course, under some judiciously chosen growth conditions, such as pressure. The (MET, X) composition may become more rapidly X-rich than that expected in a bulk (MET, X) alloy if (1) the nanoparticle has a surface composition (MET, X) and has a relatively small size; (2) the partial pressure of $R_S \equiv X$ during growth is relatively high, and (3) the temperature of the growth chamber is gradually increased during growth. This (MET, X) composition may be the eutectic alloy-like composition even at $T < T_E$, though it may be less stable at the said temperature T, which is lower than T_E. The liquid droplet thus formed may lack the characteristics of the eutectic droplet. It may not also be spherical or hemispherical. An alloy, cluster or solid solution (e.g., from MET≡Au and $R_S \equiv$Ge) may melt at a temperature T lower than the eutectic temperature T_E and even over a wider range of temperature if certain selected impurities and contaminants are introduced into it. These impurities and contaminants must cause the reduction of attractive interatomic interactions between MET and $R_S \equiv X$ of the (MET, X) alloy prompting this alloy to be non-eutectic alloy molten (semimolten) at $T < T_E$. For some $X_m Y_n$ nanomaterials growths, the nanoparticle surface composition may be $R_L \equiv$(MET, X, Y) instead of (MET, X). If it happens, the composition $R_L \equiv$(MET, X, Y) may have similar behavior as the composition $R_L \equiv$(MET, X).

12.2.3 Nanoparticle Surfaces Influenced by Various Parameters

The precursor(s) landing on nanoparticle surface, for example, for thermal or plasma-enhanced CVD, may undergo a series of reactions (interactions) with the nanoparticle. The reactions (interactions) may vary and contribute to the formation of alloy, cluster, solid solution, or ionic species up to a depth δ_{amor} of the nanoparticle surface. They may otherwise disturb the lattice structure of the nanoparticle surface (up to a depth δ_{amor}). The nanoparticle surface may consequently exhibit single or multiple phases. The materials species of this surface, called the R_L species, may be solid, liquid, quasisolid, or quasiliquid. Depending on material state, it may have grains, grain boundaries, voids, vacancies, dislocations, pits and roughness; it may be loose and have nanopores. The number of grains and grain boundaries is not arbitrary, but pre-determined. For example, for nanobelt growths, the R_L species should preferably have two grains with a single grain boundary in between the two grains. Being the top surface, the R_L species surface may experience the influence, for example, of the following:

1. Vibration, oscillation, and fluctuation;
2. Non-uniform temperature and pressure;
3. Hydrodynamic, thermodynamic, and surface tension effects, if any;
4. Charge distribution(s);
5. Electronegativity of the R_L species with respect to the R_S species; and
6. The injection of some ionic species.

If thermally annealed, this surface may have a lattice structure quite disturbed; it may be altered by annealing. In general, the smaller the size of the nanoparticle, the larger may be the alteration of its lattice structure, though it may happen up to a depth of δ_{amor}. This alteration may, otherwise, be significant. And as stated earlier, it may hence exhibit amorphicity, pits, hillocks, and nanopores and have surface melting (semi-melting).

12.2.4 Illustrations of Nanoparticle Surfaces

For the sake of clarity, we present in Figs. 12.1a, b and 12.2a, b schematic diagrams showing quasiliquid (quasisolid) nanoparticle surfaces. Figure 12.1a shows pits and hillocks formed on the nanoparticle surface due to surface treatment. These pits and hillocks have molten (semimolten) edges (shown in green). The rest of the pits and hillocks though remains solid. Figure 12.1b shows molten (semimolten) protrusions formed on the coarsened nanoparticle surface. The nanoparticle regions shown in red are solid, but the protrusions are in green. The sharp edges of these protrusions are molten (semimolten) due to size-dependent mesoscopic effect. Additionally, Fig. 12.2a shows islands of coarsened, disordered, amorphous, and porous

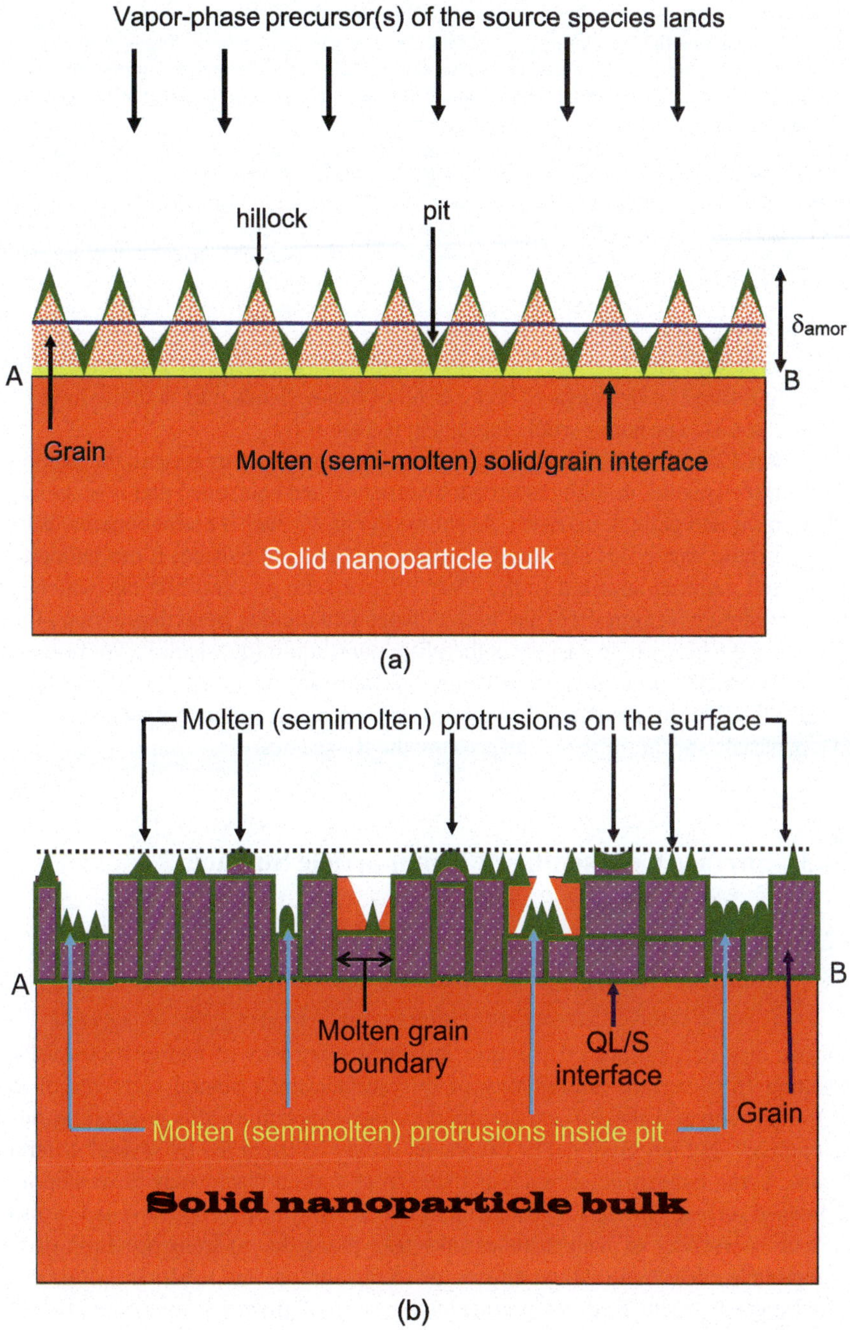
Vapor-phase precursor(s) of the source species lands
hillock
pit
δ_{amor}
A
B
Grain
Molten (semi-molten) solid/grain interface
Solid nanoparticle bulk
(a)
Molten (semimolten) protrusions on the surface
A
B
Molten grain boundary
QL/S interface
Grain
Molten (semimolten) protrusions inside pit
Solid nanoparticle bulk
(b)

◂**Fig. 12.1** **a** Nanoparticle exhibiting pits and hillocks on its surface. These are formed by the surface treatment of this surface. The blue line marks the nanoparticle surface prior to this surface treatment; δ_{amor} is the approximate depth of the disordered nanoparticle surface. The nanoparticle regions in red are crystalline solid, but the nanoparticle regions in green at the sharp edges of the pits and hillocks are molten (semimolten) due to size-dependent mesoscopic effect. Vapor-phase precursors of $R_S{\equiv}X$ and $R_S{\equiv}Y$ species lands on the nanoparticle surface. **b** Nanoparticle exhibiting molten (semimolten) protrusions, grains, and grain boundaries formed on its disordered, coarsened surface. The nanoparticle regions in red are crystalline solid, but the protrusions and grain boundaries in green, particularly, the sharp edges of the protrusions are molten (semimolten) due to size-dependent mesoscopic effect

lattice structure, and Fig. 12.2b shows amorphous, porous, and coarsened lattice structure formed on the surface. Some or all of these islands may actually have two or more grains. The nanoparticle regions in red are solid.

Figures 12.1a, b and 12.2a, b indicate that, despite surface treatment, the nanoparticle surface is solid, though disturbed, disordered, and coarsened and it is stressed. It has grains and grain boundaries. As shown in Fig. 12.3a, b it has nanopores. While some of these nanopores are straight, others are bent and twisted. It has nanopores disordered material adjacent to the sharp edges of the pits and hillocks which are however molten (semimolten) due to size-dependent melting point depression [4–6]. Overall, the nanoparticle surface is therefore quasiliquid (quasisolid). We reiterate, the nanoparticle surfaces shown in Figs. 12.1a, b and in 12.2a, b are quasiliquid (quasisolid). Due to size-dependent mesoscopic effect, the nanopores are also molten. The materials species of these surfaces are the R_L species.

12.3 Surface Coarsening of Nanoparticle Surface

12.3.1 Elements of Surface Coarsening; Tamman and Heutting Temperatures

Surface disorder, surface coarsening, and surface porosity of nanoparticle surface resulting from surface amorphicity of this surface are essential elements of the VQS mechanism. Like surface amorphicity (see Appendix), this coarsening must be optimal—but not marginal or excessive. It may be triggered by several different means. Some nanoparticles, such as those from carbon black, may have naturally coarsened surface. Sputter-deposited nanoparticles may also have coarsened surface. Surface coarsening of nanoparticles may take place due to transformation of the nanoparticle surface into alloy, cluster, or solid solution; and it happens during the initial stage of growth (e.g., the pre-nucleation stage of growth) prior to nanomaterial nucleation. As stated earlier, if FECA nanoparticle is formed on a substrate surface, it can also be coarsened by high-temperature annealing [7]. Such surface coarsening of the nanoparticle surface begins at a temperature much lower than the bulk melting

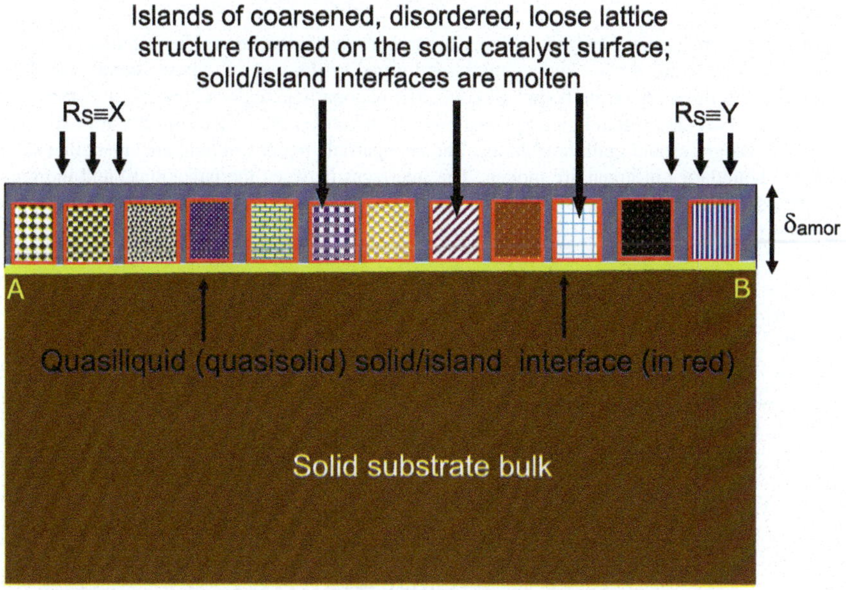
Islands of coarsened, disordered, loose lattice structure formed on the solid catalyst surface; solid/island interfaces are molten
Rs≡X
Rs≡Y
δamor
A
B
Quasiliquid (quasisolid) solid/island interface (in red)
Solid substrate bulk

(a)

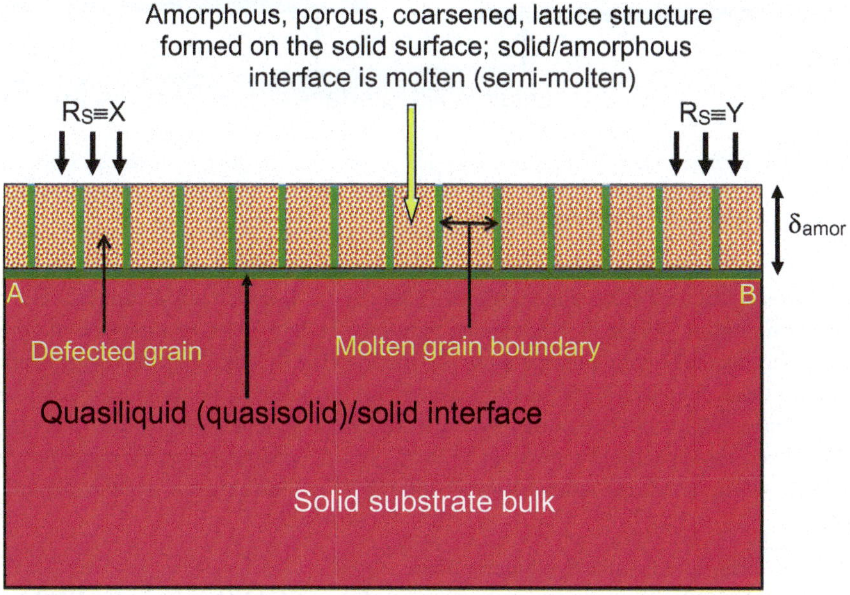
Amorphous, porous, coarsened, lattice structure formed on the solid surface; solid/amorphous interface is molten (semi-molten)
Rs≡X
Rs≡Y
δamor
A
B
Defected grain
Molten grain boundary
Quasiliquid (quasisolid)/solid interface
Solid substrate bulk

(b)

Fig. 12.2 **a** Nanoparticle, instead of being amorphous in the entire region, exhibits islands of coarsened, disordered, porous lattice structures formed on its surface. These islands may have grains, defects, disorders, grain boundaries, etc. The nanoparticle regions in brown are crystalline solid. **b** Nanoparticle exhibiting amorphous, porous, and coarsened lattice structure formed on its surface. It has grains and grain boundaries. The nanoparticle regions in pink are crystalline solid, but grain boundaries in green are molten. The interfaces between the bulk (pink) and the grain boundaries (green) are at least solid/liquid or solid/quasisolid or solid/(quasiliquid) interfaces. The grains have nanopores

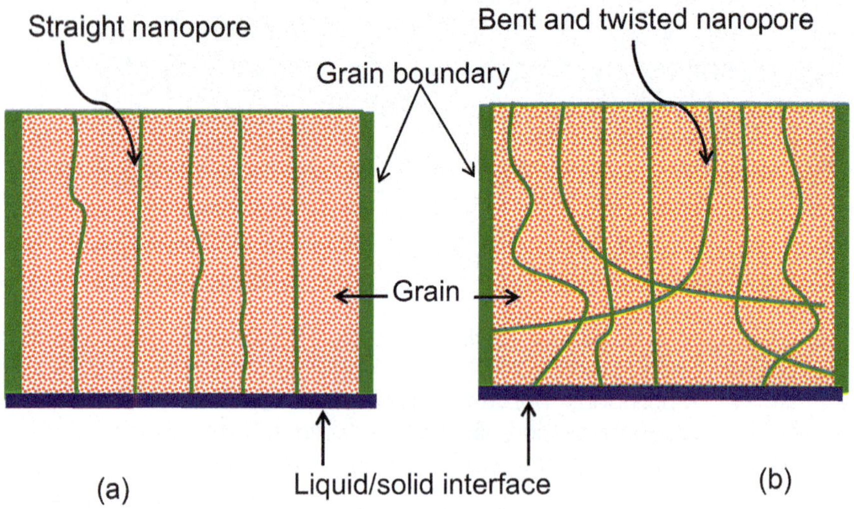

Fig. 12.3 Enlarged views of porous grains and porous, molten grain boundaries. Schematic diagrams showing nanopores formed inside the grains of the R_L species surface; **a** nearly straight nanopores, **b** bent and twished nanopores

temperature of the FECA nanoparticle. There can be Tamman temperature or Huetting temperature. While the Tamman temperature is defined as 0.5 times the melting temperature, the Huetting temperature is defined as 0.3 time the melting temperature, both in the Kelvin scale. These temperatures are actually the temperatures at which the nanoparticle atoms become mobile in the bulk and at defect sites, respectively [8]. Let us be more specific, Fe melts at 1535 °C, but the coarsening of Fe nanoparticle surface starts at a temperature as low as 269 °C. This coarsening, of course, depends on

1. The substrate/nanoparticle interactions;
2. Size-dependent melting point depression of the FECANO (e.g., both METANO and SUBSANO); and
3. The annealing atmosphere [9–11] during growth.

Owing to the effects stated above, the said temperature could even be lower than 269 °C.

12.3.2 Basics of Surface Coarsening

Surface energy of nanoparticle is higher than the surface energy of the corresponding bulk. Because of this, surface atoms easily break away from the surface of surface-treated nanoparticle. We cite some examples. Depending on the ion bombardment energy and ion fluxes on a substrate, microwave plasma causes surface etching and generates surface disorder on this substrate. Ion energy and/or ion power determine the disorder on this surface. This surface thus becomes amorphous (semi-amorphous, amorphous-like) and nanoporous. This is supported by experiments in which Fe, Au, and even diamond surfaces treated with microwave plasma, aqua regia solution, ethanol solution, etc., became disordered and amorphous (amorphous-like). Ethanol treatment created surface roughness of diamond nanoparticles [12, 13]. SEM images of nanodiamond particles [12, 13] after air (700 °C) and Ar/H_2 heating (850 °C) indicated that these nanoparticles had highly defective surfaces. Lin et al. [14] noted that, during the pre-nucleation stage of growth, the carbon black nanoparticle surface turned quite rough having many bumps with diameters ranging between 40 and 80 nm in height. The pristine surface was, on the other hand, smooth and free from roughness. No CNTs could be grown on the pristine graphite surface. But MWCNTs could be grown on the treated rough surface. Treatment with aqua regia solution for 60 s created surface roughness of Au nanoparticles [15]. Liu et al. [16] could grow SWCNT on silica, but not on Si. It is because, based on AFM height analysis, silica surface, unlike silicon surface, was coarsened; and the surface roughness of this surface was in the ranges of 2–30 nm.

12.3.3 Illustrations of Surface Roughness

An experiment by Reddy et al. [17] is very illuminating in illustrating surface roughness. In this experiment, Ni surfaces were manually roughened by sanding them with SiC paper. And three different Ni nanoparticles, thus created, had surface roughness of 415, 265, and 129 nm, respectively. It was observed that increase in surface roughness led to decrease in nanoparticle sizes, but increase in the density of CNTs and CNFs grown on these nanoparticles. Carbon atoms for these CNTs and CNFs, 20–60 nm in diameter, and grown on the said nanoparticles were from C_2H_2 precursor. The experiment was repeated also with Ni alloy containing 63 at.% Ni, 2.5 at.% Fe, 28–34 at.% Cu, 0.3 at.% C, 2.0 at.% Mn, 0.5 at.% Si, and 0.024 at.% contaminants. Three of the Ni alloy nanoparticles thus created had similar surface roughness, but of heights 415, 265, and 129 nm, respectively. The SEM images indicated again that the density of CNTs grown on them decreased with decrease in the alloy surface roughness. More recently, CNTs were grown on sputtered copper-on-silicon oxide wafers [18]. AFM measurements indicated that the RMS surface roughness of the sputtered copper-on-silicon oxide was 2.5 ± 0.2 nm. Interestingly, the CNTs grown on it were not only dense, but also vertically aligned.

12.4 Surface Looseness and the Porosity ρ_c

12.4.1 Effect of Annealing and Temperature on the Opening of Mask

Surface looseness, pore formation, and the porosity ρ_c of a nanoparticle surface occur due to the presence of defects and/or irregularities in the lattice structure at this surface. Thermal conductivity and thermal expansion coefficients of some materials are listed in Table 12.1. As apparent from this table, there is always a finite difference between the thermal expansion coefficients of two materials, for example, a substrate and a thin film (e.g., native oxide thin film) formed on this substrate. Upon annealing at high temperature, stress is generated due to the different thermal expansion coefficients of the two materials. It is the tensile stress in the thin film and the compressive stress in the substrate itself. In general, the higher the temperature, the larger is the stress at the interface. And to accommodate this stress, voids, vacancies, disorders, nanopores, pits, hillocks, etc., are generated at the interface of the

Table 12.1 List of thermal conductivity and thermal expansion coefficient of some inorganic materials

No.	Material	Linear thermal expansion coefficient (K^{-1})	Thermal conductivity W/(m K)
1	Si	2.56×10^{-6}	150
2	SiC	2.77×10^{-6}	120
3	Silver	18.00×10^{-6}	430
4	SiO_2	5.62×10^{-7}	1.1–1.4
5	Gold	14.20×10^{-6}	320
6	Si_3N_4	$1.40–3.72 \times 10^{-6}$	100
7	Iron	11.80×10^{-6}	79
8	Ge	6.12×10^{-6}	0.58
9	Inconel	$11.50–12.61 \times 10^{-6}$	15
10	Indium	33.00×10^{-6}	86
11	Alumina (Al_2O_3)	8.12×10^{-6}	50
12	Sulfur	–	0.205
13	TiO_2	11.8×10^{-6}	11.8
14	Carbon	–	140
15	ZnO	4.77×10^{-5}	0.34–1.36
16	AlN	5.63×10^{-6}	60–177
17	BN	$2.7–38.0 \times 10^{-6}$	27
18	Mica	3.0×10^{-6}	0.71
19	GaN	2.8×10^{-6}	120–140

two. Obviously, defects and irregularities are the inevitable results of surface amorphicity and surface coarsening, or vice versa. They both must be well controlled and interrelated rather than arbitrary. While the defects include humps, voids, vacancies, and dislocations, the irregularities give rise to pits, hillocks, and sharp edges. While growing InAs nanowires, Mandl et al. [19] found no openings or major irregularities in the SiO_z film produced on InP(111) and InAs(111) substrates prior to annealing at $T > 500$ °C. However, an annealing at $T = 600$ °C for 10 min during this growth led to the generation of surface disorder and the formation of a multitude of openings of sizes ranging from smaller than 100 nm up to several micrometers. The diameter distribution of nanowires is directly correlated with the size distribution of openings on the nanoparticle surface. Mandl et al. [19] observed a diameter distribution of 70 $\pm$ 22 nm for a growth of 10 s, which is a comparatively narrow size distribution. Mandl et al. [19] also found that increase in growth temperature leads to a decrease in the number of openings in the thin film. Such temperature dependence of the number of openings could not however be correlated with the nanowire growth by the SCG mechanism.

12.4.2 Migration of the R_S Source Species and of the Droplets

Most of the nanowires, produced by Koblmüller et al. [20] on sputter-deposited amorphous SiO_z film on the epi-ready side of the Si(111) substrates, were approximately 25–45 nm thick; they had nucleation directly at the interface of the substrate and the non-stoichiometric SiO_z layer. These nanowires were fairly straight and non-tapered with hexagon-shaped geometries. The results by Koblmüller et al. [20] indicated that the $R_S{\equiv}$In and $R_S{\equiv}$As source species released on the SiO_z thin film migrated to the SiO_z/Si interface, and it would not be possible without the SiO_z film being porous. Indeed, Mandl et al. [19] observed that In droplets formed on SiO_z thin film diffused freely from the top SiO_z surface to the interface between the SiO_z thin film and the InAs substrate. The migration was very fast, which would not take place without the SiO_z film being transparent to the diffusion of the In droplets. Energy-dependent measurements showed that the "free" In droplets did not disappear; only the signal was attenuated by the SiO_z layer formed on the top of the InAs substrate. They noted that the In droplets could even diffuse through the SiO_z layer prior to this layer being cracked up. The study by Park et al. [21] indicated similarly that SiO_z film was transparent to the diffusion of In droplet to the SiO_z/substrate interface. Cassell et al. [22] noted that nanoparticle surface, that exhibited metal/oxide mixed phases, was transparent to diffusion of the $R_S{\equiv}$C source species; and that there was higher yield of SWCNTs by more open pore structure of nanoparticles. The diffusion of the growth species was increased in more open pore structure. This implies that the diffusion of the source species through nanoparticle surface depends both on the nanopore volume and nanopore density. The observation of Berlanga et al. [23], in this regard, was more appealing. They found nanotubes growing on the porous non-stoichiometric side of Si_3N_4 nanoparticles, but not on the non-porous stoichiometric side of the same Si_3N_4 nanoparticles. They also found that the surface morphology of the porous side of the nanoparticle surface was rough and amorphous. It confirmed that nanoparticles can indeed be porous and that the porosity ρ_c of nanoparticles is related to coarsening and non-stoichiometric lattice structure of the nanoparticle surface.

12.4.3 *Illustrations of Nanopores Generated on Nanoparticle Surface*

O'Hern et al. [24] were able to achieve controlled, high-density nanopores in macroscopic single-layer graphene membranes. The nanopores of diameters 0.40 ± 0.24 nm had densities exceeding 10^{12} cm^{-2}. Thomas and Raja [25] also observed nanopores of radii ~0.2 nm in FECA nanoparticles. Pellen et al. [26] found that the dimensions of these nanopores can even be small, on the order of 0.1 nm. While studying oxide nanoparticles, Kelsall et al. [27] observed nanopores also in the range of ~0.1 nm. We assume oxides of these nanoparticles were non-stoichiometric. Gitis and Rothenberg [28] observed these nanopores even in ceramic membranes; and the radii of these nanopores were about 0.15 nm. More recently, Matteini et al. [29] observed nanopores of radii ~0.15 nm in oxide nanoparticles. Based on the finding made above, we argue that nanoparticle surfaces of disturbed, disordered, metastable nanostructures may all be porous. They may have nanopores of radius r_c slightly larger than, but comparable to the radius r_{as} of the diffusing R_S species.

12.5 Melting (Semi-melting) of Nanoparticle Surface

12.5.1 *Basics*

Due to surface treatments or some other effects, nanoparticle surface prior to growth becomes essentially an aggregate of nanograins and hence exhibits a high density of grain boundaries. Melting of this nanoparticle surface is a continuous and reversible process taking place at a temperature $T < T_M$. It results from the positional disturbance and disorder of the crystal surface lattice [30]. As temperature increases, nanoparticle surface becomes increasingly disordered undergoing a phase transition to a disordered state at a transition temperature, suppose $T = T_s$. The disordering is initiated at the surface, but migrates into the interior of the nanoparticle. As a result, the disordered phase intrudes between the top surface and the ordered bulk of the nanoparticle. This process is called surface-induced disorder; it causes melting and it may be quasiliquid (quasisolid). Indeed, the surface melting on an atomic level was observed by Frenken and van der Veen [31].

12.5.2 A Simple Model of Surface Melting

Consider a disordered surface layer of thickness δ_{amor} exhibiting free energy per unit area called E_{ds}. Following Lipowski [32, 33], the nanoparticle has two interfaces: (1) the interface between the disordered surface phase and the vapor called the v-d interface, and (2) the interface between the disordered surface layer and the ordered bulk state of the crystal called the d-o interface. At a temperature $T = T_s$, the disordered phase co-exists with the ordered phase and also with the vapor. The leading term of the free energy per unit area, namely E_{ds} should comprise the sum of the interfacial tension $\Sigma_{v\text{-}d}$ of the vapor-disordered interface and the interfacial tension $\Sigma_{d\text{-}o}$ of disordered-ordered interface:

$$\Sigma_{v\text{-}o} = \Sigma_{v\text{-}d} + \Sigma_{d\text{-}o} \quad (12.1)$$

There would also be direct intermolecular interactions E_{do} between the v-d and d-o interfaces, which would arise from the microscopic forces between atoms and molecules. There would be bulk free energy for both the ordered phase and the disordered phase, as well. If expressed in terms of enthalpy and entropy, it would, for the ordered phase, be

$$E_{ord} = H_{ord} - S_{ord}(T - T_s), \quad (12.2)$$

but, for the disordered phase would be

$$E_{disord} = H_{disord} - S_{disord}(T - T_s), \quad (12.3)$$

where E_{ord}, H_{ord}, and S_{ord} are the bulk free energy, enthalpy, and entropy associated with the ordered bulk phase, and E_{disord}, H_{disord}, and S_{disord} are the bulk-free energy, enthalpy, and entropy associated with the disordered phase. Assuming that $H_{ord} \approx H_{disord}$, the difference between the two bulk free energies would approximately be

$$E_{bulk} = E_{ord} - E_{disord} \approx \Delta S_{bulk}(T - T_s), \quad (12.4)$$

where $\Delta S_{bulk} = S_{ord} - S_{disord}$ for a disordered surface layer of thickness t_{ds}. Then, the free energy of the disordered surface layer of thickness t_{ds} would be

$$E_{ds} \approx \Delta S_{bulk}(T - T_s)\delta_{amor} + \Sigma_{v\text{-}o} + E_{do} \quad (12.5)$$

Roughly, it would be the measure of surface melting of the nanoparticle surface.

12.5.3 Illustrations of Surface Melting

Erlandsson et al. [34] and Eriksson et al. [35] found that, due to coarsening during cyclic exposure to O_2 and H_2, there occurred appreciable restructuring of 0.5 nm thick Pd films (melting and Huetting temperatures of 1554 and 275 °C, respectively) only at 200 °C. Platinum has a bulk melting point T_M = 1768 °C. Wang et al. [36] noted that an increase in temperature from 600 to 660 °C led to a striking changes on the Pt nanocrystal (diameter ~ 8 nm) surface. These changes resulted from surface melting of the first few atomic layers of the nanocrystal surface. Chen et al. [37] performed molecular dynamics simulations to study disordering and melting of the Ni(110) surface. They noted that, due to the generation of vacancies and the formation of an adlayer, surface disorders can take place at a temperature as low as 1180 °C. They predicted that a quasiliquid region exhibiting liquid-like energetic, structural, and transport properties sets in at this temperature. The thickness of this quasiliquid region increased with increase in temperature to the melting point T_M = 1455 °C. Levitas and Samani [38] noted that, depending on temperature $T < T_M$, the thickness of the molten surface layer of a nanoparticle can be as large as ~2 nm. Using calorimetry, temperature-programmed oxidation and Raman spectroscopy, the evolution of MET≡Fe during SWCNT growth was studied by Harutyunyan et al. [39]. (Fe, C) eutectic composition of the R_L≡(Fe, C) alloy at the eutectic temperature T_E = 1153 °C is $Fe_{0.82}C_{0.18}$. Yet they observed that liquid phase was favored for SWCNT growth at $T < T_E$, In fact, they found no SWCNT growth at $T < T_E$ if the FECA was in solid phase; the SWCNT growth almost ceased as the FECA approached solidification. The experiment by Harutyunyan et al. demonstrates almost conclusively that surface melting does occur leading to the creation of a liquid (semi-liquid) phase during nanomaterials growths.

12.5.4 Possible Causes of Surface Melting

We argue that there are at least two possible causes of surface melting (semi-melting) of a nanoparticle. These are the formation of grains and grain boundaries in the nanoparticle, and the generation of pits, hillocks, edges, etc., on the nanoparticle surface. High density of grain boundaries may indeed be a dominant cause of surface melting (semi-melting). While the grain boundaries may be defected, the grains may be free from defects. Atoms, including FECA and X atoms lying in the vicinity of grain boundaries (semi-grain boundaries, or grain boundary-like regions) of the R_L species are less tightly bound to one another than those in the absence of these boundaries. It is true also for substrate/nanoparticle and nanomaterial/nanoparticle interfaces. The melting point of materials at or near grain boundaries and interfaces is generally lower than that in the bulk. This is evident from an experiment by Howe and Saka [40], who performed transmission electron microscopy to demonstrate that, even if Si and metal are solid at a certain temperature, the interface of this Si and metal is partially molten at this same temperature.

Pits, hillocks, etc., formed on the rough surface of nanoparticle generally have sharp edges. Hence, they undergo size-dependent melting point depression [4–6]. This is illustrated in Fig. 12.1a. This figure suggests that the nanoparticle surface may hence be molten (semimolten) at least in locations of pits and hillocks, and the overall nanoparticle surface may be quasiliquid (quasisolid) with pockets of molten (semimolten) spots lying on it. Taking various features, discussed above, into account, an important element of the VQS mechanism is surface melting (semi-melting). If surface melting of the R_L species of a FECANO is denoted by ξ_m, then $\xi_m = 0$ for solid R_L species, $\xi_m = 1$ for liquid R_L species (e.g., droplet), and $0 < \xi_m < 1$ for a quasiliquid (quasisolid) R_L species.

12.6 Nanoparticle Structure and Morphology

The FECA nanoparticles for nanowire growth are shown schematically in Fig. 12.4a, b, but the FECA nanoparticles for nanotube growth are shown schematically in Fig. 12.5a, b. The R_L species of these nanoparticles are shown in Figs. 12.4b and 12.5b; it is porous and amorphous. This R_L species exists in all of the FECA nanoparticles of Figs. 12.1a, b and 12.2a, b. The FECA nanoparticles of Figs. 12.4a and of 12.5a are crystalline; they have a crystalline structure of height AB, but no R_L

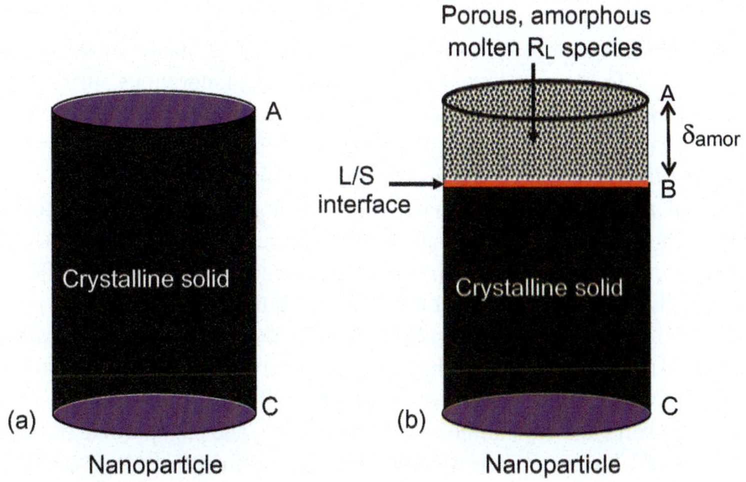

Fig. 12.4 The FECA nanoparticles for nanowire growth; **a** crystalline solid nanoparticle not suitable for nanowire growth; **b** crystalline solid nanoparticle with coarsened, amorphous, porous, molten (semimolten) R_L species of depth δ_{amor} and formed on the top of it suitable for nanowire growth. L/S interface is the liquid/solid interface; it is between the crystalline solid bulk and the molten (semimolten) surface on the top of it

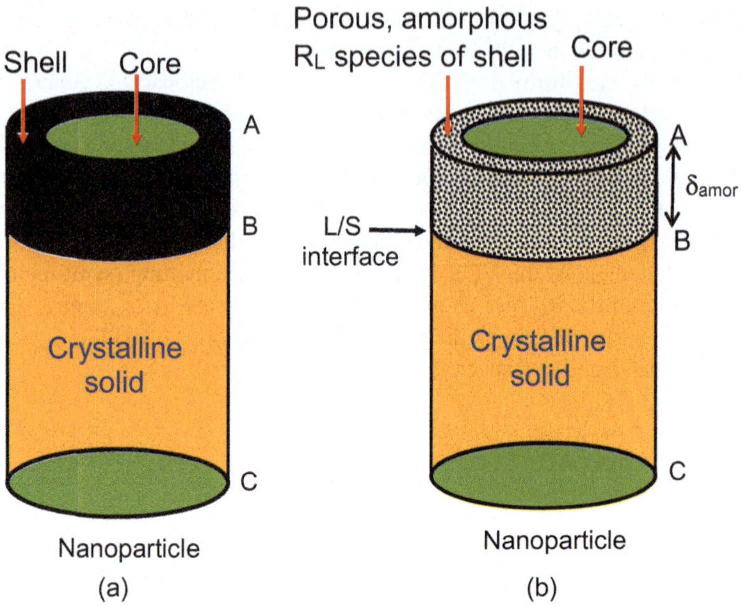

Fig. 12.5 The FECA nanoparticles for nanotube growth; **a** crystalline solid nanoparticle not suitable for nanotube growth, **b** but crystalline solid with coarsened, amorphous, porous, molten (semimolten) shell of depth δ_{amor} and on its top surface suitable for nanotube growth. L/S interface is the liquid/solid interface between the crystalline solid bulk and the molten (semimolten) shell on its top surface

species. The FECA nanoparticles of Figs. 12.4b and of 12.5b are, on the other hand, cylindrical; the structure of them exhibits disordered, amorphous surface of depth AB≡δ_{amor} and crystalline bulk of depth BC. These nanoparticles underwent surface disturbance and disorder leading to the creation of the R_L species on their surfaces. The depth of the R_L species is again AB≡δ_{amor}. It may be liquid or solid. If solid, it is disordered and amorphous with a network of nanopores molten (semimolten) due to size-dependent mesoscopic effect [4–6]. These are true for all FECA nanoparticles irrespective of their crystalline state and structure [41]. An interface called liquid/solid (L/S), liquid/quasisolid (L/QS) , or liquid/quasiliquid (L/QL) interface is, as a result, created between the solid bulk and the liquid, quasiliquid (quasisolid) R_L species on the top of this bulk. The FECA nanoparticle of Fig. 12.5a is surrounded with a shell of depth BC. It does not though have R_L species. If this R_L species at all exists, it is solid and crystalline (near-crystalline). The nanoparticle of Fig. 12.5b has a height BC, but a core surrounded with a shell of depth BC≡δ_{amor}. While the core is crystalline (near-crystalline), the shell is non-crystalline (disordered, amorphous, semi-amorphous, amorphous-like) due to disturbance of the surface. The R_L species has pits and hillocks and a network of nanopores molten (semimolten) due to size-dependent mesoscopic effect [4–6]. The L/S, L/QS or L/QL interface of this nanoparticle is created between the peripheral solid bulk and the liquid, quasiliquid (quasisolid) R_L species of the peripheral shell. The R_L species of METANO is an alloy, cluster, or solid solution, and hence the R_L≡(EMNO), R_L≡(MET, X), R_L≡(MET, Y), R_L≡(MET, X, Y), R_L≡(MET, X, Y), etc. If the nanoparticle is a SUBSANO, the R_L species of the SUBSANO is also an alloy, cluster, or solid solution, and hence the R_L≡(ESNO), R_L≡(ESNO, X), R_L≡(ESNO, Y), R_L≡(ESNO, X, Y), etc. They may have contaminant ϑ. If R_S≡Y species is highly volatile, it may not exist in any of the R_L species.

12.7 Phase Transition(s) and Phase Separation(s)

The following growth parameters greatly influence the R_L species composition, morphology, and the morphological changes during growths:

1. Reactions and reaction dynamics on the R_L species surface;
2. Concentration of the active sites on the R_L species surface;
3. The vibration, oscillation, fluctuation, relaxation, etc., of the surface atoms [42] of the R_L species;
4. Gradual, uniform or non-uniform increase in temperature and pressure during growth;
5. The presence of ambient (carrier gas) in the growth chamber, and its influence of it on the R_L species;
6. Impact of oxygen and/or other contaminant(s) intentionally or unintentionally introduced into the growth chamber;
7. Surface reconstructions, surface enrichment, and surface segregation taking place due to changes in the gaseous environment of the growth chamber.

The surface morphology may be affected by surface treatment, as well. The surface morphology may go through several stages of disorder [43–45]. There may as well occur large specific heat anomaly and anomalously large surface thermal expansion in the R_L species [46, 47]. Depending on the strength of bonds between atoms and molecules, these may be experienced more in some nanoparticle than in some other nanoparticle. They may be larger at higher temperature.

The precursor landing on the nanoparticle surface, or the precursor intermediate(s) created on it, may influence it (e.g., nanoparticle surface) by a combination of local electrostatic effect and long-range electronic perturbations in the valence charge density and in the Fermi-level local density of states [48]. The end result of all these would be phase transition and the generation of an effective amorphicity α_{amoreeff} (see Appendix) of the R_L species. Under the influence of oxygen and accompanying elements, there may be a series of phase transitions in the R_L species; there may as well be phase separation in the R_L species. The final phase of the R_L species may be a combination of phases. Some of these phases may be metastable exhibiting dislocations. This final phase may contain contaminants such as O, N, and/or some dopant atoms and other impurities. The phase separation may yield (1) a phase of a metal, an alloy, or a compound underlying (b) another phase of a metal; both of these phases may remain as components of the R_L species. While this metal may serve the primary purpose of the R_L species, the alloy (compound) underlying it may serve as a support for it. Due to strong interactions with this support and/or other component(s) of the R_L species, the said metal may have porous (pillbox-like) morphology during growth. As described in the Appendix, the highest value of the effective amorphicity $\alpha_{amoreeff}$ would be $\alpha_{amoreff0}$. The effective amorphicity $\alpha_{amoreeff}$ would be different from the surface amorphicity α_{amor} for disorder higher than the one (e.g., disorder) that corresponds to $\alpha_{amoreff0}$. High-energy sites (HETs) would also be created on the amorphous R_L species surface.

12.8 Creation of High-Energy Sites (HETs)

The VQS mechanism mandates the creation of high-energy sites (HETs) on nanoparticle surface as critical for nanomaterials growths. As elucidated earlier, the precursor landing on nanoparticle surface, or its intermediate(s) created on this surface during the pre-nucleation stage of growth may react or interact with the surface atoms and/or molecules of the FECANO (viz., METANO and and SUBSANO) nanoparticle surface. Such reaction and interaction would cause charge transfer from FECA nanoparticle to $R_S{\equiv}X$ ($R_S{\equiv}Y$) species or vice versa. Pockets of charged islands thus created on the nanoparticle surface act as high-energy sites (HETs). These HETs may be the sites of unsaturated accumulated charges, bonded oxygen atoms, bonded hydroxygenated radicals, and/or dangling bonds. They are derived from the precursor or from the nanoparticle itself. They are very effective tools in decomposing (dissociating) even at relatively low temperature the precursor(s) landing on the nanoparticle surface. Some examples of HETs created by plasma on nanoparticle surface are radicals such as CH_z, with $z < 4$, NH_y with $y < 3$, C_2H, C_2, H, etc. Gohier et al. [49] found that, during CNT growth by PECVD, sp^2 carbon layers were first segregated to the edge of FECA nanoparticles. The Raman spectra showed that disordered carbon was deposited during PECVD, but not during CVD. The propensity of the PECVD to deposit disordered carbon material had genesis in its ability to dehydrogenate the precursor without any need of increase in growth temperature. The active species created in the plasma, such as CH_z radicals had dramatic effect in accelerating the carbon structuring into MWCNTs. The experiment by Gohier et al. [49] thus provided direct experimental evidence of HETs in enhancing the decomposition of precursor(s), and in leading to the release of the R_S species on the nanoparticle surface. Previous studies by Muradov [50] indicated that dangling carbon structures of the type of bumps as observed by Lin et al. [14] could catalytically decompose hydrocarbons. Activated carbon was found to provide the highest initial activity of methane decomposition probably because defective sites were crucial [51] for decomposing the carbon precursors. The HETs (for example, those in SiO_z, ZrO_z, etc.) may otherwise be created by appropriate treatment of the substrate surface or by deposition of an appropriate material on this surface.

12.9 NP1 and NP2 Nanoparticles

12.9.1 Basic Definitions

Based on the discussions made above, and illustrations made in Figs. 12.4a, b and 12.5a, b, there can be two types of nanoparticles, namely NP1 and NP2 nanoparticles [52]. These nanoparticles can both be METANOs and SUBSANOs. Among them, the nanoparticles exhibiting crystalline or near-crystalline surface are NP1 nanoparticles. But nanoparticles exhibiting non-crystalline non-stoichiometric surface due, for example, to doping, sputtering, oxidation, oxygenation, alloying, annealing, and/or surface treatment with plasma, aqua regia, water, laser, ion implantation, etc., are NP2 nanoparticles. Unlike NP1 nanoparticles, NP2 nanoparticles have some distinct surface characteristics. They have, for example, amorphous surface possessing unsaturated dangling bonds, porosity, high-energy sites (HETs), dipole moments, etc., on this surface. They may have grain boundaries, as well. The degree of surface disorder, surface non-crystallinity and surface amorphicity may however vary, although in a judiciously controlled manner. Under no circumstances, they are arbitrary for growth by the VQS mechanism. The nanoparticles are mildly or moderately NP2 nanoparticles if the amorphicity, unsaturated dangling bonds, porosity, high-energy sites (HETs), dipole moments, etc., of them are mild. These nanoparticles are, otherwise, highly NP2 nanoparticles if they have high effective surface amorphicity $\alpha_{amoreff}$, unsaturated dangling bonds, porosity, high-energy states (HETs), dipole moments, etc.

12.9.2 Illustrations

MET≡MET1 and MET≡MET2 of bimetallic METANO = $(MET1)_{1-z}(MET2)_z$ $(z > 0)$ are almost never intimately mixed to produce long range crystal order. Similar lack of long-range crystal order may prevail in non-stoichiometric oxide (e.g., SiO_z, ZrO_z, TiO_z, etc.), trimetallic, and oxidized (oxygenated) bimetallic (trimetallic) nanoparticles. Due to the lack of long-range crystal order, they may possess internal stress and have sizable surface amorphicity, surface porosity, surface melting, and surface coarsening. They may therefore be classified as NP2 nanoparticles [52]. In fact, they may be natural NP2 nanoparticles in the sense that they possess surface disorder and surface amorphicity even without surface treatment. The carbon black nanoparticles are similarly natural NP2 nanoparticles. The surface characteristics of them may though change if they are subsequently surface treated, for example, by laser, plasma bombardment, annealing, acid, water vapor, or sputtering without though losing surface stability. The metal complexes of ferrocene, iron phthalocyanine, cobalt phthalocyanine and nickel phthalocyanine have been found to be very useful for the synthesis of nanomaterials. These metal complexes may decompose at high temperatures into gaseous species and yield very small metal clusters. The composition and characteristics of them may also change at high temperature. The NP2 character and the effective domain of this character of them are therefore functions of temperature. Depending on temperature, they can produce, for example, nanowires of varied diameters.

12.10 Evidences of Phase Transitions

High-resolution images by Sharma [53] showed that structural transformations do occur, for example, in Fe nanoparticle prior to CNT growth. These transformations include oxidation of Fe under low vacuum conditions (e.g., under a pressure of 0.001 Pa). The transformations during heating led to the formation of Fe_3O_4. Upon introduction of the C_2H_2 precursor into the CVD chamber, the face-centered cubic oxide was transformed to body-centered cubic oxide and then to ferrite (bcc α-Fe, an interstitial solid solution). The ferrite was subsequently carburized to Fe_3C (cementite), which mediated the CNT nucleation. Wirth et al. [54] also observed that Fe was carburized to cementite, and that a mixture of ferrite and cementite led to CNT nucleation. Recall that cementitle is a metastable solid, but porous [55] due plausibly to the presence of dislocations [56]. To be more explicit, the transformation was as follows:

$$\mathrm{Fe_2(CO)_9} \rightarrow \mathrm{Fe\ and\ C} \xrightarrow{\mathrm{H_2/H_2O,\ 650°C}} \mathrm{Fe_3O_4\ (fcc)} \xrightarrow{\mathrm{C_2H_2,\ 650\ °C}} \mathrm{Iron\ oxide\ (bcc)} \xrightarrow{\mathrm{C_2H_2,\ 650\ °C}}$$

$$\alpha\text{-Fe (ferrite)} + (\mathrm{CO_2{+}H_2}) \xrightarrow{\mathrm{C_2H_2,\ 650\ °C}} \mathrm{Fe_3C\ (cementite)} \xrightarrow{\mathrm{C_2H_2,\ 650\ °C}} \mathrm{Fe_3C + CNT.}$$

Notably the end result of the transformation was not an oxide. Hofmann et al. [57] noted that the active state of a metal catalyst for CNT growth is not also an oxide. No graphitic network was created on the oxidized Fe.

12.11 Evidences of Phase Transitions and Co-existence of Multiple Phases

Some important elements of the VQS mechanism are phase transitions and the co-existence of multiple phases of the R_L species prompting nanomaterial growths. These have been substantiated by a number of experiments. Cassell et al. [22] observed that metal/oxide mixed phases at nanoparticle surface give rise to porosity ρ_c of this surface. Kohigashi et al. [58] observed that CNT growths are facilitated by non-equilibrium catalyst phase mixtures. Multiple catalyst phases of this mixture co-exist and compete to influence the CNT growths.

We focus, for example, to the equilibrium (Au, Zn) phases [59]. These phases include

1. The fcc terminal solid solution with a maximum solid solubility of 33.5 at.% of Zn in Au;
2. The terminal solid solution with about 7.5 at.% solid solubility of Au in Zn;
3. Four intermediate phases in the composition range between 10 and 30 at.% of Zn, although with poorly defined phase boundaries;
4. The intermetallic compound Au_5Zn_3 that formed peritectoidal alloy at about 300 °C;
5. The β' phase with a maximum solubility range between 38 and 57 at.% of Zn at very low temperatures (lower than 100 K);
6. The δ phase with about 56 at.% of Zn, leading to the formation of peritectoidal alloy at 180 °C;
7. Three intermediate phases in the composition range between 63 and 83 at.% of Zn, although with no well-established phase boundaries; and
8. The ε phase stable below 490 °C in the composition range between ~84 and 89 at.% of Zn.

Many of the Au-catalyzed ZnO nanowires believed to be grown by the VLS mechanism were carried out at 800–900 °C. These nanowires by Campos et al. [60] were though grown at 300 to 350 °C. Au-catalyzed nanowires, thus produced on Si(100) wafers, were 20–100 nm in diameter. Synchrotron X-ray diffraction pattern indicated that the nanowire growth was catalyzed by a solid solution of γ-AuZn and β'-AuZn: namely, $R_L \equiv (\gamma\text{-AuZn}, \beta'\text{-AuZn})$. The major contribution to it was from cubic gamma brass structure, namely, γ-AuZn (lattice parameter, $a = 0.922$ nm), and the minor contribution to it was from cubic beta structure, β'-AuZn ($a = 0.314$ nm). Being in an environment of oxygen, the said solid solution might have contained also oxygen. The R_L species was thus a solid solution and amorphous (semi-amorphous, amorphous-like) with a network of nanopores. Bayer et al. [61] noted that there can be mixtures of α-Fe and γ-Fe phases of Fe nanoparticles during treatment before CNT growth. It was due to the presence of even minor carbonaceous background contaminants typically present in CVD reactors. Complex metallic Fe and iron carbide mixtures might have evolved during subsequent hydrocarbon exposure at a suitable temperature T [54]. Hu et al. [62] studied the phase changes as function of temperature of the Mg/C system. They found that during increase in temperature, a mixture of Mg and Mg_2C_3 was formed at 415.7 °C. This phase then changed to metastable MgC_2 phase at 597 °C, and finally to metastable Mg_2C_3 phase at 657 °C. Similar observation by Singhal et al. [63] was made during BNNT growth via the formation of metastable intermediate compounds such as nanostructured BN-M phases of the compound B-N-O-M (M: transition metal). All these appear to duly support the basic tenets of the VQS mechanism.

12.12 Evidence of Phase Separation

We cite two examples of phase separation of nanoparticle found necessary for SWCNT growths. He et al. [64] grew SWCNTs using $Co_zMg_{1-z}O$ nanoparticle and CO precursor. This $Co_zMg_{1-z}O$ nanoparticle is actually a solid solution. The dynamics of the nucleation of nanoparticles, and also the temporal evolution of the solid solution under the CO reduction were recorded in real time in microscope. The dynamics led to the phase separation of $Co_zMg_{1-z}O$ into Co and MgO during growth at high temperature. Co, resulting from the phase separation of $Co_zMg_{1z}O$ solution, migrated to the surface and crystallized into metal nanoparticles. MgO, on the other hand, remained as support. The migration of Co to the surface from the $Co_zMg_{1-z}O$ solid solution was though interrupted by its strong interaction with MgO of the solid solution. HRTEM bright-field image showed Co nanoparticles homogeneously distributed on the top of the MgO support. Very large lattice mismatch (~16%) between Co and MgO and strong interactions between them, led Co nanocrystal to accommodate the lattice of the MgO matrix. It was though severely strained exhibiting the metastable face-centered cubic (fcc) phase. It might have been porous as well during growth. He et al. [65] grew SWCNTs employing also $Fe_zTi_{1-z}O$ nanoparticle and CO precursor. This $Fe_zTi_{1-z}O$ nanoparticle is also a solid solution. In situ ETEM experiments recorded the reduction of the $Fe_zTi_{1-z}O$ solid solution by CO precursor at 700 °C. They observed again the phase separation of $Fe_zTi_{1-z}O$ into Fe and TiO, and the migration of small-diameter Fe particles to the top of TiO. The strong interaction between the Fe metal and the support led to the agglomeration of the metal particles under the prevailing reaction condition; but it induced also the metallic particle to adopt a pillbox-like porous morphology. These observations suggest that the SWCNT growths took place actually by the VQS mechanism.

12.13 Experimental Evidences of the Benefits of Surface Treatments

We described earlier in this chapter that nanoparticle surface characteristics are keys to the nanomaterials growths by this nanoparticle. And these surface characteristics can be realized by many different surface treatments as manifested from a number of experiments. We cite a few of them.

12.13.1 Surface Treatment Yields Surface Amorphicity and Surface Roughness

Newby et al. [66] showed that surface treatment, such as irradiation fluence of both oxidized and non-oxidized samples gives rise to surface amorphicity of these samples. The surface amorphicity of the samples as function of irradiation fluence, as observed by them (see Fig. 12.6a) can be controlled to a desired level by appropriately choosing the irradiation fluence. And this is very important for growths by the VQS mechanism. Gupta and Gupta [67] studied the effects of nitrogen partial pressure on the composition of FeN and NiFeN films. This is shown in Fig. 12.6b, which indicates that interatomic spacing of these films increase almost linearly with increase in the N partial pressure. This implies that FeN and NiFeN films become increasingly disordered due to increased N incorporation into them. This should be true for other films, as well. Fu et al. [68] carried out surface annealing of the TiNiCu film and observed that increase in annealing temperature accompanied increase in surface roughness of the film. The results of Fu et al. shown in Fig. 12.6c suggest

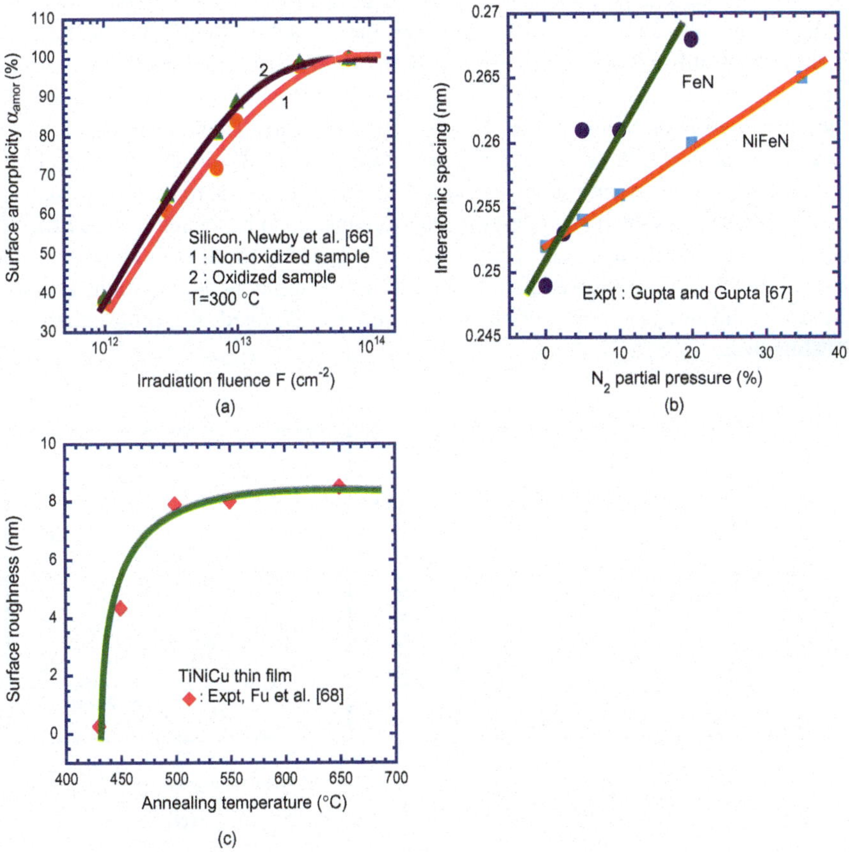

Fig. 12.6 **a** Variation of Si surface amorphicity as function of irradiation fluence; the plot is made with experimental data by Newby et al. [66]; **b** variation of interatomic distance of FeN and NiFeN as function of N_2 partial pressure; the plot is made with experimental data by Gupta and Gupta [67]; and **c** variation of surface roughness of TiNiCu film with annealing temperature which causes this roughness; the plot is made with experimental data by Fu et al. [68]

that surface roughness of a film can, in fact, be governed both by surface annealing time and surface annealing temperature.

12.13.2 Surface Treatment Yields Surface Porosity

A study by Saha et al. [69] on the surface treatment of poly(ethylene oxide) is interesting. As presented in Fig. 12.7a, this study demonstrates that surface porosity resulting from gamma ray irradiation increases and reaches a peak and then decreases with further increase in the radiation dose. It is attributed to the void (i.e., empty) spaces in the material content of the surface generated by the displacement of the atoms of the lattice. However, the displaced atoms become very mobile under the influence of high irradiation dose, and they move back to their original locations if they are highly mobile with excessively high gamma ray radiation. The surface treatment should therefore be optimal in order to yield the best results. In this context, the investigation by Sämann et al. [70] is also interesting. They made use of pulsed laser irradiation to achieve nanopores in Si film. The results depicted in Fig. 12.7b indicate that porosity and hence nanopore radius increased, reached a peak, and then decreased with increasing laser irradiation fluence. It happened in both PECVD annealed and sputtered anneal samples. And it happened for the same reason stated above. The two experiments cited above are remarkable as they demonstrate that nanopore and nanopore radius of a nanoparticle can be dictated by appropriate surface treatment(s).

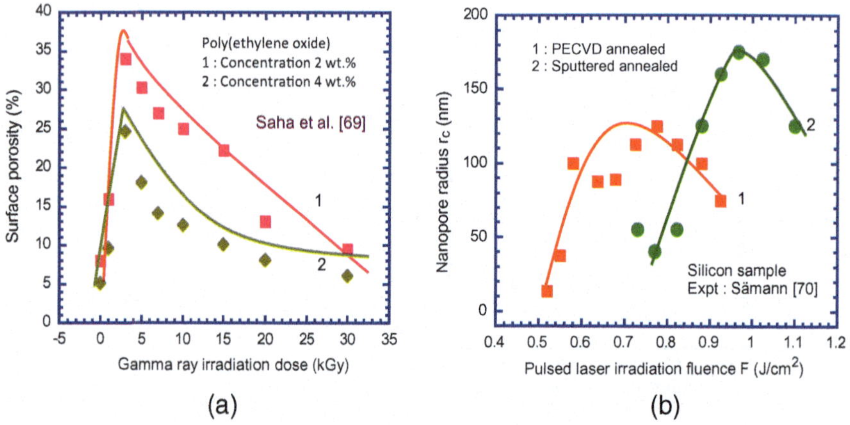

Fig. 12.7 **a** Variation of surface porosity of poly(ethylene oxide) film with gamma ray irradiation; it is drawn with experimental data by Saha et al. [69]; and **b** variation of nanopore radius of PECVD annealed and sputter annealed Si film with pulsed laser irradiation; it is drawn with experimental data by Sämann et al. [70]

12.13.3 Optimal Level of Defect and Amorphicity Essential for Nucleation and Growth

Indeed, the optimal level of surface disorder, effective surface amorphicity, surface porosity, and surface coarsening of the R_L species of nanoparticle surface is crucial for high growth of nanomaterials on this nanoparticle surface. Desired number of grains and grain boundaries on the R_L species surface and the level of defects in the grain and grain boundary may similarly determine the nanomaterial to be grown on it. Indeed, the optimal level of surface disorder, effective surface amorphicity, surface porosity, and surface coarsening of the nanoparticle surface are crucial for high growth of nanomaterials on this nanoparticle surface. We cite some examples, Li et al. [71] grew carbon nanotubes on the *as grown* and the damaged surfaces of stainless steel substrates. The damage of the stainless steel surface was caused by ion beam bombardment. A thin Fe catalyst film with a nominal thickness of 1 nm was formed on the stainless steel-supported $Al_{z'}O_z$ (z' and z are the Al and O mole fractions, respectively) layer. The CNT growth was carried out at 750 °C using a gas mixture of C_2H_4 and H_2. The results are shown in Fig. 12.8, which indicate that the CNT length increases with increase in growth time. CNTs grew far more rapidly on the damaged substrate surface than on the *as grown* surface. We argue that it happened because the Fe nanoparticle formed on the damaged substrate, rather than on the *as grown* substrate, became amorphous and porous. This is consistent with the observation by Ong et al. [72] that Ni nanoparticle formed on amorphous

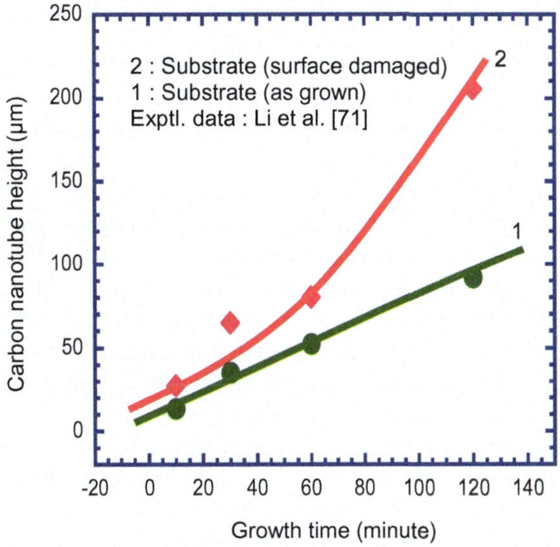

Fig. 12.8 Experimental dependence of the height of carbon nanotube on growth time. The nanotube was produced on (1) *as grown* steel substrate surface and (2) surface-damaged steel substrate. The plots are made with the experimental data by Li et al. [71]

carbon substrate became amorphous solid solution due to diffusion of carbon atoms from amorphous carbon substrate into the Ni nanoparticle. Shekari et al. [73] grew GaN nanowires on *as grown* Si(111) and porous silicon substrates using thermal evaporation method. The nanowires grown on porous silicon substrate were thinner, longer, and denser than those on *as grown* Si substrate. This confirmed the role of surface porosity and surface amorphicity in superior nanomaterials growths. And our discussions in sections 12.13.1 and 12.13.2 indicate that parameters such as surface porosity and surface amorphicity of a nanoparticle can be precisely controlled by appropriate surface treatments.

Sobanska et al. [74] reported on plasma-assisted molecular beam epitaxial growth of GaN nanowires on a thin amorphous Al_2O_3 layer produced by atomic layer deposition on Si(111) substrate. The amorphous Al_2O_3 layer dramatically enhanced the spontaneous GaN nanowire nucleation. This nucleation was slower on partially amorphous silicon nitride films, but absent on sapphire films under identical growth conditions. The finding demonstrates that an optimal level of defect density and amorphicity of the substrate (nanoparticle) surface is essential for nanowire nucleation.

12.14 Distinctive Features of Nanomaterials Syntheses by the VQS Mechanism

12.14.1 Transformation from Vapor Phase to Solid Phase

We argue that the R_S species are rarely transformed from the vapor phase to directly the solid phase for nanomaterial growth. This transformation instead takes place, for example, for the MET-mediated growths, from the vapor phase to the solid phase via the liquid (quasiliquid, quasisolid) phase. So, it involves vapor phase, liquid (quasiliquid, quasisolid) phase, and solid phase, with the liquid (quasiliquid, quasisolid phase, see Fig. 12.9) lying in between the vapor phase and the solid phase. Figure 12.9 shows possible variation of nanoparticle porosity ρ_c and the diffusivity D_c of the R_S species through this nanoparticle as function of the liquid (quasiliquid, quasiliquid) state of the nanoparticle. Note that the quasiliquid (quasisolid) state for the VQS mechanism lies in between the liquid molten state for the VLS mechanism and the solid crystalline state for the VSS mechanism for growths. Depending on growth conditions (for example, pressure), the porosity of the nanoparticle for VQS growth may vary. There may be many different pathways for the variation of porosity ρ_c. Only five pathways (pathway 1–5 in Fig. 12.9) for the variation of porosity, and hence for the diffusivity of the R_S species as function of quasiliquid (quasisolid) state of the nanoparticle have been shown. Note that, if the intermediate phase is quasiliquid or quasisolid, it may be very close to the solid phase, very close to the liquid phase, or somewhere in between the solid phase and the liquid phase. It may, for example, be 99.99% solid and 0.01% liquid. It may otherwise be 0.01% solid and

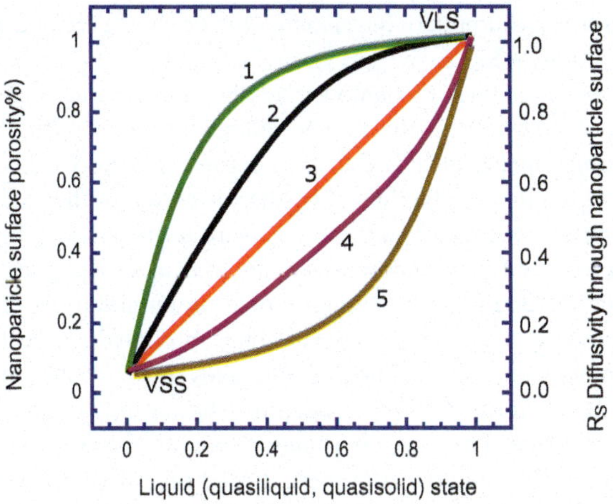

Fig. 12.9 Nanoparticle surface porosity and the R_S diffusivity through the nanoparticle surface as function of liquid–solid (quasiliquid, quasisolid) state of the nanoparticle surface. Note that the growth by the VSS mechanism corresponds to solid [or probably marginally liquid (quasiliquid, quasisolid)] state of nanoparticle; the porosity of this nanoparticle surface is almost zero and hence the R_S species diffusivity through the nanoparticle surface is also marginally low. On the other hand, the growth by the VLS mechanism corresponds to the completely liquid state of the nanoparticle droplet, and the porosity of this nanoparticle surface is high; the R_S species diffusivity through the nanoparticle surface is also high

99.99% liquid. This means the VLS and VSS mechanisms are practically the special cases of the VQS mechanism.

12.14.2 Formation of Intermediate Quasiliquid (Quasisolid) Phase

Being vapor, the R_S species generally possess very high kinetic energy at the growth temperature. They collide with each other (one another), but hardly bond to each other (one another). Hence, they have hardly an interatomic (intermolecular) distance equal to the solid-state interatomic (intermolecular) distance, which is needed for them to be transformed into solid. In order for this to happen, $R_S \equiv X$ and $R_S \equiv Y$ species must lose energy and be less mobile and less kinetic at the growth temperature and pressure. And it would be best possible if they lose some energy by diffusing through some liquid, quasiliquid, or quasisolid medium, namely the one possessed by the R_L species of the nanoparticle, as shown in Figs. 12.1a, b and 12.2a, b. The loss of their energy would be realized via collisions with the atoms and/or molecules of the R_L species. For the MET-mediated VLS growth of nanowires (nanotubes), this R_L species would be the $R_L \equiv$(MET, X) eutectic liquid alloy created at a temperature $T = T_E$. For MET-mediated non-VLS growth of nanowires (nanotubes), this R_L species would however be quasiliquid, or quasisolid alloy or cluster formed at a temperature $T < T_E$, $T = T_E$ or $T > T_E$.

We cite some examples to demonstrate the intermediate liquid (quasisolid, quasiliquid) phase formed for the transformation of vapor-phase R_S species into the solid-phase nanomaterial. Li et al. [75] argued that SLS nanowire nucleation and subsequent growth were facilitated by the formation of thermally-etched pyramidal pits on the Si substrate. And these pits were formed during high-temperature annealing prior to the onset of nanowire growth. Li et al. [75] suggested that the SLS nanowire nucleation and growth were promoted preferentially at the pit edges. He et al. [76–83] grew GaN nanowires and nanotubes through direct reaction of gallium vapor with ammonia in CVD chamber. SEM images showed amorphous GaN matrix with pits and hillocks formed during annealing. There were also platelets with polycrystalline hillocks and pits. The edges and corners of these platelets were rich in defects and nanowires emerged from them. Based on cross-sectional TEM images of *as grown* nanowires, Zi et al. [84] emphasized that pits were actually the preferred sites for nanowire growth. Cuscunà et al. [85] performed surface treatment of flat Si substrate surface prior to nanowire growth and found that, due to surface treatment, this surface was coarsened with hillocks of heights of 10–15 nm on it. Nanowires grew from them (e.g., hillocks). All the observations cited above-demonstrated nanowire growths, not on the flat substrate surface, but at the sharp tips and at edges of pits and hillocks. These sharp edges of pits and hillocks are shown schematically in Fig. 12.1a, b. Unlike the flat surface, they have small dimension. And hence they undergo size-dependent mesoscopic effect [4–6]. As shown in Fig. 11.3 of Chap. 11, they are molten on the surface, but solid underneath this surface. This tip (edge) surface becomes semimolten at relatively low temperature. As a result, a liquid or a semi-liquid phase is set in between the solid phase and the R_S vapor phase. This is very important as it demonstrates that the R_S vapor species land on the molten tip (edge) surface, diffuse through this molten tip (edge) surface to the solid tip (edge) surface, and then nucleate as nanomaterial on the solid tip (edge) surface.

12.14.3 The Need of Streamlining for Porosity and Growth

During diffusion through the liquid, or through the nanoporous quasiliquid (quasisolid) medium of the R_L species, the R_S species should be streamlined and should come close to one another so that they undergo the reaction: $mX + nY \rightarrow X_mY_n$ to produce X_mY_n molecules. To form solid nanomaterial(s), these X_mY_n molecules would next be supersaturated before being solidified into nanomaterial (e.g., nanowire, nanotube, nanodot, etc.). But, to form, for example, single-element nanomaterial, the $R_S{\equiv}X$ species must first lose kinetic energy during diffusion through the liquid, quasiliquid, or quasisolid R_L species and then be supersaturated before nucleation into nanomaterial. The supersaturation must take place at the L/S or QL/S (QS/S) interface (see Figs. 12.4b and 12.5b). The rate of diffusion of the $R_S{\equiv}X$ species, $R_S{\equiv}Y$ species, and/or X_mY_n molecules through the liquid or the nanoporous semi-liquid (quasisolid) medium of the R_L species would depend on (1) the shape and size of the $R_S{\equiv}X$ species, $R_S{\equiv}Y$ species, and/or the X_mY_n molecules, (2) the liquid (liquid-like) state of the R_L species, and (3) the porosity of the R_L species. For porous, but non-liquid R_L species, the porosity would be a function of pore radius r_c and the density n_c of nanopores, as shown in Fig. 12.10a–f. Both the pore radius r_c and the pore density n_c of this species would be governed by the composition of the R_L species and the growth parameters such as temperature and pressure [86, 87]. The porosity ρ_c would be marginally small (see Fig. 12.10a) if the R_L lattice structure is almost solid (e.g., $\xi_m \approx 0$) and has hardly any network of molten (semimolten) nanopores in it. This might occur at temperatures $T < T_E$. The porosity ρ_c of the nanoparticle of Fig. 12.10b is, on the other hand, very high as it is molten (e.g., $\xi_m = 1$). It is most suitable for the VLS-like growth. The porosity would also be high (see

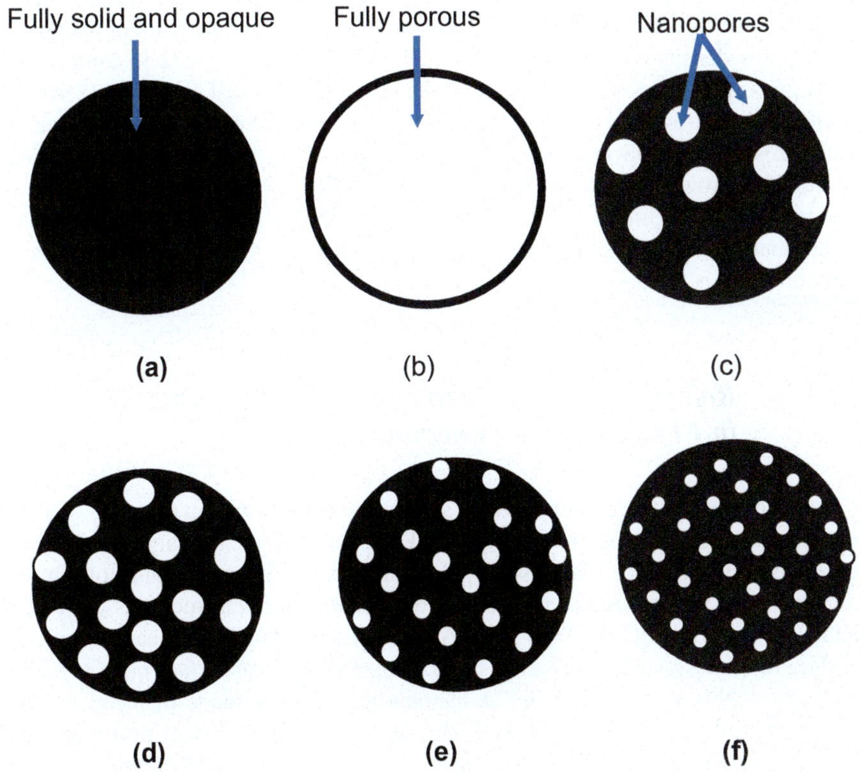

Fig. 12.10 Schematically shown nanoparticles of varied surface porosity; **a** nanoparticle with fully crystalline solid and opaque surface; **b** nanoparticle with fully porous surface; **c** nanoparticle with fewer, but large nanopores on the surface; **d** nanoparticle with relatively smaller, but relatively larger number of nanopores on the surface; **e** nanoparticle with smaller, but larger number of nanopores on the surface; **f** nanoparticle with very small nanopores, but very large number of these nanopores on the surface

Fig. 12.10f), if the R_L lattice structure of nanoparticle has very high density of finite-sized nanopores. The porosity of the R_L species of nanoparticles of Fig. 12.10c–e would however be somewhere between low and high as the R_L lattice structure of them is solid, but has a network of finite-sized molten (semimolten) nanopores. If there is a horizontal (e.g., perpendicular to the nanomaterial axis) nanostripe molten (semimolten) at the growth temperature, the R_S species (X_mY_n molecules) would enter these stripes and coalescence creating a ledge at the L/S (QL/S, QS/S) interface. The ledge is the result of mass nucleation of the R_S species (X_mY_n molecules). The semimolten nanostripe would remain intact during growth; it would be pushed up and would remain at the nanomaterial tip. The extent of applicability of the VSS, VQS , and the VLS mechanisms would depend on (1) the pore radius r_c and the pore density n_c, and, of course, on (2) the porosity ρ_c of nanoparticle surface, as depicted in Fig. 12.10a–f. Note that the VSS growth corresponds to marginally low pore density (see Fig. 12.10a) of solid crystalline lattice structure of the R_L species. This R_L species is essentially solid; it may have at the most a network of nanopores of very low density and of marginally small pore radii. So, $\xi_m \approx 0$. In contrast, the VLS growth, corresponds to very large porosity, viz., $\xi_m = 1$ (see Fig. 12.10f) of molten lattice structure of the R_L species. Depending on the pore radius, pore density and porosity, the VQS growth can be close to the VSS growth if $\xi_m \approx 0$ or close to the VLS growth if $\xi_m \approx 1$. The R_L species of Fig. 12.10f has a high density of nanopores. The regimes of applicability of the VSS and the VLS mechanisms are quite small. The regime of applicability of the VQS mechanism is, on the other hand, very broad; it is for $0 \leq \xi_m \leq 1$.

12.14.4 Role of Catalyst Support and Dipole Moment in FECA Surface Functionalization

Various features of FECA nanoparticle support were described in Chap. 4. The role of support material in functionalizing the FECA nanoparticle's catalytic activities is described in this sub-section. This support material should be such that it contributes, together with surface treatments, to the functionalization that modifies the FECA nanoparticle (both FECANO and SUBSANO) surface and facilitates the amorphicity, surface roughness, surface porosity, surface polarization, and surface melting (semi-melting) of this surface. The FECA nanoparticle may be made of metal or non-metal, or both. To serve its purpose, the support material should be porous and non-stoichiometric and have strong chemical and/or physical interactions with the FECA nanoparticle material. Amorphous (semi-amorphous) oxides such as alumina and silica would be very suitable [71, 88] as such support materials. The support material should generate functional components on the lattice structure of the FECA surface and may have heteroatoms, such as oxygen, boron, nitrogen, phosphorus, and sulfur incorporated into the FECA lattice framework. It could be achieved [89, 90], for example, by a number of means:

1. A judicious choice of the electronic structure matching of the metal and of the support;
2. Creation of bonds between the surface atoms of the FECA nanoparticle and of the support;
3. Breaking of the long-range electronic structure leading to the delocalization of charges;
4. Enhancement of local deformation, corrugation, and defect introduction into the FECA lattice structure;
5. Yielding of the morphological changes of the FECA nanoparticle due to such deformation, corrugation , and defect formation;
6. Redistribution of electron density of the FECA surface;
7. Charge transfer from support to the FECA nanoparticle or vice versa; and
8. The freeing of atoms near the adsorption sites thus increasing the dangling bonds at the FECA nanoparticle (e.g., METANO or SUBSANO) surface.

All these should take place at relatively high temperature during growth. They should ensure the role of vacancies in the interdiffusion of atoms from FECA nanoparticle into the support and vice versa. They should confirm that the composition of FECA clusters formed near the top of the FECA nanoparticle surface due to diffusion of atoms, molecules, or ions from the support are very important. The introduction of heteroatoms into the FECA nanoparticle lattice structure should generate defects and clusters, and all of them should tend to alter the FECA nanoparticle surface structure. They should also act as active sites. They should additionally help generate pore texture of the resulting mesoporous FECA nanoparticle structures. The doping with some appropriate elements can lead to some enhanced insertion of oxygen functionalities into the FECA nanoparticle structure.

12.14.5 Illustrations of the Role of Catalyst Support in FECA Nanoparticle Surface Functionalization

It was found that FECA nanoparticle formed on nanoporous zeolite support [91, 92] resulted in significantly high yields of CNTs of narrow diameter distribution. Zhu et al. [93] formed Fe and Co FECA nanoparticles on mesoporous silica and found that this mesoporous silica played a determining role in guiding the initial nanotube growth. We argue that this happened because the nanoporous silica-induced porosity was realized in the FECA nanoparticle lattice. And also various interactions, particularly, the physical interactions such as van der Waals and electrostatic forces between the support and the FECA nanoparticle were strong. They were so strong that the FECA nanoparticle movement on the support surface was prevented and the sintering of FECA nanoparticle on this surface was eliminated. The size distribution of the FECA nanoparticle was consequently stabilized during growth.

12.14.6 *Amphoteric Characteristics of Catalyst Support Should Be Preferred*

The amphoteric characteristics of the support material which give rise to both acid and base sites should promote the chemical interactions between the FECA nanoparticle and the support. There can be an increase in electron density of the FECA nanoparticle due to these interactions. And such an increase should translate into donation of electrons from FECA nanoparticle to precursor(s) enhancing the rate of precursor decomposition on the FECA nanoparticle surface. Aluminum oxide has generally been observed to be a superior support compared to silicon dioxide because it possesses both the acid and base sites and have stronger chemical interactions with FECA nanoparticles. Tripathi et al. [94] and Rümmeli et al. [95] noted that this material can even serve as both the support and FECA for superior nanomaterials growths. Binninger et al. [96] performed theoretical calculations to demonstrate that electronic equilibration between FECA metal nanoparticles (e.g., METANOs), and support material gives rise to charges at the catalytically active outer surface of FECA nanoparticles. Also, the charge transfer from the metal to FECA and vice versa is proportional to the work function difference between the FECA nanoparticle material and the support material. It is dependent as well on the shape, size, and proximity of the FECA nanoparticles.

12.14.7 *Catalyst Support Should Enhance FECA Nanoparticle Surface Disturbance and Polarity*

The contribution of the support to achieving an optimal level of surface disturbance, disorder, and amorphicity of the FECA nanoparticle surface is highly desirable. These nanoparticles should preferably be the NP2 nanoparticles. Increase of charges on the top FECA surface due to surface treatments and catalyst support gives rise to two significant layers of opposite charges aligned side-by-side on this surface [97]. A dipole moment Θ_M is thus created on the NP2 nanoparticle surface. And this dipole moment exists all over this surface for nanowire growth, but only over the shell surface for nanotube growth. It exists along the grain boundary between two neighboring grains for nanobelt growth. Note that there occurs charge transfer from the bulk to the shell surface of NP2 nanoparticle for nanotube growth. No dipole moment exists over the NP1 nanoparticle surface.

The entire NP2 nanoparticle surface for nanowire growth has (1) high density of charges, (2) a large number of dangling bonds, (3) high densities of HETs and nanopores, and (4) a large dipole moment. But due to charge transfer to peripheral surfaces for nanotube growth, only the shell formed on NP2 nanoparticle surface has high density of charges, a large number of dangling bonds, and high densities of HETs and nanopores. It also has large dipole moment. The large dipole moment enhances the landing of precursor molecules on the R_L species surface during growth. The HET reactivity resulting from unsaturated accumulated charges leads to decomposition of the precursor molecules at temperature lower than the thermal decomposition temperature. The R_S (R_S≡X and R_S≡Y) species released from this decomposition are aligned along the contour of charges created on the nanoparticle surface and undergo bulk (volume) diffusion through the nanopores for nucleation and growth.

In the absence of shell (hill) [52] formation, the entire NP2 nanoparticle surface on the top of the support possesses significant surface disturbance, porosity, amorphicity, HET concentration , and dipole moment. If the nanoparticle surface has two or three grains separated by grain boundaries, and the surface disturbance, porosity, amorphicity, HET concentration and dipole moment are present in each of these grains, two or three nanowires could grow on the nanoparticle surface. Each of these nanowires would be on each of the grains. They (e.g., amorphicity, porosity, HET concentration, dipole moment, etc.) would all promote nanowire growth. The situation changes with the formation of shell (hill) at the nanoparticle periphery. The shell (hill) is formed due to surface segregation of the R_S species from the bulk to the shell of FECA nanoparticle [52] during the pre-nucleation stage of growth. In general, the higher the surface segregation, the lower is the concentration of charges in the bulk, the lower is the dipole moment in the bulk, but the higher is the dipole moment in the shell. The formation of armchair (near armchair) SWCNTs in the shell corresponds to a significant dipole moment in the shell. On the same scale, the formation of zigzag SWCNTs in the shell corresponds to less-than-significant dipole moment in the shell. These are consistent with the results of DFT calculations [98, 99], which suggest that the C–C triple bonds in the shell are formed for armchair SWCNTs and that they correspond to a finite charge and a large dipole moment at the nanoparticle edge. The calculations could not point to any C–C triple bonds on the nanoparticle surface suitable for yielding other SWCNTs (e.g., zigzag SWCNTs and chiral SWCNTs).

12.15 Nanomaterial Growths by Low-Melting Point Metals

Metal nanoparticles may be of two different types: namely, the high melting-point metal nanoparticles and the low melting-point metal nanoparticles. In the high melting-point metal nanoparticles, the melting point T_M of the metal is generally higher than the eutectic temperature T_E of its R_L≡(MET, X) eutectic alloy [59] with the source species R_S≡X. For example, the melting point of Au is 1064 °C. But the eutectic temperatures T_E of its eutectic alloys (Au, Si) , and (Au, Ge) with R_S≡Si and R_S≡Ge are 363 °C and 361 °C, respectively. In the low melting point metal nanoparticles, the melting point T_M of the metal may be almost identical or comparable to the eutectic temperature T_E of its R_L≡(MET, X) eutectic alloy with the source species R_S≡X. The melting points of Ga, In, and Sn are 29.7 °C, 156.6 °C, and 231.9 °C, respectively. But the eutectic temperatures of (Ga, Si), (In, Si) and (Sn, Si) eutectic alloys are 29.8 °C, 156.0 °C, and 232.0 °C, respectively [59]. Table 12.2 lists R_L≡(MET, Ge) eutectic temperatures and the Ge mole fractions in the

Table 12.2 Melting point, eutectic temperature, and growth temperature of various low-melting point FECA metals (e.g., METs) used for the growth of Ge nanowires

FECA Metal	R_L≡(MET, Ge) eutectic temp (°C)	Ge mole fraction in R_L≡(MET, Ge) eutectic alloy		Observed MET mole fraction in R_L≡(MET, Ge) alloy	
		Mole fraction	References	Mole fraction	References
Ga	29.7	0.070	[100]	8.0–10.0	[100]
In	156.6	0.002	[101]		–
Sn	231.6	0.003	[101]	4.3	[104]
Sn	231.6	0.003	–	28.0	[105]
Bi	271.0	0.001	[102, 103]	2.0	[106]

(MET, Ge) eutectic alloy compositions for Ga, In, Sn and Bi MET nanoparticles [100–103], respectively. This table also lists MET incorporation in the Ge nanowires [100, 104–106] produced by the mediation of this MET. R_L≡(In, Si) eutectic alloy formed at a eutectic temperature of $T_E = 156.5$ °C is comparable to the In melting point of $T_M = 156.6$ °C. Only 0.001 at.% Si is present in this eutectic alloy. Yet Scala et al. [107] obtained $In_{1-z}Si_z$ alloy with Si atomic % $z = 0.15$. In the same way as this observation, Table 12.2 suggests that both MET and Ge contents in a (MET, Ge) alloy can be significant and that R_S≡Ge species can be adsorbed by MET if released on the MET surface. Such adsorption would convert the MET surface to R_L species surface containing both MET and Ge, namely R_L≡(MET, Ge). It would be true also for R_L≡(MET, Si). Note that a (MET, X) alloy with R_S≡Si, Ge, etc., would be eutectic only under a stringent condition of some specific temperature and component mole fractions. This condition would pertain to very low content of X and very high content of MET in the R_L≡(MET, X) alloy. An alloy with significant content of both MET and X would not obviously be eutectic; it would rather be non-eutectic. It would be solid, but a cluster or solid solution. Hence, the growth mechanism for low-temperature nanowire growths by low melting point metals would actually be the VQS mechanism, rather than the VLS mechanism.

12.16 Nanomaterials Tips

Supersaturation takes place if and when the solubility of the R_S species in the FECA nanoparticle (METANO or SUBSANO) is significantly reduced. Crystallization into the nanomaterial phase is induced by this supersaturation. Crystalline growth of nanomaterial begins and continues at the L/S, QL/S, or QS/S interface. As this interface is the only interface, the growing nanomaterial phase exhibits a pseudo-one-dimensional morphology. The one-dimensional growth continues until the delivery of the R_S species onto the FECA nanoparticle surface is discontinued. The melting point of the FECA material, the solubility of the R_S species into the R_L species, and the reactivity of the R_L species with the R_S species are the keys to the suitability of the FECA nanoparticle for nanomaterial growth. At the end of growth, FECA nanoparticle, in its original form or in a form significantly altered by the growth condition(s), may be observed at the nanomaterial tip (for example, nanowire tip). And, as shown in Fig. 12.11, it may have many different shapes and sizes and it may be flat, circular, or hemispherical. The ability, performance, and effectiveness of the

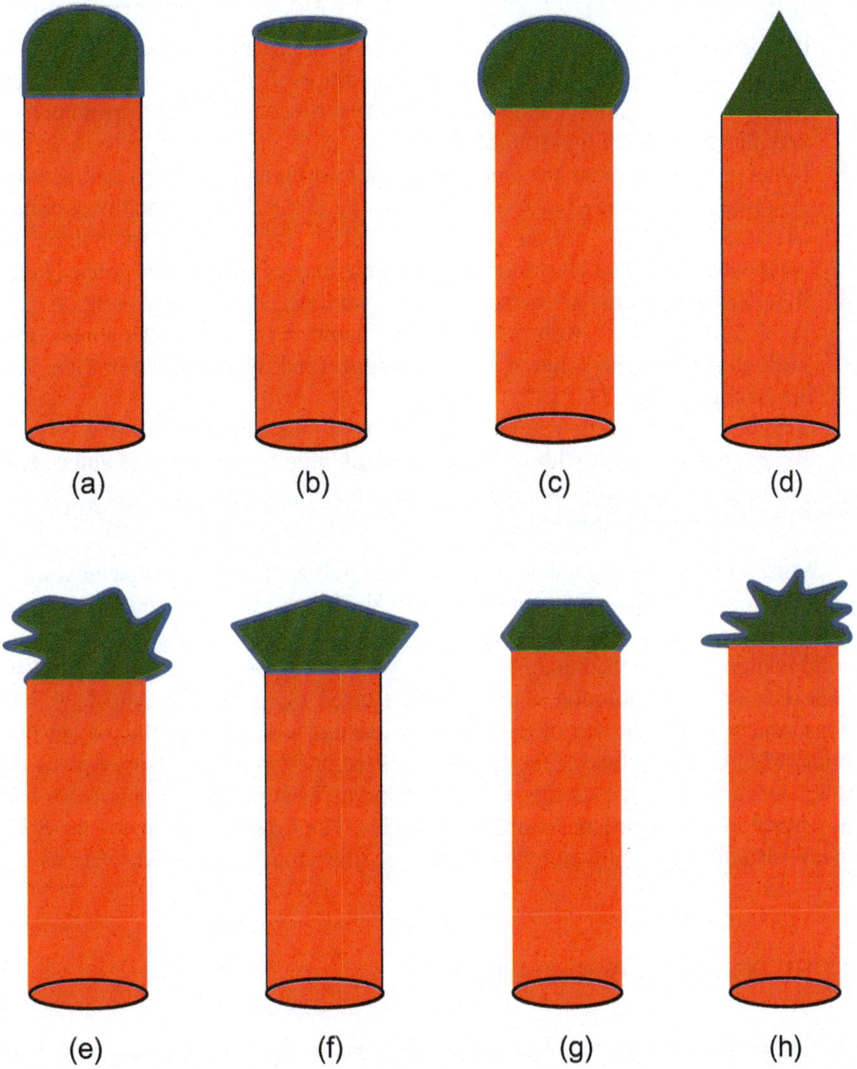

Fig. 12.11 Schematic diagrams of eight different nanowires (shown in red) of eight different shapes and morphologies of the tips (shown in green). These include tips of regular flat tips (see Fig. 12.10b) and regular hemispherical tip (see Fig. 12.10c)

R_L species in mediating nanomaterials growths may be reflected, at least in part, from the shape, size, morphology, and lattice structure of the nanomaterial tip. To repeat, this tip can have regular or irregular shape, size, and morphology. The component elements of the R_L species (viz., β_1, β_2, β_3, β_4, β_5, β_6, etc.) at the nanomaterial tip can be relatively unstable if some or all of them are unreacted.

1. If reacted, the said assemblage of materials (e.g., β_1, β_2, β_3, β_4, β_5, β_6, etc.) may yield a molten eutectic alloy, and the nanomaterial tip can be hemispherical in shape [108].
2. If unreacted or partially reacted, the same assemblage of materials (e.g., β_1, β_2, β_3, β_4, β_5, β_6, etc.), with or without the addition of contaminant(s), may yield cluster or solid solution, and the nanomaterial tip composed of them may have very flat and smooth tip [109, 110].
3. If one or more of them has a melting point lower than the growth temperature, the same assemblage of materials (e.g., β_1, β_2, β_3, β_4, β_5, β_6, etc.) may remain unreacted yielding a cluster or a solid solution. But yet the nanomaterial tip composed of the said cluster or solid solution may be of spherical shape [111].
4. If unreacted or partially reacted, the same assemblage of materials (e.g., β_1, β_2, β_3, β_4, β_5, β_6, etc.), with or without the addition of foreign element(s), may yield cluster or solid solution and the nanomaterial tip composed of them may have circular shape [112].
5. If unreacted or partially reacted, the same assemblage of materials (e.g., β_1, β_2, β_3, β_4, β_5, β_6, etc.), with or without the addition of foreign element(s), may yield cluster or solid solution and the nanomaterial tip composed of them may have very irregular shape [113].

While growing CdTe quantum wires, Wang and Buhro [114] noted that the catalyst composition was converted from Bi to $Bi_xCd_yTe_z$, and that it could be liquid or solid. They also noted that the quantum wire growth could be achieved only by the solid composition of the catalyst. Obviously, it could not be crystalline solid; we believe it was solid solution with irregular faceted morphologies. And they were consistent with the growth of the wurtzite quantum wires from solid, rather than liquid FECA nanoparticles. But they were different from the nanowire tips by the SoLS mechanism. For example, the Bi nanoparticles at the tips of nanowires had, in general, pseudo-hemispherical morphologies. The CdTe nanowires by the SoLS mechanism though exhibited high densities of zincblende-wurtzite alternations.

12.17 Concluding Remarks

The VQS mechanism and the crucial elements of this mechanism have been described. The impact of various growth parameters on this mechanism has been discussed. Attempts have been made to establish fundamental physicochemical framework for growths by the VQS mechanism.

12.17.1 Surface Energy

Surface energy plays a critical role in growths by the VQS mechanism. In many experiments, nanowires and even nanotubes produced, for example, by the CVD technique, are coated with oxide shell. This may be explained by taking into consideration the presence of more than one species not bound well to each other, while on the nanoparticle surface. These species possess individualized surface energy listed in Table 12.3. The entries in this table are obtained from varied sources of the literature and plausibly determined under diverse conditions. The data may not, therefore, be very precise and consistent. They may nevertheless provide an approximate measure of the surface energy of various elements, oxides, and semiconductors listed in Table 12.3. This table indicates that the surface energy of the metal or semiconductor oxide species is generally smaller than that of the corresponding metal or semiconductor. For example, under identical condition, the surface energy of SiO_2 is much smaller than the surface energy of Si.

Note that the nanoparticle has its own surface energy. And this surface energy is higher at the peripheral surface than in the core (bulk) of it. It best achieves its equilibrium if its surface energy is equal or at least approximately equal in every location of it. It becomes possible at least in part if the low-surface energy species, such as metal oxide populated on its surface migrates from the core (bulk) to the peripheral surface. To be specific, prior to growth and even during growth, the low-surface energy species populated on the nanoparticle or arriving at the nanoparticle migrate to its high-surface energy peripheral area. The nanomaterials (nanocrystals) thus grown on this surface may have core surrounded with oxide shell (sheath).

12.17.2 Nanomaterials Nucleation

Needless to say, the process of nucleation of nanomaterials is still controversial [115]. Many different features of this nucleation is still shrouded with confusion and misunderstanding. While some researchers consider it to take place via molten nanoparticle, others consider it to take place via solid nanoparticle. We argue that it takes place via quasiliquid (quasisolid) nanoparticle of ξ_m between 0 and 1, and that the quailiquid state nay be very close to the liquid state, for example, $\xi_m \rightarrow 0.9999$ and the quasisolid state may be very close to solid state, for example, $\xi_m \rightarrow 0001$. This is the most remarkable novelty of the growth by the VQS mechanism. Moshkalev and Verissimo [116] analyzed the synthesis of multiwalled carbon nanotubes via chemical vapor deposition employing nickel as a FECA and methane as a carbon precursor. The nucleation was viewed to be a measure of specific instability developed on the surface of metal catalyst nanoparticle supersaturated with carbon. The energy released during graphitization of carbon via the metal-carbon solution was crucial for the CNT nucleation. It was suggested that this energy released during growth might be high enough to cause substantial metal heating leading to a partial melting

Table 12.3 List of surface energy and energy bandgap of some selected inorganic materials and oxides

No.	Material	Surface energy (mJ/m^2)	Energy band gap (eV)
1	C	75–150	–
2	B	1600–2000	–
3	Ga	700	–
4	In	560	–
5	Pt	2540	–
6	Al	1450	–
7	He	2000	–
8	Cd	440–470	–
9	Zn	660	–
10	Si	1140	1.12
11	Ge	800	0.66
12	Fe_3C	2500	–
13	SiO_2	287	9.00
14	B_2O_3	211	–
15	Al_2O_3	78	6.04
16	TiO_2	68	3.25
17	MgO	1113	8.00
18	CaO	1310 ± 200	7.03
19	CaOH	1180 ± 100	–
20	SiC	3000	3.22
21	GaAs	860	1.42
22	GaP	1190	2.26
23	ZnO	1420	3.37
24	InAs	641	0.36
25	InP	700–750	1.35
26	InN	1276	0.78
27	GaN	1890	3.39
28	AlN	2210	6.20
29	BN	40	7.51
30	Calcium carbonate	230	–
31	Potassium chloride	110	–
32	Mica	120	–

of the nanoparticle. This implied that the nanoparticle could be quasiliquid (liquid-like) and hence relatively unstable during nucleation and growth. It is consistent with the concept of VQS mechanism. A recent observation by Maliakkal et al. [117] that FECA composition may change during nanowire growth makes it more plausible for the FECA to remain quasiliquid (quasisolid) rather than, molten, and eutectic of well-defined composition, during growth.

12.17.3 *Superiority of the VQS Mechanism*

The droplet for the VLS growths is molten and has no specific shape. This shape can change due to variation in pressure and temperature, and also due to mechanical fluctuations. They both can affect the VLS growth. Note that temperature fluctuation during growth influences droplet composition. This temperature may be different in different locations of a droplet and at different times during growth. At a certain time, the droplet composition may consequently be different at different locations of a droplet. The composition in the same location of a droplet may similarly be different at different times during growth. The nanomaterial (e.g., nanorod, nanowire, etc.) diameters are limited by the sizes of the FECA nanoparticles. This diameter may be non-uniform if the droplet shape and size varies during growth. Ternary or even quaternary nanowires are difficult to grow by the VLS and the SoLS mechanisms. This is not true for growths by the VQS mechanism. The FECA nanoparticles for the VQS growth can simultaneously serve as catalysts and reactants under optimized conditions. And they can thus enable the growth of ternary and quaternary alloy nanowires. Furthermore, under optimized conditions, the FECA nanoparticles for VQS mechanism can initiate the growth of near-defect-free semiconductor nanowires. There would though be temperature and pressure induced tradeoff. The temperature must not, for instance, be too low and $\xi_m \rightarrow 0$ to cause the diffusion of the R_S species through the R_L species too slow yielding defects in the lattice. The temperature should not, on the other hand, too high and $\xi_m \rightarrow 1$ to cause again the introdction of defects in the lattice. It will be demonstrated in Chap. 16 that the VSS and SoSS mechanisms are variants of the VQS mechanism. Wang and Buhro [118] found that the growth of colloidal wurtzite II–VI nanowires by FECA nanoparticles of bismuth chalcogenides and by the SoSS mechanism was nearly defect-free. Notably, the nanowire growth by the VLS mechanism proceeds via a sequence of periodic nucleation events. And in situ TEM images [119–121] demonstrated that these events take place at the FECA-nanowire interface. They give rise to planar defects and phase alternations at this interface [122, 123]. Also, the nucleation barriers determine the phase formed in a given nucleation event.

Importantly, Wang and Buhro observed that the nanoparticle morphologies during the VSS [124, 125] and the SoSS growths [126] were static. So, the liquid/solid contact angle (see Fig. 4.3 of Chap. 4) remained nearly fixed [124] during these growths. And this was different from the morphological fluctuations observed in FECA droplet during the VLS and SoLS growths. The liquid/solid contact angle varies up to 20° during these growths. A near-constant liquid/solid contact angle during nanowire growth by the VSS and SoSS mechanisms, which are variants of the VQS mechanism, is important because it removes a critical factor of fluctuations in the liquid/solid angle. It minimizes as well a significant cause of phase alternations [118].

Being in the liquid droplet state (e.g., $\xi_m=1$), the catalyst for the VLS growth is too unstable to prevent the MET particles from being incorporated into the nanomaterials that it produced. Dailey et al. [127] found the presence and configuration of MET≡Au on the sidewalls of Au-mediated Si nanowires produced by the VLS mechanism. They noted that, under certain growth conditions, the molten R_L≡(Au, Si) droplet spread from the catalyst nanoparticle at the nanowire tip to the sidewalls during growth. Eventually, it solidified creating a separate phase of small Au clusters during cooling after growth. The catalyst for the VQS mechanism can be free from this shortcoming. The possibility of sidewall cluster formation is quite remote in growths by the VQS mechanism.

References

1. F.C. Frank, The influence of dislocations on crystal growth. Discuss. Faraday Soc. **5**, 48–54 (1949)
2. W.K. Burton, N. Cabrera, F.C. Frank, Role of dislocations in crystal growth. Nature **163**, 398–399 (1949)
3. R.S. Wagner, W.C. Ellis, The vapor-liquid-solid mechanism of crystal growth and its application to silicon. Trans. Metall. Soc. AIME **233**, 1053–1064 (1965)
4. K. Dick, T. Dhanasekaran, Z. Zhang, D. Meisel, Size-dependent melting of silica-encapsulated gold nanoparticles. J. Am. Chem. Soc. **124**, 2312–2317 (2002)
5. E. Sutter, P. Sutter, Phase diagram of nanoscale alloy particles used for vapor-liquid-solid growth of semiconductor nanowires. Nano Lett. **8**, 411–414 (2008)
6. F. Gao, Z. Gu, Melting temperature of metallic nanoparticles, in *Handbook of Nanoparticles*, ed. by M. Aliofkhazraei (Springer, Cham, 2016), pp. 661–690
7. S. Pisana, M. Cantoro, A. Parvez, S. Hofmann, A.C. Ferrari, J. Robertson, The role of precursor gases on the surface restructuring of catalyst films during carbon nanotube growth. Phys. E **37**, 1–5 (2007)
8. J.A. Moulijn, A.E. van Diepen, F. Kapteijn, Catalyst deactivation: is it predictable? What to do? Appl. Catal. A: Gen. **212**, 3–16 (2001)
9. T. de los Arcos, M.G. Garnier, J.W. Seo, P. Oelhafen, V. Thommen, D. Mathys, The influence of catalyst chemical state and morphology on carbon nanotube growth. J. Phys. Chem. B **108**, 7728–7734 (2004)
10. Y. Qi, T. Cagin, W.L. Johnson, W.A. Goddard III., Melting and crystallization in Ni nanoclusters: the mesoscale regime. J. Chem. Phys. **115**, 385 (2001)
11. P.L. Hansen, J.B. Wagner, S. Helveg, J.R. Rostrup-Nielsen, B.S. Clausen, H. Topsoe, Atom-resolved imaging of dynamic shape changes in supported copper nanocrystals. Science **295**, 2053–2055 (2002)
12. D. Takagi, Y. Kobayashi, Y. Homma, Carbon nanotube growth from diamond. J. Am. Chem. Soc. **131**, 6922–6923 (2009)
13. C. Varanasi, J. Petry, L. Brunke, B.T. Yang, W. Lanter, J. Burke, H. Wang, J.S. Bulmer, J. Scofield, P.N. Barnes, Growth of high-quality carbon nanotubes on free-standing diamond substrates. Carbon **48**, 2442–2446 (2010)

14. J.H. Lin, C.S. Chen, M.H. Rümmeli, A. Bachmatiuk, Z.Y. Zeng, H.L. Ma, B. Büchner, H.W. Chen, Growth of carbon nanotubes catalyzed by defect-rich graphite surfaces. Chem. Mater. **23**, 1637–1639 (2011)
15. J.-H. Lin, C.-S. Chen, M.H. Rümmeli, Z.-Y. Zeng, Self-assembly formation of multi-walled carbon nanotubes on gold surfaces. Nanoscale **2**, 2835–2840 (2010)
16. B. Liu, D.-M. Tang, C. Sun, C. Liu, W. Ren, F. Li, W.-J. Yu, L.C. Yin, L. Zhang, C. Jiang, H.-M. Cheng, Importance of oxygen in the metal-free catalytic growth of single-walled carbon nanotubes from SiO_x by a vapor-solid-solid mechanism. J. Am. Chem. Soc. **133**, 197–199 (2011)
17. N.K. Reddy, J.-L. Meunier, S. Coulombe, Growth of carbon nanotubes directly on a nickel surface by thermal CVD. Mater. Lett. **60**, 3761–3765 (2006)
18. G. Atthipalli, Growth of aligned carbon nanotubes on copper substrates. Ph.D. thesis, University of Pittsburgh, Pittsburgh, PA (2011)
19. B. Mandl, J. Stangl, E. Hilner, A.A. Zakharov, K. Hillerich, A.W. Dey, L. Samuelson, G. Bauer, K. Deppert, A. Mikkelsen, Growth mechanism of self-catalyzed group III-V nanowires. Nano Lett. **10**, 4443–4449 (2010)
20. G. Koblmüller, S. Hertenberger, K. Vizbaras, M. Bichler, F. Bao, J.-P. Zhang, G. Abstreiter, Self-induced growth of vertical free-standing InAs nanowires on Si(111) by molecular beam epitaxy. Nanotechnology **21**, 365602 (2010)
21. H.D. Park, S.M. Prokes, M.E. Twigg, R.C. Cammarata, A.-C. Gaillot, Si-assisted growth of InAs nanowires. Appl. Phys. Lett. **89**, 223125 (2006)
22. A.M. Cassell, J.A. Raymakers, J. Kong, H. Dai, Large scale CVD synthesis of single-walled carbon nanotubes. J. Phys. Chem. B **103**, 6484–6492 (1999)
23. I. Berlanga, R. Mas-Ballesté, F. Zamora, J. González-Julián, M. Belmonte, Carbon nanotubes growth on silicon nitride substrates. Mater. Lett. **65**, 1479–1481 (2011)
24. S.C. O'Hern, J.-C. Idrobo, Y. Song, J. Kong, T. Laoui, M. Atieh, M.S.H. Boutilier, R. Karnik, Selective ionic transport through tunable subnanometer pores in single-layer grapheme membranes. Nano Lett. **14**, 1234–1241 (2014)
25. J.M. Thomas, R. Raja, Nanoporous and nanoparticle catalysts. Chem. Record **1**, 448–466 (2001)
26. M.J. Pellen, J.N. Hryn, J.W. Elam, Catalyst nanoporous membrane. US Patent No. 7,625,840 B2, 1 Dec 2009; see also, Catalytic nanoporous membrane. US Patent No. 8,518,845 B2, 27 Aug 2013
27. R. Kelsall, I.W. Hamley, M. Geoghegan, Templated nanostructures, Chap. 7.5, in *Nanoscale Science and Technology* (Wiley, New York, 2005), p. 365
28. V. Gitis, G. Rothenberg, *Ceramic Membranes, New Opportunities and Practical Applications* (Wiley, New York, 2016), p. 236
29. F. Matteini, G. Tütüncüoglu, D. Mikulik, J. Vukajlovic-Plestina, H. Potts, J.-P. Leran, W.C. Carter, A.F. Morral, Impact of the Ga droplet wetting, morphology, and pinholes on the orientation of GaAs nanowires. Cryst. Growth Des. **16**, 5781–5786 (2016)
30. J.F. van der Veen, J.W.M. Frenken, Dynamics and melting of surfaces. Surf. Sci. **178**, 382–395 (1986)
31. J.W.M. Frenken, J.F. van der Veen, Observation of surface melting. Phys. Rev. Lett. **54**, 134 (1985)
32. R. Lipowski, Critical surface phenomena at first-order bulk transitions. Phys. Rev. Lett. **49**, 1575 (1982)
33. R. Lipowski, Surface-Induced disorder and surface melting, in *Magnetic Properties of Low-Dimensional Systems*, 1st edn., ed. by L.M. Falicov, F. Mejí-Lira, J.L. Morin-Lopez. Springer Proceedings in Physics, vol. 50 (Springer, Berlin, Heidelberg, 1990), pp. 158–166
34. R. Erlandsson, M. Eriksson, L. Olsson, U. Helmersson, I. Lundstrom, L.-G. Petersson, *Fifth International Conference on Scanning Tunneling Microscopy/Spectroscopy*, vol. 9 (Boston, MA, USA, 1991), p. 825
35. M. Eriksson, L. Olsson, U. Helmersson, R. Erlandsson, L.-G. Ekedahl, Morphology changes of thin Pd films grown on SiO_2: influence of adsorbates and temperature. Thin Solid Films **342**, 297–306 (1999)

36. Z.L. Wang, J.M. Petroski, T.C. Green, M.A. El-Sayed, Shape transformation and surface melting of cubic and tetrahedral platinum nanocrystals. J. Phys. Chem. B **102**, 6145–6151 (1998)
37. E.T. Chen, R.N. Barnett, U. Landman, Surface melting of Ni(110). Phys. Rev. B **41**, 439 (1990)
38. V.I. Levitas, K. Samani, Size and mechanics effects in surface-induced melting of nanoparticles. Nat. Commun. **2**, 284 (2011)
39. A.R. Harutyunyan, T. Tokune, E. Mora, Liquid as a required catalyst phase for carbon single-walled nanotube growth. Appl. Phys. Lett. **87**, 051919 (2005)
40. J.M. Howe, H. Saka, In situ transmission electron microscopy studies of the solid–liquid interface. MRS Bull. **29**, 951–957 (2004)
41. Q. Jiang, H.M. Lu, M. Zhao, Modeling of surface energies of elemental crystals. J. Phys.: Condens. Matter. **16**, 521–530 (2004)
42. T. Matsubara, K.Z. Kamiya, Self-consistent Einstein model and theory of anharmonic surface vibration. I. One-dimensional model. Prog. Theor. Phys. **58**, 767–776 (1977); see also, T. Hesjedal, E. Chilla, H.-J. Fröhlich, Direct visualization of the oscillation of Au (111) surface atoms. Appl. Phys. Lett. **69**, 354–356 (1996)
43. J.F. der Veen, J.W.M. Frenken, Order-disorder transitions at surfaces. Surf. Sci. **251**(252), 1–5 (1991)
44. Y. Fukaya, Y. Shigeta, New phase and surface melting of Si (111) at high temperature above the (7×7)-(1×1) phase transition. Phys. Rev. Lett. **85**, 5150–5153 (2000)
45. R. Rao, N. Pierce, D. Liptak, D. Hooper, G. Sargent, S.L. Semiatin, S. Curtarolo, A.R. Harutyunyan, B. Maruyama, Revealing the impact of catalyst phase transition on carbon nanotube growth by in situ Raman spectroscopy. ACS Nano **7**, 1100–1107 (2013)
46. S. Narasimhan, M.A. Scheffler, A model for the thermal expansion of Ag(111) and other metal surfaces. Z. Phys. Chem. **202**, 253–262 (1997)
47. K. Pohl, J.-W. Cho, K. Terakura, M. Scheffler, E.W. Plummer, Anomalously large thermal expansion at the (0001) surface of beryllium without observable interlayer anharmonicity. Phys. Rev. Lett. **80**, 2853–2856 (1998)
48. J.W. Ciston, Crystallographic perturbations to valence charge density and hydrogen-surface interactions. Doctoral thesis, Northwestern University, Evanston, IL (2009)
49. A. Gohier, T.M. Minea, S. Point, J.-Y. Mevellec, J. Jimenez, M.A. Djouadi, A. Granier, Early stages of the carbon nanotube growth by low pressure CVD and PECVD. Diam. Relat. Mater. **18**, 61–65 (2009)
50. N. Muradov, F. Smith, A.T. Raissi, Catalytic activity of carbons for methane decomposition reaction. Catal. Today **102**(103), 225–233 (2005)
51. N. Muradov, Catalysis of methane decomposition over elemental carbon. Catal. Commun. **2**, 89–94 (2001)
52. S.N. Mohammad, Systematic investigation of the growth mechanisms for the synthesis of the conventional, doped, and bamboo-shaped nanotubes, primarily the carbon nanotubes. Carbon **75**, 133–148 (2014)
53. R. Sharma, *In-Situ Electron Microscopy: Applications in Physics, Chemistry and Materials Science*, Chap 6, ed. by G. Dehm, J.M. Howe, J. Zweck (Wiley VCH, Weinheim, Germany, 2012), p. 160
54. C.T. Wirth, B.C. Bayer, A.D. Gamalski, S. Esconjauregui, R.S. Weatherup, C. Ducati, C. Baehtz, J. Robertson, S. Hofmann, The phase of iron catalyst nanoparticles during carbon nanotube growth. Chem. Mater. **24**, 4633–4640 (2012)
55. M.-C. Lee, G. Simkovich, Electrical conduction behavior of cementite, Fe_3C. Metal. Trans. A **18**, 485–486 (1987)
56. Y.A. Suchikova, V.V. Kidalov, G.A. Sukach, Influence of dislocations on the process of pore formation in *n*-InP (111) single crystals. Semiconductors **45**, 121–124 (2011)
57. S. Hofmann, R. Blume, C.T. Wirth, M. Cantoro, R. Sharma, C. Ducati, M. Havecker, S. Zafeiratos, P. Schnoerch, A. Oestereich, D. Teschner, M. Albrecht, A. Knop-Gericke, R. Schlogl, J. Robertson, State of transition metal catalysts during carbon nanotube growth. J. Phys. Chem. C **113**, 1648–1656 (2009)

58. Y. Kohigashi, H. Yoshida, Y. Homma, S. Takeda, Structurally inhomogeneous nanoparticulate catalysts in cobalt-catalyzed carbon nanotube growth. Appl. Phys. Lett. **105**, 073108 (2014)
59. T.B. Massalski (ed.), *Binary Alloy Phase Diagrams*, 2nd edn., vol. 3 (American Society of Metals, Metals Park, OH, 1986)
60. L.C. Campos, M. Tonezzer, A.S. Ferlauto, V. Grillo, R. Magalhães-Paniago, S. Oliveira, L.O. Ladeira, R.G. Lacerda, Vapor-solid-solid growth mechanism driven by epitaxial match between solid AuZn alloy catalyst particle and ZnO nanowire at low temperature. Adv. Mater. **20**, 1499–1504 (2008)
61. B.C. Bayer, S. Hofmann, C. Castellarin-Cudia, R. Blume, C. Baehtz, S. Esconjauregui, C.T. Wirth, R.A. Oliver, C. Ducati, A. Knop-Gericke, R. Schlogl, A. Goldoni, C. Cepek, J. Robertson, Support-catalyst-gas interactions during carbon nanotube growth on metallic Ta films. J. Phys. Chem. C **115**, 4359–4369 (2011)
62. B. Hu, Y. Du, H. Xu, W. Sun, W.W. Zhang, D. Zhao, Thermodynamic description of the C-Ge and C-Mg systems. J. Min. Metall. Sect. B-Metall. **46**, 97–103 (2010)
63. S.K. Singhal, A.K. Srivastava, A.K. Gupta, Z.G. Chen, Synthesis of boron nitride nanotubes by an oxide-assisted chemical method. J. Nanopart. Res. **12**, 2405–2413 (2010)
64. M. He, H. Jiang, B. Liu, P.V. Fedotov, A.I. Chernov, E.D. Obraztsova, F. Cavalca, J.B. Wagner, T.W. Hansen, I.V. Anoshkin, E.A. Obraztsova, A.V. Belkin, E. Sairanen, A.G. Nasibulin, J. Lehtonen, E.I. Kauppinen, Chiral-selective growth of single-walled carbon nanotubes on lattice-mismatched epitaxial cobalt nanoparticles. Sci. Rep. **3**, 1460 (2013)
65. M. He, L. Zhang, H. Jiang, H. Yang, F. Fossard, H. Cui, Z. Sun, J.B. Wagner, E.I. Kauppinen, A. Loiseau, Fe-Ti-O based catalyst for large-chiral-angle single-walled carbon nanotube growth. Carbon **107**, 865–871 (2016)
66. P.J. Newby, B. Canut, J.-M. Bluet, S. Gomès, M. Isaiev, R. Burbelo, K. Termentzidis, P. Chantrenne, L.G. Frechette, V. Lysenko, Amorphization and reduction of thermal conductivity in porous silicon by irradiation with swift heavy ions. J. Appl. Phys. **114**, 014903 (2013)
67. R. Gupta, M. Gupta, Nanocrystallization and amorphization induced by reactive nitrogen sputtering in iron and permalloy. Phys. Rev. B **72**, 024202 (2005)
68. Y. Fu, H. Du, S. Zhang, Y.-W. Gu, Stress and surface morphology of TiNiCu thin films: effect of annealing temperature. Sci. Coat. Technol. **198**, 389–394 (2005)
69. M. Saha, M. Mukhopadhyay, R. Ray, S. Tarafdar, Impact of tailored gamma irradiation on pore size and particle size of poly [ethylene oxide] films: correlation with molecular weight distribution and microstructural study, Indian. J. Phys. **92**, 325–336 (2018)
70. C. Sämann, J.R. Köhler, M. Dahlinger, M.B. Schubert, J.H. Werner, Pulsed laser porosification of silicon thin films. Materials **9**, 509 (2016)
71. X. Li, M. Baker-Fales, H. Almkhelfe, N.R. Gaede, T.S. Harris, P.B. Amama, Rational modification of a metallic substrate for CVD growth of carbon nanotubes. Sci. Rep. **8**, 4349 (2018)
72. S.-E. Ong, S. Zhang, J.-H. Hsieh, H. Du, S.-H. Oh, Growth of carbon nanotubes via rapid thermal processing from sputtered amorphous carbon. Int. J. Nanomanuf. **2**, 40–49 (2008)
73. L. Shekari, A. Ramizy, K. Omar, H. Abu Hassan, Z. Hassan, High-quality GaN nanowires grown on Si and porous silicon by thermal evaporation. Appl. Surf. Sci. **263**, 50–53 (2012)
74. M. Sobanska, K. Klosek, J. Borysiuk, S. Kret, G. Tchutchulasvili, S. Gieraltowska, Z.R. Zytkiewicz, Enhanced catalyst-free nucleation of GaN nanowires on amorphous Al_2O_3 by plasma-assisted molecular beam epitaxy. J. Appl. Phys. **115**, 043517 (2014)
75. C.P. Li, X.H. Sun, N.B. Wong, C.S. Lee, B.K. Teo, Ultrafine and uniform silicon nanowires grown with zeolites. Chem. Phys. Lett. **365**, 22–26 (2002)
76. M. He, I. Minus, P. Zhou, S.N. Mohammad, J.B. Halpern, R. Jacobs, W.L. Sarney, L. Salamanca-Riba, R.D. Vispute, Growth of large-scale GaN nanowires and tubes by direct reaction of Ga with ammonia. Appl. Phys. Lett. **77**, 3731–3733 (2000)
77. M. He, P. Zhou, S.N. Mohammad, G.L. Harris, J.B. Halpern, R. Jacobs, W.L. Sarney, L. Salamanca-Riba, Growth of GaN nanowires by direct reaction of Ga with ammonia. J. Cryst. Growth **231**, 357–365 (2001)

78. A.M.S. El Ahl, M. He, P. Zhou, L. Salamanca-Riba, F. Felt, H. Shaw, A.K. Sharma, M. Jah, D. Lakins, T. Steiner, S.N. Mohammad, Systematic study of effects of growth conditions on the (nano-, meso-, micro)size and (one-, two-, three-dimensional) shape of GaN single crystals grown by a direct reaction of Ga with ammonia. J. Appl. Phys. **94**, 7749 (2003)
79. M. He, S.N. Mohammad, Novel chemical vapor deposition technique for the synthesis of high-quality single-crystal nanowires and nanotubes. J. Chem. Phys. **124**, 064714 (2006)
80. M. He, M.E.E. Fahmi, S.N. Mohammad, InAs nanowires and whiskers grown by reaction of indium with GaAs. Appl. Phys. Lett. **82**, 3749 (2003)
81. M. He, S.N. Mohammad, Structural characteristics of single-crystal nanowires grown by self-catalytic chemical vapor deposition method. J. Vac. Sci. Technol. B **25**, 1909 (2007)
82. M. He, S.N. Mohammad, Novelty of self-catalytic nanowire growth: a case study with InN nanowires. J. Vac. Sci. Technol. B **25**, 940 (2007)
83. M. He, A. Motayed, S.N. Mohammad, Phase separations of single-crystal nanowires grown by self-catalytic chemical vapor deposition method. J. Chem. Phys. **126**, 064704 (2007)
84. Y. Zi, S. Suslov, C. Yang, Understanding self-catalyzed epitaxial growth of III–V nanowires toward controlled synthesis. Nano Lett. **17**, 1167–1173 (2017)
85. M. Cuscunà, A. Convertino, L. Mariucci, G. Fortunato, L. Felisari, G. Nicotra, C. Spinella, A. Pecora, F. Martelli, Low-temperature, self-catalyzed growth of Si nanowires. Nanotechnology **21**, 255601 (2010)
86. X. Ma, H. Yang, L. Yu, Y. Chen, Y. Li, Preparation, surface and pore structure of high surface area activated carbon fibers from bamboo by steam activation. Materials **7**, 4431–4441 (2014)
87. S.S. Park, D.H. Park, Temperature dependence of pore size distribution of highly crystalline titanosilicate MCM-41. Solid State Phenom. **119**, 131–134 (2007)
88. N. Nagaraju, A. Fonseca, Z. Konya, J.B. Nagy, Alumina and silica supported metal catalysts for the production of carbon nanotubes. J. Mol. Catal. A: Chem. **181**, 57–62 (2002)
89. S. Campisi, C.E. Chan-Thaw, A. Villa, Understanding heteroatom-mediated metal-support interactions in functionalized carbons: a perspective review. Appl. Sci. **8**, 1159 (2018)
90. B.K. Min, A.K. Santra, D.W. Goodman, Understanding silica-supported metal catalysts: Pd/silica as a case study. Catal. Today **85**, 113–124 (2003)
91. I. Willems, Z. Konya, J.F. Colomer, G.V. Tendeloo, N. Nagaraju, A. Fonseca, J.B. Nagy, Control of the outer diameter of thin carbon nanotubes synthesized by catalytic decomposition of hydrocarbon. Chem. Phys. Lett. **317**, 71–76 (2000)
92. M. Kumar, Y. Ando, Controlling the diameter distribution of carbon nanotubes grown from camphor on a zeolite support. Carbon **43**, 533–540 (2005)
93. J. Zhu, M. Yudasaka, S. Iijima, A catalytic chemical vapor deposition synthesis of double-walled carbon nano-tubes over metal catalysts supported on a mesoporous material. Chem. Phys. Lett. **380**, 496–502 (2003)
94. N. Tripathi, P. Mishra, B. Joshi, Harsh, S.S. Islam, Catalyst free, excellent quality and narrow diameter of CNT growth on Al_2O_3 by a thermal CVD technique. Phys. E **62**, 43–47 (2014)
95. M.H. Rümmeli, F. Schäffel, C. Kramberger, T. Gemming, A. Bachmatiuk, R.J. Kalenczuk, B. Rellinghaus, B. Büchner, T. Pichler, Oxide-driven carbon nanotube growth in supported catalyst CVD. J. Am. Chem. Soc. **129**, 15772–15773 (2007)
96. T. Binninger, T.J. Schmidt, D. Kramer, Capacitive electronic metal-support interactions: outer surface charging of supported catalyst particles. Phys. Rev. B **96**, 16540 (2017)
97. S.N. Mohammad, A possible role of the dipole moment of the catalyst droplet in nanotube growth, alignment, chirality, and characteristics. Nanotechnology **23**, 085701 (2012)
98. Z. Yang, Q. Wang, X. Shan, W.-Q. Li, G.-H. Chen, H. Zhu, DFT study of Fe-Ni core-shell nanoparticles: stability, catalytic activity, and interaction with carbon atom for single-walled carbon nanotube growth. J. Chem. Phys. **142**, 074306 (2015)
99. H. Dumlich, *A path to monochiral ensembles of carbon nanotubes and their properties.* Doctoral Thesis, Freie Universität, Berlin, Germany (2013)
100. P. Pertl, M.S. Seifner, C. Herzig, A. Limbeck, M. Sistani, A. Lugstein, S. Barth, Solution-based low-temperature synthesis of germanium nanorods and nanowires. Monatsh. Chem. Chem. Mon. **149**, 1315–1320 (2018)

101. SGTE, SGTE 2017 Alloy Phase Diagrams (1176), www.crct.polymtl.ca›fact›documentation ›SGTE2017_Figs
102. C. Yan, P.S. Lee, Bismuth-catalyzed growth of germanium nanowires in vapor phase. J. Phys. Chem. C, Lett. **113**, 2208–2211 (2009)
103. R.W. Olesinski, G.J. Abbaschian, The Bi−Ge (bismuth-germanium) system. Bull. Alloy Phase Diagrams **7**, 535–540 (1986)
104. M.I. Bodnarchuk, K.V. Kravchyk, F. Krumeich, S. Wang, M.V. Kovalenko, Colloidal tin-germanium nanorods and their Li-ion storage properties. ACS Nano **8**, 2360–2368 (2014)
105. M.S. Seifner, S. Hernandez, J. Bernardi, A. Romano-Rodriguez, S. Barth, Pushing the composition limit of anisotropic $Ge_{1-z}Sn_z$ nanostructures and determination of their thermal stability. Chem. Mater. **29**, 9802–9813 (2017)
106. K. Tabatabaei, H. Lu, B.M. Nolan, X. Cen, C.E. McCold, X. Zhang, R.L. Brutchey, K. van Benthem, J. Hihath, S.M. Kauzlarich, Bismuth doping of germanium nanocrystals through colloidal chemistry. Chem. Mater. **29**(17), 7353–7363 (2017)
107. R. Scala, L. Bonanno, S. Haringer, A. Giannattasio, V. Moser, J. Samsonov, M.J. Binns, Fabrication of indium-doped silicon by the Czochralski method. US Patent No. 10,060,045 B2, 28 Aug 2018
108. For Au catalyzed Ge nanowires, see Y. Wu, P. Yang, Direct observation of vapor-liquid-solid nanowire growth. J. Am. Chem. Soc. **123**, 3165–3166 (2001)
109. For flat nanowire tip of tungsten oxide nanowires, see T. Tokunaga, T. Kawamoto, K. Tanaka, N. Nakamura, Y. Hayashi, K. Sasaki, K. Kuroda, T. Yamamoto, Growth and structure analysis of tungsten oxide nanorods using environmental TEM. Nanoscale Res. Lett. **7**, 85 (2012)
110. For very flat nanowire tip of AlN nanowires grown by catalyst-free physical vapor transport method, see G.R. Yazdi, P.O.A. Persson, D. Gogova, R. Fornari, L. Hultman, M. Syväjärvi, R. Yakimova, Aligned AlN nanowires by self-organized vapor-solid growth. Nanotechnology **20**, 495304 (2009)
111. C.-L. Cheng, Y.-F. Chen, Low temperature synthesis of ZnSe nanowires by self-catalytic liquid-solid growth. Mater. Chem. Phys. **115**, 158–160 (2009)
112. For circular nanowire tip of Au catalyzed Ge nanowires, see T.I. Kamins, X. Li, R.S. Williams, Growth and structure of chemically vapor deposited Ge nanowires on Si substrates. Nano Lett. **4**, 503–506 (2004)
113. For nanowire tip of very irregular shape of Au catalyzed ZnSe nanowires, see J. Basu, R. Divakar, J. Nowak, S. Hofmann, A. Colli, A. Franciosi, C.B. Carter, Structure and growth mechanism of ZnSe nanowires. J. Appl. Phys. **104**, 064302 (2008)
114. F. Wang, W.E. Buhro, Role of precursor-conversion chemistry in the crystal-phase control of catalytically grown colloidal semiconductor quantum wires. ACS Nano **11**(12), 12526–12535 (2017)
115. V. Jourdaian, C. Bichara, Current understanding of the growth of carbon nanotubes in catalytic chemical vapor deposition. Carbon **58**, 2–39 (2013)
116. S.A. Moshkalev, C. Verissimo, Nucleation and growth of carbon nanotubes in catalytic chemical vapor deposition. J. Appl. Phys. **102**, 044303 (2007)
117. C.B. Maliakkal, D. Jacobsson, M. Tornberg, A.R. Persson, J. Johansson, R. Wallenberg, K. A. Dick, In situ analysis of catalyst composition during gold catalyzed GaAs nanowire growth. Nat. Commun. 4577 (2019)
118. F. Wang, W.E. Buhro, Crystal-phase control by solution-solid-solid growth of II–VI quantum wires. Nano Lett. **16**, 889–894 (2016)
119. A.D. Gamalski, C. Ducati, S. Hofmann, Cyclic supersaturation and triple phase boundary dynamics in germanium nanowire growth. J. Phys. Chem. C **115**, 4413–4417 (2011)
120. S.H. Oh, M.F. Chisholm, Y. Kauffmann, W.D. Kaplan, W. Luo, M. Ruhle, C. Scheu, Oscillatory mass transport in vapor-liquid-solid growth of sapphire nanowires. Science **330**, 489–493 (2010)
121. C.Y. Wen, J. Tersoff, K. Hillerich, M.C. Reuter, J.H. Park, S. Kodambaka, E.A. Stach, F.M. Ross, Periodically changing morphology of the growth interface in Si, Ge, and GaP nanowires. Phys. Rev. Lett. **107**, 025503 (2011)

122. J. Johansson, L.S. Karlsson, C.P.T. Svensson, T. Martensson, B.A. Wacaser, K. Deppert, L. Samuelson, W. Seifert, Structural properties of <111> B-oriented III–V nanowires. Nat. Mater. **5**, 574–580 (2006)
123. F. Glas, J.-C. Harmand, G. Patriarche, Why does wurtzite form in nanowires of III–V zinc blende semiconductors? Phys. Rev. Lett. **99**, 146101 (2007)
124. S. Hofmann, R. Sharma, C.T. Wirth, F. Cervantes-Sodi, C. Ducati, T. Kasama, R.E. Dunin-Borkowski, J. Drucker, P. Bennett, J. Robertson, Ledge-flow-controlled catalyst interface dynamics during Si nanowire growth. Nat. Mater. **7**, 372–375 (2008)
125. S. Kodambaka, J. Tersoff, M.C. Reuter, F.M. Ross, Germanium nanowire growth below the eutectic temperature. Science **316**, 729–732 (2007)
126. J. Wang, K. Chen, M. Gong, B. Xu, Q. Yang, Solution-solid-solid mechanism: superionic conductors catalyze nanowire growth. Nano Lett. **2013**(13), 3996–4000 (2013)
127. E. Dailey, P. Madras, J. Drucker, Au on vapor-liquid-solid grown Si nanowires: spreading of liquid AuSi from the catalytic seed. J. Appl. Phys. **108**, 064320 (2010)

Chapter 13
Growths on METANO Surface by the VQS Mechanism

Abstract The fundamentals, applicability, and novelty of the metal-assisted vapor–quasiliquid–solid (vapor–quasisoild–solid) growth mechanism, called the VQS growth mechanism have been described. The structure, composition, and morphology of the MET-based nanoparticles most suitable for growths by the VQS mechanism have been studied. So, the formation of many different types of the MET-containing R_L species has been presented. Phase transformation and phase generation of this species have been articulated. Possible events during the pre-nucleation stage of growth via the MET-assisted VQS mechanism have been discussed. Illustrative demonstration of nanomaterials growths by the MET-containing nanoparticles has been made. The roles of surface energy in growths have been manifested. A model for the role of surface energy in the MET-mediated nanowire growth has been developed. In this regard, barrier to the exchange of materials on the METANO surface has been investigated. Taking the results from this model into account, possible reasons of Au being the most suitable catalyst for the VQS growths of Si and Ge nanowires have established. Also, attempts have been made to explain why CNT growth rates with Fe, Co, and Ni are very high.

13.1 Forwarding Note

Nanomaterials synthesis routes were narrated in Chap. 2. General definitions of FECANO, METANO and SUBSANO nanoparticles, of the EMNO surface, and of the R_L species were presented in Chap. 3. SECINI and SECINI0 were also defined in Chap. 3. Various features of nanomaterials synthesis including the pre-nucleation and pro-nucleation stages of growths were described in Chap. 4. NP1 and NP2 nanoparticles were defined in Chap. 12. The basic principles of growths by the VQS mechanism were laid down also in Chap. 12. Based on these and on the R_S source species defined in Chap. 1, the basic concepts of MET-mediated nanomaterials growths would be described in the following. They would take into account amorphicity α_{amor} and the effective amorphicity $\alpha_{amoreff}$ detailed in the Appendix. These amorphicities defined

S. N. Mohammad, *Synthesis of Nanomaterials*, Springer Series in Materials Science 307, https://doi.org/10.1007/978-3-030-57585-4_13

in terms of the parameters such as η_0, ν, and b would quantify the basic concepts of MET-mediated nanomaterials growths.

13.2 Basic Concepts

13.2.1 Formation of R_L Species

Our discussions in Chap. 12 indicated that, depending on growth parameters and growth conditions, R_L≡(MET, X) species or R_L≡(MET, X, Y) species may be eutectic at $T = T_E$. Also, the R_L≡(MET, X) species and/or R_L≡(MET, X, Y) species may be non-eutectic at $T \langle T_E, T \rangle T_E$ and even at $T = T_E$. The R_L≡(MET, Y) species, R_L≡(MET, Y, ϑ) species, R_L≡(EMNO) species, R_L≡(MET, X, ϑ) species, and R_L≡(MET, X, Y, ϑ) species, where ϑ is contaminant may also be non-eutectic and amorphous of effective amorphicity $\alpha_{\text{amoreff.}}$ The contaminant ϑ may, for example, be oxygen, native oxide, dopant atom(s), hydrogen, and/or nitrogen from the carrier gas. Unless otherwise achieved at $\alpha_{\text{amoreff}} = \alpha_{\text{amoreff0}}$, all of these R_L species may have SECINI ≠ SECINI0 (see Chap. 3). These non-eutectic R_L species may or may not also be less stable than the R_L≡(MET, X) eutectic species. In order to achieve stability, these non-eutectic R_L species may undergo processes such as precipitation, grain growth, structural transformation, compound formation, and reaction with external species during increase (decrease) in temperature during growth. These may be due to fluctuations of the R_L species, decomposition of certain elements of this R_L species, adsorption of reactants into the R_L species, and the temporal evolution of the distribution of the reactant atoms within the R_L species and in some surface areas of the nanoparticle. The reactants may be the R_S (R_S≡X and R_S≡Y) species, and some species migrated from the support or substrate, and/or the contaminants. Walser and Bené [1] and Wittmer et al. [2] found that, probably because of these, a congruently melting phase close to the single-crystal eutectic phase is initially formed. Lin et al. [3] also used state-of-the-art environmental transmission electron microscope (ETEM) to record in situ atomic-level resolution and the dynamic phase transformations caused by fluctuations in MET≡Co during SWCNT growth.

13.2.2 Phase Transformations and Generations

The concurrent phase of the R_L species, as noted by Lin et al. [3], may undergo transformation to subsequent phase(s) that could be more stable. This transformation is dictated by diffusion and nucleation that generate new phases. It is controlled as well by temperature, pressure, ambient gases, and contamination. It is influenced also by the intermediates of the precursors not fully transformed to yield the R_S≡X and $R_S \equiv Y$ species. Corresponding to each temperature (pressure), there is an activation energy at which the composition of the R_L species, the diffusion of the R_S ($R_S \equiv X$ and/or $R_S \equiv X$) species through the R_L species, the diffusion of the X_mY_n molecules through the R_L species, the transition of the R_L species from one phase to another phase, and the supersaturation of the R_L species with the R_S species (X_mY_n molecules) begin. Among them, the transformation from one phase to another phase may be complete or partial. If partial, multiple phases may co-exist at the same time. The concurrent phase may also give rise to a new phase while preserving its own original existence and identity. And all these may be caused by events during the pre-nucleation stage of growth prior to nanomaterial nucleation.

13.2.3 Possible Events During the Pre-nucleation Stage of Growth

There may be a number of events during the pre-nucleation stage of growth. First, the presence of one or more of the foreign contaminants ϑ (e.g., oxygen, native oxides, dopant atoms, nitrogen from carrier gas N_2, etc.) may almost always be present or introduced into the growth chamber. They may contribute to the amorphicity, porosity, surface roughness, and melting (semimelting) of the R_L species at a temperature $T < T_E$ even before the R_L species reaches the eutectic phase. This is the central of the VQS mechanism. The said R_L species may be one or other of $R_L \equiv$(EMNO), $R_L \equiv$(MET, X, Y), $R_L \equiv$(MET, X, ϑ), $R_L \equiv$(MET, Y, ϑ), or $R_L \equiv$(MET, X, Y, ϑ), etc. They all have lattice structure disturbed, disordered, coarsened and hence amorphous (semi-amorphous, amorphous-like). Second, foreign contaminants, if introduced into the R_L species [4], may create stress, cause pronounced thermodynamic imbalance and deactivation, and lead to a change in composition and morphology of the R_L species. They may as well be responsible for sintering and loss of porosity of the R_L species. They may be the reason of why oxygen (and probably oxidation) was found to have a deleterious effect on the nanowire growth [5] by the VLS mechanism. Third, there can be substrate–nanoparticle interface and nanomaterial–nanoparticle interface. These interfaces are almost always heterointerfaces. They suffer from lattice mismatch and mismatch of thermal expansion coefficients leading to stress, vibration, fluctuation, oscillation and hence compositional change inside the R_L species. Fourth, the miscibility (solubility) of MET in X or vice versa, and or sticking of X, *Y*, and/or ϑ to MET on the METANO surface, for example, as function of temperature, may also influence $R_L \equiv$(MET, X) species, $R_L \equiv$(MET, X, Y) species, $R_L \equiv$(MET, X, ϑ) species, and/or the $R_L \equiv$(MET, X, Y, ϑ) species. If the $R_L \equiv$(MET, X) species exists at all, the process of transition of this $R_L \equiv$(MET, X) species from non-eutectic to eutectic phase may particularly be hampered. The $R_L \equiv$(MET, X) species may never attain eutectic phase if it is transformed to $R_L \equiv$(MET, X, Y) species, $R_L \equiv$(MET, X, ϑ) species, or $R_L \equiv$(MET, X, Y, ϑ) species even at a temperature $T = T_E$. Instead, a relatively metastable intermediate phase comprising a mixture of one, two or more of the R_L≡(EMNO) species, R_L≡(MET, X) species, R_L≡(MET, X, Y) species, R_L≡(MET, X, ϑ) species, R_L≡(MET, X, ϑ) species, and R_L≡(MET, X, Y, ϑ) species may be created. This intermediate phase may be amorphous (or semi-amorphous, amorphous-like); see Appendix. It may as well be porous and coarsened. It may have pits, hillocks, dislocations, stacking faults, grain boundaries, and grain boundary-like regions at $T < T_E$. It has obviously SECINI $\neq$ SECINI0 unless its effective amorphicity is $\alpha_{amoreff} = \alpha_{amoreff0}$, The porosity of the R_L species, as function of the effective amorphicity $\alpha_{amoreff}$ would depend on all of the component elements, viz. MET, X, Y, and ϑ. Grains, grain boundaries, semi-grain boundaries, or grain boundary-like regions in the R_L species of the METANO surface may have small dimensions. Being porous and coarsened, the R_L species may become semi-molten and quasiliquid, quasisolid serving as fast channels for adatom diffusion through the R_L species at $T \langle T_E, T \rangle T_E$ T_E or $T \approx T_E$. The nanomaterial growth is consequently carried out by a mechanism which we call the VQS (vapor–quasiliquid–solid, vapor–quasisolid–solid) mechanism. We argue that the material phases (see Table 13.1) of the R_L species during growth and sometimes (but not always) after growth are determined on the basis of material(s) at the nanomaterial tip. They may confirm that this phase is different from that of the R_L≡(MET, X) eutectic species. As apparent from Table 13.1, the growth temperature for X_mY_n nanomaterials growths with various FECAs [6–23] is indeed almost always lower than the corresponding (MET, X) eutectic temperature T_E.

13.3 Illustrative Demonstration of the R_L Species

The defining of the R_L species, various means to create these R_L species, and various characteristics of these R_L species, as articulated earlier in Chap. 3, are very important for a comprehensive description of the VQS mechanism. As emphasized, the R_L species may be solid, quasisolid (quasiliquid), or liquid (droplet). They may have many different compositions. We illustrate a few of them in the following.

13.3.1 Non-eutectic R_L Species Created by Some Oxide-Assisted Growth Experiments

Tang et al. [24] grew InP nanowires. For this, they conducted reactions first of InP and In_2O_3 at a high temperature of 1150 °C to produce In, P, O, and In_2O vapors and then transported these vapors to a low-temperature (500–600 °C) deposition zone of the growth chamber. TEM studies revealed that most of the nanowires emerged from nanoparticle. EDX analysis indicated that the nanoparticles and also the nanowire tips were composed of In, P, and O with the approximate composition of 87:10:3. The R_L species of the nanoparticle was therefore R_L≡(In, P, O). While growing

Table 13.1 Growth parameters used for the growth of some selected nanowires (NWs)

No	NW	Catalyst, T_M (°C)	Growth process	T_E, T_F	Phase at eutectic temp	Growth temp, °C	Phase at NW tip	References
1	4H-SiC	Fe, 1535	Microwave-evaporation	1207	$Fe_{0.66}Si_{0.34}$	1550	Fe_2Si	Sundaresan et al. [6]
2	Si	Au, 1064	CVD	363 1239	$Au_{0.80}Si_{0.20}$	300	Au	Colli et al. [7]
3	Si	Au, 1064	PECVD	363	$Au_{0.80}Si_{0.20}$	320	Au	Mbenkum et al. [8]
4	Si	Ti, 1660	CVD	1330 1372	$Ti_{0.16}Si_{0.84}$	600–700	$TiSi_2$	Kamins et al. [9]
5	Si	Fe, 1535	CVD	1207 1474	$Fe_{0.66}Si_{0.34}$	450–500	Fe, FeSi	Wan et al. [10]
6	Si	Co, 1495	PVD	1259 1454	$Co_{0.22}Si_{0.78}$	1100	Co_2Si	Carter et al. [11]
7	Si	Cu, 1083	CVD	802 1248	$Cu_{0.70}Si_{0.30}$	450–650	Cu_3Si	Arbiol et al. [12]
8	Si	Pd, 1552	CVD	892	$Pd_{0.48}Si_{0.52}$	670	Pd_2Si	Hofmann et al. [13]
9	Si	Pt, 1772	CVD	973	$Pt_{0.33}Si_{0.67}$	500–800	PtSi, $PtSi_2$	Baron et al. [14]
10	Si	Mn, 1245	CVD	1149 1281	$Mn_{0.32}Si_{0.68}$	550–600	Mn_5Si_3	Lensch-Falk et al. [15]
11	Si	Al, 660	UHV-CVD	577 1037	$Al_{0.88}Si_{0.12}$	465	α-Al	Wang, et al. [16]
12	Ge	Ni, 1453	CVD	762	$Ni_{0.33}Ge_{0.67}$	275	Ni_2Ge	Kang et al. [17]

(continued)

Table 13.1 (continued)

No	NW	Catalyst, T_M (°C)	Growth process	T_E, T_F	Phase at eutectic temp	Growth temp, °C	Phase at NW tip	References
13	Ge	Cu, 1083	CVD	644	$Cu_{0.635}$ $Ge_{0.365}$	200	Cu_3Ge	Kang et al. [18]
14	Ge 59 nm	Au, 1064	UHV-CVD	361	$Au_{0.80}Ge_{0.20}$	265–320	$Au_{0.71}Ge_{0.29}$	Kodambaka et al. [19]
15	Ge	Sn, 232	Thermal evaporation	231	$Sn_{0.998}Ge_{0.002}$	350–400	AuSn	Ko et al. [20]
16	Ge	Fe, 1535	Laser ablation	838	$Fe_{0.32}Ge_{0.68}$	820	$FeGe_2$	Colli et al. [7]
17	Ge 40 nm	Au, 1064	CVD	361	$Au_{0.80}Ge_{0.20}$	320	γ-$Au_{0.6}Ge_{0.4}$	Kamins et al. [21]
18	Ge	Ag, 962	CVD	647	$Ag_{0.75}Ge_{0.25}$	800	$Ag_{0.875}Ge_{0.125}$	Gouveia et al. [22]
19	Ge	Mn, 1245	CVD	720	$Mn_{0.60}Ge_{0.40}$	350	$Mn_{11}Ge_8$	Lensch-Falk et al. [15]
20	Ge	Au, 1064	UHV-CVD	361	$Au_{0.80}Ge_{0.20}$	275	Au	Wang and Dai [23]

Various parameters include METANOs used to grow various nanowires, growth techniques used for these growths, the melting temperatures T_M of the METANOs used for the growths, the eutectic temperatures T_E (°C) of the (MET, X) alloyed seeds, and the material phase(s) at the nanowire tips. The material phase at the nanowire tips constituted the R_L species. The melting temperature of Si is 1414 °C and of Ge is 937 °C. The temperature T_F (in °C) is $T_F = (T_M + T_X)/2$, where T_M is the melting temperature of MET and T_X is the melting temperature of X of the (MET, X) alloy

InN nanowires on Au nanoparticle, Lan et al. [25] noted that Au nanoclusters were first formed. Oxygen and In vapors were then dissolved in Au nanoclusters to form R_L≡(Au, In, O) species during the pre-nucleation stage of growth. R_S≡In species resulting from In precursor, and R_S≡N species released from NH_3 were adsorbed onto the R_L species yielding InN nanowires during the pro-nucleation stage of growth. Klein et al. [26] used CuNi nanoparticle to study the influence of its alloy composition on the growth of vertically aligned carbon nanofibers. Carbon nanofibers were grown on this nanoparticle by plasma enhanced chemical vapor deposition. EDX analysis of carbon nanofiber tips indicated that the nanoparticle surface was composed of Ni, Cu, C, O, and Si implying that the R_L species of the nanoparticle was actually R_L≡(Ni, Cu, C, O, Si) species. It was cluster or solid solution.

13.3.2 Eutectic R_L Species Created by Some Non-oxide-Assisted Growth Experiments

Sundaresan et al. [6] claimed that the synthesis of SiC nanowires by catalyst-assisted sublimation-sandwich method was by the VLS mechanism. R_S≡Si and R_S≡C species were sublimated from SiC wafer in the high-temperature zone. The physical vapor transport (PVT) of these species to the low temperature zone was carried out for the nanowire synthesis. Metal catalysts employed for the growth at 1650–1750 °C were Fe, Ni, Pd, and Pt, respectively. Electron backscattering diffraction patterns of the nanowire cap materials indicated that they had Fe_2Si, Ni_3Si, Pd_2Si, and PtSi eutectic alloy for MET≡Fe, MET≡Ni, MET≡Pd, and MET≡Pt, respectively. The R_L species catalyzing the nanowire growth by the VLS mechanism were therefore R_L≡Fe_2Si, R_L≡Ni_3Si, R_L≡Pd_2Si, and R_L≡PtSi, respectively. The said eutectic alloys were molten enabling the R_S species to diffuse through them before supersaturation and nucleation for nanowire growths. As noted by Lehlooch et al. [27], Si can form solid solution with metal such as Fe and that such solid solution may also be molten.

13.3.3 Non-eutectic R_L Species Created by Some Non-oxide-Assisted Growth Experiments

Neumann et al. [28] produced ZnTe nanowires at 450 °C on GaAs substrate by employing molecular beam epitaxy. FECA≡Au was used for the growth. The Au-alloy catalyst was visible at the nanowire tip. An amorphous layer appeared to cover the catalyst nanoparticle surface. EDX analyses indicated that the chemical composition of the catalyst nanoparticle surface had 78 atomic % of Au, 14 atomic % of Ga, 4 atomic % of Zn, and 4 atomic % of Te. The R_L species mediating the nanowire growth was therefore R_L≡(Au, Ga, Zn, Te) species. It was amorphous (semi-amorphous, amorphous-like). Wang et al. [29] grew CdS nanowires at 800 °C by thermal evaporation of CdS powders in the presence of MET≡Au. The SAED pattern and HRTEM images indicated that the CdS nanowires were structurally uniform and single crystalline. The EDX measurements of the nanowire tips showed that they were composed of Au, Cd, and S. The R_L species catalyzing the nanowire growth was therefore R_L≡(Au, Cd, S) species. Yuan et al. [30] synthesized single-crystalline HfC nanowires at 1280 °C employing MET≡Ni. EDX data suggested that the nanowire tips contained Ni, O, Hf, and C, but the body of the nanowire contained only Hf and C. According to Ni/Hf binary phase diagram, some intermediate product of Hf and Ni could be realized at a temperature lower than their eutectic temperature $T_E = 1280$ °C. So, the catalyst droplet was possibly R_L≡(Hf, Ni, C) alloy with traces of contaminant O. The R_L species catalyzing the nanowire growth was therefore modified to R_L≡(Ni, Hf, C, O) species.

13.4 The Role of Surface Energy in the Met-Mediated Growths

13.4.1 Surface Energy Defined

Depending on growth conditions, R_S≡X and/or R_S≡Y source species released on the METANO surface may react (interact) with MET atoms yielding (MET, X), (MET, Y), and/or (MET, X, Y) alloy, cluster or solid solution. These METANO atoms at the surface have fewer nearest neighbors than those in the bulk of it. The METANO may have dangling bonds at the surface. The interatomic interactions are also lower at the surface than in the bulk. In view of these, the formation of isothermal, irreversible, and stable METANO surface is possible only via changes in free energy called the surface energy, or more specifically the surface free energy. Rather than being the total free energy of the atoms at the surface, it is the excess free energy of the atoms at the surface. The surface atoms possess this energy by virtue of being at the surface. Defined as the energy per unit area, it is the specific surface free energy.

13.4.2 METANO Surface Characteristics

If METANO and its surface are reconstructed or regenerated, the atoms at this surface are generally under non-equilibrium condition. They are also immobile and hence relatively stressed. The METANO surface is, as a result, in non-equilibrium state. It is as well inhomogeneous implying that it may have clusters, islands, grains, and grain boundaries. This surface is different from the surface of a liquid in which atoms readily attain equilibrium. It may though achieve equilibrium and a homogeneous surface structure if the bulk of it exerts sufficient force to its own surface. The

METANO size influences the excess free energy of it. Reduction of this size accompanies a decrease in the excess free energy more slowly at the surface than in the bulk of it. The surface energy of it however increases with decrease in its dimension.

We assume that the $R_S{\equiv}Y$ species may often be too volatile to stick to METANO surface, to interact with the METANO surface atoms, and to create (MET, Y) alloy on this surface. We also assume that the high density of the $R_S{\equiv}X$ species (a) released from its precursor on METANO surface, (b) not desorbed from this surface or (c) not migrated deep inside this surface, may, on the other hand, create a METANO/R_S interface or at least pockets of MET/R_S interface. Some intermediates of the precursors not fully transformed to yield the $R_S{\equiv}X$ species, may participate in creating this interface. We call them R_S intermediate species. Depending on the surface density and the sticking coefficient of the R_S species and R_S intermediate species, such an interface suffers from interface stress, affects the boundary between the two, and also affects the reaction between the MET atoms and the $R_S{\equiv}X$ species. The equilibrium of the METANO surface may thus be altered. Of course, temperature and pressure would have a role for the alteration. This role would be quite apparent in changes in the characteristic features of the phase(s) that already exist or tend to arise at the interface due to a change in temperature and/or pressure. The energetics of the reaction (interaction) at the METANO/R_S interface would simultaneously play a role in restructuring or even eliminating the interface. The phase equilibrium at the interface would also be modified. These would particularly happen when MET and $R_S{\equiv}X$ would tend to reach the eutectic phase.

13.4.3 Barrier to the Exchange of Materials on the METANO Surface

We consider METANO surface, the nanopores of which are not reduced or closed by the overcrowded R_S species and/or the R_S intermediate species. Suppose, one component of the $R_L{\equiv}$(MET, X) species at the METANO/R_S interface is solute and the other component of it is solvent. The barrier to the exchange of materials between them is determined by temperature and pressure at the interface, the characteristics of the solute, the characteristics of the solvent, and the environment in which they exist and react (interact). This barrier is determined, as well, by (1) the chemical reaction (s) taking place during the exchange of materials, (2) the bond-making and bond-breaking between (among) surface atoms of METANO and the atoms of the $R_S{\equiv}X$ species, and (3) the interruption in the MET/X bond-breaking (bond-making) by the presence of dopants, contaminants and the R_S intermediate species. Some or all of them experience gain or loss in energy with the immediate result that the barrier may be reduced or even eliminated. This may lead to the formation of new phases inside the $R_L{\equiv}$(MET, X) species, and also at the interface between METANO and $R_S{\equiv}X$ species. These new phases may altogether be different from the old phases. The equilibrium shape and size of the $R_L{\equiv}$(MET, X) species, and the alloying (clustering, solid solution) characteristics of this species may thus be altered, tailored, or modified ensuring stable crystallography at interface. The (MET, X) alloy phase diagram consequently undergoes some qualitative or even quantitative changes.

13.5 Model for the Role of Surface Energy in the Met-Mediated Growth

13.5.1 Model for the Exchange of Materials

The surface energy of METANO plays an important role in MET-mediated nanomaterials growths. For this surface energy to play a role, the barrier to the exchange of materials between the solute (e.g., $R_S{\equiv}X$ species) and the solvent (e.g., METANO) at the METANO/R_S interface must be surmounted. Suppose, E_{AF} is the activation energy needed to surmount this barrier. This activation energy, and hence the said exchange between the METANO and the $R_S{\equiv}X$ species, has genesis in the thermal energy $k_B T$. As the temperature T increases, the barrier decreases, and the exchange of materials between METANO and $R_S{\equiv}X$ increases. This means an increase in the exchange of materials between the METANO and the $R_S{\equiv}X$ species accompanies a decrease in the activation energy E_{AF}. This decrease in activation energy arises at the interface enhancing reaction between the $R_S{\equiv}X$ and the METANO species, which leads to the formation of intermediate compounds made of the $R_S{\equiv}X$ species and the atoms (molecules) of METANO. A path to the said exchange (reaction) may thus be created. For carbon nanotube growths, it would require that the $R_S{\equiv}C$ is soluble in METANO at the growth temperature T [31–34]. The solubility in the METANO nanoparticle at a temperature T may be given by

$$\chi_S = \chi_0 \exp\left[-\frac{E_{AF}}{k_B T}\right], \tag{13.1}$$

where χ_0 is the solubility in the corresponding bulk, and

$$E_{AF} = E_A\left(\frac{T_E \gamma_{bulk}}{T \gamma_{nano}}\right). \tag{13.2}$$

Note that E_A is the bulk activation energy, γ_{bulk} is the surface energy of the bulk material, but γ_{nano} is the surface energy of the METANO nanoparticle of the same bulk material. As earlier, T_E is the eutectic temperature of the $R_L{\equiv}$(MET, X) eutectic alloy. Note also that the surface energy γ_{bulk} remains unchanged, but the surface energy γ_{nano} [35] increases with decrease in METANO size. Equation (13.2) suggests that the effective activation energy E_{AF} decreases with decrease in METANO size. Similarly, the effective activation energy E_{AF} decreases with increase in temperature

T provided the eutectic temperature T_E remains unaltered. This effective activation energy E_{AF} may, therefore, have different values for exchange-induced $R_L = R_{LC}{\equiv}$(MET, X) species formed on the METANO surface. While it may have some value for the $R_L = R_{LC}{\equiv}$(Au, Si) species, it may have some other value for the $R_L = R_{LC}{\equiv}$(Fe, C) species, and some other value for the $R_L = R_{LC}{\equiv}$(Ni, In) species. It may be sensitive to the ambient gas, chamber pressure and contaminants present in the chamber. Note that χ_{SM} is the value of χ_S at the eutectic temperature $T = T_E$. Equation (13.2) then suggests that, at a temperature $T = T_E$, the effective activation energy for the bulk (e.g., $\gamma_{nano} = \gamma_{bulk}$ for the bulk) reduces to $E_{AF} \rightarrow E_A$. Equation (13.1) then transforms to

$$E_A = -k_B T_E \log\left(\frac{\chi_S}{\chi_0}\right). \tag{13.3}$$

Wilcox and Chapelle [36] may have noted that the exchange of materials due to solubility of Au in silicon obeys (13.1). In general, the effective activation energy E_{AF} for a reaction between the METANO and the $R_S{\equiv}$X is lower for higher value of the surface energy γ_{nano}. The barrier for this reaction is also lower for higher value of surface energy γ_{nano}.

This is interesting because it suggests that the higher probability of forming a $R_L = R_{LC}{\equiv}$(MET, X) non-eutectic alloy species, which has essentially the same property (e.g., melting) as that of the $R_L = R_{LCE}{\equiv}$(MET, X) eutectic alloy species formed at $T = T_E$, may be achieved even at a temperature $T < T_E$. The nature of phase changes determines if this non-eutectic alloy may or may not be subsequently transformed from the molten state to the semi-molten (quasiliquid, quasisolid) metastable state. If a transformation to the semi-molten state takes place, the said alloy may very well be metastable and solid and may still have some degree of porosity via a network of nanopores.

13.5.2 Alternative Model for the Exchange of Materials

The exchange of materials between METANO and $R_S{\equiv}$X may alternatively be modeled in terms of the surface energy γ_{nano}. If χ_{bulk} is the bulk solubility of the solute (e.g., $R_S{\equiv}$X species) in the solvent (e.g., METANO), and T_{nanob} is the bulk melting point of the bulk of the METANO nanoparticle, the Freundlich–Ostwald formula for the solubility of the $R_S{\equiv}$X species in a METANO nanoparticle would be [37].

$$\chi_{FROST} = \chi_{bulk} \exp\left(\frac{\vartheta_{nano}}{T}\right), \tag{13.4}$$

where $\chi_{bulk} = \chi_0$ is the solubility in the bulk material of the nanoparticle, and

Table 13.2 Examples of Type A1 (face-centered cubic 4d transition metal), Type A2 (body-centered cubic 4d transition metal), Type A3 (hexagonal close packed crystal), and Type A4 (diamond cubic crystal) structures

Type of structures	Examples
A1	Ag, Au, Cu, Mn, Ni, Pd, Pt, Rh, Ir, Pb, Al, Ca, Sr, Ac, Th
A2	Li, Na, K, Rb, Cs, Ba, Ra, Eu, V, Fe, Nb, Mo, Ta, W
A3	Be, Mg, Zn, Cd, Tl, Sc, Ti, Co, Y, Zr, Tc, Ru, Lu, Hf, Re, Os
A4	Si, Ge

$$\vartheta_{\text{nano}} = \frac{4\Omega_{\text{nano}}\gamma_{\text{nano}}}{\text{RD}_{\text{nano}}}, \tag{13.5}$$

R is the gas constant, and Ω_{nano} is the atomic volume. Recall that D_{nano} is the nanoparticle diameter. The surface energy of METANO may be modeled as [33]

$$\gamma_{\text{nano}} = \gamma_{\text{bulk}} + \gamma_{\text{atom}}(1 + \alpha_{\text{nano}}T)\left[1 + \left(\frac{d_{\text{atom}}}{D_{\text{nano}}}\right)\right], \tag{13.6}$$

where d_{atom} is the diameter of each atom of the nanoparticle, and α_{nano} is the coefficient of thermal stress of the METANO nanoparticle. If we define Z_b and Z_s as the coordination numbers of the bulk and the surface atoms, respectively, E_{cohb} as the cohesive energy of an atom in the bulk, a_L as the lattice constant, S_{area} is the area of the two-dimensional unit cell, and N_T as the Avogadro number, $N_T = 6.02214179 \times 10^{23}\ \text{mol}^{-1}$, then the surface energy per surface atom may be given by [38]

$$\gamma_{\text{atom}} = E_{\text{cohb}}\left[\frac{2 - (Z_s/Z_b) - (Z_s/Z_b)^{1/2}}{2N_T S_{\text{area}}}\right]. \tag{13.7}$$

Different METANO may have different crystal structures. They can thus be A1, A2, A3, and A4 type of METANOs (see Table 13.2). The values of Z_s, Z_b, and S_{area} listed in Table 13.3 are determined by the crystal type [38] of a METANO.

Table 13.3 Values of S_{area}, Z_s and Z_b for type A1, A2, A3, and A4 crystal structures and surfaces

Structure	Surface	S_{area}	Two-dimensional Brillouin zone	Z_s	Z_b
A1	(111)	$\sqrt{3}a_L^2/4$	Hexagonal	9	12
A1	(100)	$a_L^2/2$	Square	8	12
A1	(110)	$\sqrt{2}a_L^2/2$	Rectangular-P	6	12
A2	(110)	$\sqrt{2}a_L^2/2$	Rectangular-C	6	8
A2	(100)	a_L^2	Square	4	8
A2	(111)	$\sqrt{3}a_L^2$	Hexagonal	2	8
A3	(0001)	$\sqrt{3}a_L^2/2$	Hexagonal	9	12
A3	$(10\bar{1}0)$	$\sqrt{8/3}a_L^2$	Rectangular-P	16/3	12
A4	(110)	$\sqrt{2}a_L^2/5$	–	3	4

S_{area} is the area of the two-dimensional unit cell defined in terms of the lattice constant a_L. The two-dimensional Brillouin zones include rectangular prism (rectangular-P) and rectangular cuboid (rectangular-C)

13.6 Analyses of the Role of Surface Energy in the Metano-Mediated Growth

A close look of the Freundlich–Ostwald formula for the solubility, namely (13.4), would indicate that the solubility χ_{FROST} decreases with increase in temperature. And this contradicts the experimental observations. To shed light on it, we examined solubility and exchange of materials modeled by (13.1). A series of calculations was performed employing bulk surface energy γ_{bulk}, bulk cohesive energy E_{cohb}, lattice constant a_L (which, for example, is the lattice constant a of the conventional lattice constants, a, b, and c), atom diameter d_{atom}, melting temperature T_M, heat of fusion H_L, and atomic volume Ω_{nano} of some representative METANOs listed in Table 13.4.

13.6.1 Reduced Solubility of METANO Surface

The reduced solubility of the METANO surface may be given by $\Re = \chi_S / \chi_{SM}$, where the instantaneous solubility is χ_S and the highest solubility is χ_{SM}. Employing this solubility χ_S, the exchange of materials in the Au/Si system is calculated by using (13.1). Recall that Si is the solute and Au is the solvent for this system. Recall also that this Au/Si system is not much different from the Au/Ge system. This is apparent from the eutectic temperature $T_E = 363$ °C for the Au/Si system, but the eutectic temperature $T_E = 361$ °C for the Au/Ge system. Several values of the nanoparticle dimension D_{nano} were used to perform the calculations for the Au/Si system. The highest solubility χ_{SM} of Si in Au was assumed to create the R_{LC}≡(Au. Si) bulk eutectic alloy at $T_E = 363$ °C and that it was 20 atomic % (e.g., $\chi_{SM} = 0.20$);

Table 13.4 Characteristic properties of some representative FECAs (both metals and non-metals)

FECA	E_{cohb} (kJ/mol)	a_L (Å)	d_{atom} (Å)	γ_{bulk} (J/m^2)	T_M (°C)	Ω_{nano} (cm^3/mol)	H_L (kJ/mol)
Ni	428	3.58	2.48	2.38	1453	6.59	17.48
Mo	658	3.17	2.78	2.91	2617	9.40	32.00
Fe	413	2.87	2.52	2.42	1535	7.09	13.81
Co	424	2.53	2.50	2.80	1495	6.67	16.20
Ru	650	2.72	2.68	3.04	2250	8.30	24.00
Re	775	2.76	2.74	2.00	3180	8.81	33.21
Au	368	4.20	2.92	1.51	1064	13.60	36.95
Ag	284	4.18	2.88	1.25	962	10.335	11.30
Cu	336	3.66	2.56	1.79	1083	7.11	13.26
Mn	282	3.53	2.70	1.54	1245	7.35	12.05
Pt	564	4.02	2.78	2.49	1772	9.10	19.70
Pd	376	3.85	2.74	2.00	1552	8.78	17.60
Al	327	4.05	2.86	1.14	660	9.50	10.71
Ca	178	5.62	3.94	0.50	839	29.9	8.54
Sr	166	6.17	4.30	0.42	769	33.32	8.00
Li	158	3.99	3.10	0.52	180	13.12	3.00
Pb	196	5.11	3.50	0.59	327	18.27	4.79
Zn	130	2.68	2.76	0.99	420	9.22	7.35
Cd	112	3.06	3.42	0.76	321	13.10	6.30
Ta	782	3.35	2.98	2.90	2996	10.93	36.00
Na	107	4.20	3.80	0.26	98	23.7	2.60

(continued)

Table 13.4 (continued)

FECA	E_{cohb} (kJ/mol)	a_L (Å)	d_{atom} (Å)	γ_{bulk} (J/m^2)	T_M (°C)	Ω_{nano} (cm^3/mol)	H_L (kJ/mol)
K	90	5.30	4.70	0.13	64	45.3	2.33
Rb	82	5.71	4.96	0.12	39	55.90	2.19
Cs	78	6.26	5.34	0.10	29	70.0	2.09
In	241	–	3.32	0.90	157	15.72	3.26
Ti	468	2.95	2.90	1.99	1660	10.64	18.70
Bi	210	3.26	3.40	0.49	232	21.30	10.90
Sn	150	–	3.24	0.71	271	16.30	7.00
Mg	145	3.20	3.20	0.79	639	12.97	8.70
Si	446	7.71	2.64	1.14	1410	12.058	50.20
Ge	372	8.10	2.74	0.88	937	13.62	31.80
4H-SiC	612	4.36	2.23	2.50	2830	12.53	–
Diamond	800	3.57	1.82	3.70	3500	3.42	–
SiO_2	405	4.19	4.37	0.32	1830	26.35	–
GeO_2	637	4.98	4.27	0.62	1115	24.61	–
CaO	1061	4.80	5.02	1.31	2572	16.90	–
MgO	993	4.21	4.00	1.113	2852	12.80	–
TiO_2	940	3.74	4.24	0.790	2116	24.05	–

Bulk surface energy is γ_{bulk}, bulk cohesive energy is E_{cohb}, lattice constant is a_L (which, for example, is the lattice constant a of the conventional lattice constants, *a*, *b*, and *c*), atom diameter is d_{atom}, melting temperature is T_M, heat of fusion is H_L, and atomic volume is Ω_{nano}. The data are from the World Wide Web

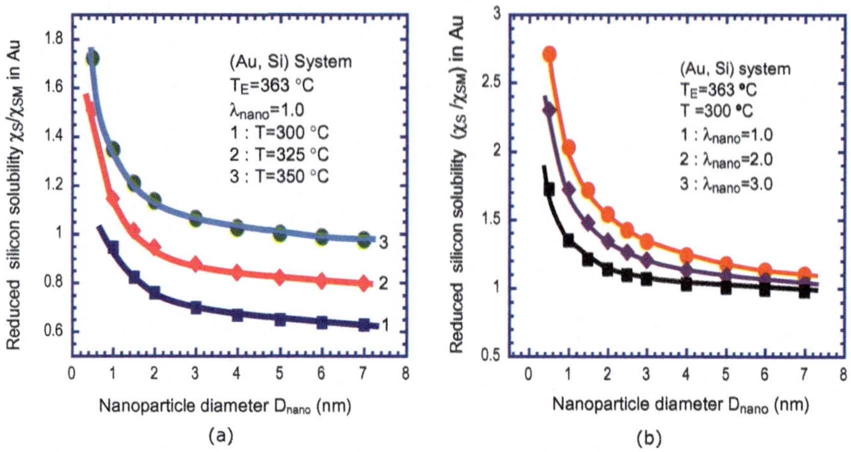

Fig. 13.1 Variation of reduced solubility, namely (χ_S/χ_{SM}) of Si in Au nanoparticle with the nanoparticle diameter D_{nano} at **a** temperatures T = 300, 325, and 350 °C, respectively, and for, **b** λ_{nano} = 1.0, 2.0, and 3.0, respectively. The R_L≡(Au, Si) species is formed on the nanoparticle at $T_E = 363$ °C

it is identical to the one observed by experiment [39]. The exchange of materials resulting from this solubility at $T_E = 363$ °C was considered to be 100%. The entire R_{LC}≡(Au, Si) alloy surface was, as a result, molten. Nevertheless, the calculations were performed for temperatures 300 °C $\le T \le$ 350 °C, and the calculated results for χ_S at various temperatures and nanoparticle dimensions were weighted with respect to the highest solubility value $\chi_{SM} = 0.20$ to determine the barrier to the exchange of materials. The reduced solubility $\Re = \chi_S/\chi_{SM}$ as function of the nanoparticle diameter D_{nano} is plotted for three different values of the growth temperature and shown in Fig. 13.1a. The reduced solubility $\Re = \chi_S/\chi_{SM}$ as function of the nanoparticle diameter D_{nano} is also plotted for three different values of λ_{nano} as shown in Fig. 13.1b. Figure 13.1a shows that the solubility increases with decrease in nanoparticle dimension D_{nano}, but decreases with decrease in temperature T. These results thus indicate that the exchange of materials on the nanoparticle surface is lower at lower temperature. This also indicates that, at a certain temperature T, the exchange of materials on the nanoparticle surface increases with decrease in nanoparticle dimension. These are corroborated with experiments by Dayeh and Picraux [40]; see Fig. 3 by these authors. Curve 1 of Fig. 13.1a is for the temperature T = 300 °C. It shows that $\Re = \chi_S/\chi_{SM}$ approaches the highest value of the reduced solubility, namely $\Re_H = 1$ for $D_{nano} \approx 0.8$ nm. The exchange of MET≡Au material and R_S≡X=Si material for this nanoparticle is therefore significantly high. It is high enough to yield non-eutectic R_{LC}≡(MET, X) alloy characteristics at T = 300 °C. Strikingly, it is suitable for producing thin Si nanowire (diameter ~0.8 nm) at this temperature, and it can do it almost identically as the eutectic R_{LCE}≡(MET, X) alloy of characteristics attained at the eutectic temperature T_E = 363 °C. The results presented above points to distinct advantage of higher non-eutectic temperature (e.g., $T > 300$ °C) growth. It suggests that the non-eutectic R_{LC}≡(MET, X) alloy characteristics at these temperatures can be suitable for producing thick Si nanowire (e.g., large diameter nanowires on large diameter nanoparticle) almost in an identical way as the eutectic R_{LCE}≡(MET, X) alloy of characteristics attained at the eutectic temperature T_E = 363 °C. Figure 13.1b shows the decrease in the reduced solubility $\mathfrak{R}$ as function of the increase in the nanoparticle dimension D_{nano}. This decrease is near exponential; it is faster for a nanoparticle of larger λ_{nano} than for a nanoparticle of smaller λ_{nano}. Although not shown in the figure, even at a certain lower temperature T [for example, 200 °C for the Au/Si system], the reduced solubility $\mathfrak{R}$ approaches the highest solubility $\mathfrak{R}_H$ for D_{nano} approaching 1 nm. And it is true for almost all values of λ_{nano}.

13.6.2 Exchange of Materials on METANO Surface

Exchange of materials on the METANO surface may be very high or very low. It may be partial or complete as demonstrated in Fig. 13.2a–f, where the small red circles correspond to MET species and small yellow circles correspond to (MET, X) alloy species. The exchange can be quite significant in METANO of smaller dimension. Figure 13.2a shows METANO surface composed only of MET species. But Fig. 13.2b shows METANO surface composed of about 95% MET species; it has only ~5% R_{LC}≡(MET, X) species. These species resulted from exchange of materials, namely the MET species and the R_S≡X species. But Fig. 13.2c shows METANO surface composed of 20% R_{LC}≡(MET, X) species and 80% MET species. Lastly, Fig. 13.2f shows METANO surface composed of 100% R_{LC}≡(MET, X) species. To reiterate, all of them involve the exchange of materials on the METANO nanoparticle surface.

Among various parameters, METANO's surface energy significantly influences the complete or even partial exchange of MET and X materials species on the METANO surface. And this exchange causes partial melting of the R_L species. The R_L species nevertheless remains largely metastable and a quasiliquid (quasisolid) due additionally to the presence of amorphicity-induced surface porosity. The quasiliquid (quasisolid) characteristics of the R_L species are nevertheless enhanced under the influence of surface energy. It appears indeed to be justified based on the experiments by Kim et al. [41], who observed METANO surface as partially filled R_{LC}≡(MET, X) alloy, and this resulted from the exchange of materials on the METANO surface. Temperature and some other parameters played key role in it. Recall that Fig. 13.1a demonstrates the exchange of materials on a nanoparticle surface and that it is dependent on nanoparticle dimension. This exchange of materials, for example, on a METANO surface of diameter D_{nano} > 10 nm could be insignificant, for example, at a temperature of 200 °C. And this prediction is corroborated also with the observation by Kim et al. [41]. It is in line, as well, with the experiment by Gamalski et al. [42], who observed Au/Ge system quite metastable and solid during nanowire growth at 220 °C $\leq T \leq$ 270 °C. It was found to be true even when exchange of materials during nanowire growth led to the creation of METANO surface composed of (a) exchange

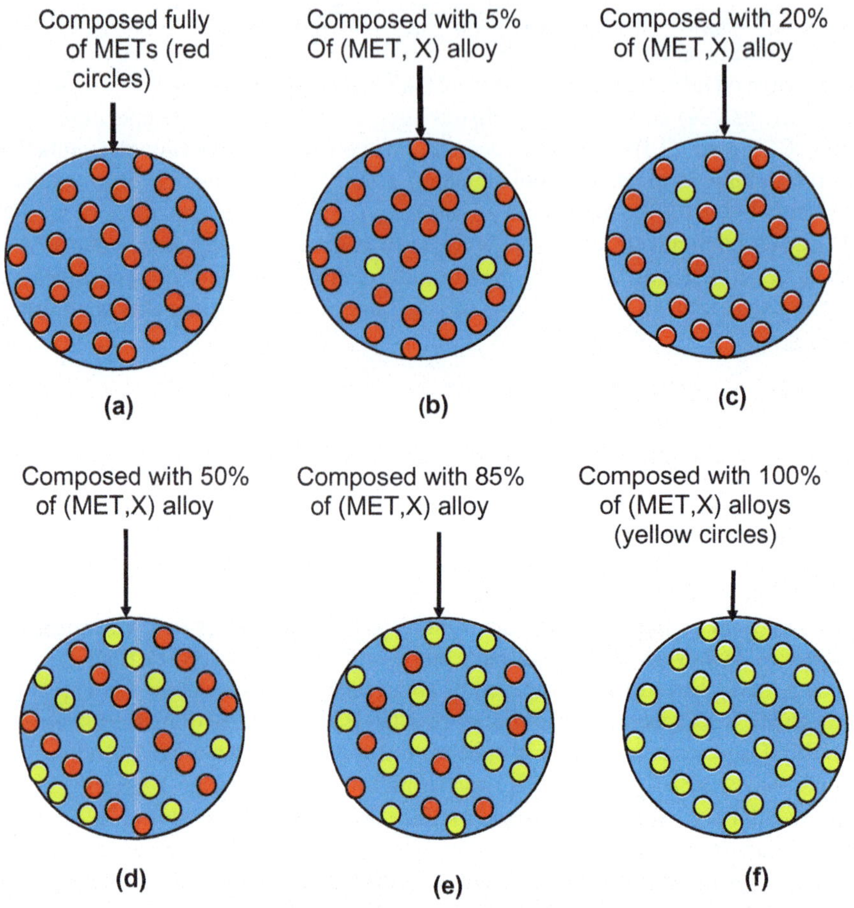

Fig. 13.2 Schematic diagrams of six different nanoparticles with six different interdiffusions and exchanges of MET and X materials leading to the formation of six different $R_L \equiv$(MET, X) species (crystallites) on the nanoparticle surface. **a** No interdiffusion and exchange of material, **b** 5% exchange of MET and X materials, **c** 20% exchange of MET and X materials, **d** 50% exchange of MET and X materials, **e** 85% exchange of MET and X materials, and **f** 100% exchange of MET and X materials, respectively

materials induced $R_{LC} \equiv$(MET, X) species and (b) the surface amorphicity-induced $R_L \equiv$(MET, X) species.

13.7 Why Are Au-Mediated Si and Ge Nanowire Growths So Successful?

13.7.1 Possible Reasons of Au Being Suitable for the VLS Growths of Si and Ge Nanowires

During the past years, many different nanowires have been synthesized. Among them, Si nanowires are very promising for new technology development. For a number of reasons, they have been most frequently grown by employing MET≡Au. These reasons include (1) non-oxidation in air, (2) high chemical stability, and (3) non-toxicity of Au. The binary phase diagram [39] shows that (Au, Si) alloy possesses simple eutectic feature even for silicon concentration as high as ~20 atomic % (e.g., $\chi_S = 0.2$) at the eutectic point. This means R_L≡(Au, Si) alloy droplet on Si substrate is rich in silicon. Low vapor pressure of Au even at high temperatures provides an added advantage. Also, high solubility of Si in Au guarantees high dissolution of Si in the R_L≡(Au, Si) alloy at the eutectic temperature $T_E = 363$ °C. Low Si pressure is required to achieve an increase in the Si concentration in Au even beyond its equilibrium value.

Non-Au metals have also been identified as METANOs for Si nanowire growths. One of these metals is Al. Wang et al. [16] demonstrated that MET≡Al can produce quite uniform, single-crystalline Si nanowires on Si(111) substrate at a temperature range 430 °C $\leq T \leq$ 490 °C, which is below the eutectic temperature $T_E = 577$ °C of the R_L≡(Al, Si) binary alloy. The (Al, Si) phase diagram [39] closely resembles the (Au, Si) phase diagram. The main problem with Al-mediated Si nanowire growth is that the eutectic point of the (Al, Si) binary alloy located at 577 °C has Si concentrations of ~12 atomic % ($\chi_S = 0.12$) and that it is much lower than that in the (Au, Si) binary eutectic alloy.

13.7.2 Reasons for Other Metals not Being Very Suitable for the VLS Growths of Si and Ge Nanowires

Pd and Pt metals used as catalysts for the Si nanowire growths have similar physical and chemical properties. However, Pd catalyst requires at least 892 °C and Pt catalyst requires at least 973 °C for the MET-mediated VLS growths of Si nanowires (see Table 13.1). These temperatures are considerably high. The silicon nanowires have nevertheless been grown at temperatures lower than the (Pd, Si) eutectic temperature via the Pd_2Si alloy nanoparticles and at a temperature lower than the (Pt, Si) eutectic temperature via the PtSi, $PtSi_2$ alloy nanoparticles. Si and Ge nanowires have been grown at sub-eutectic temperatures, as well, by other METs. And these growths were via some relatively unstable silicide or germanide alloy intermediate(s) formed during growths. There are two notable problems with these growths. First, alloy intermediates were formed, for example, during Pd- and Pt-catalyzed Si and Ge nanowire growths. In contrast, no alloy intermediates were formed during Au- and Al-catalyzed Si and Ge nanowire growths. Second, Pd- and Pt-catalyzed Si and Ge nanowire growths suffer from the formation of defects in nanowires. These nanowires obtained with Al and Au catalysts had, on the other hand, relatively fewer defects. High-resolution TEM images by Wang and Dai revealed [23] that Au-catalyzed single-crystalline Ge nanowires grown by CVD at 275 °C had indeed high purity and high crystalline quality. The nanowires, 23 nm in diameter, were single crystalline. Si nanowires grown at 500 °C with Cu catalyst also suffered greatly from stacking faults and microtwin boundaries. They had $R_L \equiv (Cu_3Si)$ silicide particle at their tips.

13.7.3 Surface Energy, Activation Energy, and Exchange of Materials on METANO Surface

Surface energy γ_{nano}, activation energy E_A at the eutectic temperature T_E, barrier E_{AF} to the exchange of materials between MET and Si at the experimental growth temperature T (see Table 13.4), and also the reduced solubility $\Re = \chi_S / \chi_{SM}$ for the growths of silicon nanowires 10 nm in diameter were calculated for some representative METANOs. The results are presented in Table 13.5. This table shows that, for MET≡Au, E_{AF} is quite small and $\Re = \chi_S / \chi_{SM}$ is quite large even at a temperature $T = 320$ °C. Notably, $\chi_{SM} = 0.38$ for MET≡Cu, $\chi_{SM} = 0.52$ for MET≡Pd, $\chi_{SM} = 0.67$ for MET≡Pt, and $\chi_{SM} = 0.68$ for MET≡Mn. However, $\Re = \chi_S / \chi_{SM}$ is larger for MET≡Au at $T = 320$ °C than for other METs even at temperatures $T > 320$ °C. This indicates that METANO from Au, unlike all other METANOs, undergoes large surface-energy-mediated partial melting of the R_L species. The temperature $T = 1100$ °C is close to the temperature $T_E = 1195$ °C of the $R_L \equiv$(Co, Si) eutectic alloy species. Probably for this reason, E_{AF} for Co-catalyzed Si nanowire growth at

Table 13.5 Calculated results for surface energy γ_{nano}, activation energy E_A at the eutectic temperature, barrier to exchange of materials between FECA χ_S / χ_{SM} and Si at the experimental growth temperature T (see Table 13.3), and the ratio for the growths of silicon nanowires, 10 nm in diameter

FECA	T (°C)	γ_{nano} (J/m^2)	χ_{SM}	E_A (eV)	E_{AF} (eV)	χ_S / χ_{SM}
Au(111)	320	1.555	0.18	0.0882	0.0972	0.7468
Al(111)	577	1.182	0.12	0.1553	0.1858	0.4485
Fe(111)	450	2.496	0.34	0.1376	0.3578	0.9428×10^{-2}
Mn(111)	550	1.585	0.68	0.0473	0.0959	0.3803
Co(0001)	1100	2.861	0.78	0.0328	0.0367	0.9398
Cu(111)	450	1.837	0.30	0.1115	0.1936	0.1490
Pd(111)	670	2.051	0.52	0.0656	0.0852	0.6739

1100 °C is very high. Table 13.5 suggests that MET≡Au is perhaps the most suitable one for Si and Ge nanowire growths. This is a crucial finding that highlights, in general, the exceptional surface-energy-mediated catalytic potential for nanomaterials growths. Unlike Au, other METs lack significant surface-energy-mediated melting. And, as a result, Si and Ge nanowire growths catalyzed by them (e.g., METs other than Au) are solely dictated by Knudsen diffusion through nanopores created on the nanoparticle surface.

13.8 Carbon Solubility in Metano

13.8.1 Reduced Solubility

We examined the solubility and exchange of materials between MET and R_S≡C species based again on (13.1)–(13.7). Calculations were performed assuming the peak solubility $\chi_0 = 1$ and employing the parameters γ_{bulk}, E_{cohb}, a_L, d_{atom}, T_{melt} (melting temperature), H_L (heat of fusion), and Ω_{nano} (atomic volume) listed in Table 13.4. Unless otherwise stated, we also used $\beta_{nano} = 10^{-5}/°C$ and $\lambda_{nano} = 1$ for the calculations. Recall that the solubility χ_S at $T = T_E$ is assumed to be the highest and is denoted by χ_{SM}. Then, the reduced solubility is again $\Re(\chi_S/\chi_{SM})$. The variation of this reduced solubility $\Re$ with the outer CNT diameter D_{NTO} is shown in Fig. 13.3a. This reduced solubility is actually the carbon solubility in Fe(111),

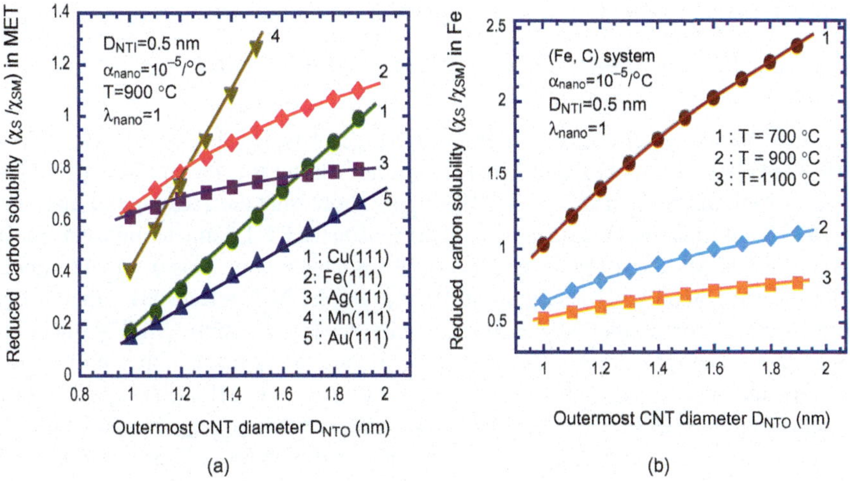

Fig. 13.3 Variation of reduced carbon solubility with the outermost CNT diameter D_{NTO} for two sets of growth temperature T: **a** with $T = 900$ °C in MET≡Cu(111), Fe(111), Ag(111), Mn(111), and Au(111), respectively; and **b** with $T = 700$ °C, 900 °C, and 1100 °C, respectively, but in MET≡Fe(111)

Cu(111), Ag(111), Mn(111), and Au(111), respectively, and it is at a temperature $T = 900$ °C. The variation of the same reduced solubility $\Re$ with the outermost CNT diameter D_{NTO} is shown in Fig. 13.3b. This variation is for carbon solubility in Fe(111) at 700 °C, 900 °C, and 1100 °C, respectively. Figure 13.3a shows a very interesting feature of reduced carbon solubility in various MET species. This reduced carbon solubility is quite high in some MET species, but quite low in some other MET species. Further, the variation in this solubility with CNT diameter is large in some MET species but small in some other MET species. As apparent from Fig. 13.3b, carbon solubility in Fe(111) increases with increase in the outermost diameter D_{OUT} but decreases with increase in temperature. Influenced by surface energy, there occurs partial exchange of materials between MET and X≡C, and this exchange leads to partial melting of the R_L species. Note that R_L≡(MET, C) alloy becomes molten if the carbon diffusion in MET reaches an optimal level. This level corresponds to about 18 atomic % of carbon diffusion (e.g., $\chi_{SM} \approx 0.18$) in Fe [39]. Figure 13.3b suggests that the exchange of material becomes lower at higher temperature. The R_L species though remains quasiliquid (quasisolid) due additionally to the presence of pits, hillocks, nanopores, etc. In view of the finding in Fig. 13.3b, this is true particularly at higher temperature. The R_L species is more amenable to bulk diffusion of the R_S species under the influence of surface energy than without the influence of surface energy. Kim et al. [41] observed that, due to exchange of materials, there can indeed be partially or fully formed (MET, X) alloy crystallites on the nanoparticle surface. Gamalski et al. [42] found R_L≡(Au, Ge) alloy to be metastable and solid at temperature between 220 and 270 °C despite some changes in its composition due possibly to exchange of materials during nanotube growth.

13.8.2 Effective Barrier to Exchange of Materials

Figure 13.4 shows effective barrier to exchange of materials between R_S≡C and MET≡Re(0001), Pt(111), and Mn(111) species on the Re, Pt, and Mn METANO nanoparticle surfaces, respectively. This figure shows that the exchange of materials on METANO nanoparticle surfaces is dependent on the CNT diameter implying that it is sensitive to METANO nanoparticle dimension. This is more sensitive to the dimension of some METANO nanoparticles [for example, MET≡Re(0001)] than to the dimension of some other METANO nanoparticles [for example, MET≡Mn(111)]. Also, the exchange of materials for $D_{nano} > 10$ nm may be marginally small.

E_{AF} and $\Re(\chi_S/\chi_{SM})$, for example, for the C species in METANO species show three different types of behavior: First, E_{AF} is small and $\Re(\chi_S/\chi_{SM})$ large for Fe, Ni, and Co, known to be highly active METANO nanoparticles for CNT growths. Second, E_{AF} is moderately large and $\Re(\chi_S/\chi_{SM})$ large for Mn, Au, Cu, and Mo known to be moderately active METANO nanoparticles for CNT growths. And third, E_{AF} is large and $\Re(\chi_S/\chi_{SM})$ large for Ag, Ti, Re, and Pt known to be lowly active METANO for CNT growths. Although not shown specifically, these are evident from carbon solubility in Ni, Co, Mn, and Ti. These are apparent also from E_{AF}, which is the barrier

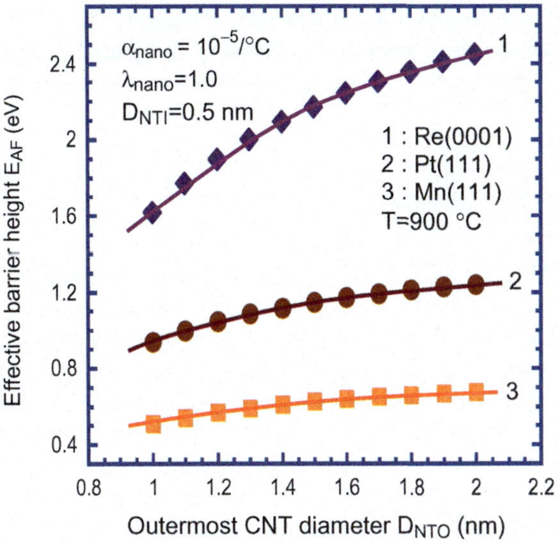

Fig. 13.4 Effective barrier E_{AF} to exchange of materials between carbon and metals such as Re (0001), Pt(111), and Mn(111) catalysts at 900 °C as function of the outermost nanotube (outer shell) diameter D_{NTO}

to the exchange of materials between carbon and Ni, Mn, and Ti. Figure 13.5a, b shed light on the barrier to exchange of materials in the R_L≡(Fe, C) system. Figure 13.5a is for three different METANO surface energies quantified by λ_{nano} and Fig. 13.5b

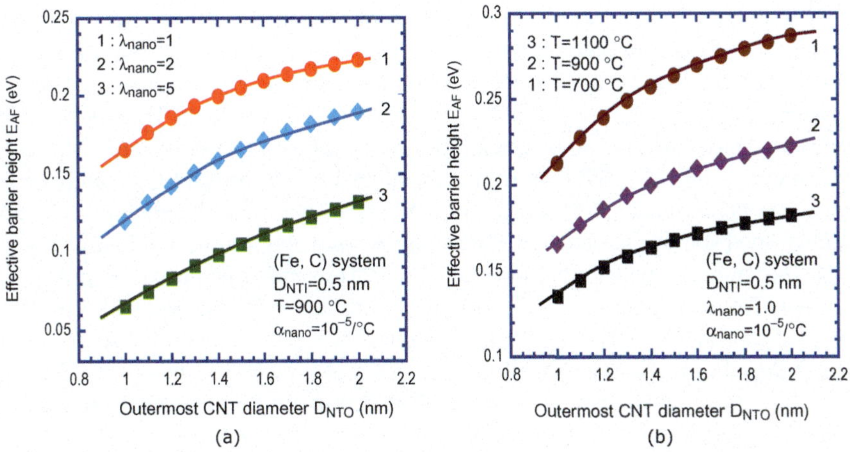

Fig. 13.5 Effective barrier E_{AF} to exchange of materials between carbon and Fe(111), at **a** $\lambda_{nano} = 1$, 2, and 5 as function of the outermost nanotube diameter (outermost shell) D_{NTO} and **b** temperatures 700 °C, 900 °C and 1100 °C

is for three different METANO temperatures. It is found that the larger the surface energy, the smaller is the barrier to the exchange of materials. Likewise, the higher the temperature, the lower is the barrier to the exchange of materials.

13.8.3 Important Revelation

Note that SWCNTs have characteristics distinctly different from the MWCNT characteristics. While outer diameter D_{OUT} is small for SWCNT, it is large for MWCNT. Figures 13.3, 13.4, and 13.5 reveal a very important fundamental feature of CNT growths. They demonstrate that both reduced solubility and exchange of materials on the nanoparticle surface are small in SWCNTs, but large in MWCNTs. This implies that the SWCNT growth is dictated more by the amorphous, disordered, defect-induced surface effects than by exchange of materials at the nanoparticle surface. In contrast, the MWCNT growth is dictated more by the exchange of materials than by the amorphous, disordered, defect-induced surface effects on the nanoparticle surface.

13.9 Why CNT Growth Rates with Fe, Co, and Ni Are Very High

13.9.1 Experimental Demonstration

The observed growth rates G_R of carbon nanotubes obtained from various experiments [43–55] are presented in Table 13.6. This table lists primarily the METANO-mediated CNT growth rates and demonstrates that these growth rates of SWCNT growths by Fe, Ni, and Co METANOs are particularly very high. The CNT growth rate on Fe based METANO by Chen [46] was 442.00 nm/s. Unfortunately, the CNT growth rates on METANOs from metals other than Fe, Ni, and Co could not be precisely determined. SEM images of CNTs grown by Esconjauregui et al. [56] though shed some light on them. These images were obtained for CNTs grown with CH_4 precursor and under atmospheric pressure. They indicated that CNTs grown for 1 min at 750 °C and under a C_2H_4 flow of 100 ml/min on Ni, Co, and Fe METANOs were longer than the CNTs grown for 15 min at the same temperature, but on Al, In, Pt, Ti, Mg, Pd, K, Cs, and Na METANOs. They (e.g., CNTs grown on Fe, Ni, and Co METANOs) were also longer than CNTs grown for 15 min at 950 °C under a CH_4 flow of 1000 ml/min on W and Mn METANOs. These images by Esconjauregui et al. [56] suggest that the CNT growth rates on Fe, Co, and Ni METANOs are much higher than those on Al, In Pt, Ti, Mg, Pd, K, Cs, Na, W, and Mn METANOs. Raman studies by Zhong et al. [57] suggested that, under some favored conditions, SWCNTs grow more spontaneously than MWCNTs. Wood et al. [58] and Lee et al. [59] found

Table 13.6 Comparison of CNT growths by the VSS mechanism and catalyzed by FECAs

Carbon precursor	FECA	Temp (°C)	CNT	Technique	Growth rate (nm/s)	References
CH_4	FeMo	900	SWCNT	CVD	2.00×10^4	Huang et al. [43]
CH_4, N_2	–	850	MWCNT	Microwave plasma CVD	–	Kumar et al. [44]
CH_4, C_2H_5OH	Fe		SWCNT	Thermal CVD	2.00×10^3	Reina et al. [45]
CH_4, C_2H_5OH	Fe		SWCNT	Random	2×10^2	Reina et al. [45]
C_2H_2, H_2	Fe	800	MWCNT	Photothermal CVD	442.00	Chen [46]
CH_4	Co	700	DWCNT	Random	6×10^2	Hiramatsu and Hori [47]
Ethanol	Fe	900	SWCNT	CVD on Si substrate	1.10×10^4	Zheng et al. [48]
C_2H_5OH	Fe	950	SWCNT	Thermal CVD	$0.45–2.24 \times 10^4$	Yao et al. [49]
C_2H_2, H_2	Ni	600	MWCNT	Plasma-assisted CVD	8.9713×10^4	Chiang and Sankaran [50]
C_2H_2, H_2	$Fe_{0.33}Ni_{0.67}$	500	MWCNT	Plasma-assisted CVD	4.81×10^2	Chiang and Sankaran [50]
C_2H_2, H_2	$Fe_{0.73}Ni_{0.27}$	500	MWCNT	Plasma-assisted CVD	3.06×10^2	Chiang and Sankaran [50]
C_2H_4	Fe, Co	700	SWCNT	Thermal CVD	4.20×10^3	Hata et al. [51]
C_2H_2, H_2	TiFe	370	MWCNT	Photothermal CVD	22.66	Shang et al. [52]
CH_4	Co	900	SWCNT	Thermal CVD	2.50×10^3	Liu et al. [53]
CH_4, H_2	a-Carbon	350–550	MWCNT	PECVD	38.88	Song and Lim [54]
CH_4, H_2	Ni	700	MWCNT	Microwave plasma CVD	1.50×10^2	Choi et al. [55]

that nanotube growth rate depends on temperature and that it increases with increase in temperature.

13.9.2 Possible Causes of Said Observations

To reveal the causes of the said observations, we relied on carbon solubility in Fe and calculated E_{AF} for Fe-catalyzed CNT growth at 700 °C, 900 °C and 1100 °C, respectively. The calculations were performed keeping the innermost diameter D_{NTI} constant, but gradually increasing the outermost diameter D_{NTO}. The results shown in Fig. 13.5b indicate that, essentially at all temperatures, E_{AF} increases with increase in the outermost diameter D_{NTO}. The activation energy E_{AF} however decreases with increase in the temperature T, which causes an increase in the exchange of materials between Fe and C. The immediate result of this is the increase in CNT growth rate with increase in temperature T. To understand the increase in E_{AF} with increase in D_{NTO}, we must note that increase in D_{NTO} keeping D_{NTI} constant transforms the nanotube from single-walled nanotube to multiwalled nanotube. We must also recall that single-walled nanotubes grow primarily from hill by the base growth mode, but multiwalled nanotubes grow primarily on shells by the tip growth mode [60]. Hills are formed at the nanoparticle periphery due to lower barrier to the segregation of $R_S{\equiv}C$ species from the bulk to the surface periphery of this nanoparticle. Shells, rather than hills, are however formed at the nanoparticle periphery due to higher barrier to the segregation of R_S species from the core to the surface periphery of this METANO nanoparticle. In general, the barrier to surface segregation of the R_S species depends on the looseness of the $R_S{\equiv}C$ species bound to the METANO surface lattice structure. This barrier is smaller for a METANO surface exhibiting less tightly bound $R_S{\equiv}C$ species to its lattice structure. The lattice structure of the METANO surface may however be more tightly bound if an increased release of the $R_S{\equiv}C$ species on the METANO surface leads to stronger adhesion of some of these species to this surface. This appears to be justified as Zhong et al. [57] observed SWCNTs grown predominantly at C_2H_2 concentration lower than 10%, but MWCNTs grown predominantly at C_2H_2 concentration higher than 10%. As manifested by Table 13.6, growth rate of the Fe-catalyzed SWCNT growth can be extremely high; it can be as high as 11,400 nm/s. The growth rates of Ni-catalyzed and Co-catalyzed SWCNT growths can also be very high, but not as high as the growth rate of Fe-catalyzed SWCNT growth.

13.9.3 Possible Causes of Discrepancy Based on Calculated Results

To explain the discrepancy stated above, we resort again to Fig. 13.5a, b. These figures show the variation of carbon solubility in Fe with variation in D_{NTO} keeping $D_{NTI} = 0.66$ nm. Figure 13.5a is for $T = 900$ °C, but $\lambda_{nano} = 1, 2$, and 3 respectively. But Fig. 13.5b is for $\lambda_{nano} = 1$, but $T = 700$ °C, 900 °C and 1100 °C, respectively. They both are for a SWCNT if $D_{NTO} = 1$ nm and the wall thickness is 0.34 nm. They are however for SWCNTs (MWCNTs) of wall thickness larger than 0.34 nm if $D_{NTO} > 1$ nm. Recall that carbon solubility in Fe is $\chi_{SM} = 0.18$ (e.g., 18 atomic %) at $T = T_E = 1153$ °C of the $R_L \equiv$(Fe, C) binary eutectic alloy [39]. Further, the surface energy γ_{nano} of METANO nanoparticle is sensitive to growth conditions, and also to surface treatments of the METANO nanoparticle surface [61–63]. It is dictated by λ_{nano}. This λ_{nano} generated by surface treatment (for example, soft plasma treatment) [61–63] can be large if this surface treatment gives rise to a large surface energy γ_{nano}. It is reflected from Fig. 13.5a, which shows different carbon solubility for the growth of different SWCNTs under different growth conditions: Calculated results for (a) the carbon solubility is $\chi_S \approx 0.25$ at $T = 900$ °C and $\lambda_{nano} = 1$; (b) the carbon solubility is $\chi_S \approx 0.19$ at $T = 800$ °C and $\lambda_{nano} = 1$; (c) the carbon solubility is $\chi_S \approx 0.11$ at $T = 700$ °C and $\lambda_{nano} = 1$, and (d) the carbon solubility is $\chi_S \approx 0.23$ at $T = 700$ °C and $\lambda_{nano} = 2$. These results are quite remarkable. A notable feature of these results is that carbon solubility in Fe can exceed the eutectic temperature value of it, viz. $\chi_{SM} = 0.18$ even at 800 and 900 °C if $\lambda_{nano} = 1$. It can exceed the eutectic temperature value of $\chi_{SM} = 0.18$ even at 700 °C, if however $\lambda_{nano} = 2$. And all these have several major and probably paramount implications on the CNT growths.

13.9.4 Implication of Higher Solubility of C in Fe, Co, and Ni for SWCNT Growths

Recall that solubility of carbon in Fe is critical for the segregation of this C species to the METANO periphery from the METANO core. The immediate result of this is the formation of shell surrounded by hill (see Fig. 13.6). Let us be more specific. Higher solubility of carbon in the METANO≡Fe translates to higher segregation of this species to the METANO periphery from the METANO core creating shell. Depending on the level of segregation, there can also be a robust hill surrounding the shell at the METANO periphery. All these can take place during the pre-nucleation stage of growth. Second, higher solubility of carbon in METANO≡Fe leads to the formation of amorphous and porous $R_L \equiv$(Fe, C) species in the shell during the pre-nucleation stage of growth. Third, higher solubility of carbon species in the METANO≡Fe leads to a highly porous shell, and higher bulk diffusion of the $R_S \equiv$C species through the R_L species of this shell. This takes place during the pro-nucleation stage of growth. It leads, as a result, to higher nanotube growth. Also, the $R_L \equiv$(Fe, C)

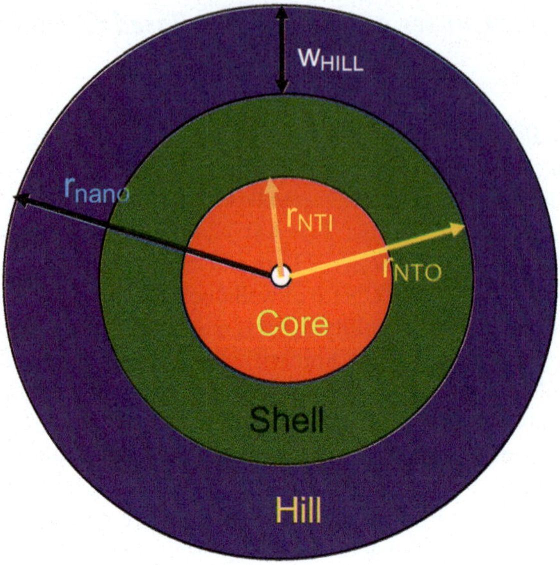

Fig. 13.6 Schematic diagram showing shell and hill formed around the core of a nanoparticle due to surface diffusion of the R_S species (for example, carbon species) from the core to the surface periphery of the nanoparticle. This schematic diagram shows also the innermost radius $r_{NTI} = D_{NTI}/2$ and the outermost radius $r_{NTO} = D_{NTO}/2$

species generated in the shell of the Fe catalyst surface can be quasiliquid (quasisolid) at $T \leq 900\,°C$, which (e.g., temperature T) is lower than the eutectic temperature $T_E =$ 1153 °C of the $R_L \equiv$(Fe, C) binary alloy. Then, there are the surface-energy-induced effects, which enhance the surface-disorder (effective surface amorphicity, surface porosity, surface melting, sutace coarsening, etc.). The surface-induced diffusion can indeed give rise to very high nanotube growth rate. We reiterate. The present results demonstrate that, due to very high surface-energy-induced effects, the CNT growth rate on Fe could be extremely high. The CNT growth rate on METANO≡Ni and METANO≡Co could also be high. However, due to lower surface energy-induced effects, the CNT growth rate on METANOs other than Fe, Ni, and Co could be low or moderately low.

13.9.5 Implication of Lower Solubility of C in Fe for MWCNT Growths

As stated earlier, $D_{NTI} = 0.66$ nm and $D_{NTO} > 1$ nm of a CNT may correspond to MWCNT. The reduced carbon solubility per unit area of a METANO would be

$$\Re_{\text{nano}} = \frac{4\Re}{\pi}\left(D_{\text{NTO}}^2 - D_{\text{NTI}}^2\right)^{-1}. \tag{13.8}$$

Equation (13.8) may be transformed to

$$\frac{\pi \Re_{\text{nano}}}{4} = \Re\left(D_{\text{NTO}}^2 - D_{\text{NTI}}^2\right)^{-1} \tag{13.9}$$

Making use of thc calculated value of $\Re$ for the METANO≡Fe, we obtained $(\pi \Re_{\text{nano}}/4) = 0.94$, 0.66 and 0.48 nm^{-2} for $D_{\text{NTI}} = 0.5$ nm, but $D_{\text{NTO}} = 1.5$ nm, 2.0 nm and 2.5 nm, respectively. Considering that volume (bulk) carbon diffusivity through a shell for carbon nanotube growth is determined by carbon solubility into this shell, this carbon diffusivity gradually decreases with increase in D_{NTO} keeping D_{NTI} constant. This implies that, under identical conditions, the MWCNT growth can be lower than the SWCNT growth.

13.10 Limit of Growth Rate

To obtain high nanomaterial growth rate is always highly desirable. However, there is a limit to it. SEM and TEM images by Hemesath et al. [64] revealed substantial incorporation of Au catalyst atoms at grain boundaries in ⟨110⟩ oriented Si nanowires. Šilhavík et al. [65] showed that foreign atom incorporation into the crystal structure of a semiconductor Si nanowires, for example, by plasma-enhanced chemical vapor deposition process, is possible just not only with Au catalyst, but also with several metal catalysts such as In, Sn, Bi, Ga, and Pb catalysts. They noted that the catalyst metals would be homogenously distributed inside nanowires or be segregated in clusters on their surface. Depending on the metal catalyst used, they could as well be in the core of the nanowires. Unless such incorporation is desired, some preventive measures can be taken to overcome the problem. We argue that the said incorporation occurs due to high energy possessed by the R_S species during their rapid diffusion through the droplet. Such a rapid diffusion leads to high growth rate, but the energetic R_S species knock the catalyst metal atoms out of the droplet, which eventually find their ways in the bulk or surface of nanowires. This happens particularly if the droplet is highly molten due to increase in temperature or there are insufficient interactions between the R_S species atoms and the MET atoms of the R_L≡(MET, X) catalyst nanoparticle transformed into the droplet. While examining impurity concentration in Si nanowires, Chen et al. [66] noted that this incorporation increases with increase in nanowire growth rate. This incorporation, shown in Fig. 13.7 indicates that both Sn impurity and In impurity resulting from catalyst incorporation in Si nanowire increase with increase in nanowire growth rate.

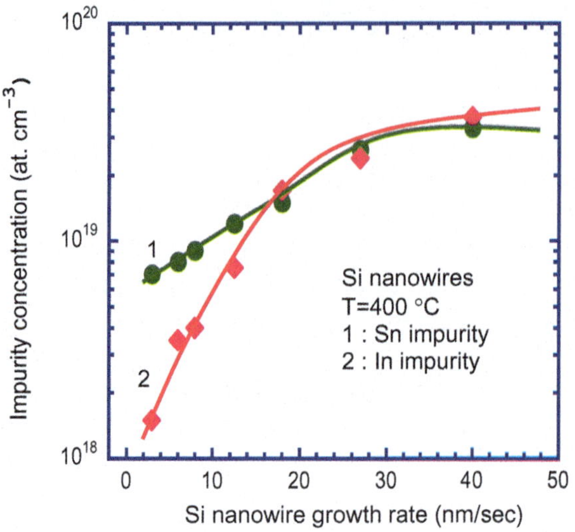

Fig. 13.7 Variation of Sn and In impurity concentration introduced in Si nanowire with silicon nanowire growth rate. The figure was obtained with experimental data by Chen et al. [66]

13.11 Conclusions

Taking the thermodynamic and kinetic viewpoints into account, the nanomaterials (nanocrystals) growths by the metal catalyst-mediated VQS mechanism can be carried out via (1) oxygen and oxide-assisted means, (2) plasma-enhanced means, (3) laser-assisted means, (4) Frank's screw dislocation induced means, (5) surface-treatment-induced means, (6) defect-induced means, and (7) water-treatment-induced means. The goal, in all the cases, as described in Chap. 3, would be to create on the metal nanoparticle surface the $R_L \equiv (\beta_1, \beta_2, \beta_3, \beta_4, \beta_5, \beta_6,$ etc.) species or the $R_L \equiv$(EMNO) species, or the $R_L \equiv (\beta_1)_z(\beta_2)_{1-z}$, where z is the mole fraction of β_1 in the R_L species. One or more of the components of the R_L species would obviously be metal. The R_L species thus created must have surface disturbance and disorder, effective surface amorphicity and dipole moment up to a depth δ_{amor} (see Chap. 12 and Appendix). They should have surface roughness of certain optimal density and height. None of them would be arbitrary. They would rather be planned, controlled, interdependent, and optimal and have fundamental physicochemical foundation. To achieve it would, no doubt, be challenging. The R_L species would satisfy other requirements, as well. These requirements would include porosity, optimal phase(s), surface melting (semimelting), etc. (see Chap. 12). While one of the seven means for growths, spelled out above, would be preferred under certain growth conditions, another one of them would be preferred under some other growth conditions. All of them should however yield the most desired products, namely the ones with the best possible crystallinity, morphology, orientation, length, diameter, uniformity, interfaces, etc.

Generally, the defect-induced means for growth is preferred. The behavior, effectiveness, and characteristics of the R_L species in mediating growths would be reflected, at least in part, from their shape, size, morphology, and lattice structure. For example, the R_L species would be porous cluster or solid solution, rather than a non-porous alloy if the tips of the nanomaterial (s) are irregular and relatively unstable. They would generally be unreacted, rather than reacted. The assemblage of materials (e.g., $\beta_1, \beta_2, \beta_3, \beta_4, \beta_5, \beta_6$, etc.) at the tips would consequently have no regular or well-defined shape and size. Thc tip would otherwise assume a spherical (hemispherical) shape if the assemblage is reacted.

We argue that there are misunderstandings in the basics of some of the above-mentioned seven means for VQS growths. We point out one of them, e.g., the water-treatment-induced means. While trying this means, the objective of Hata et al. [51] was to find a weak oxidizer suitable for selectively removing amorphous carbon from the metal nanoparticle surface on which the CNTs were produced. The objective was also to achieve it without damaging the CNTs produced by the said means. The objective was therefore worthwhile, novel, and purposeful.

Note that the amorphous carbon creates a coating on the catalyst nanoparticle surface thus impeding the activity and lifetime of the nanoparticles. The experimental data by Hata et al. [51] demonstrated that indeed water significantly promoted and preserved the catalytic activity of the nanoparticle surface. SWCNTs were grown by employing a mixture of ethylene precursor with Ar (or He) and H_2. This mixture contained a small, but controlled amount of water vapor. The relative levels of ethylene and water were balanced in order to maximize the catalytic activity and lifetime of the metal nanoparticles. Szabo et al. [67] and Patole et al. [68] subsequently studied it, rather in some details. They examined the impact of water vapor on CNT growths. We reproduce the experimental results by Szabo et al. in Fig. 13.8a, b.

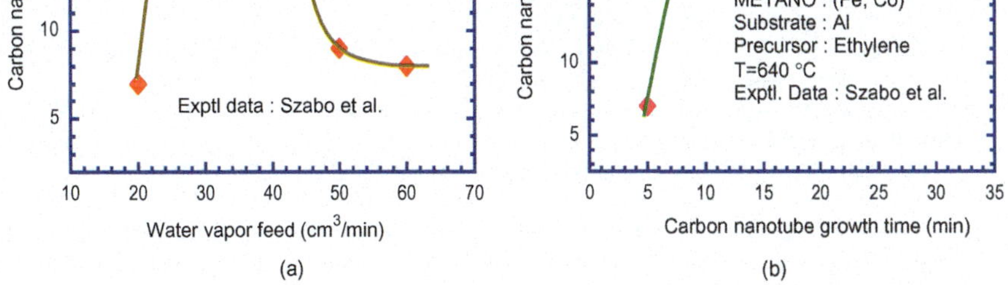

Fig. 13.8 The impact of **a** water vapor feed and **b** growth time on the height of CNTs grown by the CVD method at 640 °C. The experimental data for these plots were obtained by Szabo et al. [67]

Figure 13.8a indicates that CNT height increases, reaches a peak, and then decreases with increase in water vapor feed. Figure 13.8b indicates that CNT height increases, reaches a peak, and then decreases with increase in growth time. The said decrease in CNT height must be caused by decrease in CNT growth rate due to increase in water vapor feed and also due to increase in time. If the purpose of water vapor is to remove the amorphous carbon from the nanoparticle surface, then why would CNT height eventually decrease with increase in water vapor feed and growth time?

We believe that the observations by Hata et al. [51], Szabo et al. [67], and Patole et al. [68] would be consistent with each other and be explained well if judged from another point of view. Based on this viewpoint, the primary goal of the water vapor feed would be to generate R_L species on the nanoparticle surface. The gradual increase in water vapor would lead to gradual increase in surface amorphicity α_{amor} (see Appendix), effective surface amorphicity $\alpha_{amoreff}$, surface porosity, surface coarsening, HET concentration, and dipole moment and hence CNT growth rate until the surface amorphicity α_{amor} reaches the peak at $\alpha_{amoreff} = \alpha_{amoreff0}$ on this surface. The effective surface amorphicity $\alpha_{amoreff}$, surface porosity, surface coarsening, HET concentration, and dipole moment and hence CNT growth rate would then all begin to decrease with further increase in surface amorphicity α_{amor} beyond $\alpha_{amor} = \alpha_{amoreff0}$ (see Appendix). The CNT growth rate would also begin to decrease with increase in time after the CNT growth rate reaches the peak at a time $t = t_0$. It would happen because water vapor particles would gradually block and even close the nanopores of the R_L species after the surface amorphicity α_{amor} reaches $\alpha_{amoreff0}$ at a time $t = t_0$.

References

1. R.M. Walser, R.W. Bené, First phase nucleation in silicon-transition-metal planar interfaces. Appl. Phys. Lett. **28**, 624–625 (1976)
2. M. Wittmer, M. Nicolet, J.W. Mayer, The first phase to nucleate in planar transition metal-germanium interfaces. Thin Solid Films **42**, 51–59 (1977)
3. P.A. Lin, J.L. Gomez-Ballesteros, J.C. Burgos, P.B. Balbuena, B. Natarajan, R. Sharma, Direct evidence of atomic-scale structural fluctuations in catalyst nanoparticles. J. Catal. **349**, 149–155 (2017)
4. M.D. Argyle, C.H. Bartholomew, Heterogeneous catalyst deactivation and regeneration: a review. Catalysts **5**(1), 145–269 (2015)
5. H.D. Park, S.M. Prokes, M.E. Twigg, R.C. Cammarata, A.-C. Gaillot, Si-assisted growth of InAs nanowires. Appl. Phys. Lett. **89**, 223125 (2006)
6. S.G. Sundaresan, A.V. Davydov, M.D. Vaudin, I. Levin, J.E. Maslar, Y.-L. Tian, M.V. Rao, Growth of silicon carbide nanowires by a microwave heating-assisted physical vapor transport process using group VIII metal catalysts. Chem. Mater. **19**, 5531–5537 (2007)
7. A. Colli, A. Fasoli, P. Beecher, P. Servati, S. Pisana, Y. Fu, A.J. Flewitt, W.I. Milne, J. Robertson, C. Ducati, S. De Franceschi, S. Hofmann, A.C. Ferrari, Thermal and chemical vapor deposition of Si nanowires: shape control, dispersion, and electrical properties. J. Appl. Phys. **102**, 034302 (2007)
8. B.N. Mbenkum, A.S. Schneider, G. Schutz, C. Xu, G. Richter, P.A. van Aken, G. Majer, J.P. Spatz, Low-temperature growth of silicon nanotubes and nanowires on amorphous substrates. ACS Nano **4**, 1805–1812 (2010)

9. T.I. Kamins, R.S. Williams, D.P. Basile, T. Hesjedal, J.S. Harris, Ti-catalyzed Si nanowires by chemical vapor deposition: microscopy and growth mechanisms. J. Appl. Phys. **89**, 1008–1016 (2001)
10. Q. Wan, G. Li, T.H. Wang, C.L. Lin, Titanium-induced germanium nanocones synthesized by vacuum electron-beam evaporation: growth mechanism and morphology evolution. Solid State Commun. **125**, 503–507 (2004)
11. J.D. Carter, Y. Qu, R. Porter, L. Hoang, D.J. Masiel, T. Guo, Silicon-based nanowires from silicon wafers catalyzed by cobalt nanoparticles in a hydrogen environment. Chem. Commun. 2274–2276 (2005)
12. J. Arbiol, B. Kalache, P.R. Cabarrocas, J.R. Morante, A.F. Morral, Influence of Cu as a catalyst on the properties of silicon nanowires synthesized by the vapor-solid-solid mechanism. Nanotechnology **18**, 305606 (2007)
13. S. Hofmann, R. Sharma, C.T. Wirth, F. Cervantes-Sodi, C. Ducati, T. Kasama, R.E. Dunin-Borkowski, J. Drucker, P. Bennett, J. Robertson, Ledge-flow-controlled catalyst interface dynamics during Si nanowire growth. Nat. Mater. **7**, 372–375 (2008)
14. T. Baron, M. Gordon, F. Dhalluin, C. Ternon, P. Ferret, P. Gentile, Si nanowire growth and characterization using a microelectronics-compatible catalyst PtSi. Appl. Phys. Lett. **89**, 233111 (2006)
15. J.L. Lensch-Falk, E.R. Hewmesath, J.L. Lauhon, Syntaxial growth of Ge/Mn-germanide nanowire heterostruc-tures. Nano Lett. **8**, 2669–2673 (2008)
16. Y.W. Wang, V. Schmidt, S. Senz, U. Gösele, Epitaxial growth of silicon nanowires using an aluminum catalyst. Nat. Nanotechnol. **1**, 186–189 (2006)
17. K. Kang, D.A. Kim, H.-S. Lee, C.-J. Kim, J.-E. Yang, M.-H. Jo, Low-temperature deterministic growth of Ge nanowires using Cu solid catalysts. Adv. Mater. **20**, 4684–4690 (2008)
18. K. Kang, G.H. Gu, D.A. Kim, C.G. Park, M.-H. Jo, Self-organized growth of Ge nanowires from Ni-Cu bulk alloys. Chem. Mater. **20**, 6577–6579 (2008)
19. S. Kodambaka, J. Tersoff, F.M. Ross, Ge nanowire growth below the eutectic temperature. Science **316**, 729–732 (2007)
20. Y.-D. Ko, J.-G. Kang, G.-H. Lee, J.-G. Park, K.-S. Park, Y.-H. Jin, D.-W. Kim, Sn-induced low-temperature growth of Ge nanowire electrodes with a large lithium storage capacity. Nanoscale **3**, 3371–3375 (2011)
21. T.I. Kamins, X. Li, R.S. Williams, Growth and structure of chemically vapor deposited Ge nanowires on Si substrates. Nano Lett. **4**, 503–506 (2004)
22. R.C. Gouveia, H. Kamimura, R. Munhoz, A.D. Rodrigues, E.R. Leite, A.J. Chiquito, Germanium nanowires grown using different catalyst metals. Mater. Chem. Phys. **183**, 145–151 (2016)
23. D. Wang, H Dai, Low-temperature synthesis of single-crystal germanium nanowires by chemical vapor deposition. Angew. Chem., Int. Edn. **41**, 4783–4786 (2002)
24. C. Tang, Y. Bando, Z. Liu, D. Goldberg, Synthesis and structure of InP nanowires and nanotubes. Chem. Phys. Lett. **376**, 676–678 (2003)
25. Z.H. Lan, W.M. Wang, C.L. Sun, S.C. Shi, C.W. Hsu, T.T. Chen, K.H. Chen, C.C. Chen, Y.F. Chen, L.C. Chen, Growth mechanism structure and IR photoluminescence studies of indium nitride nanorods. J. Cryst. Growth **269**, 87–94 (2004)
26. K.L. Klein, A.V. Melechko, P.D. Rack, J.D. Fowlkes, H.M. Meyer, M.L. Simpson, Cu–Ni composition gradient for the catalytic synthesis of vertically aligned carbon nanofibers. Carbon **43**, 1857–1863 (2005)
27. A.-F. Lehlooh, S.M. Fayyad, S.H. Mahmood, Mössbauer spectroscopy study of Fe–Si solid solution prepared by mechanical milling. Hyperfine Interact. **139**(140), 335–344 (2002)
28. W. Neumann, H. Kirmse, I. Häusler, C. Zheng, A. Mogilatenko, Quantitative structure analysis of nanosized materials by transmission electron microscopy, in *Materials Research Society Symposium Proceedings*, vol. 1184, 184-GG03-02 (2009); https://doi.org/10.1557/PROC-1184-GG03-02, (Cambridge University Press: Cambridge, UK, 31 January 2011)
29. Y. Wang, G. Meng, L. Zhang, C. Liang, J. Zhang, Catalytic growth of large-scale single-crystal CdS nanowires by physical evaporation and their photoluminescence. Chem. Mater. **14**, 1773–1777 (2002)

30. J. Yuan, H. Zhang, J. Tang, N. Shinya, K. Nakajima, L.-C. Qin, Synthesis and characterization of single crystalline hafnium carbide nanowires. J. Am. Ceram. Soc. **95**, 2352–2356 (2012)
31. S.N. Mohammad, VQS (vapor-quasiliquid-solid, vapor-quasisolid-solid) mechanism for the catalyst-free and catalyst-mediated non-eutectic syntheses of single-crystal nanowires. J. Appl. Phys. **120**, 084307 (2016)
32. S.N. Mohammad, Bimetallic-catalyst-mediated syntheses of nanomaterials (nanowires, nanotubes, nanofibers, nanodots, etc.) by the VQS (vapor–quasiliquid–solid, vapor–quasisolid–solid) growth mechanism. J. Phys. D Appl. Phys. **49**, 495304 (2016)
33. S.N. Mohammad, VQS (vapor-quasiliquid-solid, vapor-quasisolid-solid) mechanism presents a unified foundation for the syntheses of nanotubes, primarily carbon nanotubes. AIP Adv. **7**, 095011 (2017)
34. S.N. Mohammad, VQS (vapor-quasiliquid-solid, vapor-quasisolid-solid) mechanism for realizing narrow distributions of chirality and diameters of single-walled carbon nanotubes (SWCNTs). J. Nanosci. Nanotechnol. **19**, 5388–5417 (2019)
35. S.N. Mohammad, Thermodynamic imbalance, surface energy, and segregation reveal the true origin of nanotube synthesis. Adv. Mater. (Weinheim, Germany) **24**, 1262–1275 (2012)
36. W.R. Wilcox, T.J. La Chapelle, *Mechanism of Gold Diffusion in Silicon*, Unclassified Document, (Defense Documentation Center for Scientific and Technical Information, Cameron Station, Alexandria, VA, USA. April 10, 1963)
37. A. Moisala, A.G. Nasibulin, E.I. Kauppinen, The role of metal nanoparticles in the catalytic production of single-walled carbon nanotubes—a review. J. Phys. Consens. Matter **15**, S3011–S3035 (2003)
38. Q. Jiang, H.M. Lu, M. Zhao, Modeling of surface energies of elemental crystals, J. Phys.: Condens. Matter. **16**, 521–530 (2004)
39. T.B. Massalski (ed.), *Binary Alloy Phase Diagrams*, vol. 3, 2nd edn. (American Society of Metals, Metals Park, OH, 1986)
40. S.A. Dayeh, S.T. Picraux, Direct observation of nanoscale size effects in Ge semiconductor nanowire growth. Nano Lett. **10**, 4032–4039 (2010)
41. B.J. Kim, C.-Y. Wen, J. Tersoff, M.C. Reuter, E.A. Stach, F.M. Ross, Growth pathways in ultralow temperature Ge nucleation from Au. Nano Lett. **12**, 5867–5872 (2012)
42. A.D. Gamalski, J. Tersoff, R. Sharma, C. Ducati, S. Hofmann, Metastable crystalline AuGe catalysts formed during isothermal germanium nanowire growth. Phys. Rev. Lett. **108**, 255702 (2012)
43. S.M. Huang, M. Woodson, R. Smalley, J. Liu, Growth mechanism of oriented long single walled carbon nanotubes using fast-heating chemical vapor deposition process. Nano Lett. **4**, 1025–1028 (2004)
44. S. Kumar, I. Levchenko, K. Ostrikov, J.A. McLaughlin, Plasma-enabled, catalyst-free growth of carbon nanotubes on mechanically-written Si features with arbitrary shape. Carbon **50**, 325–329 (2012)
45. A. Reina, M. Hofmann, D. Zhu, J. Kong, Growth mechanism of long and horizontally aligned carbon nanotubes by chemical vapor deposition. J. Phys. Chem. C **111**, 7292–7297 (2007)
46. J.S. Chen, *Selective Growth of Carbon Nanotubes Via Photo-Thermal Chemical Vapor Deposition*. Doctoral thesis (University of Surrey, England, 2015)
47. M. Hiramatsu, M. Hori, Aligned growth of single-walled and double-walled carbon nanotube films by control of catalyst preparation, in *Carbon Nanotubes—Synthesis, Characterization, Applications*, ed. by Siva Yellampalli (InTech, Rijeka, Croatia, 2011). ISBN: 978-953-307-497-9, Chapter 10
48. L.X. Zheng, M.J. O'Connelli, S.K. Doorn, X.-Z. Liao, Y.-H. Zhao, E.A. Akhadov, M.A. Hoffbauer, B.J. Roop, Q.-X. Jia, R.C. Dye, D.E. Peterson, S.M. Huang, J. Liu, Y.-T. Zhu, Ultralong single-wall carbon nanotubes. Nat. Mater. **3**, 673–676 (2004)
49. Y.G. Yao, R. Liu, J. Zhang, L.Y. Jiao, Z.F. Liu, Raman spectral measuring of the growth rate of individual single-walled carbon nanotubes. J. Phys. Chem. C **111**, 8407–8409 (2007)
50. W.-H. Chiang, R.M. Sankaran, Relating carbon nanotube growth parameters to the size and composition of nanocatalysts. Diam. Relat. Mater. **18**, 946–952 (2009)

51. K. Hata, D.N. Futaba, K. Mizuno, T. Namai, M. Yumura, S. Iijima, Water-assisted highly efficient synthesis of impurity-free single-walled carbon nanotubes. Science **306**, 1362–1364 (2004)
52. N.G. Shang, Y.Y. Tan, V. Stolojan, P. Papakonstantinou, S.R.P. Silva, High-rate low-temperature growth of vertically aligned carbon nanotubes. Nanotechnology **21**, 505604 (2010)
53. B. Liu, W. Ren, C. Liu, C.-H. Sun, L. Gao, S. Li, C. Jiang, H.-M. Cheng, Growth velocity and direct length-sorted growth of short single-walled carbon nanotubes by a metal-catalyst-free chemical vapor deposition process. ACS Nano **3**, 3421–3430 (2009)
54. W. Song, D.G. Lim, Growth of metal-free carbon nanotubes with an amorphous carbon catalyst layer. J. Korean Phys. Soc. **53**, 2142–2146 (2008)
55. Y.C. Choi, Y.M. Shin, Y.H. Lee, B.S. Lee, G.-S. Park, W.B. Choi, N.S. Lee, J.M. Kim, Controlling the diameter, growth rate, and density of vertically aligned carbon nanotubes synthesized by microwave plasma-enhanced chemical vapor deposition. Appl. Phys. Lett. **76**, 2367 (2000)
56. S. Esconjaureguia, C.M. Whelan, K. Maex, The reasons why metals catalyze the nucleation and growth of carbon nanotubes and other carbon nanomorphologies. Carbon **47**, 659–669 (2009)
57. G. Zhong, S. Hofmann, F. Yan, H. Telg, J.H. Warner, D. Eder, C. Thomsen, W.I. Milne, J. Robertson, Acetylene: A key growth precursor for single-walled carbon nanotube forests. J. Phys. Chem. C **113**, 17321–17325 (2009)
58. R.F. Wood, S. Pannala, J.C. Wells, A.A. Puretzky, D.B. Geohegan, Simple model of the interrelation between single- and multiwall carbon nanotube growth rates for the CVD process. Phys. Rev. B **75**, 235446 (2007)
59. Y.T. Lee, J. Park, Y.S. Choi, H. Ryu, H.J. Lee, Temperature-dependent growth of vertically aligned carbon nanotubes in the range 800–1100 & #xB0;C. J. Phys. Chem. B **106**, 7614–7618 (2002)
60. S.N. Mohammad, Systematic investigation of the growth mechanisms for the synthesis of the conventional, doped, and bamboo-shaped nanotubes, primarily the carbon nanotubes. Carbon **75**, 133–148 (2014)
61. R. Wolf, A.C. Sparavigna, Role of plasma surface treatments on wetting and adhesion. Engineering **2**, 397–402 (2010)
62. M.C. Kim, D.K. Song, H.S. Shin, S.-H. Baeg, G.S. Kim, J.-H. Boo, J.-G. Han, S.H. Yang, Surface modification for hydrophilic property of stainless steel treated by atmospheric-pressure plasma jet. Surf. Coat. Technol. **171**, 312–316 (2003)
63. C.-C. Wu, C.-K. Wei, C.-C. Ho, S.-J. Ding, Enhanced hydrophilicity and biocompatibility of dental zirconia ceramics by oxygen plasma treatment. Materials **8**, 684–699 (2015)
64. E.R. Hemesath, D.K. Schreiber, E.B. Gulsoy, C.F. Kisielowski, A.K. Petford-Long, P.W. Voorhees, L.J. Lauhon, Catalyst incorporation at defects during nanowire growth. Nano Lett. **12**, 1167–1171 (2012)
65. M. Šilhavík, M. Müller, J. Stuchlík, H. Stuchlíková, M. Klementová, J. Kočka, A. Fejfar, J. Červenka, Comparative study of catalyst-induced doping and metal incorporation in silicon nanowires. Appl. Phys. Lett. **114**, 132103 (2019)
66. W. Chen, L. Yu, S. Misra, Z. Fan, P. Pareige, G. Patriarche, S. Bouchoule, P.R. Cabarrocas, Incorporation and redistribution of impurities into silicon nanowires during metal-particle-assisted growth. Nat. Commun. **5**, 4134 (2014)
67. A. Szabó, E. Kecsenovity, Z. Pápa, T. Gyulavári, K. Németh, E. Horvath, K. Hernadi, Influence of synthesis para-meters on CCVD growth of vertically aligned carbon nanotubes over aluminum substrate. Sci. Report **7**, 9557 (2017)
68. S.P. Patole, P.S. Alegaonkar, H.-C. Lee, J.-B. Yoo, Optimization of water assisted chemical vapor deposition parameters for super growth of carbon nanotubes. Carbon **46**, 1987–1993 (2008)

Chapter 14
Growths on SUBSANO Surface by the VQS Mechanism

Abstract The fundamentals, applicability, and novelty of the metal-free vapor–quasiliquid–solid (vapor–quasisoild–solid) growth mechanism, called the VQS growth mechanism have been described. The structure, composition, and morphology of the MET-free nanoparticles (e.g., SUBSANOs) most suitable for growths by the VQS mechanism have been presented. A critical look of the SUBSANOs suggests that the MET-free R_L species can have many different compositions and morphologies. They can, for instance, be solid solution (clustered) islands, coarsened islands of substrates, defective islands of metallic substrates, and defective islands of nonmetallic substrates. They can have phase transformation and phase generation during growths. Nanomaterials growths on surfaces have then be articulated. These surfaces can be single or composite and have several different components. These surfaces would depend on both the surface treatment and surface functionalization. They would primarily be the surfaces of the R_L species of SUBSANOs. They should have characteristics suited for significant catalytic potential for growths. They would be (a) an alloy, cluster, or solid solution, (b) amorphous material exhibiting an effective surface amorphicity $\alpha_{amoreff,}$ (c) paracrystalline material, (d) stepped material, (e) material influenced by synergy, and/or (f) material inundated with defects and dislocations. The events such as Knudsen diffusion, low-temperature decomposition of gaseous precursors, and catalyst poisoning, and the impacts of catalyst template on supersaturation during the MET-free growths by the VQS mechanism have been described.

14.1 Forwarding Note and Basic Concepts

14.1.1 Forwarding Note

The basic principles of growths by the VQS mechanism were laid down in Chap. 12. General definitions of FECANO, METANO, and SUBSANO nanoparticles, of ESNO substrate surface, and of the R_L species were made in Chap. 3. Note that

S. N. Mohammad, *Synthesis of Nanomaterials*, Springer Series in Materials Science 307, https://doi.org/10.1007/978-3-030-57585-4_14

FECANO is the FECA nanoparticle, which can be METANO (e.g., metal nanoparticle) or SUBSANO (e.g., substrate nanoparticle). SECINI and SECINI0 were also defined in Chap. 3, the pre-nucleation and pro-nucleation stages of growths were defined in Chap. 4, but NP1 and NP2 nanoparticles were defined in Chap. 12. Based on these and on the R_S source species defined in Chap. 1, the basic concepts of substrate-mediated nanomaterials growths by the VQS mechanism would be laid down in the following.

14.1.2 A Critical Look at SUBSANOs

Realization of SUBSANOs is central to the MET-free growths by the VQS mechanism. It should be recalled that METANOs result from the fragmentation of a metal film formed on a substrate (or on a support formed on the substrate). But SUBSANOs are created on metallic or nonmetallic substrate (or on the support formed on this substrate). A thin film of gold is a metallic substrate, but a thin film of graphite is a nonmetallic substrate. SUBSANOs are actually the pockets of disturbed, disordered, amorphous, coarsened lattice structure created on the substrate surface. They appear as islands, about 1–100 nm in diameter, on substrate surface due to surface treatment such as oxygenation, oxidation, acid treatment, plasma treatment, sputtering, aqua regia treatment, laser ablation. The SUBSANOs are porous and amorphous (semi-amorphous, amorphous-like) possessing HETs and network of nanopores [1–5]. Surface amorphicity (semi-amorphicity) α_{amor} and effective surface amorphicity $\alpha_{amoreff}$ of the SUBSANOs and hence of the R_L species of the SUBSANOs follow properties described in the Appendix. The R_L species of this SUBSANO may actually be ESNO: R_L≡(ESNO) species. The R_L species of the SUBSANO may also be R_L≡(ESNO, X) species, R_L≡(ESNO, Y) species, R_L≡(ESNO, X, Y) species, R_L≡(ESNO, X, ϑ) species, R_L≡(ESNO, Y, ϑ) species, R_L≡(ESNO, ϑ) species, and/or R_L≡(ESNO, X, Y, ϑ) species. They may all have small dimensions.

14.2 Illustrative Demonstrations of the R_L Species

In the following, we present some illustrations of the R_L species for metal-free nanomaterials (nanocrystals) growths. It is hoped that they would provide clarity of the discussions subsequently made in the chapter.

14.2.1 *Nanomaterials Growth on Solid Solution (Clustered) Islands*

Ma et al. [6] synthesized multiwalled BN nanotubes from a $B_4N_3O_2H$ precursor. The morphological and structural features of the BN nanotube tips suggested that they were grown from boron oxynitride nanoclusters containing silicon, aluminum, and calcium as impurities. The R_L species plausibly responsible for BN nanotube growth was therefore $R_L \equiv$(B, N, O, Si, Al, Ca) cluster or some of its derivative. HRTEM investigation of catalyst-free growth of GaN nanowires on Si substrates by Stoica et al. [7] showed small clusters made of Ga and N formed at the interface of oxidized Si substrate and an amorphous layer formed on this substrate. The formation of amorphous $R_L \equiv$(Ga, N, Si, O) species was plausibly responsible for nanowire growths. Wu and Liu [8] grew highly oriented ZnO nanorods on several different substrates. While growing these nanorods on Si (100) substrate, they observed an amorphous SiO_z interfacial layer formed between ZnO nanorods and Si(100) substrate. We believe that the $R_L \equiv$(Si, O) species, or $R_L \equiv (SiO_z)$ $1 \le z < 2$, or more likely the $R_L \equiv$(Si, Zn, O) species was responsible for the ZnO nanorod growths.

14.2.2 *Nanomaterials Growth on Coarsened Substrates*

Matteini et al. [9] grew GaAs nanowires by Ga-assisted self-catalytic method on Si (111) substrate. Surface roughness of the disordered SiO_z $(1 \le z < 2)$ layer formed on the Si substrate was crucial for this growth; it possessed pinholes and pits necessary for GaAs nanowire growth. It would not be rough without being disordered and amorphpus. In general, the lower the surface roughness, the higher was the temperature and the Ga injection rate needed for realizing comparable nanowire density, and vice versa. It happened because the higher the surface roughness, the higher was the surface disorder, the higher was the HET density, and the higher was the release of the R_S source species from the decomposition of their precursor by the HETs. The pinholes (pits) were created by the adsorption of Ga into the non-stoichiometric oxide. They generated NP2 nanoparticle and the $R_L \equiv$(Si, Ga, O) species on this NP2 nanoparticle surface. Morral et al. [10] achieved GaAs nanowire growth on GaAs wafers coated with SiO_2; this growth was similarly possible by gallium adatoms interacting with SiO_2 and creating NP2 nanoparticles and non-stoichiometric islands of $R_L \equiv$(Si, O, Ga) nanocraters on this nanoparticle surface. HETs generated on this nanocraters served as tools to release the R_S species for nanowire growths.

The sputter-deposited SiO_2 film is markedly different from the thermally grown SiO_2 film. The sputter-deposited film generally has disturbed, disordered, amorphized (semi-amorphized), coarsened, porous, islands (clusters) exhibiting grains, grain boundaries, and metastable nanostructures of NP2 nanoparticles. HETs are generated in pockets of accumulated charges and unsaturated dangling bonds on its surface. The thermally grown SiO_2 film generally lacks such islands (clusters);

they yield NP1 nanoparticles, which have a few HETs, if any, on their surfaces. The islands (clusters) on the sputter-deposited SiO_2 surface are therefore NP2-type SUBSANOs exhibiting R_L≡(Si, O) species or R_L≡(SiO_z), $1 \leq z < 2$, species. These R_L species could as well be R_L≡(Si-based ESNO, O) species. These species enabled Liu et al. [11, 12] to grow SWCNTs on the NP2-type SUBSANO surface. Generated on ZrO_2 films, similar SUBSANOs were composed of R_L≡(Zr, O) species or R_L≡(Zr-based ESNO, O) species. They enabled Steiner et al. [13] to grow, in the same way, SWCNTs on the ZrO_2 surface. While the growth by Liu et al. could be attributed to the SiO_z ($1 \leq z < 2$) surface composition and the HETs generated on the SiO_z surface, the growth by Steiner et al. could be attributed to ZrO_z ($1 \leq z < 2$) surface composition and the HETs generated on the ZrO_z surface. Carbon black surface and defect-rich graphite surface are not different in this regard. Both of them have natural NP2-type SUBSANOs of R_L species of similar characteristics as those of NP2-type SUBSANOs formed on sputter-deposited SiO_2 and ZrO_2 substrates. Lin et al. [14, 15] could consequently produce MWCNTs on the SUBSANO surface.

14.2.3 Nanomaterials Growth on Metallic Substrates

The illustrations made above-exemplified R_L species created on nonmetallic substrates. Lin et al. [16] showed that these R_L species can be created also on SUBSANOs of NP2-type characteristics formed on metallic substrate. They produced self-assembled carbon nanostructures (e.g., multiwalled carbon nanotubes, carbon nanofibers, etc.) at 750–900 °C on Au substrate by thermal CVD. Prior to the synthesis of carbon nanostructures at 850 °C, the substrate surface was purified at room temperature for a period of 4 h using nitric acid. It was followed by surface treatment with aqua regia solution for 60 s. And such a surface treatment created islands (clusters) of disturbed (disordered) lattice structure on the substrate surface. EDS analysis suggested that these islands (clusters) were composed of carbon, gold, and copper implying that the islands (clusters) were NP2-type SUBSANOs composed of R_L≡(Au, Cu, C) solid solutions. These SUBSANOs could alternatively be composed of R_L≡(Au-based ESNO, C, O) solid solutions. This solid solution contained probably other contaminants and was created during the pre-nucleation stage of growth. It was amorphous (semi-amorphous, amorphous-like) and had HETs. We argue that the R_L species mediating the nanowire growths had plausibly the composition: R_L≡(MET-based ESNO, C, ϑ) solid solution and/or cluster. This MET can be Au and some other metal of metallic substrate.

14.2.4 Nanomaterials Growth on Nonmetallic Substrates

Huang et al. [17] deposited 1-μm-thick SiO_2 film on a clean Si wafer in order to grow CNTs. No catalyst was deposited on the SiO_2 surface; instead the SiO_2 surface was scratched with a diamond blade. The lattice structure of the SiO_2 surface was, as a result, disturbed (disordered). In another experiment, the SiO_2 surface was treated with HF aqueous solution in air and at a temperature of 1000 °C for 1 h. The HF treatment disrupted, and even destroyed, some atomic (molecular) bonds. It displaced (and even freed) some of the atoms (molecules) of the SiO_2 surface converting it to a disordered SiO_2 layer. Both the processes created NP2-type SUBSANOs and R_L≡(Si oxide-based ESNO) species on these SUBSANOs. Chemical vapor deposition of the C species released from CH_4 precursor was carried out at 900 °C. CNTs were consequently produced on these NP2-type SUBSANOs of R_L≡(Si oxide-based ESNO). These ESNOs could actually be SiO_z, $1 \leq z < 2$, transformed from SiO_2.

The experiment by Kim et al. [18] to perform CVD growth of single-crystal Si and Ge nanowires was similar to that of Huang et al. [17]. A piece of hydrogen-terminated silicon wafer, rather than a SiO_2 wafer, was simply etched with ultrapure water to generate reactive silicon-rich oxide SUBSANOs of the R_L≡(SiO_z, $1 \leq z < 2$) species on the substrate surface. The nanowire growth at 490 °C employing SiH_4 and GeH_4 precursors yielded Si and Ge nanowires, respectively. Interestingly, they were selectively grown on NP2-type SUBSANOs of the R_L≡(SiO_z, $1 \leq z < 2$) species, but not on the SiO_2 film containing NP1-type SUBSANOs. They were though both formed on Si wafer. This happened because non-stoichiometric SiO_z-based SUBSANOs had disturbed, disordered, amorphized lattice structure and HETs, but stoichiometric SiO_2-based SUBSANOs lacked all of them. The R_L species responsible for nanowire growths was again R_L≡n(Si-based ESNO, O) cluster (solid solution) or R_L≡(Ge-based ESNO, O) cluster (solid solution); they had HETs for the release of the R_S species from the decomposition of their precursors.

The experiment by Liu et al. [11] to grow CNTs is not much different from those of Huang et al. [17] and Kim et al. [18]. In this experiment, a clean Si/SiO_2 wafer was scratched by sharp tip of another Si/SiO_2 wafer to ensure that the process was completely free from metal. The R_L species responsible for this was R_L≡(silicon oxide-based ESNO). And this ESNO could actually be non-stoichiometric SiO_z transformed from SiO_2. Examples cited above point to the nonmetallic substrates serving as SUBSANOs for growths by the VQS mechanism if these SUBSANOs have the composition of non-stoichiometric semiconductor or metal oxide.

14.3 Nanomaterials Growths on Surfaces

14.3.1 Basics

In the previous section, we described SUBSANOs and nanomaterials growth on SUBSANOs. Recall that these SUBSANOs may be formed on substrate, or on support produced on substrate. The materials structure of the islands, which may finally be SUBSANOs, may be different from those of the substrate or support. If formed on a support, there may, at the growth temperature T, be migration of atoms, ions, etc. from the substrate to the islands or vice versa. Some of the atoms loosely bound to the islands may also be freed. There may also be charge transfer. We cite an example. Due to the loss of oxygen atoms, a SUBSANO formed on a stoichiometric oxide surface may hence be converted to a non-stoichiometric oxide. Due to the migration of impurity metal atoms from substrate (support) to the SUBSANOs formed on a nonmetallic support produced on this substrate, may have metallic components in the SUBSANO lattice structures.

In this section, we, therefore, deal with surfaces, which should be single or composite, and have components shown in Fig. 14.1. They should have charac-

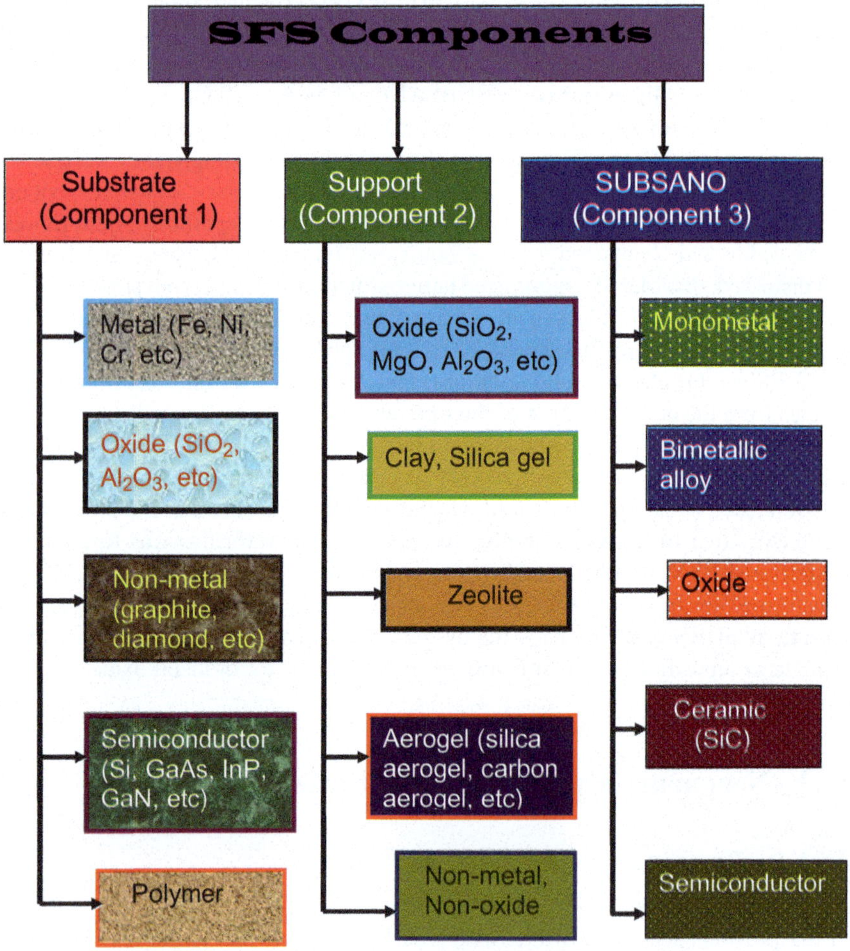

Fig. 14.1 Diagrammatic presentation of the components of various surfaces of the functional substrates (SFS). The components are the substrate, support, and the catalyst nanoparticle and the materials yielding these components

teristics suited for significant catalytic potential for growths. They would require a rational choice of the substrates and supports, and of the materials structures of these substrates and/or supports. The material characteristics of the island surfaces and also of subsurfaces would again depend on both the surface treatment and surface functionalization. For the sake of convenience, they (e.g., island surfaces) may be called the surfaces of functional substrates (SFSs), which would comprise substrate, support, and SUBSANO. The topmost layers of the SFSs possessing catalytic potential would actually be those of the SUBSANOs. Various possible components of SFSs shown in Fig. 14.1 may be of five or more different types, as listed in this figure.

The growths of nanomaterials (nanowires, nanotubes, nanobelts, nanodots, nanosheets, nanofibers, grapheme, etc.) on SUBSANOs must be controlled. It would happen if the said islands possess optimal level of lattice disorder and dangling bonds, and have novel properties as determined by the effective surface amorphicity α_{amoreff} (see Sect. 12.2 of Chap. 12 and also the appendix). The nanomaterials should, as a result, grow aligned; they should rest on the SUBSANO surfaces and have desired dimension, morphology, and characteristics.

14.3.2 Illustrations of SFSs

Table 14.1 shows that, depending on metal, semiconductor, oxide, polymer, etc. SFSs can be single-layered or composite-layered structures. If composite-layered, they can have multiple layers, one above the other, needed to facilitate the SFS's catalytic action for growth. We cite some examples, Yamada et al. [19] grew silicon nanowires on Si(100) substrate with and without the formation of a thermal oxide layer on it. They carried out rf magnetron sputtering of Si in Ar/H_2 before forming a thin Au layer on the top of both silicon and SiO_2, and before performing the growth. Randomly oriented, polycrystalline Si nanowires typically 20 μm in length and 350 nm in diameter were grown on both Si and SiO_2, after 60 min of growth at 700 °C. While the Si substrate served as catalyst support for one set of growths, the SiO_2 layer served as catalyst support for another set of growths, both the growths on Au catalyst. Had this Au catalyst been generated as an island on the Au film, it would be SUBSANO.

SFS can probably be illustrated well by the research of Bartness et al. [20]. This research for GaN nanowire growth by the MBE technique widely employed Si-AlN-GaN composite layer with Si as the substrate, AlN as the buffer (support) layer, and GaN as the SUBSANO layer. The AlN layer, which was the 30–120 nm thick layer, separated the top GaN layer from the bottom Si substrate layer. The GaN nanowire growth was carried out at a temperature 800–840 °C. While the linear thermal expansion coefficient of GaN is 2.8×10^{-6}/K, the linear thermal expansion coefficient of AlN is 5.63×10^{-6}/K. There occurred large mismatch of thermal expansion of AlN and GaN at the growth temperature, and islands of disturbed (disordered) lattice structure were formed on GaN to relieve the stress resulting from

Table 14.1 Illustrations of some possible SFS compositions and components

No	SFS components and compositions
1	Just a nonmetallic substrate (e.g., Si), which serves as nanoparticle catalyst for growth
2	Just a metallic substrate (Al, Cu, Inconel, stainless steel, etc.), which serves as nanoparticle catalyst for growth
3	A nonmetallic substrate, and a nonmetallic non-oxide support on this substrate; this support serves as nanoparticle catalyst for growth
4	A nonmetallic substrate, and an oxide support formed on this substrate. This oxide support serves as nanoparticle catalyst for growth
5	Oxide substrate (e.g., silica, Al_xO_y) and a monometallic nanoparticle formed on this substrate. This monometallic nanoparticle serves to catalyze the growth
6	An oxide substrate (e.g., silica, Al_xO_y) and a bimetallic nanoparticle formed on this substrate. This nanoparticle serves to catalyze the growth
7	A metallic substrate (e.g., Cu, stainless steel) and a nonmetallic non-oxide support (e.g., Si, Ge) formed on this substrate; this support serves as nanoparticle to catalyze the growth
8	A metallic substrate and a nonmetallic oxide support (for example, SiO_z, $1 \leq z < 2$) formed on this substrate. This support serves as nanoparticle to catalyze growth
9	An oxide substrate and a nonmetallic non-oxide support formed on the oxide substrate. This support serves as nanoparticle to catalyze growth
9	A nonmetallic non-oxide substrate (e.g., Si) and a nonmetallic support (e.g., Al_2O_3) formed on it. This support serves as catalyst to mediate growth
10	A metallic substrate, an oxide support (e.g., Al_2O_3) on it, and then a metallic nanoparticle (e.g., Ni, Fe, etc.) on this support; the metal nanoparticle serves as catalyst to mediate growth
11	Non-oxide substrate (e.g., Si) and bimetallic nanoparticle formed on this substrate. The nanoparticle serves to catalyze the growth

this mismatch. These islands serve as SUBSANOs for growth. Another illustration of SFS [21] would be a 50-nm-thick $Si_{0.7}Ge_{0.3}$ layer deposited by CVD on Si (001) wafers after the growth of a thin Si buffer layer on the Si(001) wafer. To relieve the stress resulting from the lattice mismatch between Si and Ge, islands were generated on the SiGe layer formed on the top of thin Si wetting layer (e.g., buffer layer). The heights of the islands varied between 20 and 50 nm. These islands were subsequently implanted with carbon ions (energy 30 keV, dose of 3×10^{16} cm^{-2}) to induce damage creating amorphous SUBSANOs formed on the SiGe surface. The SiGe top layer, the Si buffer layer, and the Si substrate were the constituents of the SFS.

Hernadi et al. [22] investigated the catalytic growth of carbon nanotubes. They found that catalyst nanoparticle formed (created) only on the external (outermost) surface of the porous support could catalyze carbon nanotube growth. This observation is similar to that by Zhu et al. [23], who synthesized double-walled carbon nanotubes by CVD over metal-containing SUBSANOs obtained from $Fe(CH_3COO)_2$ and $Co(CH_3COO)_2$ salts. These SUBSANOs were formed on mesoporous silica, which obviously served as substrate and also as catalyst support. Very uniform sizes of the nanotubes grown on the SUBSANOs endorsed mesoporous silica support as important element for initiating nanotube growth. The observation is similar also to that by Su et al. [24] who attributed high SWCNT growths to the strong interactions between the Al_2O_3 aerogel and the $Fe_{1-z}Mo_z$catalyst formed on this aerogel. Note that aerogel is a high-surface area, high-porosity, ultralow-density substrate prepared by sol-gel method. Obviously, the aerogel served also as catalyst support.

Ngo-Duc et al. [25] grew ZnO nanowires in a 2 cm inner-diameter quartz tube furnace at 550 °C for 5 to 30 min. A (FeCrAl) metal-alloy substrate with atomic compositions of 76 atomic % Fe, 21 atomic % Cr, and 3 atomic % Al was used to grow these nanowires. They were vertically aligned ZnO nanowires on FeCrAl metal substrates formed on a self-forming, thin pseudo-buffer layer by annealing the FeCrAl metal substrate prior to growth. It was a thin sapphire-like aluminum oxide surface comprising only Al and O formed on the FeCrAl metal substrate. The substrate annealing, carried out prior to growth, resulted in a preferential segregation of Al toward the surface yielding this Al and O-based buffer layer within ~25 nm underneath the substrate surface. We believe it was a non-stoichiometric AlO_z layer sapphire-like in chemical composition (Al and O). No wonder Sobanska et al. [26] could produce self-assembled GaN nanowires on a non-stoichiometric AlO_z sapphire-like layer of chemical components Al and O. Obviously, it had porous, amorphous islands that served as SUBSANOs. Prabaswara et al. [27] grew III–V nitride nanowires by plasma-assisted MBE on silica substrate coated with indium tin oxide support layer. The nanowires grew perpendicular to the substrate plane with n-polar polarity. The lateral size of the nanowires followed the grain sizes of the indium tin oxide support layer. This implies that SUBSANOs plausibly of the $R_L \equiv$(In, Sn, Si, O) species were formed as islands on the support layer, and GaN nanowires grew only on these SUBSANOs. The growth required no global lattice and thermal matching between the nanowires and the substrate.

Oxide layers such as SiO_2, Al_2O_3, TiO_2, and Er_2O_3 deposited, for example, by Huang et al. [17] and Cai et al. [28] on Si substrate could serve as nanoparticles and could catalyze the SWCNT growths. These oxide layers might actually had been nanoporous ESNOs of compositions SiO_z, Al_2O_z, TiO_z, and Er_2O_z, respectively. They served as SUBSANOs. There are more. Silva et al. [29] did not even produce an oxide or non-oxide layer on a substrate. Instead they employed an Inconel (72 atomic % Ni, 16 atomic % Cr, and 8 atomic % Fe) metallic substrate, which upon annealing in air yielded R_L species of SUBSANOs plausibly of the solid solution composition: $R_L \equiv$(Ni, Cr, Fe, O). This R_L species-mediated the growth of vertically aligned carbon nanotubes by thermal CVD. This solid solution could have composite layers of Fe_3O_4, Ni, and Cr_2O_3, one above the other. No catalyst addition was needed for the CNT growth on the Inconel substrate. The substrate surface acted both as the catalyst and as the support for growth. SEM, TEM, and Raman spectroscopy confirmed that the CNTs produced on these nanoparticles had high quality. Unlike Silva et al. [29], Mathur et al. [30] made CVD growth of CNTs, not on metal substrate, but on defect-rich carbon fiber substrate. The routes activating the growths were obviously via surface amorphicity. These routes gave rise to porous SUBSANOs on the substrate surface with oxygenated functional groups that exhibited pits and hillocks (e.g., nanobumps) on this substrate surface again during the pre-nucleation stage of growth. And they acted as nucleation centers for nanotube growth. The CNT growths on carbon fiber surfaces were essentially the same as the CNT growths on carbon black [14] and bulk Au [16] substrates. The carbon nanobugs formed on the aqua regia-treated Au substrate probably were actually SUBSANOs; they had one or more of the following cluster or solid solution compositions: $R_L \equiv$(Au-based ESNO, C), $R_L \equiv$(Au-based ESNO, C, O). $R_L \equiv$(Au-based ESNO, C, H), etc. The most novel feature of the findings by Silva et al. [29] and by Mathur et al. [30] is that the carbon nanotube growth could be carried out directly on defect-rich substrate via CVD. Composite-layer SFSs could especially entail the most desired characteristics of SUBSANOs.

14.3.3 SFS Types and Defects Generation in These Types of SFSs

The illustrations made above manifest the following:

(1) SFSs containing component 1 materials (e.g., metallic or nonmetallic substrate; see Fig. 14.1), and called type-I SFSs, can serve as catalyst nanoparticles for nanomaterials growth.
(2) SFSs containing component 2 materials [e.g., nonmetallic oxide, nonmetallic non-oxide, and supports; see Fig. 14.1] formed on substrates, and called type-II SFSs, can also serve as nanoparticles catalyzing nanomaterials growth.
(3) SFSs containing component 3 materials, and called type-III SFSs, can serve as well as catalysts for nanomaterials growths. These materials are metal-based SUBSANOs (see Fig. 14.1) generated on nonmetallic support (s), which are formed on substrate.

SUBSANO nanoparticles which are formed on both metallic and nonmetallic substrates, and which catalyze growths by the VQS mechanism, should though have an effective amorphicity α_{amoreff}. It should result from loose, coarsened, disordered surface lattice structures. All these could be realized by surface functionalization induced by the substrate and/or the support. They should as well be generated by surface treatment such as plasma bombardment with ions and electrons altering the chemical composition of the surface. They could also be produced, for example, by sputter deposition. The supports can have single layer. They can as well be composite layer with two or more consecutive nonmetallic dissimilar layers formed one above the other for growth. The oxide materials serving as SUBSANO components of the SFSs and catalyzing growth by the VQS mechanism may almost always have amorphous non-stoichiometric surface lattice structure.

The defects generated in the substrate, oxide support, or even metallic SUBSANOs include point defects, cluster defects, grain boundaries, stacking faults, screw dislocations, and/or edges which also behave as defects. Among them, point defects occur predominantly within the plane in the form of lattice vacancies and impurity atoms. The impurities of them can be in the substitutional or the interstitial sites. They can also be isotopic with impact on perturbing the phonon spectra. If the impurities are foreign species (e.g., dopants), they can introduce local electric fields and strain fields because of their local charges. Functional groups may be attached to the surface via hydrogen bonding, π–cation, π–π stacking, π–anion electrostatic forces, hydrophobic interactions, and/or van der Waals forces interactions. Covalent functionalization may result in stronger interactions and may hence cause structural alterations (defects) in the catalyst materials lattice. There may also be changes in the electronic, mechanical, and thermal properties of the said materials. The immediate consequence of all of these would be (1) structural imbalance, (2) compositional imbalance, (3) band gap imbalance due to variation from larger band gap to smaller band gap, (4) defect-rich to defect-free transition, and (5) surface energy transition in them (e.g., substrate, support and/or metal nanoparticle).

The dominant effect of defect introduction in them could be the formation of oxides. These oxides could be localized oxides and could include families such as silica, alumina, clays, zeolites, TiO_2, ZnO, and ZrO_2. They could cover as well nanoporous and mesoporous metal oxides, polyoxometallates (POMs), the phosphates family [e.g., VPO, $FePO_4$, silica phosphoric acid (SPA)], multicomponent mixed oxides (molybdates, antimonates, tungstates, perovskites, hexa-aluminates), etc. And they could have surfaces containing defects, kinks, steps, terraces, etc., which impart significant impact on the catalytic role of SUBSANOs. These surfaces would not always be able to homogeneously accommodate all cations in their lattices. They would not also be easily altered when exposed to external environment. They may then have segregated phase(s). This would particularly happen in complex mixed oxides, which may experience enhanced transport of species through their nanopores leading to a concentration gradient within the nanopores.

14.3.4 Heterointerfaces and Charge Transfers in SFSs

Depending on the overall structure and components of SFSs, multiple heterointerfaces can be formed, for example, between the surface and bulk of the same substrate in type-I SFSs, between substrate and support formed on this substrate in type-II SFSs, and between SUBSANO and oxide support in type-III SFSs. At the growth temperature, they may be promoted (altered), at least in part, by vibrations, fluctuations, and oscillations of the surface. There may consequently be charge redistribution at the heterointerfaces. Local charge redistribution driven by differences in electronegativities between two dissimilar material components (e.g., SUBSANO and underlying oxide) of an SFS may though be affected by the surface properties of one (e.g., the oxide) or the other, or the both. It may take place within the first few atomic layers at the heterointerface. As electrons are strongly localized and the production and diffusion of ionic defects are limited, insulating (semi-insulating) region(s) or insulating oxides (e.g., Al_2O_3, MgO, SiO_2, etc.), if formed in the system, may give rise to domains. The probability that an electron occupies an energy state in a material is given by the Fermi–Dirac function:

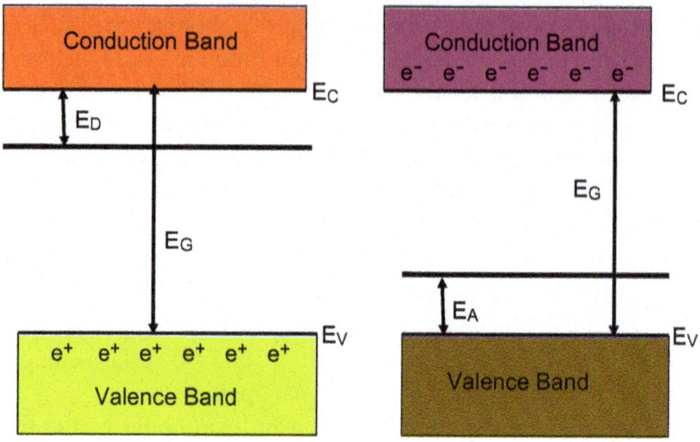

Fig. 14.2 Schematic diagrams of *n*-type and *p*-type doped nanomaterials. E_C is the energy level; of the conduction band edge, E_V is the energy level of the valence band edge, E_F is the Fermi level, and E_G is the energy band gap

$$F(E) = \frac{n_e}{N} = \frac{1}{\exp(E - E_F/RT)}, \tag{14.1}$$

where n_e is the number of electrons, N is the number of available energy states, and E_F is again the Fermi level energy. If this material is an intrinsic semiconductor, the number of electrons in the conduction band of it would equal the number of holes in the valence band of it, which would be given by:

$$n = n_e = n_h = n_0 \exp\left(\frac{E_G}{2k_B T}\right), \tag{14.2}$$

where n_0 is a constant, n_h is the number of holes, and E_G is the energy band gap. The energy band of donor-doped and acceptor-doped semiconducting materials is shown in Fig. 14.2. In this figure, E_C is the conduction band edge and E_V is the valence band edge; also E_D is the donor level energy state and E_A is the acceptor level energy state. We believe these energy levels exist in disturbed (disordered, amorphous, semi-amorphous) regions of type-I and type-II SFSs, where impurities in disturbed (disordered) lattice structures have high contents and also the energy band gap E_G is larger. In the oxides of type-III SFSs, there can also be a large band gap E_G and high contents of impurities. Extrinsic defects generated in these regions give rise to localized energy states within the band gap. They may be electrically active. Recall that the donor-type defect energy level is just below the conduction band edge E_C, and the acceptor-type defect energy level is just above the valence band edge E_V. Electrons donated by the donor level to the conduction band increase *n*-type conductivity, and holes donated by the acceptor level to the valence band increase *p*-type conductivity. The total number of charge carriers in a material exhibiting, for example, donor levels may be given by.

$$n_{\text{total}} = n_{\text{e}}(\text{dopant}) + n_{\text{e}}(\text{intrinsic}) + n_{\text{h}}(\text{intrinsic}) \tag{14.3}$$

$$n_{0\text{D}} \exp\left(-\frac{E_{\text{D}}}{k_{\text{B}}T}\right) + 2n_0 \exp\left(-\frac{E_{\text{G}}}{2k_{\text{B}}T}\right). \tag{14.4}$$

They create local electron density. This local electron density, together with the presence of structural imperfections (defects) of the surfaces, induces (generates) interactions in all SFSs (e.g., type-I SFSs, type-II SFSs, and type-III SFSs). Functional groups on the surfaces of them act as preferential sites for these interactions and as electron-donating and/or electron-withdrawing substituents. The catalytic activity and selectivity are intensified by the interactions and also by charge redistributions. The electric fields generated at the heterointerfaces increase the diffusion of ionic defects and strengthen various interactions. The catalytic activity, for example, of the metal nanoparticle in type-III SFSs surges by the alteration (modification) of the SUBSANO structure into cluster, and by the defects generated in the support. Yoon et al. [31] observed that MgO(111) support became defect-rich due to charge transfer, and that Au_8 clusters formed on such defect-rich support were catalytically active. The same metal cluster Au_8 was however inert when deposited on a virtually defect-free MgO surface. The charge transfer, together with charge redistribution, brings about three effects: (i) electronic effects, (ii) geometric effects, and (iii) dispersion effects. Among them, the electronic effects originate from the differences between the Fermi level E_{F} of the two adjacent dissimilar components of the SFS, which result in an electron transfer needed to equalize the Fermi level E_{F} at the interface of these components. The lattice mismatch between them, if any, induces changes in the shape of the SUBSANO. Due to surface treatment, in particular, the geometric effect in type-I and type-II SFSs enhances the formation of heterointerface(s). Kuznicki et al. [32] noted that, due to ion implantation and annealing of a silicon film, two Si phases were generated and coexisted in it in the form of well-defined layers separated by sharp heterointerfaces. Swiatek et al. [33] found that structural changes due to internal stress of an ion-implanted Si film yielded an amorphous-like layer in it. The region underneath the heterointerfaces, thus formed, can appear as the support of the region above the interfaces.

14.3.5 Influence of the Layer Thicknesses of Type-I and Type-II SFSs on Catalytic Activity

Cao et al. [34] performed carbon nanotube growths on Si/SiO_2 surface. They observed strong dependence of nanotube growth on the SiO_2 layer thickness. There was no detectable nanotube growth on this SiO_2 layer if the thickness of this layer was smaller than about 5–6 nm. If the thickness of this layer was between 6 and 24 nm, the nanotube growth rate increased monotonically with increasing oxide thickness and then saturated as SiO_2 approached 50 nm. Mattila et al. [35] synthesized InAs(P) nanowires on Si substrate using in situ In droplets. The thin substrate native oxide SiO_2 was found to play a crucial role in the nanowire growth. No nanowire growth was observed on silicon without the native oxide or on a thick SiO_2 layer. This observation for InAs(P) nanowires is similar to that by Cao et al. [34]. Both of them demonstrate that certain optimal thickness of SiO_2 layer was essential for growth. But the growth becomes marginally small if this thickness exceeds a certain maximum limit.

We present the explanation of the possible reasons of the observations noted above. For both the nanotube and nanowire growths, the SFS was type-II SFS with SiO_2 layer serving as catalyst. The catalyst could be active only if its top surface was disordered and had dipole moment with electronic charges aligned at or near its surface. Such a dipole moment is necessary for the attraction and decomposition of the precursor(s) and for the release of R_S species for growth. However, charge transfer and the formation of aligned charge layer become difficult if the top surface of the catalyst is far off from the Si/SiO_2 heterointerface. Migration of defects from Si to SiO_2 top surface and/or vice versa through the Si/SiO_2 heterointerface is needed for the SiO_2 top surface to successfully catalyze growth. However, the migration of defects and/or impurities to the SiO_2 surface becomes increasingly difficult if the SiO_2 layer becomes increasingly thick, and this SiO_2 layer is far off from the Si/SiO_2 heterointerface. If the SiO_2 layer is very thin, it may be too thin to accomodate charge transfer and impurity species migration into it, on the other hand. That surface disorder of the catalyst surface should be significant for this catalyst to enhance growth is exemplified by an experiment of Mor et al. [36]. While trying to grow carbon nanotubes using a type-I SFS [e.g., Si/Al_2O_3 structure], Shawat et al. formed a Fe reservoir layer underneath the thin-film alumina catalyst layer. During growth at high temperature, the Fe atoms migrated through the Al_2O_3 layer to its own top surface and increased the disturbance (disorder) of this surface. CNTs became taller when grown with Fe reservoir than when grown without Fe reservoir. This observation is similar to that by Liu et al. [37], who performed annealing of SiO_2 substrates in H_2 at high temperature to generate defects on their surfaces, and these defects promoted CNT growths.

14.3.6 Influence of the Metallic SUBSANO Layer Thickness of Type-III SFSs on Catalytic Activity

Hamadinez had [38] employed PECVD to synthesize silicon nanowires on silicon substrates coated with gold. Si(111) substrate was coated with several different layers of thin film Au, using radio frequency magnetron sputtering. Although not mentioned in the report, such sputtering generated islands of loose lattice structure that served as SUBSANOs for growth. FESEM, XRD, EDX, and TEM analyses were performed to characterize the morphology, compositions, and structures of nanowires. These nanowires were thus found to be single-crystalline Si nanowires of diameters ranging from 40 to 160 nm and length up to 3 μm. The nanowire diameter increased with increasing Au layer thickness. We can probably safely assume that the nanowire diameter increased with increasing SUBSANO diameter. Raman spectra peaks were narrow and asymmetric at 518 cm^{-1}, indicating that these nanowires had high crystalline composition. Nevertheless, the nanowire features were dependent on the Au layer thickness. Al-Taay et al. [39] also grew silicon nanowires on glass substrate coated with indium tin oxide (ITO). They employed pulsed plasma CVD to perform the growth. Note that ITO is a ternary compound of indium, tin, and oxygen of varying proportions. Depending on the oxygen content, it can be a ceramic or alloy. The thin film ITO, as deposited on the glass substrate by thermal evaporation, had several different thicknesses ranging between 10 and 100 nm. We argue that pulsed plasma CVD generated islands on the film at the growth temperature, and that these islands served as SUBSANOs for growth. The effect of the thickness of the ITO and hence of the SUBSANOs on the morphology of the silicon nanowires was studied by STEM, which indicated that the nanowire diameter increased with increase in the SUBSANO thickness. The nanowire diameter ranged between 70 and 80 nm for a 10-nm-thick SUBSANO. The nanowire diameter was of 190–200 nm for a SUBSANO thickness of 100 nm.

Zhao et al. [40] noted that the diameter distributions of nanomaterial [nanowire, nanotube, nanosheet, nanofiber, etc.) are commonly realized by controlling the thickness of the deposited catalyst substrate film during the thermal pre-nucleation stage of growth. If it is thick, it yields large metal clusters (SUBSANOs) at relatively high temperature and long annealing. The increase in size of the metal clusters is caused by increased surface migration of metal atoms. This is apparent from Fig. 14.3. by Jang

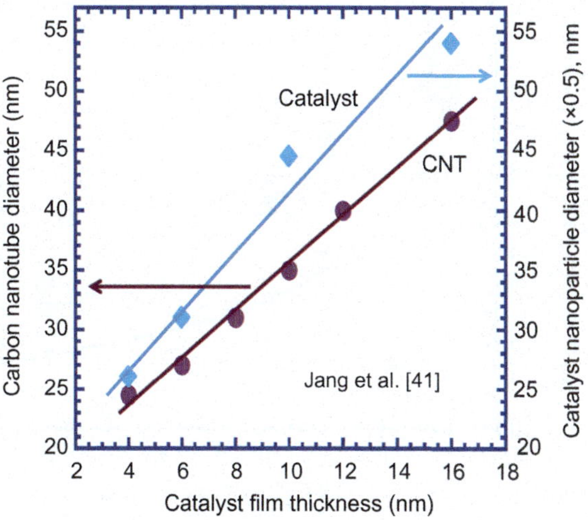

Fig. 14.3 Variation of CNT diameter and nanoparticle diameter with the catalyst film thickness. Catalyst nanoparticles were formed from the catalyst thin film. The plots are made with experimental data by Jang et al. [41]

et al. [41]. This figure shows that the size of the SUBSANOs and even METANOs, and the diameters of CNTs produced on them increased by increasing the thickness of the Co catalyst film. The Co catalyst film was deposited on SiO_2 thermal oxide formed on Si substrate. The SFSs thus produced were therefore type-III SFSs.

Jang et al. [41] also studied the variation of CNT length with the thickness of Co catalyst layer and growth time. They tried 4 different thicknesses of the Co layer, viz. 3, 6, 10, and 16 nm, respectively for this. Using the experimental data by Jang et al., we plotted the variation of CNT growth rate with CNT growth time for the said catalyst thicknesses. Using the same experimental data by Jang et al., we also obtained the variation of CNT growth rate with catalyst thicknesses for three different growth times. These are presented in Figs. 14.4, and 14.5. Figure 14.4 indicates that the growth rate decreases with growth time, but Fig. 14.5 indicates that the growth rate increases with catalyst thickness, attains a saturation, and then decreases with further increase in catalyst thickness. As growth time increased, increasing number of carbonaceous particles was embedded on the R_L species surface formed on catalyst nanoparticle. The diffusion of the $R_S{\equiv}C$ species through the R_L species was therefore increasingly hindered by the deposited carbonaceous particles. Yet the said diffusion was needed for supersaturation and nucleation. The CNT growth rate though decreased with increase in growth time (Fig. 14.5).

Substantial charge transfer should take place [42] when catalyst nanoparticle adatoms or the clusters formed in the R_L species of the catalyst nanoparticle directly

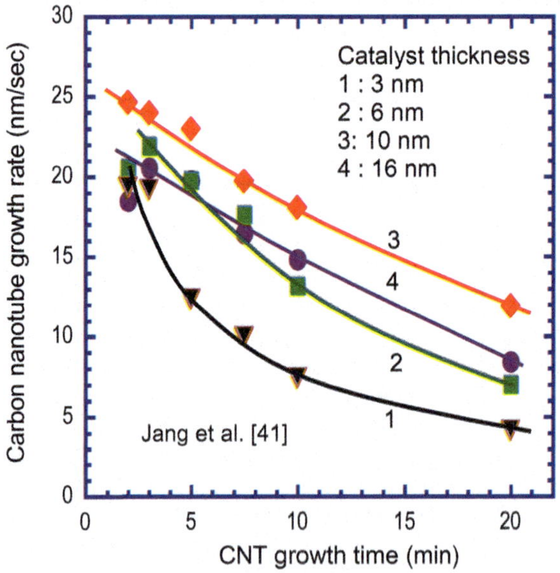

Fig. 14.4 Variation of CNT growth rate with growth time for several different catalyst film thicknesses. CNT growth took place on catalyst nanoparticles, which were formed from the catalyst thin film. The plots are made with experimental data by Jang et al. [41]

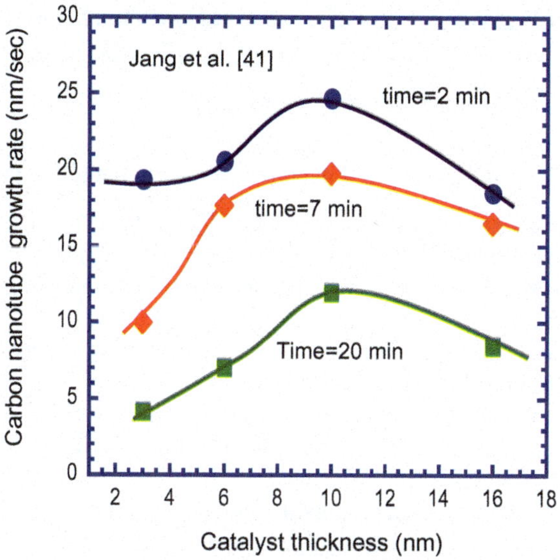

Fig. 14.5 Variation of CNT growth rate with catalyst film thickness for three different growth times. Catalyst nanoparticles were created by fragmenting the catalyst thin film by the means of annealing or by some other means. The plots are made with experimental data by Jang et al. [41]

interact with those of the support surface on which they are formed. Such interactions are enhanced by the presence of defects and cations in the support at or near the catalyst/support interface. Electron transfer thus taking place leads to a partial oxidation or reduction of the catalyst nanoparticle. There also occurs a long-range charging of the catalytically active outer surface of the catalyst nanoparticles. The electronic interactions between the catalyst and the support due to charge transfer from the support to this catalyst surface, charge reorganization at this catalyst surface, defect generation at this catalyst surface, and lattice disturbance of this catalyst surface depend on the shape, size, and proximity of the top surface of the catalyst nanoparticle. There is an optimal catalyst thickness for all of them, which should be sufficiently large, and generate the highest possible catalyst activity. This catalyst activity should gradually increase with catalyst thickness, reach a peak and then decrease with further increase in catalyst thickness. Note that the optimal catalyst thickness corresponds to an optimal distance of the top catalyst surface from the catalyst/support (catalyst/substrate) interface. This is corroborated well with the results of Fig. 14.5. Huang et al. [43] found that, under identical conditions, Ni yielded the highest CNT growth rate, the largest diameter, and the thickest wall. Co, on the other hand, resulted in the lowest growth rate, smallest diameter and thinnest wall. This happened simply because the electronic catalyst/support interactions due to charge transfer to the top surface, charge reorganization at this surface, defect generation at this surface, and lattice disturbance of this surface were the highest for Ni and the lowest for Co.

14.4 High Catalytic Activities of Catalyst Surfaces

14.4.1 Key Catalyst Activities

It is important to recall that catalyst nanoparticles serve the following purposes:

(1) They possess significant sticking capability for the selective sticking, absorption, and/or adsorption of the precursor(s) of the source species.
(2) They catalyze the decomposition of the precursors of the R_S ($R_S{\equiv}X$ and $R_S{\equiv}Y$) source species.
(3) They possess capability for the selective desorption of the by-products of the decomposition of the precursor(s).
(4) They adsorb the R_S source species, which stick to its surface.
(5) They serve as medium of diffusion (a) of the source species or (b) of the intermediates of the precursors of the source species. They enable chemical interaction (reaction) of these species during diffusion through them. The intermediates of the precursor(s) are reduced to $R_S{\equiv}X$ and/or $R_S{\equiv}Y$ during the diffusion. In order for chemical reaction (interaction) to take place, the reacting (interacting) particles lose excessive kinetic energy. Thus, they come close to each other, which is needed for their reactions. For this to haven, the depth δ_{amor} of the disturbed, disordered, amorphous surface should be optimal. It should be such the R_S species, for example, $R_S{\equiv}B$ and $R_S{\equiv}N$ are streamlined during diffusion through it and react to form $B + N \rightarrow BN$ molecule for BN nanotube growth. The said diffusion can be bulk diffusion, for example, for nanowire growths; but bulk diffusion is followed by surface diffusion for nanotube growths.
(6) They provide a nanoscale template for the nucleation and growth of nanomaterials. This template, called the liquid/solid (L/S) interface or the quasiliquid/solid (QL/S) interface or the quasisolid/solid (QS/S) interface, may be formed somewhere inside the catalyst nanoparticles or at the interface of substrate/nanoparticle or support/nanoparticle.
(7) Depending on the R_S species, the diffusive tendency of the R_S species, and the growth conditions, they provide peripheral reactive rim (shell and hill; see Fig. 13.6 of Chap. 13), for example, for nanotube and nanoring growths, This rim surrounds the core of the catalyst nanoparticle.
(8) With the generation of the shell and hill, they yield a platform for the tip growth or base growth of CNTs.
(9) With adequate pre-growth planning, they create an environment for the catalyst doping of nanomaterials.

All these are elaborated in the following.

14.4.2 *Catalyst Surface Characteristics for Effective Catalytic Activities*

To best achieve the goals stated above, and for growths by the VQS mechanism, the R_L species of catalyst nanoparticle surface must exhibit a number of material characteristics. As stated in previous chapters, these can be accomplished by surface treatment. These can also be realized by the support-induced surface functionalization. The goal of these is to generate loose lattice structure of the R_L species of the catalyst nanoparticle surface. This lattice structure can have significant sticking capability and be loose if it is

1. an alloy, cluster, or solid solution,
2. amorphous exhibiting an effective surface amorphicity $\alpha_{amoreff,}$
3. paracrystalline,
4. stepped,
5. influenced by synergy, and/or
6. inundated with defects and dislocations.

14.4.2.1 Alloy, Cluster, Solid Solution-Based Surface Structure

The lattice structure of the R_L species of catalyst nanoparticle surface may be loose with weakened interatomic interactions, if it is an alloy, a solid solution or a cluster. Recall that an alloy is a metal composed of a mixture of elements, including nonmetallic elements. A solid solution is however a mixture of crystalline solids. In contrast, a cluster is an ensemble of bound atoms or molecules that is intermediate in size between a molecule and a bulk solid. Notably an alloy can be formed in SUBSANOs produced on metallic substrate. There can also be migration of metal atoms into nonmetallic SUBSANO, if these atoms are present in the chamber, substrate or support as contaminants. So, no matter if the R_L species is an alloy, a solid solution or a cluster, it has interatomic (intermolecular) bondings weakened; some of them even broken in such a way that the surface lattice structure is loose. Also, the R_L species is porous. The R_S species smoothly diffuse through it. We cite some examples. Kratzer et al. [44] investigated the impact of Au-catalysis on GaAs nanowire growth. They found that, apart from the formation of the compound Au_7Ga_2, Ga formed also surface alloy with Au, which catalyzed the GaAs nanowire growth. Wang et al. [45] grew ZnS single-crystal nanowires in bulk quantities by a simple and high-yield method. They observed that the Au film, serving as catalyst, reacted with ZnS vapor to form $R_L \equiv$(Au, Zn, S) alloy, cluster, and/or solid solution. The EDS analysis showed that the R_L species of the nanoparticles catalyzing the growth contained Au, Zn, and Wang et al. [46] obtained large-scale CdS nanowires by thermal evaporation of CdS powders in the presence of Au. EDS measurements made on the nanoparticle and on the nanowire tip indicated that the R_L species of the nanoparticles were composed of Au, Cd, and S. It was confirmed by observation of Au-CdS alloy (cluster, solid solution) at the nanowire tips. Wang et al. [47]

grew high-purity BNNTs by CVD. The XRD analysis indicated that the nanotube product was pure, hexagonal BNNT. EDS used to characterize the composition of the nanoparticles encapsulated on the BNNT tips indicated that the R_L species for the growth of the BNNTs was R_L≡(Mg, Fe, O) species (alloy, cluster, solid solution). EDS analysis indicated that B and N were dissolved in Mg-Fe-O nanoparticles. It was confirmed by the line-scan energy-dispersive X-ray spectra.

14.4.2.2 Amorphous (Semi-amorphous) Surface Structure

The impact of amorphous (semi-amorphous) surface on growth was discussed earlier. We elaborate it here pointing to a slightly different scenario. Prabaswara et al. [48] demonstrated the growth of InGaN/GaN nanowires directly onto an amorphous quartz substrate using TiN/Ti interlayer. The nanowire growth by MBE on quartz substrate was catalyst-free under nitrogen-rich condition. The commercially available quartz substrate (thickness ~500 μm) was amorphous. Before growth, 20-nm-thick Ti layer was deposited on this amorphous substrate using electron beam evaporation. After loading the substrate into the growth chamber at a temperature of 650 °C, the substrate surface was exposed to a nitrogen plasma to partially convert the Ti interlayer into TiN interlayer. TiN interlayer thus formed had a linear thermal expansion coefficient of 9.35×10^{-6} K^{-1}. But the linear thermal expansion coefficient of quartz was extremely small, about 5.55×10^{-7} K^{-1}. Therefore at the substrate temperature of 650 °C prior to growth, there could be a substantial mismatch in thermal expansion at the quartz/InN interface. The thin InN layer thus experienced tensile stress and the thick quartz substrate experienced compressive stress. The immediate result of this was the formation of islands in the thin InN interlayer in order to relax the compressive stress of the quartz substrate. The loosely bound atoms of the amorphous quart layer underwent migration into the islands causing it disordered and loose. These islands served as SUBSANOs for the growth of GaN nanowires. Johar et al. [49] also grew GaN nanowires on amorphous glass substrate by MOCVD. We believe the basics of this growth were similar to those by Prabaswara et al. [48].

We cite another example of slightly different amorphous surface. The lattice structure of the R_L species of this surface is non-crystalline lacking long-range crystal order. It does not have any well-defined melting point. However, its melting point is lower than that of its single-crystal counterpart, implying that it has weakened interatomic (intermolecular) bondings. Some of those bonds may even be broken in such a way that the R_L species has voids, dislocations, etc. It is loose and even porous. The R_S species smoothly diffuse through it. To be specific, diamond is a very strong, nearly isotropic and tetrahedral sp^3 bonded material with no ability to decompose carbon precursors. Koga et al. [50] discovered that bulk diffusion of carbon in diamond is negligible as compared to that in metals such as Fe. Yet Takagi et al. [51] grew CNTs from diamond particles in the solid phase, and not the liquid phase. It was possible because the nano diamond surface, in the CVD environment was transformed from the sp^3-type surface to a sp^2-type π-bonded relaxed graphite surface.

This graphite was not etched from the nanodiamond surface due to the depletion of atomic hydrogen. So small domains of graphite islands were formed on the π-bonded relaxed diamond surface. It possessed effective surface amorphicity $\alpha_{amoreff}$, and also catalytic activities. Liu et al. [12, 52] grew SWCNTs from R_L species on solid, amorphous, porous SiO_z ($1 \leq z < 2$) nanoparticle surface. It was not different from CNT growth by Steiner et al. [13] who observed defects from the extraction of oxygen to play critical role in the catalytic activity of ZrO_2 nanoparticles.

14.4.2.3 Paracrystalline Surface Structure

Carbon black is a form of paracrystalline carbon composed of fine particles consisting mainly of carbon. Further, carbon blacks have varying degrees of chemisorbed oxygen complexes (i.e., carboxylic, quinonic, lactonic, phenolic groups, and others) on their surfaces. These surface oxygen groups are however quite volatile. Although carbon blacks possess short- and medium-range ordering in their lattice, they lack crystal-like long-range ordering in one or more directions. So, carbon atoms are not in a predictable lattice, as in a highly ordered, perfectly crystalline single-crystal material. Carbon blacks therefore tend to have a high degree of porosity [53]. They have weak interatomic (intermolecular) interactions; some of these interactions may be too weak to create bonds. And hence numerous SUBSANOs may naturally exist on the carbon black surface. The R_L species of these SUBSANOs formed on carbon black surface have loose lattice structures; they are generally porous. The R_S species smoothly diffuse through them. Lin et al. [14] used carbon black substrate to produce open-tip MWCNTs at 800 °C. They were self-assembled on the R_L species of its porous (20 to 80 nm in dimension), fullerene-like, paracrystalline, defected structure which decomposed ethylene precursor.

14.4.2.4 Stepped Surface Structure

Surface treatmentsuch as plasma bombardment involves high-speed stream of plasma of an appropriate gas mixture shot at a substrate in order to generate SUBSANOs on the substrate surface. The plasma source can either be charged ions or neutral atoms, molecules and/or radicals. During the process, the plasma generates volatile products even at room temperature due to chemical reactions between the elements of the substrate material and the reactive species of the plasma. The plasma also breaks interatomic (intermolecular) bonds between (among) some of the atoms (molecules) at and near the surface of the substrate. The substrate surface thus becomes stepped surface. The lattice structure at and near the surface of the substrate thus becomes relatively loose, porous stepped, and coarsened. Similar situation arises during energetic ion bombardment of a substrate. Such bombardment leads particularly to the emission of secondary electrons creating unsaturated accumulated charges at and near the stepped substrate surface. The secondary electron emission coefficient for a metal is less than 0.1, while that of an oxide is higher. Secondary electron emission from electron bombardment is much higher than that from ion bombardment. The R_S species hence smoothly diffuse through the R_L species thus created. Kumar et al. [54] grew CNTs on Si substrate. The six different experimental combinations were used. They took into account different reactive plasma/gas environments and the locations of the plasma relative to that of the substrate. No nanotube nucleation took place in neutral gas environment, and also on smooth and/or patterned substrates. Using remote plasma, no CNTS grew also on smooth and patterned surfaces. But long (up to several hundred μm), vertically aligned CNTs grew at a very high growth rate only if the plasma was in direct contact with the Si substrate surface. This happened simply because the plasma, being in direct contact with the Si substrate, yielded loose, porous, coarsened R_L species on a steppend surface.

14.4.2.5 Synergy Influenced Surface Structure

Interaction of two or more metals leading to the generation of a catalytic effect far greater than the sum of their separate effects is very attractive for nanomaterials growths. No wonder bimetallic alloy substrates and thin films well-suited for the formation of SUBSANOs by surface treatment have been grown during the past years [55–58]. Bimetallic nanoparticles produced from them have been successful for the growths of nanomaterials [58, 59]. Being an alloy or cluster, the bimetallic (trimetallic) nanoparticles have some interatomic (intermolecular) bonds significantly weakened; some interatomic (intermolecular) bonds are broken, but some new bonds are generated in such a way that the surface lattice structure of them is loose and the R_L species of the SUBSANOs are coarsened, semi-molten, and porous. They possess hence significantly increased catalytic activity. The R_S species smoothly diffuse through this R_L species. Chiang and Sankaran [59] studied the synergistic effect of $Fe_{1-z}Ni_z$ alloy on the CNT growths. They measured the activation barrier against the catalytic activity and hence against carbon diffusion through the catalyst as function of the Ni mole fraction z. They observed that the activation barrier decreased with increase in z, reached the minimum at $z \approx 0.42$, and then increased with increase in z. This implies that the surface lattice structure of the $Fe_{1\text{-}z}Ni_z$ alloy attained the greatest looseness for the Ni mole fraction of $z \approx 0.42$. We believe that this results from surface and near-surface restructuring of the alloy as function of component element mole functions and is true for other bimetallic and trimetallic alloy catalysts, as well. And this looseness contributes heavily to the synergistic catalytic effect of these alloy catalysts. In another investigation, Chiang and Sankaran [60] studied CNT growth rate as function of $Fe_{1-z}Ni_z$ alloy composition. They found that the CNT growth rate by this $Fe_{1\text{-}z}Ni_z$ alloy catalyst gradually increases, reaches a peak at a optimal composition of $z = 0.67$, and then gradually decreases for z exceeding 0.67. Magrez et al. [61] studied the catalytic potential of $Fe_{1-z}Co_z$ alloy for the growth of MWCNTs. They found that the highest growth, the narrowest diameter distribution, and the lowest defect density in the MWCNT structures were achieved for $z = 0.33$ of the $Fe_{1-z}Co_z$ alloy. Remarkably, the SUBSANO efficiency increased almost 100 times by changing z from 0 to 0.25 for MWCNT growth at 700 °C. The SUBSANO efficiency however rapidly decreased for $z > 0.33$ and became comparable to that of Fe for $z = 0$. The experiment by Magrez et al. [61] thus demonstrated that the Fe_2Co alloy had far more catalytic potential as SUBSANO than both Fe and Co, which yield METANO. We argue that the compositional effect of catalytic activity of bimetallic and trimetallic alloys is plausibly related to increased structural looseness and surface coarsening of them and hence increased diffusion of the R_S species through them.

14.4.2.6 Defects and Screw Dislocations Inundated Structure

We stated in Sect. 14.3.3 that type-II SFSs, serving as nanoparticles and catalyzing nanomaterials growths are nonmetallic nanoparticles formed on substrate. They yield SUBSANOs. These SUBSANOs should have loose, defected, disordered surface lattice structures. These could be realized by surface functionalization induced by substrate. These could be realized also by surface treatment such as plasma bombardment with ions and electrons altering the chemical composition of the SUBSANO surface. They could also be produced, for example, by sputter deposition of the SUBSANO. The defects generated in the substrate can be point defects, cluster defects, grain boundaries, stacking faults, screw dislocations, etc. Some or all of them can permeate into the SUBSANO disturbing the SUBSANO surface. This surface can thus be loose, defected and coarsened. Among various defects, axial screw dislocations can not only disturb the SUBSANO surface lattice structure, but also provide self-perpetuating steps crucial for nanomaterials growths.

Mathew et al. [62] noted that indeed nanomaterial growth is driven by screw dislocations. They found that nanoparticles with highly defected surface served well as SUBSANOs. Morin and Jin [63] intentionally generated screw dislocations in GaN substrate and utilized these dislocations for epitaxial growth of ZnO nanowires on GaN substrate. Atomic force microscopy confirmed that screw dislocations were present on the native GaN substrate surface, and that ZnO nanowires grew directly from dislocation etch pits of heavily etched GaN surface. Further, transmission electron microscopy validated the existence of axial dislocations. These results established a relationship between loose, defected nanoparticle (SUBSANO) surface inundated with screw dislocations for nanowire growths.

14.4.3 Knudsen Diffusion, Interstitial Diffusion, and Substitutional Diffusion

Recall that the bulk diffusion (volume diffusion) of the R_S species atoms may be much faster than that of the atoms of the solid nanoparticle host. While the former can be Knudsen diffusion or interstitial diffusion, the latter can be substitutional diffusion. Both the Knudsen diffusion and interstitial diffusion depend on the geometry of the host lattice and are faster in more open structures. In general, the larger the pore dimension and pore density, the larger is the Knudsen diffusion. The activation energy for interstitial diffusion of the $R_S{\equiv}C$ species is 1.53–1.57 eV in close packed structure of fcc Fe; it is only 0.83 eV in the loose open structure of the bcc Fe [64, 65] for Knudsen diffusion. The diffusion coefficient depends inversely upon solid solubility. It should be smaller for single-crystal metals than for solid solutions or clusters. Interestingly, Pd, Ni, and α-Fe that are all fcc metals with close melting points. The diffusion coefficient of $R_S{\equiv}C$ through them inversely follows the order of carbon affinity in them, which is the largest for Pd, intermediate for Ni, and the smallest for Fe. When a precursor molecule is adsorbed on a catalyst nanoparticle surface, the activation barrier for its dissociation is lowered. And the probability of this adsorption may be higher in a metal-containing and metal-free solid solution and cluster than in a metal. Even for metal, the activation barrier is reduced upon adsorption. For example, the barrier for the dissociation of acetylene on Ni(111) was measured to be 1.4 eV, while it is 5.58 eV for self-decomposition [66]. If the surface bonds of the SUBSANO are too strong, the reaction intermediates generally remain on the surface and block the adsorption of new reactant molecules. Forming adsorbate-nanoparticle bonds of intermediate strength is important for catalyst. And it may be best achieved by a SUBSANO exhibiting loose lattice structure as in solid solution or cluster composition.

14.4.4 Low-Temperature Decomposition of Gaseous Precursors

Low-temperature decomposition of precursors is important for nanomaterials growths. SUBSANOs catalyze the decomposition of the gaseous precursors of the R_S ($R_S{\equiv}X$ and $R_S{\equiv}Y$) source species. Being disturbed, disordered, amorphous, and coarsened exhibiting pits and hillocks, the SUBSANO surface has unsaturated charges accumulated at disordered locations. It has dangling bonds. The unsaturated charges give rise to HETs. There occurs charge transfer from the HETs to the precursor molecules and vice versa. Such charge transfer breaks, for example, the C–C π bond of hydrocarbons and leads to instability of these molecules. Each dangling bond orbital of the unsaturated charges has only one electron. It has hence tendency to accept a second (spin-paired) one from precursor molecules. Losing electrons, the precursor molecules become charge imbalanced and unstable. They are eventually dissociated (decomposed). The release of the R_S species and also of some intermediates from the precursor molecules, on the SUBSANO surface thus follows. The high-surface energy of the SUBSANO surface additionally promotes the formation of temporary intermediate adsorbates and accelerates specific reactions. The products of these reactions diffuse as well through the SUBSANO. The most notable feature of HETs serving as tools to decompose the precursors of the R_S species is that they release the R_S species and other intermediates at a temperature lower than the thermal decomposition temperatures of the precursor. We argue that these are the primary reasons of why Busetto et al. [67] could decompose methanol at a very low temperature by employing Cu–Zn–Al mixed oxides. This is also the reason of why He et. al [68]. could grow SWCNTs only at 600 °C by employing FeCu/MgO nanoparticles. Due to the absence of HETs, the catalytic function, for example, of SiO_2 for the decomposition of hydrocarbon is very limited. CH_4 has a high thermal stability. Hence, the decomposition of CH_4 on the stoichiometric SiO_2 surface is very difficult. In contrast, this decomposition on the non-stoichiometric SiO_z surface, which has high concentration of HETs, is easier. And that is why CNTs [52] were yielded on the SiO_z surface, but not on the SiO_2 surface.

14.4.5 Is Catalyst Poisoning Real?

It is argued that metal-containing SUBSANO nanoparticles coated with amorphous carbon may be poisoned, and hence they are not able to decompose hydrocarbons. We believe that the said arguments are not fully justified. Experimental observations indicate that METANOs and metal-containing SUBSANOs are still able to decompose hydrocarbons. For example, oxides do catalyze the formation of graphene [69] and CNTs [70, 71] without the assistance of any METANO and metal-containing SUBSANO. This means these nanoparticles (e.g., METANOs and metal-containing SUBSANOs) are not always essential for decomposing the hydrocarbon precursors. We suggest that the decomposition of hydrocarbons is carried out by HETs, irrespective of whether they are composed of metal (s) or nonmetals. These HETs are generated at the tips of pits and hillocks, which contain high concentration of unsaturated accumulated charges and dangling bonds. Notably the tips of the pits and hillocks have certain heights (depths); they remain exposed and not covered with amorphous carbon even when catalyst surfaces are covered with amorphous carbon. HETs lead to charge transfer and breaking of carbon-hydrogen and carbon–carbon bonds of hydrocarbons. The breaking of these bonds obviously forms free-radicals during hydrocarbon pyrolysis. Each of the fragments thus created keeps one electron and forms two radicals. The presence of radical(s) in a hydrocarbon molecule leads to rapid rearrangement of carbon bonds. They explain why there occurs nucleation of CNT on SUBSANOs (e.g., unstable nanohumps) formed on graphitic surfaces [15, 16].

14.4.6 Catalyst Template Effects for Supersaturation

The template effects of catalyst nanoparticle on the shape and size of the growing nanomaterials are very important. If this nanoparticle is a SUBSANO, it produces one-dimensional nanomaterials (nanowires, nanotubes, etc.). If this catalyst is a continuous film, it produces two-dimensional nanomaterials such as grapheme. This growth always requires the R_S species to undergo supersaturation at the liquid/solid (quasiliquid/solid, quasisolid/solid) interface. And this is where new epitaxial atomic (molecular) layers come about. It must hence be flat, stable, and inert to ensure that it has much larger accommodation coefficient of the reactants [72, 73] than that at the solid–vapor interface. It is also important that the R_S species, or molecules formed from them, are properly solubilized by the liquid (quasiliquid, quasisolid) medium in such a way that the nanoparticle is envisioned as defining the diameter of nanomaterial (e.g., nanowire) thus grown. To achieve this goal, the nanoparticle must not react with the R_S species (or the molecules formed from them), and must not decay during growth at high temperature, particularly at growth temperature. Note that the interface energy is about an order of magnitude smaller than the surface energy. It is smaller than even the difference between solid and liquid surface energies. Such a small interface energy must not interfere the supersaturating process of the R_S species. There is a general rule: the interfacial energy depends on the size of the interface which determines the nanomaterial diameter. The interfacial energy thus dictates the diameter of the nanomaterial product. Controlled unidirectional growth of nanomaterials (e.g., nanowires, nanotubes, etc.) is always preferred. It can be best achieved by defining the liquid/solid (quasiliquid/solid, quasisolid/solid) interface from the point of view of crystallographic orientation. In other words, the solid of the liquid/solid (quasiliquid/solid, quasisolid/solid) interface must have desired crystal orientation.

14.4.7 Membrane Template Effects

SUBSANOs on the templates of anodic aluminum oxide (AAO) membrane and of other membranes have been used to synthesize one-dimensional nanomaterials (nanorods, nanowires, nanobelts, and nanotubes). They proved to be very successful in producing aligned arrays of very long nanowires of high aspect ratios [74, 75]. These SUBSANOs do satisfy a number of requirements: First, they have desired pore dimension, pore density, and morphology. Second, the template materials are compatible with the processing conditions. One of them is the use of electrical insulator as template for electrochemical deposition. Third, the template materials are chemically and thermally inert during synthesis and also during the processing steps that follow. Fourth, the material or the solution being deposited wets the internal pore walls. Fifth, the deposition process, for example, for the nanowire or nanorod synthesis, starts from the bottom of the nanopore or from one end of the template channel to the other end of it. The deposition process for the growth of nanotubules, however, starts from the pore wall and proceeds inwardly. Unfortunately, inward growth may lead to nanopore blockade. This problem, common during solid nanowire and nanorod growths, must be carefully avoided. Kinetically, the optimal level of surface relaxation enables the highest possible packing density. This is insured by resorting to a diffusion-limited process. The ease of release of nanowires and nanorods from the templates and the ease of handling their growth during experiments should be taken into consideration.

14.5 Conclusions

Our discussions made in the previous sections of this chapter indicate that, in consideration of thermodynamics and kinetics, the metal-catalyst-free VQS mechanism for nanomaterials (nanocrystals) growths could be possible via (1) self-catalytic growths, (2) oxygen and oxide-assisted growths, (3) plasma-enhanced growths, (4) laser-assisted growths, (5) Frank's screw dislocation-induced growths, (6) surface treatment-induced growths, and (7) defect-induced growths. The goal, in all the cases, as described in Chap. 3, would be to create on the substrate (support) surface the $R_L \equiv (\beta_1, \beta_2, \beta_3, \beta_4, \beta_5, \beta_6$, etc.) species or the $R_L \equiv$(ESNO) species, or the $R_L \equiv (\beta_1)_z(\beta_2)_{1--z}$, where z is the mole fraction of β_1 in the R_L species. It may contain metal if the substrate is a metal substrate instead of nonmetal substrate. It may contain metal if metal impurities from the substrate or support migrate into the SUBSANO. The R_L species must have judiciously controlled effective surface amorphicity, surface roughness, and dipole moment on its surface, and up to a depth δ_{amor} (see Chap. 12 and Appendix). There may be other requirements, such as porosity, surface melting (semimelting), etc., as well (see Chap. 12). Which one of the 7 different means for growths spelled out above, would be preferred? It would depend on the growth conditions yielding the most desired products, namely the products of the best possible crystallinity, morphology, and physicochemical characteristics. Generally, the defect-induced means for growth is preferred. We cite some examples of this means. Kohno et al. [76] employed Au and S to grow Si nanoneedles at 1230 °C, but Ishiyama et al. [77] employed only S as catalyst to grow Si nanowires at 1200 °C on Si substrates. Sulfur produced vapor-phase Si sulfides by etching the Si substrate, and the silicon sulfide vapor served as the source gas species for Si nanoneedle growths. S and Au also created an $R_L \equiv$(Au, S, Si) or possibly $R_L \equiv$(Au, S, Si, O) species as catalyst for growths. Au and Si react to produce molten AuSi alloy at 950 °C, which is lower than the growth temperature adopted by Kohno et al. [76]. So, depending on the components and the composition, the R_L species of Kohno et al. [76] might not have been fully molten, but instead had molten nanoporous cluster even at $T \approx 950$ °C or $T > 950$ °C. The nanowire tips by Ishiyama et al. [77] contained Si, S, and O implying that the $R_L \equiv$(Si, S, O) species was a solid solution or cluster that catalyzed the nanowire growths. Most of the metal-free growths of Si nanowires made use of SiO_z ($z \leq 2$) substrate (support). The study by Ishiyama et al. is significant in the sense that it made use of some material other than SiO_z ($z \leq 2$) for the R_L species composition, and that it is permissible as long as this material creates desired amorphicity and porosity in it needed for growths. An indication of the R_L species being a porous cluster or solid solution, rather than a non-porous alloy is that the tips at the nanomaterial (s) are irregular in shape. They are as well unreacted, rather than reacted. So, assemblage of materials (e.g., $\beta_1, \beta_2, \beta_3, \beta_4, \beta_5, \beta_6$, etc.) at the tips assumes no regular well-defined shape if unreacted, but assumes a spherical (hemispherical) shape if eutectically reacted.

References

1. S.N. Mohammad, VQS (vapor-quasiliquid-solid, vapor-quasisolid-solid) mechanism for the catalyst-free and catalyst-mediated non-eutectic syntheses of single-crystal nanowires. J. Appl. Phys. **120**, 084307 (2016)
2. S.N. Mohammad, Bimetallic-catalyst-mediated syntheses of nanomaterials (nanowires, nanotubes, nanofibers, nanodots, etc.) by the VQS (vapor-quasiliquid-solid, vapor-quasisolid-solid) growth mechanism. J. Phys. D: Appl. Phys. **49**, 495304 (2016)
3. S.N. Mohammad, VQS (vapor-quasiliquid-solid, vapor-quasisolid-solid) mechanism presents a unified foundation for the syntheses of nanotubes, primarily carbon nanotubes. AIP Adv. **7**, 095011 (2017)
4. S.N. Mohammad, VQS (vapor-quasiliquid-solid, vapor-quasisolid-solid) mechanism for realizing narrow distributions of chirality and diameters of single-walled carbon nanotubes (SWCNTs). J. Nanosci. Nanotechnol. **19**, 5388–5417 (2019)
5. S.N. Mohammad, Thermodynamic imbalance, surface energy, and segregation reveal the true origin of nanotube synthesis. Adv. Mater. (Weinheim, Germany) **24**, 1262–1275 (2012)
6. R. Ma, Y. Bando, T. Sato, K. Kurashima, Growth, morphology, and structure of boron nitride nanotubes. Chem. Mater. **13**, 2965–2971 (2001)
7. T. Stoica, E. Sutter, R.I. Meijers, R.K. Debnath, R. Calarco, D. Grützmacher, Interface and wetting layer effect on the catalyst-free nucleation and growth of GaN nanowires. Small **4**, 751–754 (2008)
8. J.-J. Wu, S.-C. Liu, Catalyst-free growth and characterization of ZnO nanorods. J. Phys. Chem. B **106**, 9546–9551 (2002)
9. F. Matteini, G. Tütüncüoğlu, D. Rüffer, E. Alarcón-Lladó, A.F. Morral, Ga-assisted growth of GaAs nanowires on silicon, comparison of surface SiO_x of different nature. J. Cryst. Growth **404**, 246–255 (2014)
10. A.F. Morral, C. Colombo, G. Abstreiter, J. Arbiol, J.R. Morante, Nucleation mechanism of gallium-assisted molecular beam epitaxy growth of gallium arsenide nanowires. Appl. Phys. Lett. **92**, 063112 (2008)
11. B. Liu, W. Ren, L. Gao, S. Li, S. Pei, C. Liu, C. Jiang, H.-M. Cheng, Metal-catalyst-free growth of single-walled carbon nanotubes. J. Am. Chem. Soc. **131**, 2082–2083 (2009)
12. B. Liu, D.-M. Tang, C. Sun, C. Liu, W. Ren, F. Li, W.-J. Yu, L.C. Yin, L. Zhang, C. Jiang, H.-M. Cheng, Importance of oxygen in the metal-free catalytic growth of single-walled carbon nanotubes from SiO_x by a vapor-solid-solid mechanism. J. Am. Chem. Soc. **133**, 197–199 (2011)
13. S.A. Steiner, T.F. Baumann, B.C. Bayer, R. Blume, M.A. Worsley, W.J. MoberlyChan, E.L. Shaw, R. Schlogl, A.J. Hart, S. Hofmann, B.L. Wardle, Nanoscale zirconia as a nonmetallic catalyst for graphitization of carbon and growth of single- and multiwall carbon nanotubes. J. Am. Chem. Soc. **131**, 12144–12154 (2009)
14. J.H. Lin, C.-S. Chen, H.-L. Ma, C.-W. Chang, C.-Y. Hsu, H.-W. Chen, Self-assembling of multiwalled carbon nanotubes on a porous carbon surface by catalyst-free chemical vapor deposition. Carbon **46**, 1619–1623 (2008)
15. J.H. Lin, C.S. Chen, M.H. Rümmeli, A. Bachmatiuk, Z.Y. Zeng, H.L. Ma, B. Büchner, H.W. Chen, Growth of carbon nanotubes catalyzed by defect-rich graphite surfaces. Chem. Mater. **23**, 1637–1639 (2011)
16. J.-H. Lin, C.-S. Chen, M.H. Rummeli, Z.-Y. Zeng, Self-assembly formation of multiwalled carbon nanotubes on gold surfaces. Nanoscale **2**, 2835–2840 (2010)

17. S. Huang, Q. Cai, J. Chen, Y. Qian, L. Zhang, Metal-catalyst-free growth of single-walled carbon nanotubes on substrates. J. Am. Chem. Soc. **131**, 2094–2095 (2009)
18. B.-S. Kim, T.-W. Koo, J.-H. Lee, D.S. Kim, Y.C. Jung, S.W. Hwang, B.L. Choi, E.K. Lee, J.M. Kim, D. Whang, Catalyst-free growth of single-crystal silicon and germanium nanowires. Nano Lett. **9**, 864–869 (2009)
19. I. Yamada, Y. Hirano, K. Nishimura, Y. Takao, K. Eriguchi, K. Ono, Silicon nanowire growth on Si and SiO_2 substrates by rf magnetron sputtering in Ar/H_2, Appl. Phys. Express **8**, 066201 (2015)
20. K.A. Bertness, A. Roshko, L.M. Mansfield, T.E. Harvey, N.A. Sanford, Mechanism for spontaneous growth of GaN nanowires with molecular beam epitaxy. J. Cryst Growth **310**, 3154–3158 (2008)
21. T. Uchino, K.N. Bourdakos, C.H. de Groot, P. Ashburn, M.E. Kiziroglou, G.D. Dilliway, D.C. Smith, Metal catalyst-free low-temperature carbon nanotube growth on SiGe islands. Appl. Phys. Lett. **86**, 233110 (2005)
22. K. Hernadi, Z. Konya, A. Siska, J. Kiss, A. Oszko, J.B. Nagy, I. Kiricsi, On the role of catalyst, catalyst support and their interaction in synthesis of carbon nanotubes by CCVD. Mater. Chem. Phys. **77**, 536 (2002)
23. Z. Zhu, M. Yudasaka, S. Iijima, A catalytic chemical vapor deposition synthesis of double-walled carbon nanotubes over metal catalysts supported on a mesoporous material. Chem. Phys. Lett. **380**(5–6), 496–502 (2003)
24. M. Su, B. Zheng, J. Liu, A scalable CVD method for the synthesis of single-walled carbon nanotubes with high catalyst productivity. Chem. Phys. Lett. **322**, 321–326 (2000)
25. T. Ngo-Duc, K. Singh, M. Meyyappan, M.M. Oye, Vertical ZnO nanowire growth on metal substrates. Nanotechnology **23**, 194015 (2012)
26. M. Sobanska, S. Fernández-Garrido, Z.R. Zytkiewicz, G. Tchutchulashvili, S. Gieraltowska, O. Brandt, L. Geelhaar, Self-assembled growth of GaN nanowires on amorphous Al_xO_y: from nucleation to the formation of dense nanowire ensembles. Nanotechnology **27**, 325601 (2016)
27. A. Prabaswara, J.W. Min, M.R. Tangi, R.C. Subedi, D. Priante, T.K. Ng, B.S. Ooi, Growth of GaN nanowire on indium-tin-oxide coated fused silica for simultaneous transparency and conductivity, in *Proceedings SPIE, volume 10918, Gallium Nitride Materials and Devices XIV*, 109180B (8 March, 2019)
28. Q. Cai, Y. Hu, Y. Liu, S. Huang, Growth of carbon nanotubes from titanium dioxide nanoparticles. Appl. Surf. Sci. **258**, 8019–8025 (2012)
29. R.M. Silva, A.C. Bastos, F.J. Oliveira, D.E. Conte, Y. Fan, N. Pinna, R.F. Silva, Catalyst-free growth of carbon nanotube arrays directly on Inconel substrates for electrochemical carbon-based electrodes. J. Mater. Chem. A **3**, 17804 (2015)
30. R.B. Mathur, S. Chatterjee, B.P. Singh, Growth of carbon nanotubes on carbon fiber substrates to produce hybrid/phenolic composites with improved mechanical properties. Composite Sci. Technol. **68**, 1608–1615 (2008)
31. B. Yoon, H. Häkkinen, U. Landman, A.S. Wörz, J.-M. Antonietti, S. Abbet, K. Judai, Charging effects on bonding and catalyzed oxidation of CO on Au_8 clusters on MgO. Science **307**, 403–407 (2005)
32. Z.T. Kuznicki, J. Thibault, F. Chautain-Mathys, S. De Unamuno, Towards ion beam processed single-crystal Si solar cells with a very high efficiency, in *E-MRS Spring Meeting Proceedings Strasbourg, France, Firs Polish–Ukrainian Symposium, New Photovoltaic Materials for Solar Cells* (Krakow, Poland, 1996), 21–22 Oct
33. Z. Swiatek, J.T. Bonarski, R. Ciach, Z.T. Kuznicki, I.M. Fodchuk, M.D. Raransky, P. Gorley, Investigation of inhomogeneous structures of near-surface-layers in ion-implanted silicon. Thin Solid Films **319**, 39–43 (1998)
34. A. Cao, P.M. Ajayan, G. Ramanath, Silicon oxide thickness-dependent growth of carbon nanotubes. Appl. Phys. Lett. **84**, 109 (2004)
35. M. Mattila, T. Hakkarainen, H. Lipsanen, H. Jiang, E.I. Kauppinen, Catalyst-free growth of InAs(P) nanowires on silicon. Appl. Phys. Lett. **89**, 063119 (2006)

36. E.S.V. Mor, L.O.Y. Fleger, C.L. Pint, G.D. Nessim, What is below the support layer that affects carbon nanotube growth: an iron catalyst reservoir yields taller nanotube carpets. Nanoscale **6**, 1545–1551 (2014)
37. H. Liu, D. Takagi, S. Chiashi, Y. Homma, The growth of single-walled carbon nanotubes on a silica substrate without using a metal catalyst. Carbon **48**, 114–122 (2010)
38. H. Hamidinezhad, Thickness effect of catalyst layer on silicon nanowires morphology and features. Appl. Surf. Sci. **364**, 484–489 (2016)
39. H.F. Al-Taay, M.A. Mahdi, D. Parlevliet, P. Jennings, Controlling the diameter of silicon nanowires grown using a tin catalyst. Mater. Sci. Semicond. Process **16**, 15–22 (2013)
40. B. Zhao, D.N. Futaba, S. Yasuda, M. Akoshima, T. Yamada, K. Hata, Exploring advantages of diverse carbon nanotube forests with tailored structures synthesized by super growth from engineered catalysts. ACS Nano **3**, 108–114 (2008)
41. Y.-T. Jang, J.-H. Ahn, Y.-H. Lee, B.-K. Ju, Effect of NH_3 and thickness of catalyst on growth of carbon nanotubes using thermal chemical vapor deposition. Chem. Phys. Lett. **372**, 745–749 (2003)
42. T. Binninger, T.J. Schmidt, D. Kramer, Capacitive electronic metal–support interactions: outer surface charging of supported catalyst particles. Phys. Rev. B **96**, 16540 (2017)
43. Z.P. Huang, D.Z. Wang, J.G. Wen, M. Sennett, H. Gibson, Z.F. Ren, Effect of nickel, iron and cobalt on growth of aligned carbon nanotubes. Appl. Phys. A **74**, 387–391 (2002)
44. P. Kratzer, S. Sakong, V. Pankoke, Catalytic role of gold nanoparticle in GaAs nanowire growth: a density functional theory study. Nano Lett. **12**, 943–948 (2012)
45. Y. Wang, L. Zhang, C. Liang, G. Wang, X. Peng, Catalytic growth and photoluminescence properties of semiconducting single-crystal ZnS nanowires. Chem. Phys. Lett. **357**, 314–318 (2002)
46. Y. Wang, G. Meng, L. Zhang, C. Liang, J. Zhang, Catalytic growth of large-scale single-crystal CdS nanowires by physical evaporation and their photoluminescence. Chem. Mater. **14**, 1773–1777 (2002)
47. L. Wang, T. Li, L. Ling, J. Luo, K. Zhang, Y. Xu, H. Lu, Y. Yao, Remote catalyzation for growth of boron nitride nanotubes by low pressure chemical vapor deposition. Chem. Phys. Lett. **652**, 27–31 (2016)
48. A. Prabaswara, J.-W. Min, C. Zhao, B. Janjua, D. Zhang, A.R.M. Albadri, A.Y. Alyamani, T.K. Ng, B.S. Ooi, Direct growth of III-nitride nanowire-based yellow light-emitting diode on amorphous quartz using thin Ti interlayer. Nanoscale Res. Lett. **13**, 41 (2018)
49. M.A. Johar, H.-G. Song, A.W. Mostafa, I.V. Bagal, Y.H. Cho, S.W. Ryu, Universal and scalable route to fabricate GaN nanowire-based LED on amorphous substrate by MOCVD. Appl. Mater. Today **19**, 100541 (2020)
50. K.T. Koga, M.J. Walter, E. Nakamura, K. Kobayashi, Carbon self-diffusion in a natural diamond. Phys. Rev. B **72**, 024108 (2005)
51. D. Takagi, Y. Kobayashi, Y. Homma, Carbon nanotube growth from diamond. J. Am. Chem. Soc. **131**, 6922–6923 (2009)
52. B. Liu, W. Ren, C. Liu, C.-H. Sun, L. Gao, S. Li, C. Jiang, H.-M. Cheng, Growth velocity and direct length-sorted growth of short single-walled carbon nanotubes by a metal-catalyst-free chemical vapor deposition process. ACS Nano **3**, 3421–3430 (2009)
53. J.H. Atkins, Porosity and surface area of carbon black. Carbon **3**, 299–303 (1965)
54. S. Kumar, I. Levchenko, K. Ostrikov, J.A. McLaughlin, Plasma-enabled, catalyst-free growth of carbon nanotubes on mechanically-written Si features with arbitrary shape. Carbon **50**, 325–329 (2012)
55. M.Y. Khaywah, New Ultrasensitive Bimetallic Substrates for Surface Enhanced Raman Scattering, Doctoral Thesis, Universite de Technologie de Troyes, Troyes, Cedex, France, 2014
56. K.N. Tu, Interdiffusion and reaction in bimetallic Cu-Sn thin films. Acta Metallurica **21**, 347–354 (1973)
57. D. Su, M. Yu, G. Zhang, S. Jiang, Y.F. Qin, M.-Y. Li, Highly thermally stable Au–Al bimetallic conductive thin films with a broadband transmittance between UV and NIR regions. J. Mater. Chem. C **8**, 2852–2860 (2020)

58. D. Hardeman, S. Esconjauregui, R. Cartwright, S. Bharadwaj, L. D'Arsie, D. Oakes, J. Clark, C. Cepek, C. Dukati, J. Robertson, The synergistic effect in the Fe-Co bimetallic catalyst system for the growth of carbon nanotube forests. J. Appl. Phys. **117**, 044308 (2015)
59. W.-H. Chiang, R.M. Sankaran, Synergistic effects in bimetallic nanoparticles for low temperature carbon nanotube growth. Adv. Mater. **20**, 4857–4861 (2008)
60. W.-H. Chiang, R.M. Sankaran, Relating carbon nanotube growth parameters to the size and composition of nanocatalysts. Diam. Relat. Mater. **18**, 946–952 (2009)
61. A. Magrez, J.W. Seo, C. Mikó, K. Hernadi, L. Forró, Growth of carbon nanotubes with alkaline earth carbonate as support. J. Phys. Chem. B **109**, 10087–10091 (2005)
62. S.J. Matthew, J. Bierman, S.A. Morin, A new twist on nanowire formation: screw-dislocation-driven growth of nanowires and nanotubes. J. Phys. Chem. Lett. **1**, 1472–1480 (2010)
63. S.A. Morin, S. Jin, Screw dislocation-driven epitaxial solution growth of ZnO nanowires seeded by dislocations in GaN substrates. Nano Lett. **10**, 3459–3463 (2010)
64. C.J. Smithell, in *Smithells Metals Reference Book*. ed. by E.A. Brandes, G. Brook (Butterworth-Heinemann, Oxford, 1992)
65. A.D. Le Claire, in *Landolt-Börnstein—Numerical Data and Functional Relationships in Science and Technology*, vol. 26, ed. by H. Mehrer (Springer, Berlin, 1992), p. 471.
66. S. Hofmann, G. Csanyi, A.C. Ferrari, M.C. Payne, J. Robertson, Surface diffusion: the low activation energy path for nanotube growth. Phys. Rev. Lett. **036101**, 95 (2005)
67. C. Busetto, G. Del Piero, G. Manara, Catalysts for low-temperature methanol synthesis. Preparation of Cu-Zn-Al mixed oxides via hydrotalcite-like precursors. J. Catal. **85**, 260–266 (1984)
68. M. He, A.I. Chernov, E.D. Obraztsova, H. Jiang, E.I. Kauppinen, J. Lehtonen, Synergistic effects in FeCu bimetallic catalyst for low temperature growth of single-walled carbon nanotubes. Carbon **52**, 590–594 (2013)
69. M.H. Rümmeli, A. Bachmatiuk, A. Scott, F. Börrnert, J.H. Warner, V. Hoffman, J.-H. Lin, G. Cuniberti, B. Büchner, Direct low-temperature nanographene CVD synthesis over a dielectric insulator. ACS Nano **4**, 4206–4210 (2010)
70. A. Bachmatiuk, F. Börrnert, M. Grobosch, F. Schäffel, U. Wolff, A. Scott, M. Zaka, J.H. Warner, R. Klingeler, M. Knupfer, B. Büchner, M.H. Rümmeli, Investigating the graphitization mechanism of SiO_2 nanoparticles in chemical vapor deposition. ACS Nano **3**, 4098–4104 (2009)
71. S. Botti, C.R.L. Asilyan, L.D. Dominicis, F. Fabbri, S. Orlanducci, A. Fiori, Carbon nanotubes grown by laser-annealing of SiC nanoparticles. Chem. Phys. Lett. **400**, 264–267 (2004)
72. O. Leroy, J. Perrin, J. Jolly, M. Péalat, M. Lefebvre, Thermal accommodation of a gas on a surface and heat transfer in CVD and PECVD experiments. J. Phys. D : Appl. Phys. **30**(4), 499–509 (1997)
73. F.M. Wanlass, H. Eyring, Sticking coefficients. Adv. Chem. **33**, 140–145 (1961)
74. W. Lee, S.-J. Park, Porous anodic aluminum oxide: anodization and templated synthesis of functional nanostructures. Chem. Rev. **114**, 7487–7556 (2014)
75. G. Cao, D. Liu, Template-based synthesis of nanorod, nanowire, and nanotube arrays. Adv. Colloid. Interface Sci. **136**, 45–64 (2008)
76. H. Kohno, S. Takeda, Silicon nanoneedles grown by a simple thermal treatment using metal-sulfur catalysts. Jpn. J. Appl. Phys. **41**, 577–578 (2002)
77. T. Ishiyama, S. Nakagawa, T. Wakamatsu, Growth of epitaxial silicon nanowires on a Si substrate by a metal-catalyst-free process. Sci. Report **6**, 30608 (2016)

Chapter 15
Simple Theoretical Model for Growth by the VQS Mechanism

Abstract Taking the porosity of the FECA nanoparticle surface into account, a simple theoretical model has been developed for the nanomaterials growths by the VQS mechanism. This porosity of the nanoparticle surface has been first defined, and a mathematical form of it has then been given in terms of γ_{nano}, the surface energy of the FECA nanoparticle and in terms of γ_{bulk}, the surface energy of the FECA bulk from which the FECA nanoparticle is created. Theoretical models for porosity, pore radius, and high-energy sites (HETs) in terms of the effective surface amorphicity $\alpha_{amoreff}$ (see Appendix) have also been presented. Formulas for Knudsen diffusivity and Knudsen permeability have been presented, as well. Molecular and Knudsen diffusion have been compared, and the impact of nanoparticle surface roughness on the Knudsen diffusion has been elucidated. Finally, a simple growth model in terms of Knudsen diffusivity has been used to study the variation of CNT growth rate with CNT diameter, temperature-dependent variation of CNT growth rates, CNT growth rates dependent on growth duration, and CNT growth rates dependent on precursor flow rate. The formula for the growth rates is simple and approximate. It involves a number of empirical parameters. Nevertheless, the calculated results for the CNT growth rates compare well with the available experimental results for them.

15.1 Forwarding Note

The basic principles of growths by the VQS mechanism were laid down in Chap. 12. Various features of FECA-assisted growths by the VQS mechanism were described in Chap. 13. Various features of FECA-free growths by the VQS mechanism were however described in Chap. 14. General terminologies such as FECANO (e.g., FECA nanoparticles), METANO (e.g., metal nanoparticles) and SUBSANO (e.g., substrate nanoparticles), and also the R_L species were defined in Chap. 3. On the other hand, the pre-nucleation and pro-nucleation stages of growths were laid down in Chap. 4, and NP1 and NP2 nanoparticles, high energy sites (HETs), and surface melting were defined in Chap. 12. Based on these and on the R_S ($R_S \equiv X$, $R_S \equiv Y$) source species

S. N. Mohammad, *Synthesis of Nanomaterials*, Springer Series in Materials Science 307, https://doi.org/10.1007/978-3-030-57585-4_15

defined in Chap. 1, simple theoretical model for nanomaterials growths by the VQS mechanism would be described in the following.

15.2 FECA Nanoparticle Porosity for Nanowire Growths

Porosity of nanoparticles (both METANOs and SUBSANOs) is essential for growths by the VQS mechanism. It is the result of the void or empty spaces in the lattice structure of the R_L species of a FECA nanoparticle, which we call FECANO. This FECANO can be METANO; it can also be SUBSANO. Porosity is the fraction of the volume of voids over the total volume of the lattice structure of the FECA nanoparticle. It can vary between 0 and 1. Depending on its surface morphology, it extends over the entire surface of the R_L species of FECA nanoparticle (diameter D_{nano}). As a result, both $R_S \equiv X$ and $R_S \equiv Y$ species diffuse through it for nanomaterials growths. If k_B is the Boltzmann constant, P is the pressure, and d_c is the average pore diameter ($d_c = 2r_c$) of the R_L species, then the mean free path of the diffusing species is

$$\lambda = \frac{k_B T}{\sqrt{2}\pi d_c^2 P}. \tag{15.1}$$

We argue that the pressure P on the R_S species, which migrate through nanopores of the R_L species of nanoparticle surface, is generated, at least partly, by the surface tension force. In fact, this pressure tends to increase the porosity of the R_L species. Rayner et al. [1] noted that there occurs increase in dimensionless mass transfer, for example, through a membrane due to increase in flux of the incoming species. The R_L species can be one such membrane. And we believe this happened because there occurred increase in porosity of the R_L species membrane. If n_c is the pore density (e.g., number of pores per unit area, e.g., nm^2 or μm^2), the porosity of the R_L species for nanowire growth would be

$$\rho_c = 4n_c\left(\frac{r_c\gamma_{nano}}{D_{nano}\gamma_{bulk}}\right)^2, \tag{15.2}$$

where γ_{nano} is again the surface energy of the FECA nanoparticle, and γ_{bulk} is the surface energy of the FECA bulk from which the FECA nanoparticle is created. They belong, therefore, to the same material.

15.3 FECA Nanoparticle Porosity for Nanotube Growths

The FECA nanoparticle for nanotube growth has a shell region, as described in Chap. 4. This shell region, rather than the core region of the nanoparticle, must have porosity for growths. Note that this shell region has R_L species; see Fig. 12.4 of Chap. 12. And hence, the porosity of the nanoparticle is, in reality, the porosity of the R_L species. It may be defined as the ratio of the total area of nanopore in the shell and the total area of the shell itself. Recall that the total number of the nanopores is u_c. The radius of each of these nanopores is r_c as defined in the Appendix (A.7). Then the total area of all the nanopores together is

$$A_{\text{pore}} = \pi u_c r_c^2. \tag{15.3}$$

We suggest in the following that the nanopore radius, if created, for example, by structural void, but not by the influence of the surface amorphicity α_{amor}, would actually be r_{c0} and not r_c. The total area of the shell would, on the other hand, be

$$A_{\text{shell}} = \frac{\pi}{4}\left[D_{\text{NTO}}^2 - D_{\text{NTI}}^2\right], \tag{15.4}$$

where D_{NTO} is the outermost diameter and D_{NTI} is the innermost diameter of the nanotube. The local chemical composition of the shell is important in this regard. This composition is heterogeneous; it contains metastable nanostructures of lowly coordinated and weakened interatomic (intra-atomic) interactions. It has nanopores generated in regions of weakened interatomic (intra-atomic) interactions. And it depends on surface conditions evolved at least partly from the surface energy of the shell surface. The surface energy γ_{nano}, as compared to the surface energy γ_{bulk} of the corresponding bulk of a material, is different under different surface conditions. We assume that the surface energy of a shell is identical to the surface energy γ_{nano} of the nanoparticle at the peripheral surface of which it is formed. Taking the effect of surface energy γ_{nano} into account, the porosity may be given by

$$\rho_c = \left[\frac{A_{\text{pore}}\gamma_{\text{nano}}}{A_{\text{shell}}\gamma_{\text{bulk}}}\right]. \tag{15.5}$$

With the help of (15.3) and (15.4), the (15.5) would be rewritten as

$$\rho_c = 4u_c\left[\frac{r_c^2\gamma_{\text{nano}}}{(D_{\text{NTO}}^2 - D_{\text{NTI}}^2)\gamma_{\text{bulk}}}\right] \tag{15.6}$$

If, for example, the surface topography is locally reentrant so as to constitute a fine pore of atomic/molecular dimension, then the adsorption coefficient of the R_S species in the R_L species would probably be more than doubled. The nanotube wall thickness is $D_{\text{NT}} = D_{\text{NTO}} - D_{\text{NTI}}$, where D_{NTO} the outermost diameter of the nanotube is also the outermost diameter of the shell. Similarly, D_{NTI}, the innermost diameter of the nanotube is also the innermost diameter of the shell. If the hill formed around the shell is taken into account (see Fig. 13.6 of Chap. 13), and the width of the hill is $w_{HILL} = (r_{nano} - r_{NTO})$, the porosity of the R_L species of the shell would be given by

$$\rho_c = 4u_c \left[\frac{r_c^2 \gamma_{nano}}{\left\{ (D_{NTO} + w_{HILL})^2 - D_{NTI}^2 \right\} \gamma_{bulk}} \right]. \tag{15.7}$$

15.4 Theoretical Models for Porosity, Pore Radius, and HET in Terms of Amorphicity

15.4.1 Surface Amorphicity

We emphasize again that surface amorphicity of the nanoparticle (both METANO and SUBSANO) surface is crucial for nanomaterials synthesis. This amorphicity is a constructive amorphicity, rather than a descructive amorphicity, as detailed in the Appendix. Sobanska et al. [2] performed plasma-assisted molecular beam epitaxial growth of GaN nanowires on Si (111) substrates. They noted that the presence of a thin amorphous layer on the nanoparticle surface significantly enhanced spontaneous nucleation of GaN nanowires. Slower nucleation was observed on partially amorphous nanoparticle surface. No growth of nanowires was possible on the nanoparticle surface when it was nanocrystalline and had low amorphicity and insignificant density of defects. The finding by Sobanska et al. [2] suggests that tuning of nanoparticle surface is essential, as advocated earlier (see Chaps. 12–14), for an efficient self-induced nucleation of nanomaterials on nanoparticle surface. No nucleation on nanocrystalline surface was possible if this surface had no nanopores of appreciable dimension.

Surface amorphicity α_{amor} and HET concentration δ_{HET} of the R_L species surface increase with increase in defect (disorder) in this R_L species surface. There occurs also an increase in the porosity ρ_c. The porosity ρ_c and the pore number u_c are functions [3] of the amorphicity α_{amor}. A gradual increase in HET concentration δ_{HET} causes increasing decomposition of precursor. Unless adsorbed or desorbed, the increasing number of R_S species resulting from increasing decomposition of precursor is piled up on the R_L species surface and hence leads to further increase in the amorphicity α_{amor}. This process continues until α_{amor} reaches the maximum (e.g., $\alpha_{amor} = 1$). However, as shown in Fig. 15.1, beyond a certain optimal limit, suppose $\alpha_{amor} = \alpha_{amoreff0}$ ($\alpha_{amoreff0} \leq 1.0$), the R_L species surface is increasingly covered with the accumulated R_S species. The nanopores of the R_L species surface are thus increasingly obstructed and even closed by the accumulated R_S species. If

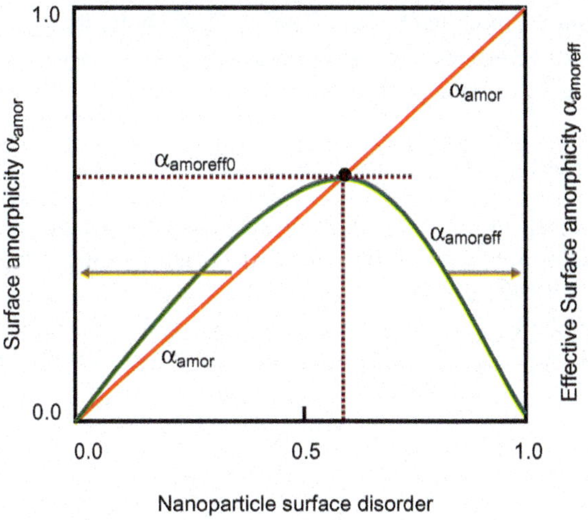

Fig. 15.1 Schematic diagram showing the dependence of the effective amorphicity α_{amoreff} on the amorphicity α_{amor} of the nanoparticle surface

partially closed, the nanopore radius is decreased. If fully closed, nanopore density is decreased. This implies that the porosity increases with increase in α_{amor} until $\alpha_{\text{amor}} = \alpha_{\text{amoreff0}}$, but decreases with increase in α_{amor} beyond $\alpha_{\text{amor}} = \alpha_{\text{amoreff0}}$. This is consistent with the observation of Chhowalla et al. [4]. So, there is an effective amorphicity α_{amoreff}. In general, $\alpha_{\text{amor}} \approx \alpha_{\text{amoreff}}$ below the peak value α_{amoreff0}. But $\alpha_{\text{amor}} \neq \alpha_{\text{amoreff}}$ beyond the peak value α_{amoreff0}. The porosity is not hence a function of the amorphicity α_{amor} beyond the peak value α_{amoreff0}. It is rather a function of the effective amorphicity α_{amoreff}, as indicated in Fig. 15.1. Our discussions made above therefore suggest that the effective amorphicity α_{amoreff} is the amorphicity which governs the porosity of the R_L species surface. This effective amorphicity α_{amoreff} increases with increase in α_{amor} until $\alpha_{\text{amor}} = \alpha_{\text{amoreff0}}$. It reaches a peak at $\alpha_{\text{amor}} = \alpha_{\text{amoreff0}}$ but gradually decreases with increase in α_{amor} beyond $\alpha_{\text{amor}} = \alpha_{\text{amoreff0}}$ (see Fig. 15.1).

15.4.2 *Pore Radius, Porosity, and HET Reactivity Defined in Terms of Effective Amorphicity*

The pore radius r_c determines the porosity ρ_c. We indicated in the Appendix (A.7) that the pore radius r_c did not exist before the nanoparticle surface was amorphized, and that it is created by the effective amorphicity α_{amoreff}. We argue that there may also be pore radius called r_{c0} of the pores, which it is not created by amorphicity, as indicated in the Appendix (A.7), but rather created by structural voids or by some other means. This pore radius r_{c0}, which is the one not created or affected by amorphicity, may be modified by the effective amorphicity α_{amoreff}. The pore radius r_{c0} may therefore be proportional to the effective amorphicity α_{amoreff}. If ζ_{cn} is the proportionality constant, the pore radius r_{cmod} would be

$$r_{c\,\text{mod}} = \zeta_{cn} r_{c0} \alpha_{\text{amoreff}}. \tag{15.8}$$

A comparison of (15.8) with (A.7) of the Appendix would indicate that r_c of (A7) may actually be $r_{c\text{mod}}$, but only for one certain specific value of r_{c0}, namely $r_{c0} = r_{cn}/\zeta_{cn}$. For this specific value of r_{c0}, $r_{c\text{mod}} = r_c$.

We suggest that the porosity ρ_c is also proportional to the effective amorphicity α_{amoreff}. If c_{bn} is the proportionality constant, then the porosity ρ_c may be modified to

$$\rho_{c\,\text{mod}} = c_{bn} \rho_c \alpha_{\text{amoreff}}. \tag{15.9}$$

The highest value of $\rho_{c\text{mod}}$ corresponding to the highest value α_{amoreff0} of α_{amoreff} is $\rho_{c\text{mod0}}$. Similarly, the HET reactivity δ_{KND} is proportional to the effective amorphicity α_{amoreff}. If c_{mn} is the proportionality constant, then the modified HET reactivity may be given by

$$\delta_{\text{HETMOD}} = c_{mn} \delta_{\text{HET}} \alpha_{\text{amoreff}}, \tag{15.10}$$

where δ_{HET} is the HET reactivity not impacted by the amorphicity α_{amoreff}. This reactivity δ_{HET} may be a measure of the number of HET sites. It may hence be a measure of the HET concentration in the R_L species. The highest value of δ_{HETMOD} corresponding to the highest value α_{amoreff0} of α_{amoreff} would be δ_{HETMOD0}. We found at least one experiment to support and confirm that porosity $\rho_{c\text{mod}}$ is proportional to the effective amorphicity α_{amoreff}, as presented in Fig. 12.5a of Chap. 12. This figure produced with experimental data by Newby et al. [5] shows that the porosity ρ_c as function of irradiation fluence is almost linear. Irradiation fluence may be a measure of effective amorphicity implying that (15.9) is at least approximately justified. Newby et al. confirmed that surface porosity is a function of surface amorphicity, and that the surface porosity increases with increase in surface amorphicity.

15.5 Knudsen Diffusivity

15.5.1 Formulation of Knudsen Diffusivity

Knudsen formula for diffusivity becomes essential for the diffusion of the R_S species if the mean free path λ of the diffusing R_S species is larger than the pore diameter d_c. Such diffusion is appropriate in an environment in which the R_L species is a porous medium and the permeability of the R_L species is low implying that the nanopore radii are small. It suits also under low-pressure condition at which the mean free path is large. The Knudsen diffusion may be prevalent in nanoporous R_L species if the pore size of this species approaches the mean free path of the diffusing species (atoms or molecules).

Knudsen diffusion of the R_S species or of molecules formed from these species is a selective transport process, as the diffusion flux during this process is inversely proportional to the atomic (molecular) mass of the diffusing species. This means the mean velocity of the species decreases with increasing species weight. It occurs due to an increase in the flux of the diffusing species with increase in temperature. This is the consequence of inverse proportionality of the concentration of an ideal gas to temperature. Hence, the mean diffusing species velocity (e.g., molecular velocity) increases with increase in the square root of temperature. For these diffusing R_S species for diffusion through the nanopores of the R_L species, the Knudsen diffusion coefficient D_{KND} may be given by [6]

$$D_{KND} = \frac{\rho_{c\,mod}\, r_c \vartheta_c}{6\tau_c} \left(\frac{8RT}{\pi M_c} \right)^{1/2}. \quad (15.11)$$

Recall that the nanopore radius r_c is the radius of each of the nanopores of the R_L species which (e.g., nanopores) are assumed to be identical. If it is assumed that all the nanopores have the same value of the nanopore radius, which did not exist before the nanoparticle surface was amorphized, this nanopore radius is modified and has different value at different amorphicity, which is given by (A7) of the Appendix. It may otherwise be given by (15.8). Also ϑ_c is the shape factor, which is unity for uniform straight pores normal to the planar surface of the nanoparticle, τ_c is the tortuosity factor, δ_{amor} is the depth AB of the disturbed (disordered), porous, amorphous R_L species of nanoparticle [see Fig. 12.4b of Chap. 12], and R is the gas constant ($R = 8.314$ J/mol/K).

15.5.2 Experimental Support for the Knudsen Diffusivity

A binary alloy such as $Fe_{1-z}Ni_z$ lacks long-range crystal order. Depending on its mole fraction z, it exhibits, as a result, surface amorphicity α_{amor} and the effective surface amorphicity $\alpha_{amoreff}$. Based on (15.9), the modified porosity ρ_{cmod} varies linearly with the effective amorphicity $\alpha_{amoreff}$. Based on (15.11), the diffusion coefficient D_{KND} varies directly and linearly with the modified porosity ρ_{cmod}. Hence, the diffusion coefficient D_{KND} varies directly and linearly with the effective amorphicity $\alpha_{amoreff}$. The variation of the experimental diffusion coefficient of carbon with the mole fraction z of the $Fe_{1-z}Ni_z$ binary alloy is shown in Fig. 15.2. This variation was obtained by Cermak and Mehrer [7]. There are only four experimental data points. Nevertheless, Fig. 15.2 shows that the effective surface amorphicity $\alpha_{amoreff}$ increases with increase in the surface amorphicity α_{amor}, attains a peak, and

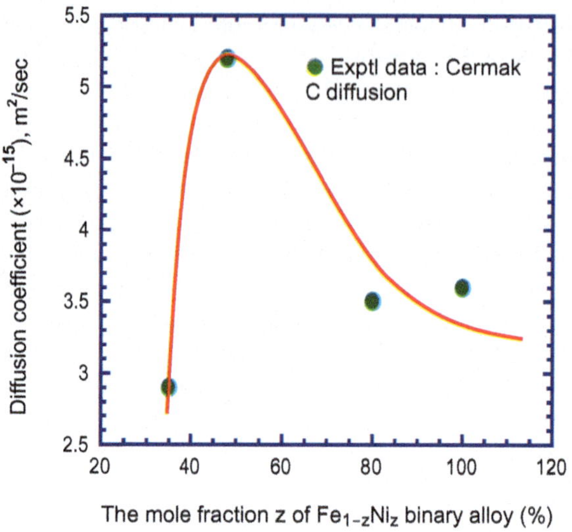

Fig. 15.2 Experimental variation of the carbon diffusion coefficient with the mole fraction z of the $Fe_{1-z}Ni_z$ binary alloy. The experimental data are by Cermak and Mehrer [7]

then decreases with further increase in the surface amorphicity α_{amor}. Note that the surface amorphicity α_{amor} increases with the mole fraction z, and hence, Fig. 15.2 shows actually the variation of the experimental diffusion coefficient of carbon with the surface amorphicity α_{amor} of the $Fe_{1-z}Ni_z$ binary alloy. This figure indicates that the diffusivity increases with increase in the surface amorphicity α_{amor}, attains a peak, and then decreases with further increase in the surface amorphicity α_{amor}. This is interesting as it demonstrates that the diffusivity varies linearly with the modified porosity $\rho_{c\text{mod}}$ as per (15.11). And the modified porosity $\rho_{c\text{mod}}$ varies directly and linearly with the effective surface amorphicity α_{amoreff} as per (15.9). The experimental result of Fig. 15.2 thus confirms again that the diffusivity depends linearly on the modified porosity $\rho_{c\text{mod}}$. And the modified porosity $\rho_{c\text{mod}}$ depends linearly on the effective surface amorphicity α_{amoreff}, which is a function of the surface amorphicity α_{amor}, and also of the mole fraction z.

15.6 Molecular and Knudsen Diffusion

The transport of the R_S species (or of molecules formed from these species) through a porous R_L species (solid solution, cluster) is a complex phenomenon. Under a constant system pressure, this transport is diffusive in nature. It may involve ordinary molecular diffusion, Knudsen diffusion, and surface migration. However, only molecular diffusion and Knudsen diffusion may be seriously taken into account, as the surface migration takes place only if the diffusing species are adsorbed in a mobile layer, and that this diffusion is negligible in most cases. Note that the molecular diffusion is essentially ordinary diffusion of the diffusing species which emerges from movements of the diffusing species. The kinetic theory of gases suggests that the forces of attraction and repulsion between the diffusing species (atoms or molecules) should be considered also the same as that for ordinary diffusion.

15.7 Knudsen Permeability

Permeability is how easily the R_S diffusing species passes through the porous R_L species. It is a measure of the rate of flow of the R_S species (atoms or molecules) through the porous R_L species. The Knudsen permeability $\wp_{KND}$ per unit pressure may then be given by [8]

$$\wp_{KND} = \frac{2\rho_{c\,\mathrm{mod}}\, r_c \vartheta_c M_c}{3RT\delta_{\mathrm{amor}}}\left(\frac{8RT}{\pi M_c}\right)^{1/2}. \tag{15.12}$$

Note that (15.12) takes the impact of the effective surface amorphicity $\alpha_{\mathrm{amoreff}}$ into account via $\rho_{c\mathrm{mod}}$ and r_c. Also, M_c is the atomic (molecular) weight of the atomic (or X_mY_n molecular) species. It is, for example, $M_c = 12.0107$ gm/mol for X_mY_n≡C, $M_c = 24.8184$ gm/mol for X_mY_n≡BN, $M_c = 28.084$ gm/mol for X_mY_n≡Si, $M_c = 72.61$ gm/mol for X_mY_n≡Ge,$M_c = 81.408$ gm/mol for X_mY_n≡ZnO, $M_c = 83.73$ gm/mol for X_mY_n≡GaN, and $M_c = 145.792$ gm/mol for X_mY_n≡InP. The diffusion rate is determined by the diffusion coefficient. It is the diffusion flux per concentration gradient. The diffusion rate increases with increase in the value of diffusion coefficient. The higher the diffusion rate, the higher is the nanomaterial growth rate G_R. Recall that, for this nanomaterial growth, ρ_c is the fraction of the bulk volume of the porous surface exhibiting nanopores. The tortuosity factor τ_c may be inversely proportional to the porosity ρ_c. With T and M_c kept fixed, D_{KND} depends only on $\rho_{c\mathrm{mod}}$ and r_c.

15.8 Knudsen Diffusion Through Rough R_L Species

Mesoporous R_L species have pore sizes generally between 2 and 50 nm. We argue again that the Knudsen diffusion is the predominant transport mechanism for the diffusion of the R_S species through such R_L species. This is also the predominant transport mechanism for the diffusion of the X_mY_n molecules, which are formed from the reaction of R_S≡X and R_S ≡X species on the R_L species surface and/or during their diffusion through the R_L species. As stated earlier, such diffusion involves collisions of the R_S species with the pore walls, rather than collisions of the R_S species among themselves. This means the said diffusion is dictated in large part by the pore characteristics, particularly the roughness of the pore wall [9]. And such roughness does exist. Experiment confirms that internal surface of many amorphous catalysts has fractal roughness [10, 11] of finite size. Smoluchowski [12] showed that pore shape, in general, has an effect on Knudsen diffusion. Santra and Sapoval [13] quantified the interactions of the diffusing species with the nanopore walls. Analytical formulas for fractal pores by Coppens and Froment [14] and by Liang and Li [15] predicted a considerable impact of nanopore surface roughness on the Knudsen diffusivity. Geier et al. [16] showed that tortuosity factor for Knudsen diffusion through pores of irregular shape and size may be significantly larger than that in the bulk regime. Malek and Coppens [17] noted, on the other hand, that diffusivity is governed by roughness of the pore wall surface. If τ_{c0} is the tortuosity factor without pore surface roughness and σ_{tor} is the effect of pore surface roughness on the tortuosity factor, then τ_c may be expressed as

$$\tau_c = \frac{\tau_{c0}}{(1 - \sigma_{tor})}. \tag{15.13}$$

The diffusivity through nanopores may then be modified to

$$D_{KND} = \frac{\rho_c r_c \vartheta_c}{6\tau_{c0}}(1 - \sigma_{tor})\left(\frac{8RT}{\pi M_c}\right)^{1/2}, \tag{15.14}$$

where τ_{c0} may be assumed to be 1.

15.9 Nanomaterial Growth Rate

If r_{as} is the radius of the R_S species ($r_{as} = 0.067$ nm for C, $r_{as} = 0.11$ nm for Si, $r_{as} = 0.09$ nm for Ga), and ζ_{KND} is a suitable growth parameter, the nanomaterial growth rate G_R would be a function of D_{KND}. It would approximately be given by [18, 19]

$$G_R = \frac{\zeta_{KND} D_{NM} D_{KND}}{r_{as}^2}, \tag{15.15}$$

where ζ_{KND} is a parameter. Note that $D_{NM} = D_{NW}$, the nanowire diameter for the growth of nanowire, but $D_{NM} = D_{NT}$, the nanotube wall thickness for the growth of nanotube; $D_{NT} = D_{NTO} - D_{NTI}$, where D_{NTI} is again the diameter of the innermost wall and D_{NTO} is the diameter of the outermost wall of this nanotube.

15.10 Diffusivity and Permeability for Growths

15.10.1 Calculated Results for Diffusivity and Permeability

Unless specifically mentioned, we disregarded, for the present study, the specific dependence of ρ_{cmod} on r_c or r_{c0}, which are given by (15.2) and (15.7), respectively. We disregarded the specific dependence of ρ_{cmod} also on the effective amorphicity $\alpha_{amoreff}$, as indicated in (15.8) and (15.9). And instead the porosity ρ_{cmod} was assumed to vary, for example, between 0.2 and 0.9. The Knudsen diffusion coefficient D_{KND} as function of ρ_{cmod} and r_c was thus obtained for the $R_S{\equiv}C$ source species and presented in Fig. 15.3a. Similarly, diffusivity of BN through the R_L species was also presented in Fig. 15.3b. Note that BN is the result of reactions between B and N on the R_L species surface or during their diffusion through the $R_L$species. Both Figs. 15.3a, b indicate that, for a certain pore radius r_c, the higher the porosity, the higher is the diffusivity D_{KND} through the nanopores. Figures 15.3a, b indicate that indeed the diffusion of the R_S species through nanopores at the growth temperature, T depends on both r_c and ρ_{cmod}. The temperature for carbon nanotube growth is generally lower than that for BN nanotube growth. But the calculations for all of them were carried out for the temperature $T = 900$ °C. These calculations suggest that nevertheless the Knudsen diffusion coefficient of the $R_S{\equiv}C$ is much higher than that of BN; it is primarily because the mass of BN is much larger than that of C. Figures 15.4a, b show the temperature-dependent diffusivity of C and BN, respectively. The variations in these figures are linear, which indicate that the diffusivity of both C and BN increases with increase in temperature, and that the diffusivity is less sensitive to the temperature T than to the porosity ρ_{cmod}.

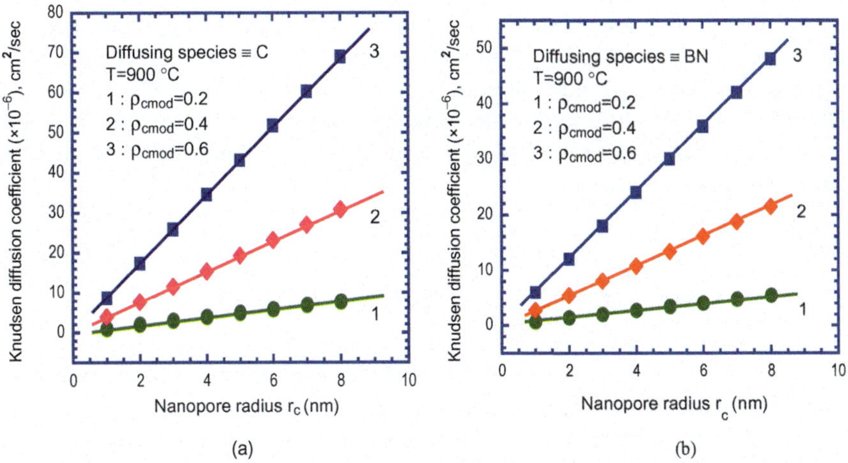

Fig. 15.3 Knudsen diffusivity of **a** carbon, **b** BN varying with nanopore radius for the porosity $\rho_{cmod} = 0.2, 0.4, 0.6$, and at temperature of $T = 900$ °C

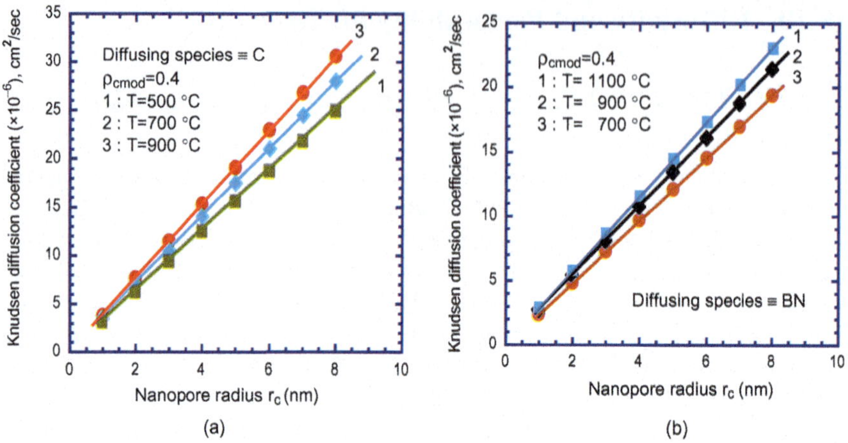

Fig. 15.4 Knudsen diffusivity of **a** carbon varying with nanopore radius for temperature T=500 °C, 700 °C and 900 °C, and **b** BN varying with nanopore radius for temperature T = 700 °C, 900 °C and 1100 °C, and for both of them the porosity $\rho_{cmod} = 0.4$

Notably, the diffusion coefficient is higher for larger pore radius. It is more sensitive to temperature for pores of larger dimension than for pores of smaller dimension. Keeping this in mind, the observations made above are in line with the experiment by Hsu et al. [20], who found ZnO nanowire growth highly dependent on the pore dimension of the porous silicon substrate. It is consistent with also the observation of Cassell et al. [21] who found strong influence of pore structures on the diffusivity of the R_S species through the R_L species. They obtained higher yield in SWCNTs by more open pore structure, and it was due to more rapid diffusion of the R_S species through these pore structures. Permeability of C and BN through the R_L species of three different porosities is shown in Fig. 15.5a, b. Again, the higher the porosity, the higher is the permeability. And this permeability for C is quite comparable to that for BN. The temperature-dependent permeabilities of C and BN are shown in Fig. 15.6a, b. Like diffusivities, the permeabilities appear to be less sensitive to temperature than to porosity. But unlike diffusivities, the permeabilities decrease with increase in temperature.

15.10.2 Surface Roughness Effects on Diffusivity

To understand the impact of surface roughness parameter σ_{tor} on diffusivity (see 15.14), we assume that the parameter τ_{c0} (see 15.13) varies inversely with ρ_{cmod}; $\tau_{c0} = 1/\rho_c$. Under this circumstance, the diffusivity may decrease, and the permeability may increase due to surface roughness of the wall. Considering that the R_S species or the molecules from these species do not interact with each other or with one another, they move in straight path through nanopores. Upon collision with

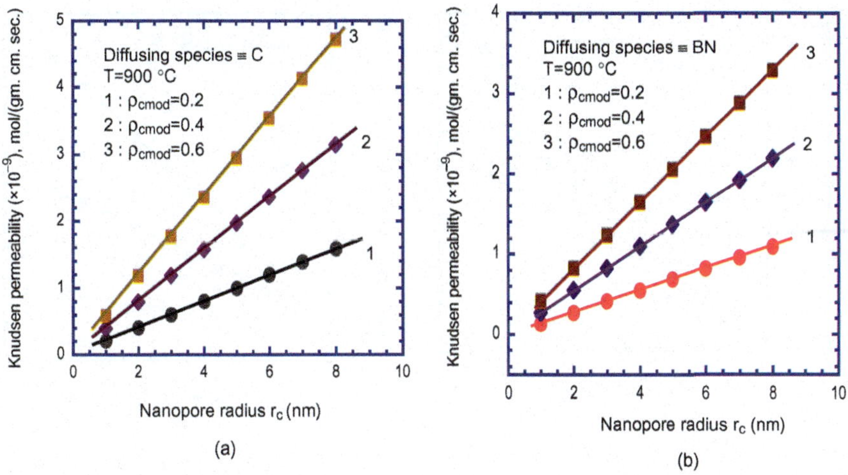

Fig. 15.5 Knudsen permeability of **a** carbon and **b** BN varying with nanopore radius for porosity $\rho_{cmod} = 0.2, 0.4, 0.6$, and temperature $T = 900$ °C

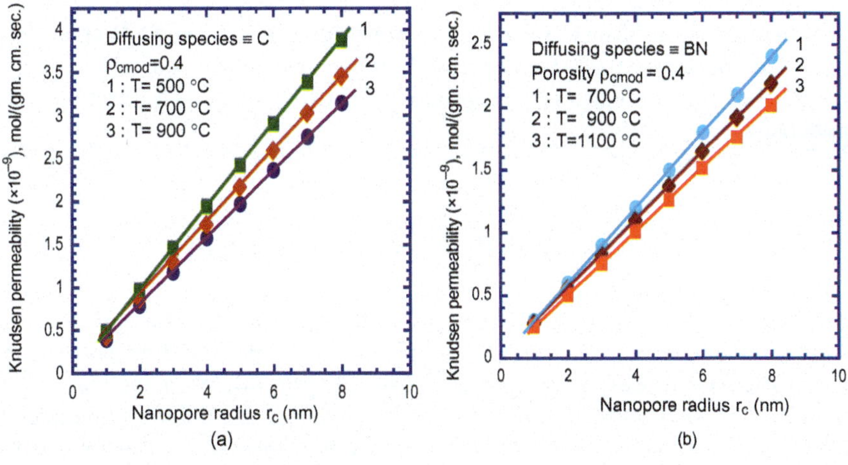

Fig. 15.6 Knudsen permeability of **a** carbon and **b** BN varying with nanopore radius for porosity $\rho_{cmod} = 0.4$, and for three different temperatures

the pore surface, they are adsorbed for a short while, but continue to move in their original direction or altered original direction. The time of physical adsorption on the surface is probably negligible as compared to the time of diffusion through the pores, and Malek and Coppens [17] demonstrated that Knudsen diffusion through amorphous R_L species is influenced by roughness of the pore surface. To understand the impact of nanowall roughness, we calculated D_{KND} using (15.13) and (15.14), respectively. The results of D_{KND} versus r_c plots are presented in Fig. 15.7a for

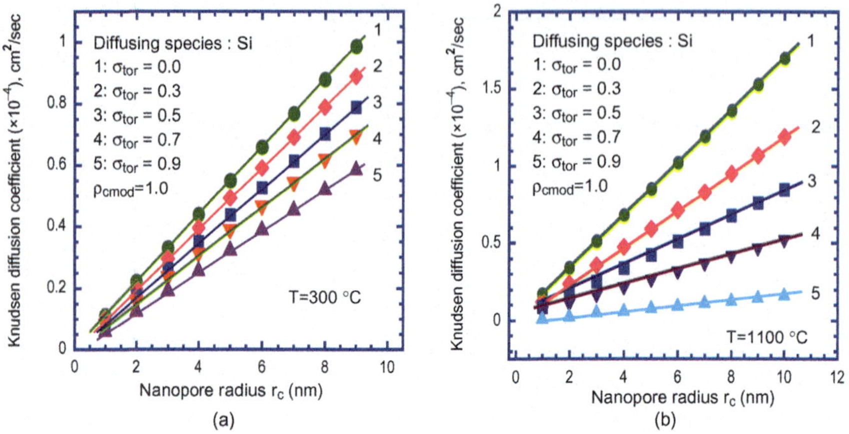

Fig. 15.7 Impact of surface roughness of the R_L species on the Knudsen diffusivity of $R_S{\equiv}C$ species through it. Knudsen diffusivity of $R_S{\equiv}C$ varying with nanopore radius is for **a** temperature $T = 300$ °C and **b** temperature $T = 1100$ °C. Curves 1 to 5 correspond to $\sigma_{tor} = 0.0, 0.3, 0.5, 0.7$, and 0.9, respectively. Also, the porosity for all the curves is $\rho_{cmod} = 1.0$

growth at $T = 300$ °C and in Fig. 15.7b for growth at $T = 1100$ °C. Both of these figures are for $\sigma_{tor} = 0.0$–0.9 and $\rho_c = 1.0$. It is remarkable to note that the higher the value of σ_{tor}, the higher is the depression (reduction) of D_{KND} as function of the pore radius r_c.

15.10.3 Surface Amorphicity Effects on Diffusivity

Needless to say, surface amorphicity α_{amor} of the nanoparticle surface gives rise to surface roughness α_{rgh}. The surface amorphicity α_{amor} may have both positive and negative effects (see Appendix) on nanomaterials growths. We argue that, unless influenced by some other parameters, surface roughness α_{rgh} follows surface amorphicity α_{amor}. There can be an effective surface roughness α_{rgheff} that follows the effective amorphicity $\alpha_{amoreff}$. In order to understand the influence of this, we modify (15.11) to

$$D_{KND} = \frac{\rho_{c\,mod}\, r_{c\,mod}\, \vartheta_c}{6\tau_c}\left(\frac{8RT}{\pi M_c}\right)^{1/2}, \tag{15.16}$$

where r_{cmod} is given by (15.8). Also, we assume that $\tau_c = \tau_{c0} = 1/\rho_c$. Recall that the nanopore radius r_{cmod}, as function of r_{c0}, did exist even before the nanoparticle surface was amorphized, and that it bears the consequence of structural void.

The variation of the diffusivity D_{KND} with the surface amorphicity α_{amor} for four different values of the nanopore radius r_{c0} is presented in Fig. 15.8. As this figure

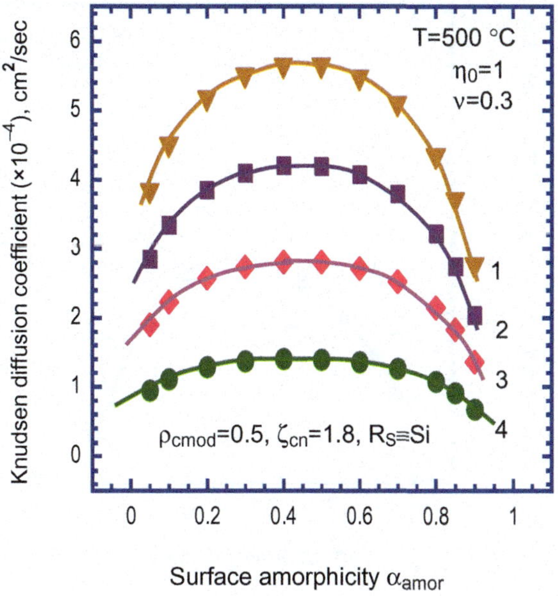

Fig. 15.8 Variation of Knudsen diffusion coefficient with surface amorphicity for four different nanopores of four different nanopore radii. Curves 1, 2, 3, and 4 are for nanopore radius $r_{c0} =$ 2, 4, 6, and 8 nm, respectively, and for $\zeta_{cn} = 1.8$. Being dependent on amorphicity, the modified nanopore radius $r_{c\text{mod}}$ is a function of the nanopore radius r_{c0} as given by (15.8)

shows that the diffusivity increases, reaches a peak, and then decreases with increase in the pore radius r_{c0}. It happens simply because there occurs increase in surface roughness with increase in amorphicity, and nanopores become increasingly curved and twisted due to increase in surface roughness. Figure 15.8 suggests that there is a limit in surface roughness, and any surface beyond an optimal level of roughness experiences decrease in diffusivity. This finding is consistent with that by Malek and Coppens [17]. Depending on the means of surface treatment, the surface roughness may slowly or rapidly increase. Figure 15.9 shows that such rapidity of increase or decrease in surface roughness, as represented by ν (see Appendix, for the definition of ν) should be avoided in practical demonstrations.

We emphasize that the Knudsen diffusion coefficient is larger for larger porosity and larger pore radius $r_{c\text{mod}}$. This pore radius is governed by the effective amorphicity α_{amoreff} implying that the effective amorphicity α_{amoreff} effectively influences the nanomaterial growth rate. This is quite apparent from a study by Chhowalla et al. [4], who examined the impact of C_2H_2 precursor concentration ϕ on the CNT growth rate at 750 °C. They observed that, beyond a certain limit of C_2H_2 concentration, any increase in C_2H_2 concentration led to a gradual decrease in CNT growth rate. It happened because increase in C_2H_2 concentration caused an increase in C

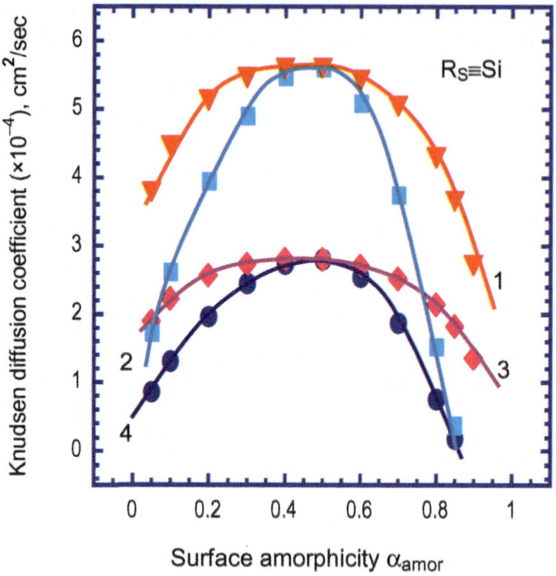

Fig. 15.9 Knudsen diffusion coefficient D_{KND} of Rs≡Si varying with the nanoparticle surface amorphicity α_{amor}. Curves 1 and 3 are for nanopore radius $r_{c0} = 4$ and 8 nm, respectively, but for $\nu = 0.3$ Curves 2 and 4 are for nanopore radius $r_{c0} = 4$ and 8 nm respectively, but for $\nu = 0.6$. The temperature $T = 500$ °C, the porosity $\rho_{cmod} = 0.5$, and the parameter $\eta_0 = 1$. Being dependent on amorphicity, the nanopore radius r_{cmod} as function of r_{c0} is given by (15.8) with $\zeta_{cn} = 1.8$

species. And beyond a certain limit of C_2H_2 concentration, there occurred an excessive concentration of C species accumulated on the METANO surface. Similar situations arose due to excessive increase in growth time. We attribute the gradual decrease in CNT growth rate to gradual decrease in the effective amorphicity $\alpha_{amoreff}$. Both Figs. 15.8 and 15.9 show that, irrespective of the nanopore dimension, the Knudsen diffusion coefficient increases with temperature. Consequently, CNT growth rate increases with increase in temperature.

Variation of diffusivity with nanopore radius is shown again in Fig. 15.10. This figure is for two different temperatures $T = 500$ °C and $T = 700$ °C and for two different amorphicities $\alpha_{amor} = 0.5$ and $\alpha_{amor} = 0.8$, respectively. Irrespective of how large or small the amorphicity is, the diffusivity always increases with increase in temperature. The increase in diffusivity is though lower for $\alpha_{amor} = 0.8$ than for $\alpha_{amor} = 0.5$ implying that excessive surface roughness due to excessive surface amorphicity adversely affects the diffusivity. Note that the effect of surface amorphicity on diffusivity has been modeled through r_{c0} which is modified to r_{cmod}. So, we show in Fig. 15.11 the dependence of r_{cmod} on r_{c0} for only $\alpha_{amor} = 0.5$. We found that, depending on temperature, r_{c0} can significantly influence r_{cmod}. For example, for $r_{c0} = 9$ nm, r_{cmod} can be larger than 25 nm. And a larger r_{cmod} can lead to larger diffusivity and consequently to larger growth rate.

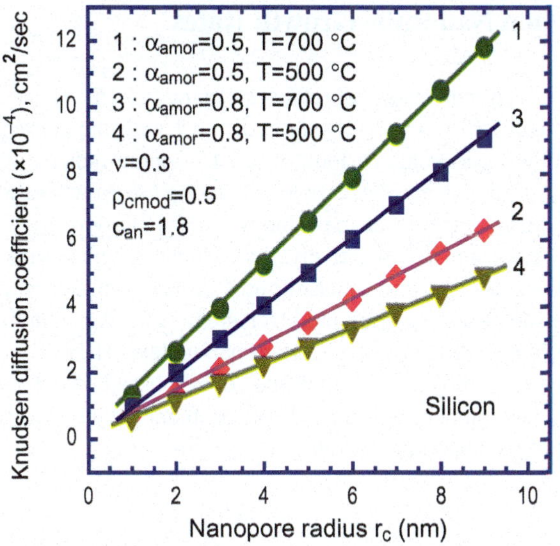

Fig. 15.10 Knudsen diffusion coefficient D_{KND} of $R_S \equiv$Si varying with nanopore radius r_{c0} for two different values of the amorphicity $\alpha_{\mathrm{amor}} = 0.5$ and $\alpha_{\mathrm{amor}} = 0.8$, and for two different temperatures: $T = 500$ °C and $T = 700$ °C, respectively.

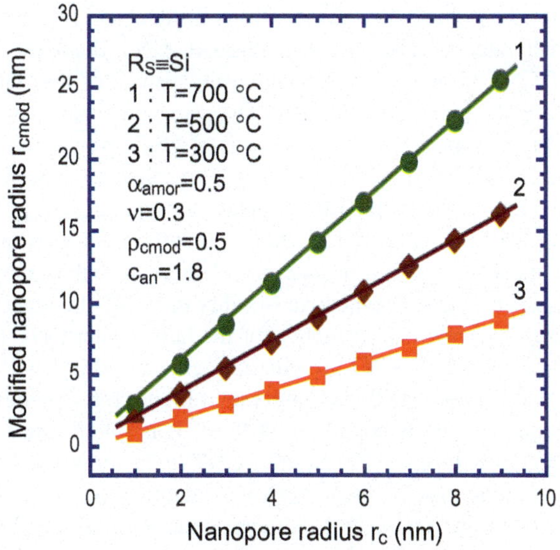

Fig. 15.11 Modified nanopore radius r_{cmod} varying as function of the nanopore radius r_{c0} for three different temperatures $T = 700$ °C, 500 °C, and 300 °C, respectively and for $\zeta_{cn} = 1.8$

15.11 Carbon Nanotube Growth Rates

The nanotube growth rates were calculated by using (15.15). These growth rates are compared in Figs. 15.14, 15.15, 15.16, and 15.17 with the corresponding growth rates obtained by experiments. A look of (15.15) would indicate that the Knudsen diffusivity D_{KND} is the primary parameters dictating the growth rate G_{KND}. And this diffusivity D_{KND}, as function of the effective amorphicity $\alpha_{\mathrm{amoreff}}$ depends on the parameters ρ_c, ν, η_0, and b, respectively. All of them, together with ζ_{KND} for calculated growth rates, are shown in these figures or stated in their captions. They are optimized ones arrived at by trial and error method. It was found that the optimal values of all of them are the most important parameters. They dictate the calculated results to be close to the experimental ones over a broad, if not the whole, experimental range. The temperatures and nanotube diameters used for the calculations were identical to the experimental ones.

15.11.1 Variation of CNT growth rate with CNT diameter

First Givargizov [22] and then Schmidt et al. [23] demonstrated that Si nanowire growth rate, particularly of relatively small-diameter nanowires, increases with increase in nanowire diameter. More recently, Sakurai et al. [24] showed that carbon nanotube growth rate increases with increase in CNT diameter. We performed calculations of CNT growth rates for increasing values of the outer CNT diameter D_{NTO} keeping the inner CNT diameter D_{NTI} constant. The results are presented in Fig. 15.12. We also performed calculations of CNT growth rate as function of the CNT diameter width $D_{\mathrm{NT}} = D_{\mathrm{NTO}} - D_{\mathrm{NTI}}$. The calculations were performed for three different values of the nanoparticle surface amorphicity α_{amor}. The results are presented in Fig. 15.13. Figure 15.12 indicates that the CNT growth rate increases, reaches a peak, and then gradually decreases with increase in the nanoparticle surface amorphicity α_{amor}. It is true irrespective of what the CNT diameter is. The rate of increase of the growth rate is also higher for the larger outer CNT diameter D_{NTO}. Figure 15.13 indicates that the CNT growth rate increases linearly with increase in the CNT diameter width D_{NT}. And this growth rate is higher for higher effective surface amorphicity $\alpha_{\mathrm{amoreff}}$ of the nanoparticle surface. It is interesting to note that the variation of CNT growth rate as function of CNT diameter width D_{NT}, as obtained by the present calculations, is linear. And it is identical to the linear variation of CNT growth rate as function of CNT diameter width D_{NT} obtained by experiments of Sakurai et al. [24]. This growth rate is higher for higher effective amorphicity $\alpha_{\mathrm{amoreff}}$ because higher effective amorphicity leads to higher porosity, higher HET concentration, and larger surface roughness.

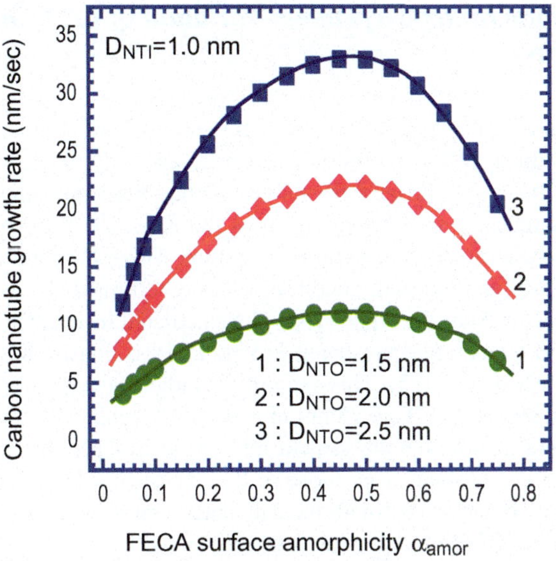

Fig. 15.12 Carbon nanotube growth rate varying with the nanoparticle surface amorphicity α_{amor} for three different outermost CNT diameter D_{NTO}, but keeping the innermost CNT diameter D_{NTI} fixed. Various parameters used for the calculations are $\rho_{\text{cmod}} = 0.5$, $\nu = 0.5$, $\eta_0 = 1.0$, carbon atom radius $r_A = 0.06$ nm, $T_0 = 300$ °C, $T = 650$ °C, $b = 2.0$, $\zeta_{\text{KND}} = 4.8191 \times 10^{-19}$, and $r_c = r_{cn}\alpha_{\text{amoreff}}$, where $r_{cn} = 0.15$ nm

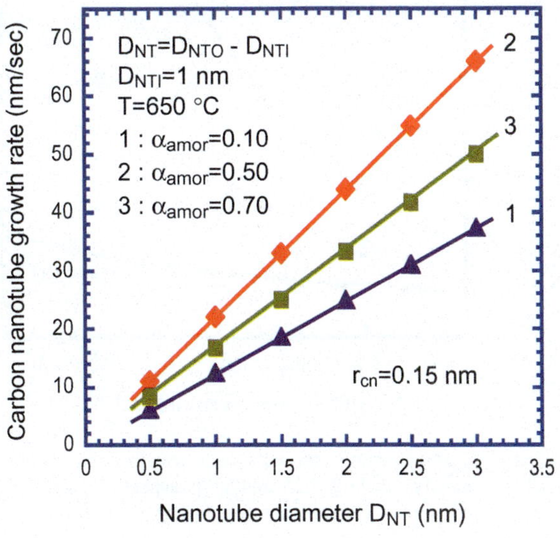

Fig. 15.13 Variation of carbon nanotube growth rate with the outermost CNT diameter D_{NTO} for three different values of nanoparticle surface amorphicity α_{amor}. Various parameters used for the calculations are $\rho_c = 0.5$, $\nu = 0.5$, $\eta_0 = 1.0$, carbon atom radius $r_A = 0.06$ nm, $T_0 = 300$ °C, $T = 650$ °C, $b = 2.0$, $\zeta_{\text{KND}} = 4.8191 \times 10^{-19}$, and $r_c = r_{cn}\alpha_{\text{amoreff}}$, where $r_{cn} = 0.15$ nm

15.11.2 Temperature-Dependent Variation of CNT Growth Rate

We studied the effect of temperature on CNT growth rate. The calculations were performed for three different values of the parameter *b*. A look at Figs. 15.4a, b would indicate that the Knudsen diffusion, which has direct implication on nanotube growth rate, is sensitive to temperature; both the Knudsen diffusion and the CNT growth rate increase with increase in temperature. The rate of increase is, though, different under different growth conditions. It is determined at least in part by the parameter *b* [see (A.5) of the Appendix]. The calculated increase in the growth rate with increase in temperature presented in Fig. 15.14 is consistent with that of the experiment by Chen et al. [25]. This experiment indicates that the MWCNT growth rate is 0.15 nm/min at 280 °C, 0.25 nm/min at 350 °C, and 0.40 nm/min at 570 °C. The experimental results for the growth rate are also dependent on experimental conditions; and it is evident from those of the experiment by Lee et al. [26]. The results by Lee et al. agree well with the calculated result for only an optimal value of *b*, namely $b = 5.25$, but not for $b = 4.00$ or $b = 6.00$.

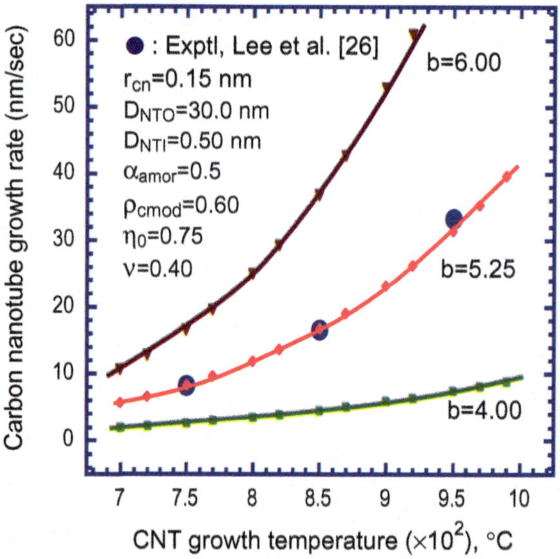

Fig. 15.14 Dependence of carbon nanotube growth rate on growth temperature *T* for three different values of the parameter *b*, namely $b = 4.00$, 5.25, and 6.00, respectively, and for $\zeta_{KND} = 0.2365 \times 10^{-21}$. Experimental data of Lee et al. [26] are presented as small solid circles

15.11.3 CNT Growth Rates Dependent on Growth Duration

Experimental studies of CNT growth rate as function of growth time were carried out in the past by Geohegan et al. [27] and by Lin et al. [28]. We performed theoretical study of the CNTgrowth rate as function of growth time. Calculations were performed for growths with two different sets of growth parameters. These parameters included growth temperatures and CNT diameters. While, for one growth, growth temperature, outer CNTdiameter and inner CNT diameter were identical to those by Geohegan et al. [27], for the other growth, growth temperature, outer CNT diameter and inner CNTdiameter were identical to those by Lin et al. [28]. Almost all other parameters for the two growths were identical. The results are presented in Figs. 15.15 and 15.16. They may suggest that the calculated results are in good agreements with the available experiments.

It would be noted that the initial growths were slow in both the cases. There was no discernible elongation of CNTs during the initial stages. And this was due to an incubation period during which the surface amorphicity of the nanoparticle surface was very low meaning that the nanoparticle surface was essentially non-porous with a few HETs formed on this surface. Decomposition of the precursor on this surface was consequently slow. However, increases in growth rates were faster during the next stage of growths. These were followed by very rapid increases in growth rates, which led the growths to have peaks. The growths during these periods were due to higher effective surface amorphicity, higher porosity of the surface, and higher density of HETs on this surface. Incorporation/attachment of carbon adatoms was therefore

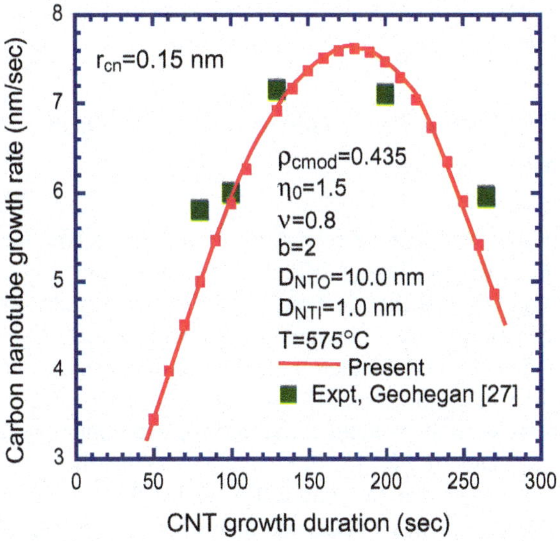

Fig. 15.15 Dependence of carbon nanotube growth rate on growth duration for $\zeta_{\mathrm{KND}} = 3.2694 \times 10^{-20}$. Experimental data of Geohegan et al. [27] are presented as small solid rectangles

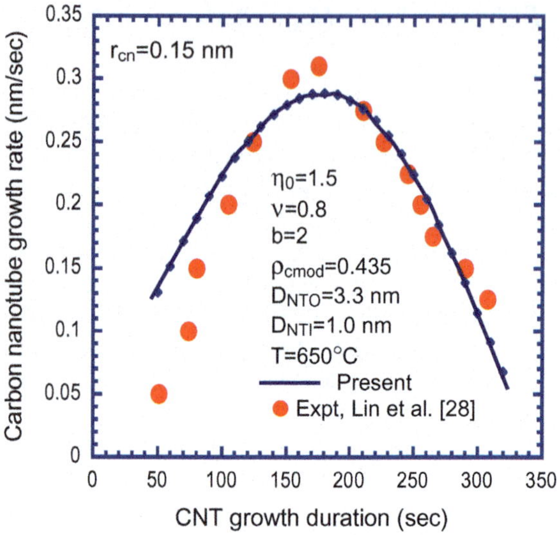

Fig. 15.16 Dependence of carbon nanotube growth rate on growth duration for $\zeta_{KND} = 0.3632 \times 10^{-20}$. Experimental data of Lin et al. [28] are presented as small solid circles

higher. Following the peak growths, there were reductions in the growth rates, which may be attributed to surface poisoning. This surface poisoning was actually the accumulation of the carbon adatoms on the nanoparticle surface to the extent that this accumulation blocked (closed) the nanopores on the nanoparticle surface. As a result, there was gradual decrease in growth rate with increase in growth time.

15.11.4 CNT Growth Rates Dependent on Precursor Flow Rate

Recall that the R_S ($R_S \equiv C$) species released from their precursor are responsible for the CNT growths. We calculated CNT growth rates as function of the precursor flow rate. We assumed that the precursor flow on nanoparticle surface is directly related to surface amorphicity of the nanoparticle surface [e.g., C_2H_2 concentration $= 10^2 \times \alpha_{amor}$) and performed calculations at the same growth temperature as the one for experiment by Chhowalla et al. [4]. The outer and inner CNT diameters chosen for the calculations were also identical to those for experiment by Chhowalla et al. The results are presented in Fig. 15.17. One may note that these results are in good agreement with the experiment by Chhowalla et al. [4]. To be more specific, the CNT growth rate increases, reaches a peak, and then decreases with gradual increase in the precursor flow.

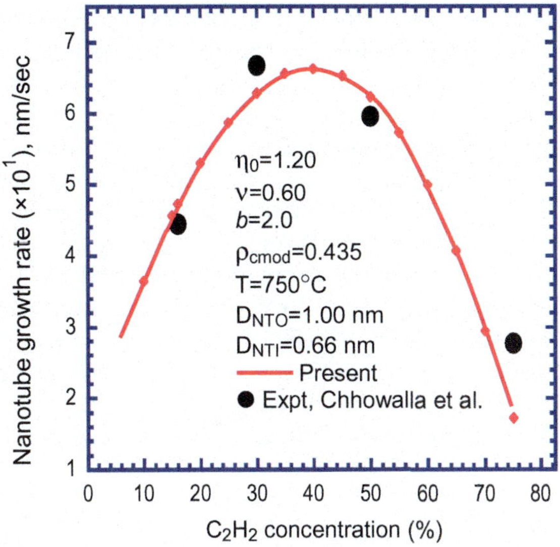

Fig. 15.17 Dependence of carbon nanotube growth rate on the C_2H_2 precursor flow rate for $\zeta_{KND} = 4.0215 \times 10^{-17}$. The experimental data of Chhowalla et al. [4] are presented as small solid circles

15.11.5 Inference

Figures 15.10 and 15.11 highlight the impact of nanoparticle surface amorphicity on the nanopore radius r_c and also on the Knudsen diffusion coefficient D_{KND}. Taking the impact of nanoparticle surface amorphicity on the Knudsen diffusion coefficient D_{KND} into account, we studied CNT growth rates under several different growth conditions and as functions of growth parameters such as CNT diameter, growth temperature, growth duration, and precursor flow rate. The said growth rates depend obviously on parameters such as growth temperature, chamber pressure, ambient, precursor flow rate, etc., which are different for growths of different nanotubes. These parameters significantly influence the growth rates. All theoretical models, including the first principles *ab initio* and molecular dynamics models, depend on parameters. And some of these parameters are empirical or semiempirical. These parameters (e.g., η_0, ν, b, and ρ_c) for the present calculations are also empirical. Yet, as apparent from Figs. 15.14, 15.15, 15.16, and 15.17, only one single set of these parameters could yield theoretical results in agreement with experiments over broad experimental range. For example, calculated CNT growth rates presented in Fig. 15.14 are in agreement with experiment [26] over the entire range of experiment. And these calculated growth rates were obtained by a single set of parameters. Similarly, calculated CNT growth rates, presented in Fig. 15.17, are in agreement with experiment [4] over the entire range of experimental C_2H_2 concentration (%); these growth rates were obtained by one single set of parameters. All these would not happen without the model taking into account the mechanistic basics of experimental nanotube growths, or unless the theoretical set of parameters represents, at least approximately, the corresponding experimental set of parameters.

A close look of (15.15) would indicate that it is an extremely simple equation. The primary element of this equation is the Knudsen diffusion coefficient D_{KND}. Taking the effective amorphicity $\alpha_{amoreff}$ in terms of parameters such as ν, η_0, and b into account, this D_{KND} probably correctly tends to suit the prevailing condition of the nanoparticle surface during growth. And as a result, D_{KND} probably governs the growth rate G_{KND}. A close correspondence of the calculated results with the experimental ones does not certify the novelty of (15.15). Instead it suggests the appropriateness of the Knudsen diffusion equipped with the concept of effective surface amorphicity, rather than any other diffusion, of the R_S species through nanopores of the R_L species for growth. It demonstrates that the R_L species of the nanoparticle must be nanoporous in order to correctly guide the nanomaterials growth.

References

1. M. Rayner, G. Tragardh, C. Tragardh, The impact of mass transfer and interfacial expansion rate on droplet size in membrane emulsification processes. Colloids Surf. A: Physicochem Eng. Aspects **266**, 1–17 (2005)
2. M. Sobanska, K. Klosek, J. Borysiuk, S. Kret, G. Tchutchulasvili, S. Gieraltowska, Z.R. Zytkiewicz, Enhanced catalyst-free nucleation of GaN nanowires on amorphous Al_2O_3 by plasma-assisted molecular beam epitaxy. J. Appl. Phys. **115**, 043517 (2014)
3. S. Pisana, M. Cantoro, A. Parvez, S. Hofmann, A.C. Ferrari, J. Robertson, The role of precursor gases on the surface restructuring of catalyst films during carbon nanotube growth. Physica E **37**, 1–5 (2007)
4. M. Chhowalla, K.B.K. Teo, C. Ducati, N.L. Rupesinghe, G.A.J. Amaratunga, A.C. Ferrari, D. Roy, J. Robertson, W.I. Milne, Growth process conditions of vertically aligned carbon nanotubes using plasma enhanced chemical vapor deposition. J. Appl. Phys. **90**, 5308–5317 (2001)
5. P.J. Newby, B. Canut, J.-M. Bluet, S. Gomès, M. Isaiev, R. Burbelo, K. Termentzidis, P. Chantrenne, L.G. Frechette, V. Lysenko, Amorphization and reduction of thermal conductivity in porous silicon by irradiation with swift heavy ions. J. Appl. Phys **114**, 014903 (2013)
6. D. Lee, Studies on hydrogen selective silica membranes and the catalytic reforming of CH_4 with CO_2 in a membrane reactor, *Doctoral Thesis, Chemical Engineering* (Virginia Polytechnic Institute and State University, Blackesburg, Virgina, 2003), Chap. 3
7. J. Cermak, H. Mehrer, Tracer diffusion of ^{14}C in austenitic NiFeCr alloys. Acta Metall. Mater. **42**, 1345–1350 (1994)
8. H. L. Lira, R. Paterson, New and modified anodic alumina membranes Part III. Preparation and characterization by gas diffusion of 5 nm pore size anodic alumina membranes. J. Membr. Sci. 206, 375–387 (2002)
9. D.H. Davis, Monte Carlo calculation of molecular flow rates through a cylindrical elbow and pipes of other shapes. J. Appl. Phys. **31**, 1169–1176 (1960)
10. D. Avnir, D. Farin, P. Pfeifer, Molecular fractal surfaces. Nature (London) **308**, 261–263 (1984)
11. M.-O. Coppens, Characterization of fractal surface roughness and its influence on diffusion and reaction. Colloid Surf., A **187–188**, 257–265 (2001)
12. M. von Smoluchowski, Zur kinetischen theorie der transpiration und diffusion verdünnter gase. Ann Phys. (Leipzig) **4F**, 1559–1570 (1910)

13. S.B. Santra, B. Sapoval, Interaction of ballistic particles with irregular pore walls, Knudsen diffusion, and catalytic efficiency. Phys. Rev. E **57**, 6888 (1998)
14. M.-O. Coppens, G. F. Froment, Diffusion and reaction in a fractal catalyst pore—II. Diffusion and first-order reaction. Chem. Eng. Sci. **50**, 1027–1039 (1995)
15. T. Liang, Q. Li, Accurate modeling of Knudsen diffusion in nanopores using a physical-based boundary model. J. Appl. Phys. **126**, 084304 (2019)
16. O. Geier, S. Vasenkov, J. Kärger, Pulsed field gradient nuclear magnetic resonance study of long–range diffusion in beds of NaX zeolite: evidence for different apparent tortuosity factors in the Knudsen and bulk regimes. J. Chem. Phys. **117**, 1935 (2002)
17. K. Malek, M.-O. Coppens, Knudsen self- and Fickian diffusion in rough nanoporous media. Appl. Phys. Lett. **119**, 2801–2803 (2003)
18. S. Noor Mohammad, VQS (vapor-quasiliquid-solid, vapor-quasisolid-solid) mechanism for the catalyst-free and catalyst-mediated non-eutectic syntheses of single-crystal nanowires. J. Appl. Phys. **120**, 084307 (2016)
19. S. Noor Mohammad, Bimetallic-catalyst-mediated syntheses of nanomaterials (nanowires, nanotubes, nanofibers, nanodots, etc) by the VQS (vapor–quasiliquid–solid, vapor-quasisolid-solid) growth mechanism. J. Phys. D: Appl. Phys. **49**, 495304 (2016)
20. H.-C. Hsu, C.-S. Cheng, C.-C. Chang, S. Yang, C.-S. Chang, W.-F. Hsieh, Orientation-enhanced growth and optical properties of ZnO nanowires grown on porous silicon substrates. Nanotechnology **16**, 297 (2005)
21. A.M. Cassell, J.A. Raymakers, J. Kong, H. Dai, Large scale CVD synthesis of single-walled carbon nanotubes. J. Phys. Chem. B **103**, 6484–6492 (1999)
22. E.I. Givargizov, Fundamental aspects of VLS growth. J. Cryst. Growth **31**, 20–30 (1975)
23. V. Schmidt, S. Senz, U. Gösele, Diameter dependence of the growth velocity of silicon nanowires synthesized via the vapor-liquid-solid mechanism. Phys. Rev. B **75**, 045335 (2007)
24. S. Sakurai, M. Inaguma, D.N. Futaba, M. Yumura, K. Hata, A fundamental limitation of small diameter single-walled carbon nanotube synthesis—a scaling rule of the carbon nanotube yield with catalyst volume. Materials **6**, 2633–2641 (2013)
25. M. Chen, C.-M. Chen, S.-C. Shi, C.-F. Chen, Low-temperature synthesis of multiwalled carbon nanotubes by microwave plasma chemical vapor deposition using CH_4-CO_2 gas mixture. Jpn. J. Appl. Phys. **42**, 614–619 (2003)
26. C.J. Lee, J. Park, Y. Huh, J.Y. Lee, Temperature effect on the growth of carbon nanotubes using thermal chemical vapor deposition. Chem. Phys. Lett. **343**, 33–38 (2001)
27. D.B. Geohegan, A.A. Puretzky, I.N. Ivanov, S. Jesse, G. Eres, In situ growth rate measurements and length control during chemical vapor deposition of vertically aligned multiwall carbon nanotubes. Appl. Phys. Lett. **83**, 1851–1853 (2003)
28. M. Lin, J. P. Y. Tan, C. Boothroyd, K. Ping Loh, E. S. Tok, Y.-L. Foo, Direct observation of single-walled carbon nanotube growth at the atomistic scale, Nano Lett. 6, 449–452 (2006)

Chapter 16
The General, Versatile Growth Mechanism

Abstract A general, versatile growth mechanism for the growths of nanomaterials has been proposed. Basics of the material phases and of the multiple phases participating in nanomaterials growths have been investigated a priori. Then important hallmarks of nanomaterials growth, namely factors influencing nanomaterials growths, optimal phase for SUBSANO-mediated growths, optimal phase for METANO-mediated growths, uniqueness of the nanoparticle phase, and characteristics of SECINI formed during growths at a temperature T have been described. Finally, the basic foundation of the general, universal mechanism called the αQS mechanism has been laid down. This mechanism is essentially the generalized VQS mechanism applicable also to the solution phase and the solid phase growths. It has been systematically demonstrated why the VLS mechanism for growth at the eutectic temperature T_E is a special case of the αQS mechanism, VSS mechanism for high-temperature ($T > T_E$) nanomaterials growths is a special case of the αQS mechanism, VSS mechanism for low- temperature ($T < T_E$) nanomaterials growths is a special case of the αQS mechanism, the MET-free VS mechanism for nanomaterials growth is a special case of the αQS mechanism, the solution phase SoLS and SFLS mechanisms are special cases of the αQS mechanism, the solid phase SLS mechanism is a special case of the αQS mechanism, the oxide-assisted growth mechanism is a special case of the VQS mechanism, and the self-catalytic growth mechanism is a special case of the αQS mechanism. The SoSS and SFSS mechanisms, which are the solution-phase analogs of the VSS mechanism, are also the special cases of the αQS mechanism. Finally, it has been shown that BN nanotube growths and the growths of semiconductor nanobelts are actually by the VQS mechanism.

16.1 Generality of Growth Mechanism

The basic principles of growths by the VQS mechanism were laid down in Chap. 12. General terminologies such as FECANO, METANO, and SUBSANO for nanoparticles, and of the R_L species were defined in Chap. 3. While FECANO is the FECA nanoparticle, METANO is the metal nanoparticle, and SUBSANO is the substrate

S. N. Mohammad, *Synthesis of Nanomaterials*, Springer Series in Materials Science 307, https://doi.org/10.1007/978-3-030-57585-4_16

nanoparticle. SECINI and SECINI0 were also defined in Chap. 3, pre-nucleation and pro-nucleation stages of growths were defined in Chap. 4, but NP1 and NP2 nanoparticles were defined in Chap. 12. All these, together with the R_S source species defined in Chap. 1, will be used in the following, while trying to describe the possibility of a general growth mechanism for the growths of nanomaterials (nanocrystals).

16.1.1 Basics of Material Phases

A material can be in the solid, liquid, or vapor phase (macrostate) [1]. While in the solid phase, the atoms or molecules of the material are closely bound to one another by molecular forces. As a result, it has a shape and a volume. The volume is fixed by the shape. While in the liquid phase, the molecular forces in the material are weaker than that in the solid phase. So, the liquid takes the shape of its container. While in the gas phase, the molecular forces of the material are very weak. As a result, the gas filling its container takes both the shape and the volume of the container. In addition to the three normal phases (macrostates) mentioned above, the material can also have another phase, called the plasma phase (macrostate), which takes place at very high temperature. At this temperature, the atoms break down with the electrons stripped off from their orbit(s) around the nucleus leaving a positively charged ion behind. There occurs thus a mixture of neutral atoms, free electrons, and charged ions, which is called plasma.

16.1.2 Multiple Phases Participating in Nanomaterials Growths

We noted that the most essential need for the X_mY_n nanomaterials growth is to first create the source species (e.g., $R_S{\equiv}X$ and $R_S{\equiv}Y$ species). They may be created in one of the six phases (e.g., macrostates), viz. $\wp 1$ (vapor), $\wp 2$ (liquid), $\wp 3$ (quasiliquid or quasisolid), $\wp 4$ (solution or fluid), $\wp 5$ (supercritical fluid), and $\wp 6$ (crystalline solid) macrostates (see Chap. 1). The source species are currently created in one of the four phases (macrostates), namely the vapor phase, solid phase, solution (fluid) phase, and supercritical fluid phase. This phase may be called the source species phase, source phase, or the α phase (macrostate). This α phase may be different from the said vapor ($\wp 1$), solution ($\wp 4$), supercritical fluid ($\wp 5$) or solid ($\wp 6$) phase. It may, for example, be quasiliquid or quasisolid ($\wp 3$) phase. It may otherwise be plasma phase or quasi-vapor phase. Quasi-vapor phase is a phase in which vapor coexists with liquid. This liquid may even be just traces of liquid. Such a phase may be suitable for dipolar molecules (precursors) to be easily physisorbed on a solid, quasisolid, or quasiliquid surface. It may have other advantages. Irrespective of whether the source species phase (macrostate) is solution (fluid), solid, vapor, quasi-vapor, or plasma, the source species are always the R_S species; for example, $R_S \equiv X$ and $R_S \equiv Y$ species. Different phase (macrostate, medium) for source species is used just differently to yield the same $R_S \equiv X$ and $R_S \equiv Y$ species. We reiterate. The R_S source species are created keeping in mind that the procedure to create the source species may be different in different phases (media). This procedure to create the source species in the vapor phase is, for example, vastly different from that in the solid phase. The precursor(s), if any, used for this may also be different for different phases. We noticed that the R_L species of both the METANO and SUBSANO may be solid, liquid, or quasisolid (quasiliquid). Among them, the quasisolid (quasiliquid) METANOs (SUBSANOs) have the structure and properties described in Sect. 12.2 of Chap. 12. Various features of the VQS mechanism have been described in Sects. 12.1–12.15 of Chap. 12. It is widely believed that the R_L species is fully solid (e.g., $\xi_m = 0$), for example, for growths by the VS and the VSS mechanisms, and that it is fully liquid (e.g., $\xi_m = 1$) for growths by the VLS, SLS, and SoLS mechanisms. We argue nevertheless that, from practical point of view, the R_L species would rarely be fully solid or fully liquid. For instance, R_L species would be quasisolid, and not fully solid, even if $\xi_m = 0.00001$. It would again be quasiliquid, and not fully liquid, even if, for example, $\xi_m = 0.99999$. This means we may safely assume that the R_L species would always be quasisolid (quasiliquid). And hence, as shown schematically in Fig. 16.1, any of the growth mechanisms may, in general, correspond to α quasiliquid–solid phases, or in other words, to α quasisolid–solid phases, for growth. The general mechanism for growth could hence be called the αQS mechanism, where α stands for the phase of the R_S source species, for example, vapor, solid, or solution phase, or for the R_S species themselves regardless of how and in which phase they are created. In other words, α may stand for the R_S source species itself. As apparent from Fig. 16.1, the proposed αQS mechanism would hence be true even for the VS mechanism.

16.1.3 General Pathway for Nanomaterials Growths

Figure 16.1 shows the R_S and R_Y species generated only in three different media (phases, macrostates), namely the vapor medium, solid medium, and the solution medium. We stated earlier in Sect. 16.1.2 that there may also be other media suitable for the generation of these R_S and R_Y species. While the R_S and R_Y species generated in the vapor medium follow the $\aleph 1$ pathway, the R_S and R_Y species generated in the solid medium follow the $\aleph 2$ pathway, and the R_S and R_Y species generated in the solution (liquid) medium follow the $\aleph 3$ pathway. They all reach the point $\Im$. Then they approach $\wp 1$ or $\wp 2$ pathway to enter either the MET-generated quasiliquid (quasisolid) medium or to substrate-generated quasiliquid (quasisolid) medium for supersaturation, nucleation, crystallization and growth. Recall that, based on conventional belief, the MET-generated quasiliquid (quasisolid) medium is the liquid medium exhibiting droplet, and the substrate-generated (MET-free) quasiliquid (quasisolid) medium is the solid medium exhibiting no droplet. The figure presented above provides a general picture of the growth pathway in the sense that, irrespective of whether the R_S and R_Y species are generated in the vapor medium, solid medium, solution medium, or some other medium, they may enter the MET-generated R_L species via the $\aleph 1 \rightarrow \Im \rightarrow \wp 1$ pathway, $\aleph 2 \rightarrow \Im \rightarrow \wp 1$ pathway, $\aleph 3 \rightarrow \Im \rightarrow \wp 1$ pathway or some other similar pathway for nucleation and growth.

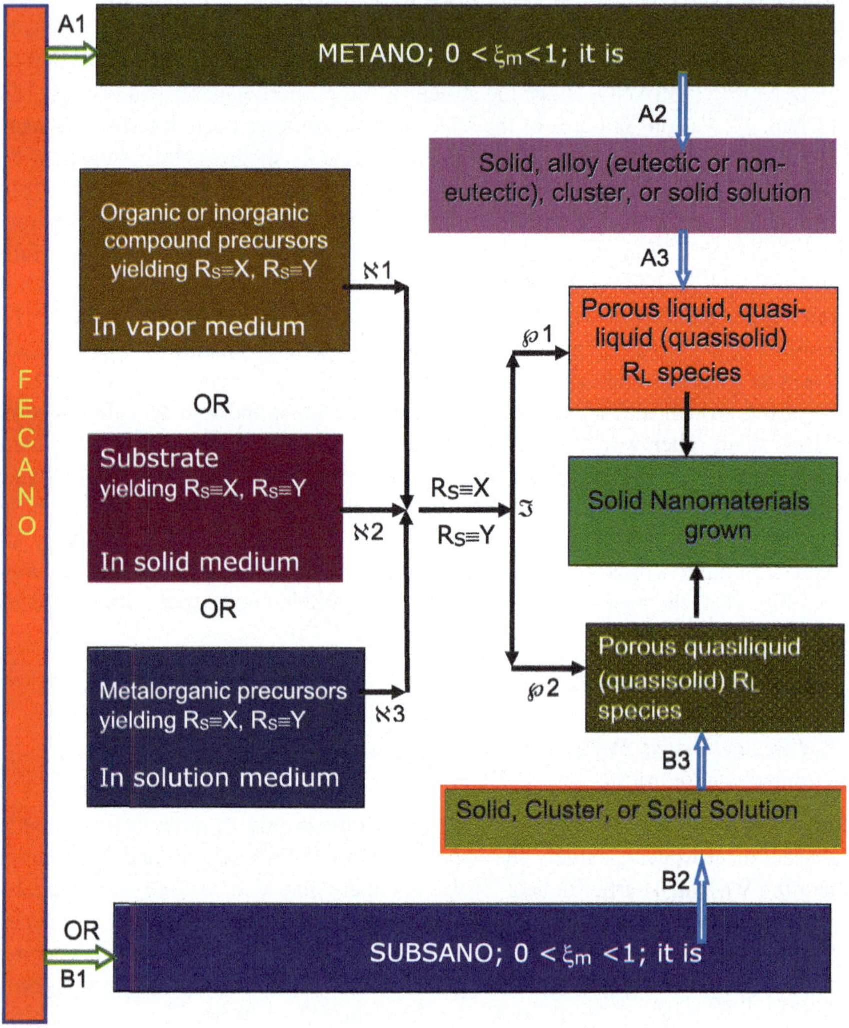

Fig. 16.1 Proposed general mechanism for nanomaterials (nanocrystals) growths applicable to all possible mechanisms including the VS mechanism. It may be called the αQS mechanism. Note that the pathways A1 → A2 → A3 lead to the formation of MET-based R_L species, but the pathways B1 → B2 → B3 lead to the formation of substrate-based R_L species. The substrate-based R_L species may be free from metal(s). This substrate-based R_L species will contain metal if this substrate is a metal substrate

They may otherwise enter the substrate-generated R_L species via the $\aleph 1 \rightarrow \Im \rightarrow \wp 2$ pathway, $\aleph 2 \rightarrow \Im \rightarrow \wp 2$ pathway, $\aleph 3 \rightarrow \Im \rightarrow \wp 2$ pathway or some other similar pathway for nucleation and growth.

16.2 Important Hallmarks

16.2.1 Factors Influencing Nanomaterials Growths

Temperature is a crucial factor for nanomaterials growth in growth chamber. (1) Precursor(s) of the source species are introduced into the growth chamber as the temperature of this chamber is gradually increased for growth on nanoparticle (METANO or SUBSANO) surface. (2) The thickness t_{nano} of the nanoparticle is judiciously optimized before growth. An R_L species is thereafter created on the nanoparticle surface during the pre-nucleation stage of growth. Depending on the choice of growth mechanism, the composition of this R_L species would vary. It would, for example, be R_L≡(MET, X) eutectic alloy molten (e.g., $\xi_m = 1$) at the eutectic temperature $T = T_E$. The depth (thickness) of the R_L species would be δ_{amor}. Depending on the nanoparticle thickness, it would be $\delta_{amor} \leq t_{nano}$. (3) Under certain growth condition (pressure, ambient, doping, fluctuation, and vibration of the nanoparticle surface, etc.), the increase in temperature during growth may accompany a change in the material phase of the R_L species. With all other growth parameters unaltered, different temperature may hence bring about different phase of the R_L species. Such a phase may be a combination of a number of phases. (4) During increase in temperature T, different phase(s) of the R_L species formed at different temperature may have different composition, morphology, and materials properties. They may nearly be solid (e.g., $\xi_m = 0.00001$) for both METANO and SUBSANO. They may otherwise be liquid (e.g., ξ_m=1) for METANO; cluster, solid solution, or non-eutectic alloy for SUBSANO, but cluster, solid solution, eutectic alloy, or non-eutectic alloy for METANO.

16.2.2 Optimal Phase for SUBSANO-Mediated Growths

If the nanoparticle is a SUBSANO, and the R_L species is a cluster and/or solid solution, there is a certain temperature T during gradual heating of the chamber during growth at which the phase of the R_L species is optimal and hence most suitable for nanomaterials growth. There is a certain suitable temperature T_{opt} for it to be stable. It may lose its SUBSANO characteristics if the temperature $T > T_{opt}$. The above-mentioned phase has (a) disturbed (disordered), amorphous (semi-amorphous) surface of effective amorphicity $\alpha_{amoreff}$; grains and grain boundaries may or may not be present on the surface; (b) rough, coarsened surface, (c) high concentration of HETs instrumental for precursor decomposition at a temperature lower than that required for thermal decomposition, (d) voids, vacancies, dislocations, etc., and hence, hillocks, pits, sharp edges, etc. It may have dipole moment. It is porous and quasiliquid (quasisolid) (e.g., $0 < \xi_m < 1$) with the highest possible porosity, best possible melting condition [even if it is quasiliquid (quasisolid)], and the largest possible density of the most effective HETs. All these are not arbitrary. They are judiciously planned, organized, controlled, optimized, and interdependent; they commensurate with the effective amorphicity $\alpha_{\mathrm{amoreff}}$. The nanoparticle is therefore the NP2 nanoparticle. We emphasize that the average depth of the nanopores from the top surface may be δ_{amor}, and in general, $\delta_{\mathrm{amor}} < t_{\mathrm{nano}}$, although under certain growth conditions, it may be possible to achieve $\delta_{\mathrm{amor}} \approx t_{\mathrm{nano}}$. The characteristics of the phase of the R_L species at the growth temperature T as described above would depend on the composition and morphology of the said cluster, alloy, or solid solution at the said temperature T.

16.2.3 Optimal Phase for METANO-Mediated Growths

If the nanoparticle is a METANO, the phase of the R_L species would exhibit the same characteristics as stated above for SUBSANO. In order for it to be most suitable for nanomaterials growth,

1. The R_L species of the METANO would though be a quasiliquid (quasisolid) non-eutectic alloy realized at a temperature $T < T_E$.
2. This R_L species may, otherwise, be liquid (e.g., molten) eutectic alloy ($\xi_m = 1$) created at a temperature higher than $T = T_E$. The eutectic alloy would be formed if there are preferably no contaminants present in the R_L species composition or there is an appropriate mole fractions of MET and $R_S = X$ species needed for the formation of $R_L \equiv$(MET, X) eutectic alloy at the eutectic temperature T_E.
3. This R_L species may, otherwise, be quasiliquid (quasisolid) non-eutectic alloy (e.g., cluster or solid solution) formed at the temperature of $T = T_E$. It may have two grains separated by a grain boundary. It may have multiple grains separated by grain boundaries. If it has two crystalline, non-defected grains separated by a defected grain boundary, it may produce nanobelt along the grain boundary. If it has multiple porous, amorphous grains separated by non-defected grain boundaries, it may produce several nanowires, each of them in a grain.
4. This R_L species may, otherwise, be a solid-core molten-shell core–shell non-eutectic alloy formed at a temperature $T < T_E$, $T \approx T_E$, or $T > T_E$. Such R_L species would produce nanotubes.
5. This R_L species may, otherwise, be quasisolid-core molten-shell core–shell non-eutectic alloy formed at a temperature $T < T_E$, $T \approx T_E$, or $T > T_E$. Such R_L species would produce nanotube filled with nanowire.

6. This R_L species may, otherwise, be quasiliquid-core molten-shell core–shell non-eutectic alloy formed at a temperature $T < T_E$, $T \approx T_E$, or $T > T_E$. Such R_L species would produce nanotube filled with nanowire.
7. This R_L species may, otherwise, be quasiliquid-core solid-shell core–shell non-eutectic alloy formed at a temperature $T < T_E$, $T \approx T_E$, or $T > T_E$. Such R_L species would produce nanowires of dimension smaller than that of the R_L species itself.
8. The R_L species may be a non-eutectic alloy (e.g., cluster or solid solution) formed at a temperature $T > T_E$ if the presence of contaminants in the R_L species composition and/or a lack of appropriate mole fraction of (h1) MET, (h2) $R_S{\equiv}X$ species and/or $R_S{\equiv}Y$ species in it did not permit the formation of molten (MET, X) eutectic alloy at the eutectic temperature T_E.
9. The growth temperature for MET-mediated growth can be $T > T_E$ if the substrate/METANO interface is not smooth. The stress at this interface permeates deep into the METANO nanoparticle causing disorders in the nanoparticle. The R_L species of the nanoparticle is consequently affected. It is influenced, as well, by thermodynamic imbalance (particularly during annealing).

The voids, vacancies, dislocations, etc., in the R_L species may give rise to hillocks, pits, sharp edges. This R_L species is generally a non-eutectic alloy, cluster, or solid solution. In this context, it may be noted that Kodambaka et al. [2] observed both VLS and VSS growth taking place at the same temperature and on the same nanoparticle if the pressure and the thermal history of the catalyst are different in different growth environments.

16.2.4 Uniqueness of the Nanoparticle Phase

We remind that the phase of the R_L species of a nanoparticle under a certain growth condition (temperature, pressure, thermodynamics, ambient, contamination, etc.) can be a MET-containing eutectic alloy, MET-containing non-eutectic alloy, or MET-free non-eutectic alloy. And it would be unique. While the eutectic alloy can be droplet, the non-eutectic alloy can be porous cluster, solid solution, etc. The characteristics of a phase are reflected from the composition and morphology of the alloy of this phase under the said growth condition, particularly temperature and pressure. They together dictate and determine the value of SECINI and ξ_m. The characteristics of this phase for non-eutectic alloy include disturbed (disordered) lattice, effective surface amorphicity, surface coarsening, surface porosity, and surface melting of the R_L species surface. They also include HETs generated on this surface. There may also be dipole moment on this surface. The SECINI and ξ_m of the phase can have distinctly different values if one or more of the following is different for the phase. These are lattice structure, surface disorder, effective surface amorphicity, surface coarsening, surface porosity, surface melting, dipole moments, HETs, etc.

16.2.5 Characteristics of SECINI Formed During Growths

SECINI≡SECINI0 and $\xi_m = 1$ if the R_L species (e.g., droplet) is stable and has eutectic phase. Also this phase has the highest porosity for the highest, smoothest, and uninterrupted diffusion of the $R_S = X$ and $R_S = Y$ species through this R_L species (e.g., droplet). SECINI≡SECINI0, $\xi_m \approx 1$, and the porosity is close to unity if the phase of the R_L species is a non-eutectic phase, and this phase corresponds to disturbed (disordered) lattice, effective surface amorphicity, surface coarsening, surface porosity, surface melting, HETs, and possibly dipole moment. Also, (1) the HETs cause the most effective (efficient) low-temperature decomposition of the precursor(s) releasing the $R_S = X$ and $R_S = Y$ species on the R_L species surface, (2) the surface disturbance yields surface roughness of the optimal density and height, (3) surface melting ensures quasiliquid (quasisolid) state of the R_L species, (4) dipole moment provides the highest level of reactivity for the adsorption of the $R_S = X$ and $R_S = Y$ species on the R_L species surface, and (5) they all together cause the highest, smoothest, and uninterrupted diffusion of the $R_S = X$ and $R_S = Y$ species through the R_L species.

Depending, for example, on temperature and/or pressure, the SECINI and ξ_m of an existing non-eutectic phase of the R_L species can be such that, they increase with increase in temperature (pressure), reache the peak, correspond to SECINI=SECINI0 and $\xi_m = 1$ at a certain optimal temperature (pressure), and then decrease with further increase in temperature (pressure). The droplet of a eutectic phase can be unstable at high temperature, e.g., at a temperature $T > T_E$. Because of this, SECINI and ξ_m of also the eutectic phase of an R_L species can be such that, they increase with increase in temperature (pressure), reach the peaks yielding SECINI≡SECINI0 and $\xi_m = 1$ at a certain optimal temperature (pressure), and then decrease with further increase in temperature (pressure). This means there is an optimal and plausibly narrow window for the fully uninterrupted diffusion of the $R_S = X$ and $R_S = Y$ species through the R_L species. There can hardly be any nanomaterials growth out (e.g., below or above) of this window.

16.3 Illustrations of Hallmarks

The growth of GaN nanowires takes place generally at 900 °C. Employing alloys of In, Co, Ni, Fe, $Ni_{1-z}Co_z$, and $Fe_{1-z}Co_z$, or the metal complexes of ferrocene, iron phthalocyanine, cobaltphthalocyanine, and nickel phthalocyanine, Chen et al. [3] grew GaN nanowire at ~910 °C. These alloys and metal complexes formed clusters at the growth temperature and acted as catalyst for growth. Obviously, these clusters were non-eutectic alloys or solid solutions. The EDX analysis confirmed it. It showed that nanowires contained only gallium and nitrogen, but the nanowire tips contained the catalyst metal, gallium, and nitrogen. The EDX data thus suggested that the metal catalysts were miscible with Ga and N yielding complex alloys and clusters at the

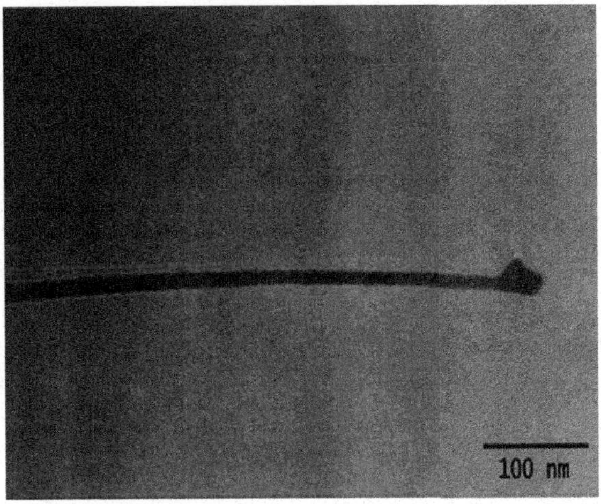

Fig. 16.2 TEM image of a single GaN nanowire reproduced from Chen et al. [3] with copyright permission from the publisher. It shows an irregular-shaped nanoparticle terminated on the tip of a nanowire prepared by using cobalt phthalocyanine as a catalyst and had a diameter of about 16 nm

growth temperature of 910 °C. It also suggested that, due to lower melting point (~29.76 °C), Ga together with N, had good miscibility with the catalyst atoms; other elements from well-dispersed mixture of constituents participated in the cluster. The cluster thus became semimolten probably due to low melting point of Ga. The melting (semi-melting) did not result from the formation of eutectic effect. It was indeed true as evidenced by the irregular non-spherical shape and the morphology of the material content present near the nanowire tip. It is shown in Fig. 16.2. A molten eutectic alloy droplet had not been formed. It is evident from the nanowire tip having no hemispherical shape. Although the growth was believed to be carried out by the VLS mechanism, it was actually carried out by the VQS mechanism. Chen et al. [3] could not grow these nanowires at $T < 850$ °C and at $T > 1050$ °C simply because the said non-eutectic alloys and alloy complexes did not fulfill the requirements of VQS growth at these temperatures, as elaborated in Chap. 12.

Surface coarsening, porosity, and melting (semi-melting) of non-eutectic alloy are, at least approximately, related to one another and to the effective amorphicity α_{amoreff}. While the porosity, together with surface coarsening, as function of liquid (quasiliquid, quasisolid) state may be the lowest (suppose, zero) for a solid nanoparticle ($\xi_m = 0$), they may be high (suppose, close to unity, see Fig. 12.5) for a certain quasiliquid (quasisolid) condition of nanoparticle with ξ_{m} between 0 and 1, and the highest for molten condition (e.g., $\xi_m = 1$) optimally favored for growth. The porosity and melting condition of a eutectic alloy are also the highest for growth. Depending on the composition and morphology of the non-eutectic alloy of a certain phase, they are in between 0 (the lowest) and 1 (the highest) for this alloy.

16.4 The VLS Mechanism is a Special Case of the VQS Mechanism

We argue that the VLS mechanism is a special case of the VQS mechanism. And it would be best illustrated with the help of Au-catalyzed Si nanowire growth. The Au/Si binary phase diagram shown in Fig. 5.2a of Chap. 5 is of the simple eutectic type [4]; it has a eutectic point at the Si composition of about 20 at. % Si at a temperature of 363 °C. This eutectic temperature is lower than the melting point of pure Au by about 700 °C. It is lower than the melting point of pure Si by about 1050 °C. It is clear from Fig. 5.2a that the phase within the V-shaped region is the liquid phase obtained only with the appropriate amount of Si supply to Au. To be molten (e.g., $\xi_m = 1$), the composition of the (Au, Si) eutectic alloy must follow the liquidus line, and be on the Si side on the phase boundary, which is on the right-hand side of the liquid phase. And it would depend on the temperature T. While on this liquidus line, Si would, at a certain temperature T, continue to be dissolved into the (Au, Si) alloy until the composition reaches the end of the liquidus line. The composition may however reach beyond the liquidus line. Even though it would occur kinetically, it might be thermodynamically unstable. To overcome it, the alloy would tend to re-establish the stable composition of Si and Au on the liquidus line either through decrease in temperature or increase in Si content. The alloy would otherwise settle as a solid solution. Thus, if all other parameters, including pressure, are judiciously set, the (Au, Si) alloy would have $\xi_m = 1$ for the Si content of ~20 at.% at $T = 363$ °C, ~23 at.% at $T = 462$ °C, ~27 at.% at $T = 565$ °C, and ~32 at.% at $T = 684$ °C [4]. If all other parameters, including pressure, are not judiciously set, a stable (Au, Si) alloy, for example, at 684 °C, would have Si content less than ~32 at.%. And the (Au, Si) alloy thus formed would be non-eutectic and hence solid. Alternatively, if all other parameters, including pressure, are not judiciously set, a stable $R_L \equiv$(Au, Si) alloy exhibiting, for example, a Si content of ~32 at.% at $T < 684$ °C, would be non-eutectic and hence solid. The examples, cited above suggest that there is a very narrow window for the molten eutectic alloy to be formed for nanomaterials growths. $R_L \equiv$(MET, X) species would be eutectic with $\xi_m = 1$ only under some very specific, very stringent growth condition(s). It would otherwise be a non-eutectic alloy, cluster, or solid solution. An important artifact of the VLS mechanism is that the temperature needed for the release of the R_S species from their precursor on a certain METANO surface is generally higher than the lowest eutectic point of the (MET, X) alloy. Such an alloy may be eutectic and hence molten if the alloy re-establishes its stable composition through decrease in temperature. The alloy may otherwise be non-eutectic and hence a solid solution if it achieves stability without any reduction in temperature. We cite another example. Al/Si binary phase diagram resembles the Au/Si binary phase diagram. The lowest eutectic point of the Al/Si binary system corresponds to the Si composition of ~12 at.% at a temperature of 577 °C. Osada et al. [5] demonstrated that, using silane as Si precursor, the CVD growth of Al-catalyzed growth of Si nanowires by the VLS mechanism can be achieved at temperatures of 580–700 °C. Obviously, the (Al, Si) alloy with a Si content of 12 at.% was eutectic and molten at 580 °C; the (Al, Si) alloy with a Si content of 21 at.% is though eutectic and molten at 700 °C.

16.5 The VSS Mechanism, for Low-Temperature ($T < T_E$) Growth, is a Special Case of the VQS Mechanism

16.5.1 Basics

Unfortunately, the fundamentals of the VSS mechanism are still obscured and poorly understood. This is evident from Table 6.1 of Chap. 6, which lists nanowire growth rates by various mechanisms, including the VSS mechanism. A close look of this table would indicate that some of the nanowire growth rates by the VSS mechanism are marginally low, while others are remarkably high. The Ge nanowire growth rate by Kodambaka et al. [2] by the VSS mechanism is about 10–100 times lower than the VLS growth rate at the same temperature and pressure. In contrast, the GaAs, GaP, and InP nanowire growth rates by Dick et al. [6] by employing the VSS mechanism are comparable with the VLS growth rates of these nanowires and at essentially the same temperature. Recall that, the R_S species for growth by the VLS mechanism diffuse through molten catalyst droplet before supersaturation at the L/S (liquid/solid) interface. This diffusion is fast. In contrast, the R_S species, for growth by the VSS mechanism, diffuse through solid MET catalyst nanoparticle before supersaturation at the L/S interface. This diffusion is obviously slow. The nanowire growth rate by the VSS mechanism must therefore be substantially lower than that by the VLS mechanism. The data by Kodambaka et al. [2] and by Dick et al. [6] contradict this logic. And this highlights the weakness of fundamental physicochemical foundation of the VSS mechanism. The strong physicochemical foundation of the VQS mechanism, as described in Chap. 12, demonstrates that the results by Kodambaka et al. and by Dick et al. both correspond to growths by the VQS mechanism, and the difference between the two can be attributed to the difference in the values of SECINI and ξ_m. Interestingly, essentially all of the nanomaterial growths carried out in the past and attributed to the VSS mechanism may be explained by the VQS mechanism. We illustrate some of them in the following.

16.5.2 Illustration with ZnO Nanowire Growths

The Au-catalyzed ZnO nanowire growth by Campos et al. [7] was carried out at ~350 °C. Taking the Au–Zn phase diagram, temperature measurement effects, and temperature size effects into account, the ZnO nanowire growth was conjectured to have a genesis in the possible epitaxial relationship between the ZnO nanowires and the γ-$Au_{1-z}Zn_z$ nanoparticle that catalyzed the growth. This relationship prompted nanowire growth by the diffusion of the R_S species driven by preferential oxidation of the Zn inside the catalyst nanoparticle. As noted by Campos et al. [7], this catalyst nanoparticle is not a solid; it has R_L species composition of γ-$Au_{1-z}Zn_z$, which is an alloy and a solid solution, and which exhibits characteristics much different from those of solid. The Ge nanowire growth by Kang et al. [8] was performed at a growth temperature of about 200 °C below the eutectic temperature of 644 °C, which corresponds to the composition of $Cu_{0.635}Ge_{0.365}$ of the R_L≡(Cu, Ge) alloy. The nanowire tips indicated that the R_L species of the catalyst nanoparticle responsible for the growth was orthorhombic Cu_3Ge alloy, which is non-eutectic solid solution. R_L≡(Ti, Si) alloy possesses a eutectic temperature $T_E = 1330$ °C adjoining the pure Si side of the Ti/Si binary phase diagram [4]. Si nanowire growth at temperatures $T <$ 1330 °C therefore proceeded via the phase neighboring the pure Si side of the liquidus line. And this phase was $TiSi_2$. A close look of the Ti/Si binary phase diagram [4] indicates that a gradual decrease in temperature from 1330 to about 1000 °C causes the transformation of the R_L species of the nanoparticle to the Ti_5Si_3 alloy. A gradual and further decrease in temperature leads the transformation of Ti_5S_{i3} first to Ti_5Si_4, next to TiSi, and finally to $TiSi_2$. The CVD growth of Si nanowire at 640–670 °C by Kamins et al. [9] was therefore catalyzed by R_L≡($TiSi_2$). This growth temperature was about 700 °C lower than the (Ti, Si) eutectic temperature of 1330 °C. And the R_L species of this catalyst nanoparticle was a non-eutectic alloy; it was a porous, coarsened, solid solution with characteristics much different from those of a solid. Various intermetallic compounds formed at temperatures between 250 and 1200 °C in the Ni/Ga system are shown in Table 16.1. A look of this table would indicate that Ni_3Ga and NiGa eutectic alloys are formed at a temperature as high as ~1200 °C. This table also indicates that all the intermetallic compounds formed in the temperature range of 580–620 °C would be peritectoid compounds (porous, semi-molten, solid solutions). Ni-catalyzed GaAs nanowire growth by Han et al. [10] was therefore by an R_L species composed of Ni-based solid solution. These growths could take place only by the VQS mechanism.

Table 16.1 Literature information on the intermetallic compounds formed in the Ni–Ga system

Compound	Composition of Ga	Melting or transition type	Temp (K)	References
α'-Ni_3Ga	0.23–0.295 (873 K)	Peritectic	1485	Feschotte [20]
δ-Ni_5Ga_3	0.37	Peritectoid	958	Helner [21]
γ-Ni_3Ga_2	0.36–0.41 (973 K)	Peritectoid	1213	Helner [21]
γ'-$Ni_{13}Ga_9$	0.40–0.42 (873 K)	Polymorphic	958	Helner [21]
β-NiGa	0.47–0.55 (873 K)	Congruent	1493	Helner [21]
Ni_3Ga_4	0.565–0.575 (673 K)	Peritectoid	815 ± 1	Feschotte [20]
β'-Ni_2Ga_3	0.60	Peritectic	1168	Helner [21]
β-$NiGa_\varepsilon$	0.80	Peritectic	525	Helner [21]
Ni_3Ga, NiGa	0.268	Eutectic	1480	Yu/an et al. [22]

16.5.3 Illustration with GaN and InN Nanowire Growths

We grew GaN and InN nanowires [11–18] in our laboratory without and with the assistance of MET (e.g., Ni and Au). Ga matrix formed on Si, substrate mediated the GaN nanowire growth. In matrix formed on Si substrate mediated the InN nanowire growth. Ammonia served as precursor of the N atoms, and the flow rate of this ammonia was 20 sccm for GaN nanowire growth, but 50 sccm for InN nanowire growth. The R_L species for GaN nanowire growth was $R_L \equiv$(Ni, Ga), but for InN nanowire growth was $R_L \equiv$(Au, In). The R_L species could have contaminants. The dimension and location of these nanowires were essentially the same as the dimension and location of the R_L species formed at the growth temperature T on the substrate surface of the growth chamber. This growth temperature was 1000 °C for GaN nanowire growth employing MET≡Ni. But this growth temperature was 525–550 °C for InN nanowire growth employing MET≡Au. There were also some MET-free InN nanowires growths, but at a temperature of 520 °C. The GaN nanowires were 4–20 nm in diameter, but the InN nanowires were lower than 50 nm in diameter. The InN nanowires grown in MET-free environment had amorphous In at the tip. The InN nanowires grown in Au-mediated environment had, on the other hand, amorphous AuIn at the nanowire tip. The amorphous material at the nanowire tip suggested that the R_L species responsible for nanowire growth by both MET≡Ni and MET≡Au were amorphous. They had obviously disturbed, disordered, porous lattice structure.

16.5.4 Role of Intermediate Phases in Growths

Pratt and Bird [19] identified Ni_3Ga, Ni_5Ga_3, Ni_3Ga_2, $Ni_{13}Ga_9$, NiGa, and Ni_2Ga_3 intermediate phases resulting from Ni and Ga reactions (interdiffusions). Feschotte and Eggimann [20], Helner [21], and Yuan et al. [22] found other intermediate phases resulting also from Ga reactions with Ni and/or from Ga interdiffusions into Ni (see Table 16.1). These intermediate phases may actually be close derivatives of the phases identified by Pratt and Bird [19]. Interestingly, all of them are non-eutectic, and hence different from the eutectic phases Ni_3Ga and NiGa, respectively, identified by Yuan et al. [22]. The eutectic temperature for both of them is $T_E \approx 1207$ °C, which is about 200 °C higher than the growth temperature for GaN nanowire growths in our laboratory. The phase of the $R_L \equiv$(Au, Ga) species mediating GaN nanowire growths in our laboratory was therefore non-eutectic alloy (solid solution).

A close look of the Au/In binary phase diagram [23] would indicate that $R_L \equiv$(Au, In) formed in the (Au, In) system can have three stable phases: $R_L \equiv Au_{0.76}In_{0.24}$ phase at 446.5 °C, $R_L \equiv Au_{0.72}In_{0.28}$ phase at 454.3 °C, and $R_L \equiv Au_{0.58}In_{0.42}$ phase at 495.4 °C, respectively. There can also be other phases such as $R_L \equiv (AuIn_2)$ at 539 °C, but $R_L \equiv$(AuIn) at 506 °C, both resulting from interdiffusion of In and Au. However, the nanowire growths were plausibly catalyzed by some close derivative of AuIn and/or $AuIn_2$, rather than by $R_L \equiv (AuIn_2)$ and $R_L \equiv$(AuIn) themselves. The entities contributing to the generation of these derivatives were the influences of temperature variation and temperature fluctuation in the growth chamber, annealing, and/or the presence of foreign elements, particularly oxygen which is often present in chamber during nanowire growth. This appears to be confirmed by the non-existence of Au at the tips of nanowires grown in our laboratory. The R_L species for the Au-mediated GaAs nanowire growth was $R_L \equiv AuGa_2$. This is a solid solution. Based on HRTEM by Cai et al. [24], $R_L \equiv AuGa_2$ had grains and grain boundaries. Hofmann et al. [25] performed Pd-catalyzed Si nanowire growth employing Si_2H_6. HRTEM indicated that the R_L species for this growth was amorphous, non-stoichiometric $Pd_{1-z}Si_z$ (more specifically, Pd_2Si) and hence a solid solution.

16.5.5 *Influence of Dopants, Contaminants and Stresses on Growths*

Liu et al. [26] showed that dopants, contaminants, and oxygen atoms do influence the lattice structure of the R_L species. They employed MET≡Fe to perform the 3C-SiC nanowire growth by the VLS mechanism. High-angle HAADF image of the R_L species and the EDX mapping of this R_L species revealed that the inner core of it had Fe, but the outer shell of it had silicon, oxygen, and iron. Iron was non-uniformly distributed throughout the outer shell, creating stress in the shell and also in the core. The R_L species had obviously disturbed, disordered lattice structure; it was amorphous, and hence a solid solution. Kamins et al. [27] grew Ge nanowires along the $\langle 111 \rangle$ direction at about 320 °C employing MET≡Au and GeH_4 precursor. HRTEM images showed that the tips of the nanowires were composed of γ-phase $Au_{0.6}Ge_{0.4}$ alloy, which is an intermediate alloy different from the eutectic phase $Au_{0.8}Ge_{0.2}$ alloy formed at a eutectic temperature of $T = T_E = 361$ °C. The intermediate γ-phase $R_L \equiv Au_{0.6}Ge_{0.4}$ alloy was created at 320 °C long before the $R_L \equiv$(Au, Ge) alloy could reach the eutectic phase; it was a solid solution.

Employing MET≡Ni, Cheze et al. [28] could grow GaN nanowires on sapphire substrate at $T = 730$ °C. The QMS and the RHEED patterns indicated that several crystalline intermediate phases were involved in the growth. There was also structural changes in METANO and phase transition in the $R_L \equiv$(Ni, Ga) alloy probably under the influence of an additional Ni–O–Al phase. It created stress. RHEED pattern, in particular, indicated that Ga was incorporated into Ni creating a disordered α'-Ni_3Ga phase. It was a solid solution with Ga content increased from 23.1 to 30 at.%. Nanowire growth was catalyzed by this solid solution. A body-centered cubic (bcc) phase with $a = 2.9 \pm 0.1$ Å and a hexagonal lattice with $a = 4.1$ Å and $c = 5.0$ Å resulted from the incorporation of more Ga into Ni during the subsequent stage of growth. Thermodynamically stable with Ga between 30 and 58 at.%, this phase was β-$Ni_{1-z}Ga_z$ phase. It had a lattice parameter of 2.896 Å. Three different Ni–Ga compounds belonged to the β-$Ni_{1-z}Ga_z$ phase. These compounds are γ-$Ni_{13}Ga_7$ ($a = 4.005$ Å, $c = 5.018$ Å), ε-Ni_3Ga_2 ($a = 3.995$ Å, $c = 4.980$ Å), and the trigonal or hexagonal β'-Ga_3Ni_2 ($a = 4.054$ Å, $c = 4.882$ Å). The β'-Ga_3Ni_2 phase corresponded to equilibrium between NiGa and Ga_3Ni_2 phases; it was liquidic at 895 °C and hence responsible for GaN nanowire growths.

Using MET≡Au, Dick et al. [6] grew GaAs and GaP nanowires at 380 °C $\leq T \leq$ 530 °C. The growths of these nanowires were attributed to the VSS mechanism. EDX micrograph suggested that R_L≡(Au, Ga) species was responsible for these nanowire growths; it was a solid solution with 5–10 at.% Ga and 90–95 at.% Au. Both GaAs and GaP remained stable in equilibrium with Au during growth. The (Au, As) phase diagram [4] indicated that no stable R_L≡(Au, As) species was formed at temperatures $T < 636$ °C. No stable R_L≡(Au, P) species was also formed at a temperature $T \leq$ 636 °C. This R_L≡(Au, P) species was formed during heat treatment of a mixture of P- and Ga-containing compounds, and hence would be stable if in equilibrium at a temperature $T \leq 440$ °C. All these imply that the R_L species for GaAs nanowire growth had no As atom(s), and the R_L species for GaP nanowire growth had no P atom(s), both at a temperature $T < 520$ °C. Instead, the R_L species for both GaAs and GaP nanowire growths had Au and Ga (~5 to 10 at.% Ga, α phase). It was a solid solution creating R_L≡(Au, Ga) at a temperature $T \leq 450$ °C. As noted by Hiraki et al. [29], it could be formed at a temperature as low as 27 °C if sufficient time was allowed for it to take place in an open system. It could be R_L≡Au_7Ga_2 (β, β' phase with 20.4–23.2 at.% Ga), R_L≡(AuGa) phase with 50 at.% Ga, and/or R_L≡($AuGa_2$) phase with 66.7 at.% Ga at a temperature of $T \leq$ 470–500 °C and under appropriate heat treatment of the Au-GaP alloy system [30–32]. Other R_L≡(Au, Ga) species [33, 34] have been found to be formed during annealing of GaAs in the presence of Au at a temperature $T \leq 470$ °C. This annealing probably alters R_L composition, R_L characteristics, and R_L dynamics. Harmand et al. [35] found that temperature, the duration of the increase (decrease) of this temperature, and the rapidity of increase (decrease) of this temperature during nanowire growth can have profound impact on the composition of the R_L≡(Au, Ga) species. This composition can be zincblende, orthorhombic, or cubic; and also the atomic % of Au in it [e.g., R_L≡engineered $Au_{1-z}Ga_z$ (e.g., EMNO)] can vary. The partial pressure of group V element (As or P) was controlled during nanowire growth by Dick et al. [6] at $T \leq 475$ °C in closed chamber environment. Because of this, the growth was not probably catalyzed by the solid solution of Au and Ga (5–10 at.% Ga). In fact, this solid solution is not at all formed and does not exist in closed environment. It is not also stable at a temperature $T > 450$ °C. The growth was not therefore catalyzed by R_L≡Au_7Ga_2 (β, β', 20.4–23.2 at.% Ga), R_L≡(AuGa) with 50 at.% Ga, and R_L≡($AuGa_2$) with 66.7 at.% Ga, as well. They were rather formed at a temperature $T > 470$ °C. Also, they had Ga component higher than the one (~10 at.% Ga) observed by Dick et al. [6]. The growths under these circumstances could be carried out only by the VQS mechanism.

16.5.6 Discrepancy in Growth Rates Explained

Table 6.1 of Chap. 6 indicates that the growth rates of GaAs, InAs, InP, and GaP nanowire, as reported by Dick et al. [6], are, in general, comparable to the VLS growth rates of InAs, GaN, Si, and ZnO nanowires. This, within the framework of the VQS mechanism, would be possible only if the catalyst for nanowire growths by Dick et al. [6] was solid, but exhibited high density of quasiliquid (quasisolid) nanopores at temperature $T < T_E$, had SECINI approaching SECINI0 and ξ_m approaching unity. This ξ_m for Ge nanowire growth by Kodambaka et al. [2] was, on the other hand, marginally small, for example $\xi_m \rightarrow 0.0001$. Figure 16.3a depicts the presumed feature of the VSS process, but Fig. 16.3b shows the real feature of the VSS process. As shown in Fig. 16.3b, the catalyst (METANO, SUBSANO) surface is disordered, for example, by plasma bombardment, aqua regia treatment, etc., up to a depth of

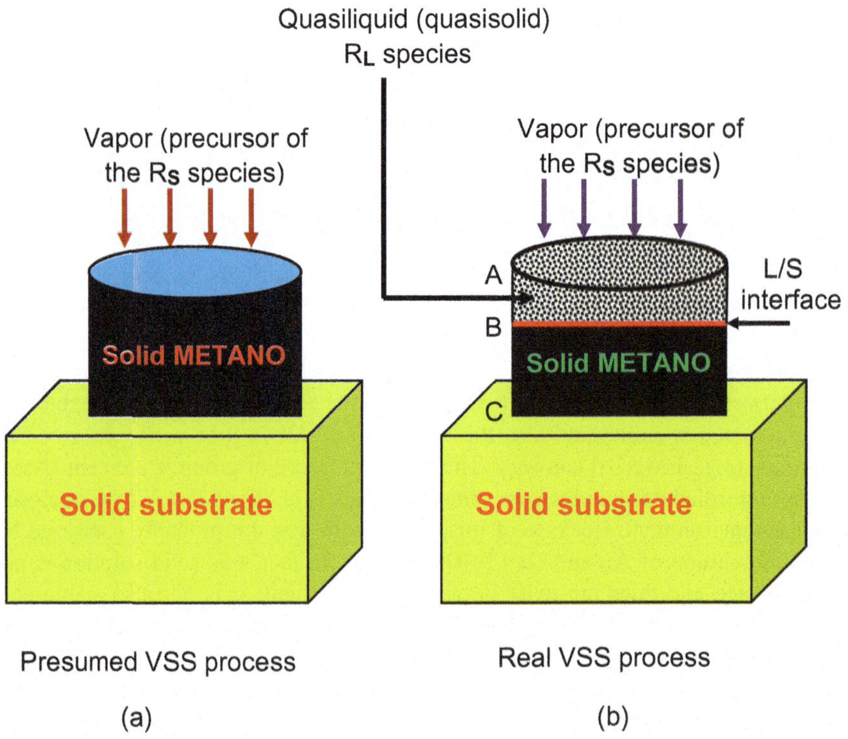

Fig. 16.3 Schematic illustration showing the differences of the presumed VSS mechanism from the real VSS mechanism taking place for growth. This real mechanism involves the diffusion of the R_S species through a disordered, coarsened, amorphous (semi-amorphous), porous R_L species formed on the top surface of the METANO. This R_L species on the top surface has grains, molten (liquidic) grain boundaries, and molten nanopores (not shown); it has a finite thickness of δ_{amor}. It is believed that the real VSS mechanism, which is a variant of the VQS mechanism, involves supersaturation and nucleation at the liquid/solid interface

AB≡δ_{amor} leaving the bulk of the depth BC still unaffected. The disordered surface appears to satisfy all the requirements for growths by the VQS mechanism: It is amorphous and has a network of nanopores; it may be molten (semimolten) due to one or more of the effects described in Sect. 12.2 of Chap. 12. An QL/S (QS/S) interface leads to the supersaturation for growths. It is the VSS mechanism only to the extent that the nanopores have very small pore radii and that the density of these nanopores is also very low. This is probably the reason of why the VSS growth rate of Au-catalyzed Ge nanowires is 10–100 times lower than the VLS growth rate of Au-catalyzed Ge nanowires at the same source pressure and temperature [2]. One primary reason for this is probably much lower solubility of the R_S [R_S≡X and R_S≡Y] growth species in a solid MET nanoparticles (e.g., Au) than in a quasiliquid (quasisolid) METANO cluster or (MET, X) droplet. VSS mechanism is, in fact, a special case of the VQS mechanism exhibiting a narrow window of applications (see Fig. 12.8 of Chap. 12).

A list of several nanowires, METANOs used to grow these nanowires, growth technique used for this growth, the melting temperature T_M of the METANO metal(s) used for the growth, the eutectic temperature T_E of the alloyed R_L species, and the material phase at the nanowire tips can be found in Table 13.1 of Chap. 13. A look into this table suggests that the R_L species for the METANO-mediated nanowire growth at a temperature $T \neq T_E$ (e.g., $T < T_E$ or $T > T_E$) can have METANO composition quite different from the eutectic composition of it. The R_L composition for growths at a temperature $T < T_E$ and even at a temperature $T > T_E$ can be non-eutectic; it can be a solid solution and hence metastable and nanoporous; it can be amorphous exhibiting HETs; see Chap. 12. As elucidated in Chap. 12, there occur phase changes in the material content of the R_L species during gradual increase in temperature during METANO-mediated growths. And even before the temperature is raised to the eutectic temperature T_E, a material phase suitable for growth can be achieved. This phase exhibits disturbed (disordered) lattice, surface amorphicity, surface coarsening, surface porosity, surface melting, and HETs; it is a cluster or solid solution and hence quasiliquid (quasisolid). Furthermore, during gradual increase in temperature during METANO-mediated growths, no eutectic phase may be achieved at a temperature $T = T_E$ or even at a temperature $T > T_E$ if the R_L species does not have proper material composition. In this case, a material phase suitable for growth at a temperature $T > T_E$ and even at a temperature $T \gg T_E$ is achieved. This phase exhibits as well disturbed (disordered) lattice, surface amorphicity, surface coarsening, surface porosity, surface melting, and HETs; it is a cluster or solid solution, and hence quasi-liquid (quasisolid).

The CNT growth rates under several different growth conditions have been presented in Figs. 15.14–15.17 of Chap. 15. One of them by Lee et al. [36] and shown in Fig. 15.14 was achieved with Fe nanoparticles. Though the eutectic temperature of the R_L≡(Fe, C) binary alloy [4] is 1153 °C, the CNT growth by Lee et al. [36] was achieved at a temperature as low as 750–950 °C. The basic tenet of the simple theoretical model for this growth was based on diffusion of the R_S species through nanopores. Figure 15.14 shows that the calculated results agree well with the experimental results. It is true also for results shown in Figs. 15.15–15.17 of Chap. 15. They all suggest that nanopores created on nanoparticle surface even at temperature $T < T_E$ and the Knudsen diffusion of the R_S species through these nanopores may primarily be responsible for nanomaterial growths at temperature $T < T_E$.

Table 16.2 displays the characteristic features of CNTs [37–42] synthesized by employing various METANOs. A close look of this table indicates that the CNT growth temperature T can almost always be lower than the (MET, C) eutectic temperature T_E. It is true irrespective of what is the MET catalyst employed for the growth and what are the substrate and carbon precursor(s) used for the growth. Sadeghian and Rashidi [43] demonstrated that it is true even when the CNT diameter is as large as 85 nm. It is true also for nanotube growths at relatively low temperature without the assistance of METANO. These would not happen without quasiliquid (quasisolid), porous medium playing decisive roles in nanotube growths at relatively low temperature. There is a wide disparity in growth rates by the VSS mechanism (see Table 6.1 of Chap. 6). And it can be explained in the framework of the VQS mechanism. If these nanopores are bent, twisted, small in size or in number, the nanomaterial growth rate is marginally small. If, on the other hand, these nanopores are straight, large in size and in number, the nanomaterial growth rate is significantly large.

16.6 The VSS Mechanism, for High-Temperature ($T > T_E$) Growth, is a Special Case of the VQS Mechanism

We showed in Sect. 16.4 of this chapter that many of the MET-mediated growths of nanomaterials take place at $T < T_E$. We show here that the MET-mediated growth of nanomaterials can take place also at $T > T_E$. The Fe-mediated 4H-SiC nanowire growth by Sundaraesan et al. [44] was carried out at a temperature $T = 1550$ °C. This temperature is far beyond the $R_L \equiv$(Fe, Si) eutectic temperature $T_E = 1207$ °C (see entry 1, Table 13.1 of Chap. 13). Yet the authors claimed that it was carried out by the VLS mechanism. Supported by the electron backscattering diffraction patterns of the nanowire tip materials, the R_L species was shown to be $R_L \equiv Fe_2Si$. However, based on binary phase diagram [4], the composition of the R_L species at the temperature $T = 1550$ °C is non-existent on the liquidus line. In fact, this temperature is higher than the melting temperature $T_M = 1538$ °C of Fe and also the melting temperature $T_M = 1377$ °C of FeO. It is though slightly lower than the melting temperature $T_M = 1565$ °C of Fe_2O_3, but much lower than the melting point $T_M = 1597$ °C of the FeO–Fe_2O_3 mixture. Lehlooch et al. [45] found that $Fe_{1-z}Si_z$ with $z < 0.10$ has the A2-type disordered structure. While performing the EDX scan, Sanderesan et al. detected nitrogen on Fe nanoparticle surface. We argue that, due to the presence of this very nitrogen, $R_L \equiv$(Fe, Si) was contaminated into $R_L \equiv$(Fe, Si, N). Such contamination impeded a gradual increase in temperature during growth causing the $R_L \equiv$(Fe, Si) species to be eutectic and molten, e.g., $R_L \equiv Fe_2Si$ at the eutectic temperature $T_E = 1207$ °C. Instead it remained minimally nanoporous, and the nanowire growth could not take place until the $R_L \equiv$(Fe, Si, N) species

Table 16.2 List of METANO nanoparticles, (MET, C) eutectic characteristics, and the CVD growth parameters of CNTs by these METANO nanoparticles

METANO	Melting temp (°C)	Source	Substrate	Growth temp T (°C)	References	Eutectic characteristics	
						T_E (°C)	Eutectic composition
Re	3180	Methane	–	950–1100	[37]	2500	$Re_{0.993}C_{0.007}$
Al	660	Ethylene	Si/SiO_2	700–850	[38]	–	–
Mn	1245	Ethylene	Si/SiO_2	600–750	[38]	1210	$Mn_{0.991}C_{0.009}$
Mo	2617	Ethylene	Si/SiO_2	600–1000	[38]	2221	$Mo_{0.825}C_{0.175}$
Ir	4163	Ethylene	Si/SiO_2	600–1000	[38]	2290	$Ir_{0.80}C_{0.20}$
W	3422	Ethylene	Si/SiO_2	600–800	[38]	2720	$W_{0.78}C_{0.22}$
Ti	1660	Ethylene	Si/SiO_2	750–800	[38]	1627	$Ti_{0.987}C_{0.013}$
Au	1064	Ethanol	Si	850	[38]	1027	$Au_{0.996}C_{0.004}$
Cu	1083	Ethanol	Al-hydroxide	850	[39]	1070	$Cu_{0.9999}C_{0.0001}$
Fe	1535	Ethanol	Al-hydroxide	850	[39]	1153	$Fe_{0.82}C_{0.18}$
Ni	1453	Ethanol	Al-hydroxide	850	[39]	1297	$Ni_{0.902}C_{0.098}$
Co	1495	Ethanol	Al-hydroxide	850	[39]	1301	$Co_{0.9111}C_{0.0889}$
Pt	1772	Ethanol	Al-hydroxide	850	[39]	1738	$Pt_{0.988}C_{0.012}$
Pd	1552	Ethanol	Al-hydroxide	850	[39]	–	–
Ag	962	Ethanol	Al-hydroxide	850	[40]	1330	$Ag_{0.999}C_{0.001}$
Cr	3192	Acetylene	SiO_2	850	[41]	1534	$Cr_{0.86}C_{0.14}$
Ru	2250	Methane	SiO_2	800–1000	[42]	1953	$Ru_{0.975}C_{0.025}$

could be sufficiently porous at $T = 1550$ °C. Based on discussions made earlier, it could happen because SECINI, for this growth, could reach SECINI0 and ξ_m could approach unity only at a temperature as high as $T = 1550$ °C. We believe that the electron backscattering diffraction patterns of the nanowire cap materials, which showed $R_L \equiv Fe_2Si$, were obtained after cooling at the end of experiment. It was not probably the correct composition of the R_L species catalyzing the nanowire growth.

Appropriate partial pressure of $R_S \equiv X$ is very important for the formation of $R_L \equiv$(MET, X) eutectic alloy. While trying to make Au-catalyzed VLS growth of GaN nanowires at 900 °C, Hou et al. [46] found that the partial pressure of the $R_S \equiv X =$ Ga species in the growth chamber dictates the composition of the $R_L \equiv$(Au, Ga) species. Under low Ga pressure, this alloy had a very low gallium content of 1–2 at.%. And hence the composition of the $R_L \equiv$(Au, Ga) species was different from the composition of the $R_L \equiv$(Au, Ga) eutectic alloy species. According to the binary phase diagram [4], the composition of the eutectic alloy at the temperature $T_E =$ 361 °C is $Au_{0.69}Ga_{0.31}$. As the growth temperature was gradually increased during growth, the $R_L \equiv$(Au, Ga) species remained non-eutectic and non-porous even at T_E = 361 °C. And it happened due to the lack of appropriate quantity of Ga present in the chamber. So, a gradual increase in temperature was needed until this temperature reached 900 °C at which $R_L \equiv$(Au, Ga) species was still non-eutectic, but molten and porous. The binary phase diagram [4] shows that, at a temperature $T = 900$ °C, the $R_L \equiv$(Au, Ga) species alloy can have 1–2 at.% of gallium and hence be in the solid phase, but not in the liquid phase. In fact, HRTEM image of the $R_L \equiv$(Au, Ga) alloy species showed that this seed had no regular shape; it was essentially non-spherical (non-hemisherical) implying that it was a solid solution. This means the GaN nanowires were actually grown at a temperature higher than the eutectic temperature $T_E = 361$ °C. And it was done by the VQS mechanism. The average nanowire diameters at a temperature $T > T_E$ were larger in high and intermediate Ga pressure, but small in low Ga pressure. The nanowire crystal quality was slightly inferior to that under high and intermediate Ga pressures.

The solubility of Ga in an ultrasmall Au catalyst particle is significantly reduced. Because of this, the eutectic growth of Au-mediated GaAs nanowires was evident from the presence of $AuGa_2$ alloy in droplets at the tips of some large-diameter GaAs nanowires. But the non-eutectic growth of Au-mediated GaAs nanowires was evident from the presence of $R_L \equiv$(Au, Ga) solid solution at the tips of some small-diameter GaAs nanowires [47]. The growth temperature of all these nanowires was set to be 530 °C. The Au/Ga binary phase diagram [4] shows that annealing at $T \approx 530$ °C does not yield an alloy of Au and Ga. Instead Au and Ga exist in the $R_L \equiv$(Au, Ga) compound as solid solution. The Ga content in this solid solution is less than 10 at.%. It is true for the R_L species of all sizes. So, we argue that the GaAs nanowire growth at $T \approx 530$ °C was catalyzed by the $R_L \equiv$(Au, Ga) solid solution. However, during cooling after growth Ga diffused more in large-sized nanoparticle than in small-sized nanoparticle yielding $AuGa_2$ eutectic alloy at a temperature of 339.4–348.9 °C.

16.7 The VS Mechanism is a Special Case of the VQS Mechanism

16.7.1 Exceptional Roles of Surface Porosity and HETs in Growths

Metal-catalyzed nanomaterials may contain high concentration of metal impurities. Also, purification routes for nanomaterials are less effective due to some specific physical and chemical properties of metals. Electromagnetic and microwave absorbing properties of nanomaterials remain obscured due to the leftover metal catalyst in nanomaterials. To overcome this obstacle, metal-free growth of nanomaterials is essential. The VS mechanism is one such metal-free growth mechanism. However, like the VSS mechanism, it is not yet well-understood. Nevertheless, some FECA-free growths of carbon nanotubes by the VS mechanism are presented in Table 16.3. A comparison of CNT growth rates on oxide-based SUBSANOs [48–50] is presented in Table 16.4. A close look of Tables 16.3 and 16.4 would indicate that CNT growth rate by thermal CVD is quite low, but the CNT growth rate by plasma CVD is quite high. Let us be more specific. Almost in an identical way, Kim et al. [50], Rao et al. [51], and Kumar et al. [52] carried out CNT growths on Si substrate. Rao et al. [51] created Si pillar and grew CNTs on these pillars. They observed CNT growth rate dependent on CNT chirality. Kumar et al. [52] cleaned and then scratched the substrate with diamond scriber [52]. They then exposed the substrate to microwave plasma before carrying out CVD growth on it at a temperature of ~850 °C. As usual, no CNT was grown in a neutral gas environment on smooth or patterned surface. No CNT was grown on also smooth or patterned surface treated with remote plasma. These surfaces belonged to NP1 type SUBSANOs. Long (up to several hundred mm), vertically aligned high-density (~1000 nanotubes per mm^2) CNTs were however grown on porous $R_L \equiv$(Si-based ESNO) species of SUBSANOs [52], which were actually NP2 type SUBSANOs. The most striking feature of this experiment was the exceptionally high and exceptionally low CNT growth rates, as evident from Table 16.4. Such growth rate by thermal CVD on SiO_2-based ESNO was only 8.30 nm/s [48], and on Al_2O_3 based ESNO was only 3.33 nm/s [49]. In contrast, an almost 100 times increase in growth rate due to an additional usage of plasma would not happen had the solid substrate been responsible for CNT growth. This increase can however be readily explained if two important parameters, namely (1) increase in SUBSANO porosity and (2) increase in the density and effectiveness of HETs by plasma bombardment, which are considered to be primarily responsible for growth. Both of them constitute the basic tenets of the VQS mechanism (see Chap. 12).

Table 16.3 Characteristics of MET-free growth of carbon nanotubes

No.	Nanotube	Substrate	Source	Growth temp (°C)	Technique	Melting temp (°C)	Comment	References
1	CNT	Si	Methane	850	PECVD	1414	Affected by plasma	[52]
2	MWCNT	Carbon black	Ethylene	800	CVD	3550	Naturally porous structure	[53]
3	SWCNT	Graphite	Ethylene	850	CVD	4489	Has defected surface	[54]
4	MWCNT	Au	Ethylene	750–900	CVD	1064	Treated with aqua regia	[55]
5	SWCNT	Diamond	Ethanol	850	CVD	4440	Has oxygenated surface	[56]

The melting temperatures listed in the table are the ones for substrates

Table 16.4 Comparison of growth rate of CNTs produced on oxide substrates by catalyst-free means

Carbon precursor	Substrate	Temp (°C)	CNT	Technique	Growth rate (nm/sec)	References
CH_4	SiO_2	900	SWCNT	Thermal CVD	8.30	Liu [48]
C_2H_5OH	Al_2O_3	850	SWCNT	Thermal CVD	3.33	Liu [49]
CH_4	SiO_2	850	MWCNT	Plasma CVD	8.33×10^2	Rao [51]

16.7.2 Exceptional Role of Surface Disorder in Growths

We go back to Table 16.3, which shows characteristic features of SUBSANO-mediated CNT growths. These growths were catalyzed by carbon black ($T_M \approx 3550$ °C) [53], defect-rich graphite ($T_M \approx 4489$ °C) [54], and substrates such as Au ($T_M \approx 1064$ °C) [55]. They were catalyzed also by Si ($T_M \approx 1410$ °C) [50], diamond ($T_M \approx 4440$ °C) [56], Cu ($T_M \approx 1083$ °C) [57], Ni ($T_M \approx 1453$ °C) [58, 59], and Fe ($T_M \approx 1535$ °C) [60] substrates, respectively. These growths are all attributed to the VS mechanism, and to the diffusion of the R_S species through solid substrate. However, surface treatments of SUBSANOs formed on substrate surface, for example, by ion bombardment, sputtering, and microwave plasma were carried out prior to all of these growths. Such surface treatments, probably with contaminants present in growth chamber, gave rise to surface disorder, surface amorphicity (effective surface amorphicity), surface porosity, surface coarsening, surface melting, and HETs on the SUBSANO surface. These are evident, for example, from the SEM images of aqua regia treated SUBSANOs on Au substrate surface [55] and of ethanol treated SUBSANOs on diamond substrate surface [56]. Among various SUBSANOs, carbon black has natural surface disorder, surface amorphicity, and surface porosity. As Kurk et al. observed [61], the surface morphology of carbon black is porous and defective. It has surface coarsening, and hence HETs at fullerene-like positions and/or at nanoscale curvatures [53].

16.7.3 Exceptional Role of Oxygen Contaminants in Growths

Kusunoki et al. [62] noted that contaminants such as O gives rise to surface disorder, surface amorphicity, and surface coarsening of SUBSANOs. This is probably why CNT growth by them on SiC substrates took place only in the presence of O contaminants. In fact, Rümmeli et al. [63] noted that SWCNT growth could not be realized without the presence of O contaminants. Li et al. [64] found that BNNTs could similarly be produced in only the presence of water vapor and probably also in the presence of O and H. These analyses indicate that the R_L species compositions on the surface of the substrates and responsible for the said CNT growths were not single-crystalline solid. They had rather oxygenated, disturbed, disordered, amorphous, and porous lattice structure. They were solid solutions. Lin et al. [53] adopted several activation routes to ensure that the R_L species of the surface of graphite used for CNT growths was defective and oxygenated. This R_L species became porous due to surface treatments. The R_L species undergoing such treatments was found to be rough [54]; it had many bumps with diameters ranging between 40 and 80 nm. All these belonged to NP2 type SUBSANOs. The R_L species of the pristine surface was, on the other hand, smooth and free from them. This surface belonged to NP1 type SUBSANOs. No CNT growth could be catalyzed by the R_L species of the pristine graphite surface. But MWCNT growth could be catalyzed by the R_L species of the treated substrate surface. Previous studies by Muradov [65] suggest that dangling carbon structures of the bumps observed by Lin et al. [53] can catalytically decompose hydrocarbons. Activated carbon was found to provide the highest initial activity of methane decomposition probably because HETs generated in the defected sites were crucial [66] for decomposing the carbon precursors.

16.7.4 The Role of Contaminant Assemblages in Growths

Lin et al. [55] observed that MWCNTs emerged at pockets (islands) having disturbed and exhibited assemblage of short-range crystalline carbon atoms on the Au film surface. Such assemblages generated R_L≡(Au, C) solid solutions during the pre-nucleation stage of growths. The carbon atoms for such assemblages were from the decomposition of the carbon precursor on the Au surface. And it happened presumably because this layer had graphic-like islands of larger lattice distance (viz. 0.352 nm) as compared to that of graphite (e.g., 0.343 nm). These islands resulted from adsorption of carbon into the substrate or from the lack of desorption of carbon from this substrate. They were the sites of the R_L species of SUBSANOs. They were metastable suitable for playing an important role in the growth of CNTs. R_L species of SUBSANOs, such as these, were observed in previous investigations by Liang et al. [67] during ZnO nanotube growths by using plasma-treated substrates. The pits were generated in these R_L species in the peripheral shell of the SUBSANO. They acted as the nucleation sites during the pro-nucleation stage of growths. During this stage of growths, the diffusion of the R_S species through the nanopores of the R_L species yielded ring-like structures finally giving rise to nanotubes. These are all incompatible to the VS mechanism, but compatible to the VQS mechanism.

16.7.5 Impact of Substrate Scratching on Growths

The experiment by Huang et al. [68] suggested that CNT growths on SiO_2 surface could be possible by creating nanosized SiO_2 domains (islands, pockets) on this surface. These domains could be induced simply by scratching. In the experiment by Liu et al. [69], CNTs grew on the sputter-deposited SiO_2 surface, but not on the thermally grown SiO_2 surface. Very thinly dispersed SWCNTs were obtained on a 10-nm-thick sputter-deposited SiO_2 film as compared to those on 30-nm- and 100-nm-thick films. This happened because the sputter-deposited SiO_2 films had the characteristics of NP2 nanoparticle, but the thermally grown SiO_2 films had the characteristics of NP1 nanoparticles. Due to scratching or during sputter-deposition, stoichiometric SiO_2 probably lost some oxygen and was transformed to non-stoichiometric SiO_z. The thickness of the sputter-deposited SiO_2 film mattered for CNT growth because the size and characteristics of nanoparticle depend on the thickness of this film. Liu et al. [69] observed SWCNTs growing in the scratched wafer but not in the unscratched wafer. CNT growth on scratched wafer was rather straightforward if it could be carried out at a precisely pre-defined location. Scratched substrates yielding CNTs had NP2 type characteristics. If the CNT growth could be carried out on unscratched substrates of NP1 type characteristics, and by the VS mechanism, then there would be no need of pre-defined, properly engineered location of some specific composition for this growth.

In a subsequent experiment, Liu et al. [70] could grow SWCNT on silica, but not on Si. Based on in situ TEM micrographs, silica, unlike silicon, was actually solid and amorphous having SiO_z ($1 \le z < 2$) composition. Liu et al. found that the R_L species of the SiO_z nanoparticles remained amorphous and solid during the entire growth process. Raman analysis confirmed that the R_L species of the SiO_z nanoparticles was indeed amorphous rather than crystalline. AFM height analysis suggested that it was coarsened; the surface roughness was in the ranges of 2–30 nm, which was not required for growth by the VS mechanism, but required for growth by the VQS mechanism.

To summarize, the studies by Liu et al. demonstrate that defect, disorder, amorphicity, and surface roughness are important ingredients of catalytic activities of the R_L species of nanoparticle surface. And this R_L species could be quasiliquid and porous. All these can be realized by scratching the nanoparticle surface or by oxygen deficiency in the non-stoichiometric oxide. This conclusion seems to corroborate with the observations by Steiner et al. [71], who made use of zirconium oxide as catalyst nanoparticles for CNT growths. These catalyst nanoparticles were active because they were neither reduced to Zr nor carbonized into ZrC during growth. Instead they remained actually as oxygen-deficient zirconium oxide [ZrO_z, $1 \le z < 2$] and hence more active as catalyst than ZrO_2. Defects due to the extraction of oxygen from ZrO_2 played a crucial role in generating R_L species and the catalytic activity of zirconium oxide nanoparticles for CNT growth.

The SUBSANO activities of Liu et al. [69, 70] for CNT growth are consistent with the SUBSANO activities of Kim et al. [50] for nanowire growth. Prior to Si nanowire growth, Kim et al. [50] performed hydrogen-terminated silicon wafer etching with ultrapure water to generate a reactive silicon-rich, oxygen-deficient, amorphous oxide (e.g., SiO_z) layer on the substrate surface. What it means may probably be explained by Fig. 16.4a, b. While Fig. 16.4a depicts the presumed feature of the VS process, Fig. 16.4b shows the real feature of the VS process. Figure 16.4b indicates that the

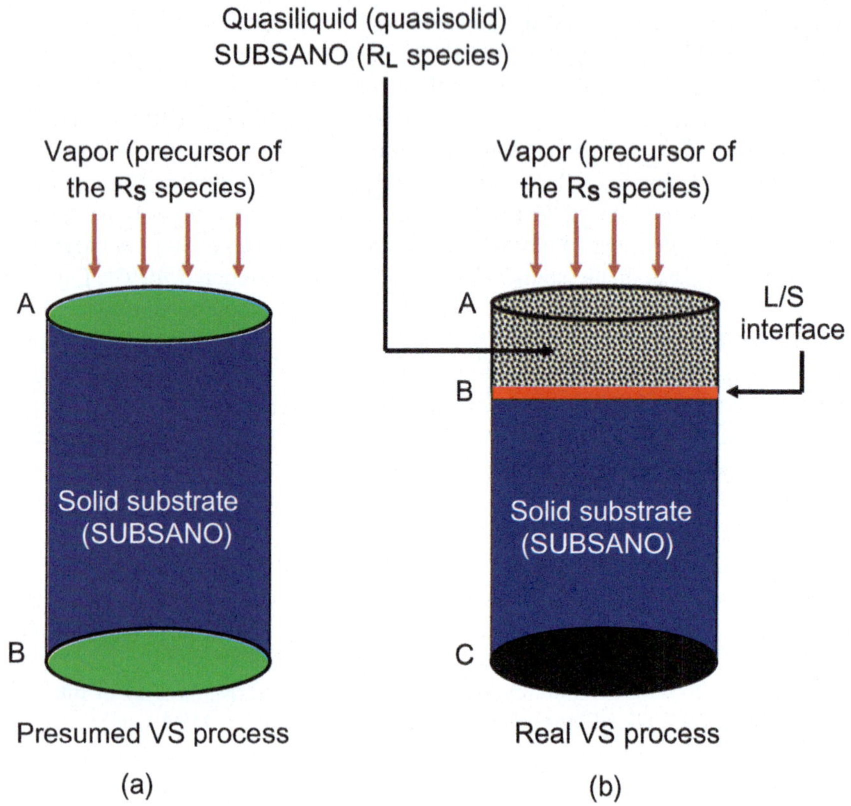

Fig. 16.4 Schematic illustration showing the differences of the presumed VS mechanism from the real mechanism taking place for growth. The real VS mechanism involves the diffusion of the R_S species through a disordered, coarsened, amorphous (semi-amorphous), porous R_L species formed on the top surface of the SUBSANO. This R_L species on the top surface has grains, molten (liquidic) grain boundaries and molten (liquidic) nanopores (not shown); it has a finite thickness of δ_{amor}. It is believed that the real VSS mechanism, which is a variant of the VQS mechanism, involves supersaturation and nucleation at the liquid/solid interface

R_L species of SUBSANO surface is actually disordered, for example, by plasma bombardment, aqua regia treatment, etc., up to a depth of $AB \equiv \delta_{amor}$ (see Figs. 12.3b and 12.4b of Chap. 12) leaving the bulk of its depth BC still unaffected. The disordered surface is amorphous (semi-amorphous, amorphous-like) and porous; it has a network of nanopores typical of a quasiliquid (quasisolid) surface, as depicted in Fig. 16.4b. The SUBSANO surface can otherwise be reconstructed. Gomes et al. [72] observed that substrate surface is reconstructed by thermal annealing. They noted that islands resulting on the reconstructed surface, if kept at a low temperature of about 400 °C for 20 min, led to the adsorption of In and As modifying the island composition from Si to $R_L \equiv$(In, As, Si) alloys. This low temperature is important to preserve the compositional integrity of the $R_L \equiv$(In, As, Si) alloys. These alloys are possibly solid solutions of (In, As, Si); they may contain some other contaminants. So, they were R_L species of disturbed (disordered) lattice structure resulting from lattice mismatch between InAs and Si. We argue that similar R_L species of disturbed, disordered, amorphous lattice structure were generated during the CNT growths by Liu et al. [69, 70], Kim et al. [50], and Steiner et al. [71]. The SUBSANOs were transformed to the NP2 type SUBSANOs and that these growths were carried out actually by the VQS mechanism rather than by the VS mechanism. The VS mechanism is a special case of the VQS mechanism only for the R_L species of negligibly small nanopores, porosity, coarsening, melting, and HETs.

16.8 Nanobelt Growth is by the VQS Mechanism, not by the VS or the VLS Mechanism

16.8.1 Preliminary Note

Quasi-one-dimensional nanostructure-based nanobelts (see Fig. 1.1 of Chap. 1) are single-crystalline nanomaterials with specific oriented surfaces. The rectangular cross section, uniform thickness, and ribbon belt-like morphology make these nanobelts ideal candidates for varied applications. They exhibit unusual optical, electronic, and/or mechanical properties as compared to those from bulk materials. They are crucial for the generation of renewable energy and for serving as sensors. They offer unique opportunities for interfacing biological systems providing platforms for individual molecules to serve at a nanoscale for biomolecular device structures. They have other important advantages such as good high-temperature stability, oxidation resistance, and stable electric properties. They have therefore attracted much attention in recent years. They are wide band gap semiconductors if made, for example, from III-V nitrides such as GaN and AlN, and from transparent conducting oxides.

16.8.2 Nanobelt Synthesis and Conflicts and Contradictions in This Synthesis

A fascinating endeavor during the past years has been to synthesize nanobelts as nanoscale building blocks of arbitrary dimensions, morphologies, desired functions, and controllable electric-mechanical properties. Although still debatable [73], the nanobelts are widely believed to be synthesized by the VLS and/or the VS growth mechanisms. These growths are extensively performed by thermal evaporation of generally pure oxide powders with or without any MET catalyst nanoparticle. Growth parameters such as evaporation time, chamber pressure, and carrier gas flow rate are carefully selected and closely monitored during growth. The evaporation temperature is determined based on the melting point of the R_S species. It varies from materials to materials. After evaporation, the vapors are transported by the carrier gas from the high-temperature zone to the low-temperature zone of the growth chamber where crucible and/or substrate are placed for the growth to take place or for the growth products to be collected. The as-deposited products are characterized and analyzed by tools such as XRD, SEM, TEM, and EDS.

While growing α-Fe_2O_3 nanostructures, including nanowires and nanobelts, Wen et al. [74] observed that these nanostructures grown in the O_2 (25 sccm) atmosphere were primarily nanobelts. Song et al. [75] reported that the partial pressure of oxygen plays a key role in determining the morphology of ZnO nanostructures. They observed that the growth of ZnO nanowires was facilitated by controlled decrease in the oxygen partial pressure of the reaction chamber. This means the growth of ZnO nanobelts was prompted by controlled increase in the oxygen partial pressure of the reaction chamber. Dalal et al. [76] and Lee et al. [77] attributed higher oxygen partial pressure to the growth of ZnO and Ga_2O_3 nanobelts, rather than to the growth of ZnO and Ga_2O_3 nanowires, respectively.

While growing SnO_2 nanostructures, Mondal et al. [78] found no nanowires grown at 750 °C. Only a very few nanorods of diameter 50–60 nm and length around 200 nm were observed. A large number of nanowires of diameter 150–200 nm, and length 6–8 μm, were however observed at higher temperature, which was about 850 °C. Gold was clearly visible at the tips of these nanowires indicating that VLS mechanism might have governed the growth of these nanowires. SnO_2 nanobelts of more than 16 μm in length and 100–110 nm in width were also produced at much higher temperature, namely $T = 950$ °C. Several branched nanostructures had also been observed at this growth temperature. And these were clearly the SnO_2 nanobelts of length 80 μm to a few mm, all produced at a temperature of 1000 °C. These observations suggest that nanowires grow at lower temperature, but nanobelts grow at higher temperature, both under identical conditions. Note that higher pressure allows higher vapor needed for higher nanomaterial growth, and thus would be possible at higher temperatures. In other words, higher temperature would be needed for growth under higher pressure. The finding by Mondal et al. [78] thus translates to the fact that nanowire growth is preferred under low partial oxygen pressure, but nanobelt growth is preferred under high partial oxygen pressure. This was contradicted by Nam et al. [73], who grew GaN nanobelts at a lower temperature of 921–970 °C, but nanowires at a higher temperature of 1000–1034 °C.

16.8.3 Illustrative Demonstration of Nanobelts Being Growths by the VQS Mechanism

16.8.3.1 SnO Nanobelt Synthesis

Employing carbothermal evaporation process, Orlandi et al. [79] synthesized SnO nanostructures in a sealed tube furnace. The SnO_2 powder, for this synthesis, was mixed with carbon black at a molar ratio ($[SnO_2]/[C]$) of 1.5. An aluminum boat containing the mixture of SnO_2 and C (e.g., carbon black) was placed in the hot zone of a tube furnace. The synthesis was carried out at 1210 °C and 1260 °C, respectively, for 2 h under a N_2 gas flow rate of 40 cm^3/min. EDX analysis was then performed of the SnO nanobelts produced in the chamber. The analysis indicated that the product had a composition close to that of SnO. While trying to understand the growth mechanism of SnO nanobelts, the following chemical reactions were assumed to take place during the carbothermal reduction process. Note that this process took place during the pre-nucleation stage of growth:

$$SnO_2(s) + C(s) \rightarrow SnO(g) + CO(g) \tag{16.1}$$

$$SnO(g) + CO(g) \rightarrow Sn(l) + CO_2(g) \tag{16.2}$$

Note also that the reaction (16.1) occurred in the alumina boat in the hot zone of the tube furnace. N_2 gas then carried various reactants downstream to a lower-temperature zone. Reaction (16.2) taking place in this zone yielded CO_2 which eventually desorbed.

One of the products of reaction (16.2) is liquid metallic tin, which has a melting point of 231.9 °C. The EDX analysis of the tin present at the nanobelt tips indicated an oxygen concentration in the range of 2–5 at.%. And this concentration is higher than the solubility limit concentration of oxygen in metallic tin. It implies that the R_L species of the nanoparticles responsible for growth was a solid solution: $R_L \equiv$(Sn, O). Residual carbon not participating in the reaction(s), together with some other contaminant ϑ, could also be present in the said solid solution. The R_L species was therefore actually $R_L \equiv$(Sn, C, O, ϑ) species. Obviously, the R_L species of the nanoparticles was amorphous, porous, and molten. It was molten plausibly because Sn as a constituent of it was molten at the growth temperature $T \gg 231.9$ °C. Following the formation of the R_L species, the pro-nucleation stage of growth began. $R_S \equiv$Sn and $R_S \equiv$O vapor source species created at the growth temperature were then adsorbed onto the R_L species surface. They underwent diffusion through the R_L species creating Sn + O $\rightarrow$ SnO molecules. Upon supersaturation at the liquid/solid interface, these molecules yielded nanobelts. This means the entire process of growth appeared to follow the MET-free VQS mechanism.

16.8.3.2 ZnO Nanobelt Synthesis

High-purity ZnO nanobelts synthesis by Chen et al. [80] involved thermal evaporation of ZnS powder in an environment of hydrogen–oxygen gas mixture and at a temperature of 1050 °C. Again, the synthesis followed the two basic stages: the pre-nucleation stage and the pro-nucleation stage. The source materials loaded into an Al_2O_3 crucible at the center of a quartz tube (diameter of 30 cm, length of 100 cm) were placed in a horizontal tube furnace. The white film-like growth products were deposited on a Si substrate located downstream of the gas flow inside the furnace. The source species employed for the growth was high-purity sphalerite. The temperature employed for the synthesis was lower than the sublimation point of about 1200 °C of sphalerite. It was lower as well than the previously reported growth temperature of 1350 °C. The sublimation of sphalerite during growth could therefore be ruled out. Various possible reactions for growth performed in the presence of H_2 and O_2 could be described as follows:

$$ZnS(fcc) + H_2(g) \rightarrow Zn(g) + H_2S(g) \tag{16.3}$$

$$2H_2S + 3O_2 \rightarrow 2H_2O(g) + 2SO_2 \tag{16.4}$$

$$2Zn(g) + O_2 \rightarrow ZnO\,(hcp) \tag{16.5}$$

As indicated above, the reaction (16.3) between the ZnS powder and H_2 at 1050 °C yielded Zn and H_2S vapors. This H_2S vapor was then oxidized to produce H_2O and SO_2. Obviously, ZnS was not fully consumed during the reaction (16.3). There was also residual ZnS powder dissociated into Zn and S atoms, or converted into small non-stoichiometric ZnS_z or $Zn_{z'}S_z$ (z and z' are the mole fractions, $z' \neq z$) clusters at the growth temperature. O atoms and some other by-products could also have migrated into these clusters. The nanobelt growth required nanoparticles for growth. We believe ZnS_z or $Zn_z{}'S_z$ (z and z' are the mole fractions, $z \neq z'$) clusters precipitated as small nanoparticles. The melting point of ZnS is 1185 °C; so these non-stoichiometric nanoparticles were solid at the growth temperature of 1050 °C. The R_L species of these nanoparticles was porous and amorphous of effective amorphicity $\alpha_{amoreff}$.

We also present an alternative possibility. It had been noted [81] that H_2S (and other sulfur impurities) can adsorb almost irreversibly to high coverage onto various substrates and/or metal catalysts. Sulfur had been found [82] also to adsorb in poly-crystalline and supported Ni catalysts. Together with Zn, O, and other contaminants, H_2S and/or S could have produced clustered nanoparticles. The melting point of sulfur is 115.2 °C and melting point of zinc is 419 °C. At the growth temperature of 1050 °C, they were both molten, and the R_L species of the said clustered nanoparticles composed of them was porous, amorphous, and molten (semimolten). During the pro-nucleation stage of growth, $R_S{\equiv}Zn$ and, $R_S{\equiv}O$ source species vapors diffused through the porous R_L species of the nanoparticles. Even at relatively low temperature, they formed polar ZnO molecules. Upon supersaturation at the interface of molten nanoparticle and solid Si substrate which lied underneath the nanoparticle, ZnO molecules were eventually crystallized into ZnO nanobelts. The growth was obviously performed by the VQS mechanism.

ZnO nanobelts by Zhang et al. [83] were prepared by thermal evaporation of $ZnCl_2$ powder in oxygen atmosphere, but without the presence of catalyst. The experimental setup was essentially the same as stated earlier. It consisted of a horizontal tube furnace and quartz tube reactor. The $ZnCl_2$ powders were placed at the center of an alumina crucible, and the crucible was then inserted into the quartz tube reactor in a horizontal tube furnace. The temperature of the crucible was rapidly increased to 700 °C from room temperature and was kept at 700 °C for 1 h under a constant flow of oxygen and argon gas mixture. The Ar:O_2 ratio was 10:1. Following the evaporation, the products were deposited onto an Al_2O_3 substrate placed at the downstream end of the quartz tube. White wool-like products formed on the substrate surface and on the walls of the crucible were found to be ZnO nanobelts. Note that they were produced by the evaporation of just $ZnCl_2$ in the presence of oxygen, and no other material was used for the growth.

During the pre-nucleation stage of growth, $ZnCl_2$ decomposed into Zn and chlorine at the high growth temperature of 700 °C. Some of the chlorine adsorbed on the alumina surface and on the wall of crucible creating islands. Scanning tunneling electron microscopy and photoemission spectroscopy together with *ab initio* simulations [84] showed that Cl easily adsorbs, for example, on the clean rutile TiO_2 surface even at room temperature. In fact, Cl adsorbs and binds to the Ti atoms in an on-top configuration. There was energetically more favorable adsorption on reduced surfaces, where bridging oxygen atoms created vacancies. STEM measurement, XPS analysis, and the simulations by density functional theoretical model [85] demonstrated together that chlorine interacts with Au(111) substrate (catalyst) creating different chlorine coverage at different levels of interactions. At low Cl coverage at the initial stage, Au atoms are reconstructed and released to relieve surface stress. Continuous chlorine adsorption leads Cl atoms to bind on top of Au layer, ultimately leading to the formation of a Cl overlayer. At higher coverage, more Au atoms are removed from the surface and a surface chloride compound containing Au atoms are formed. We argue that, in the present case, zinc and oxygen vapor atoms, together with some contaminant ϑ, were also adsorbed in the said islands creating an R_L species: $R_L \equiv$(Zn, O, Al, Cl, ϑ) species clusters and/or solid solutions on the nanoparticle surfaces. This R_L species was porous and amorphous. Due to the melting point of zinc being 419 °C, it was not solid; it was rather molten (semimolten).

There were plenty of Zn and O vapor atoms in the growth chamber during the pro-nucleation stage of growth. They landed on the nanoparticle surface and diffused through it yielding ZnO molecules. Upon supersaturation, these molecules nucleated into nanobelts. They were produced on the surface of the alumina plate at the lower-temperature zone of the furnace. We argue that the ZnO nanobelt growth was carried out by the MET-free VQS mechanism.

16.8.3.3 GaN Nanobelt Synthesis

Xiang et al. [86] synthesized GaN nanobelts employing Ga powder placed on a Ni-coated silicon substrate. The growth was conducted inside a horizontal tube furnace at 950 °C. The NH_3 flow (flow rate: 50 ml/min) into the chamber was maintained during the growth of 60 min. After the reaction, the furnace was cooled down. Typical SEM images revealed that the products were GaN nanobelts. They had the width ranging between 80 and 200 nm, and the thickness ranging between 10 and 30 nm, respectively. The nanobelt lengths were up to several tens of mm. EDS measurements indicated that the atomic Ga/N composition ratio of the nanobelts was close to 1:1.

To examine the catalytic effect of nickel on growth, the experiment was repeated under identical growth conditions, but without the presence of Ni catalyst. It was found that, under these conditions, only a few GaN nanobelts grew. This finding was a confirmation of the critical catalytic role played by the Ni catalyst during the nucleation and growth of GaN nanobelts. It was hence thought that the growth was carried out by the VLS mechanism, although no Ni was found at the nanobelt tips.

We argue that nanoparticles were formed on the Si substrate during the pre-nucleation stage of growth. The nickel/gallium system has been investigated by Hellner [21] and by Katayama et al. [87]. They suggested that the nickel solid solution extended to 28 at.% gallium, while the lattice parameter increased to 3.591 Å at the solubility limit. In the course of a systematic investigation of nickel alloy lattice parameters, Pearson [88] found a Ni_3Ga phase, which took place at 1189 °C. This alloy had an ordered phase of a relatively small homogeneous range. On the nickel-rich side, a wide heterogeneous range was though found at low temperatures. A considerable difference of lattice parameters of the two coexisting cubic phases was also observed.

Recently, Jones [89] studied Ni/Ga phases by allowing the nickel and gallium to melt by diffusion. He observed that, at an annealing temperature of 600 °C, the constituent atoms of the melt organized themselves into thermodynamically stable structures. These structures corresponded to Ni_3Ga (75 at.% Ni) α' phase, Ni_5Ga_3 (62.5 at.% Ni) δ phase, and NiGa (50 at.% Ni) β phase. They corresponded also to Ni_2Ga_3 (40 at.% Ni) phase and Ni_3Ga_7 (30 at.% Ni) phase.

To determine the crystal structures of various products, powder XRD analysis [89] was carried out. The database diffraction patterns were combined with the lattice parameters obtained via Rietveld refinement. The diffraction pattern for Ni_5Ga_3 showed peaks from Ni_5Ga_3, Ni_3Ga, and NiGa, indicating that this alloy (viz., Ni_5Ga_3) was a solid solution of three phases, namely Ni_5Ga_3, Ni_3Ga, and NiGa phases, respectively. According to Rietveld refinement, the composition was 36.3 wt% Ni_5Ga_3, 10.0 wt% Ni_3Ga, and 53.7 wt% NiGa. The diffraction pattern for Ni_3Ga_7 indicated that it had a body-centered cubic lattice and that it was a mixture of Ni_3Ga_7 (74.7 wt%) and Ni_2Ga_3 (25.3 wt%) phases. Ni_3Ga and NiGa were pure phases, but Ni_5Ga_3 and Ni_3Ga_7 were mixed solid solution phases. These phases were probably created by localized rearrangement of the Ni and Ga atoms during annealing or milling.

The experimental data listed in Table 16.5 include those by Jones [89], by Okamoto [90], and by Ingerly et al. [91]. Like the data of Table 16.1, they point to the formation of unconfirmed eutectic (Ni, Ga) alloy at the growth temperature of 950 °C. However, Rizel et al. [92] suggested that the Ni_2Ga_3 phase stable at 895 °C, as observed by Ingerly et al. [91], was actually the eutectic phase. Notably it occurred at a temperature much lower than the nanobelt growth temperature of 950 °C. So, the nanoparticles formed on the Si substrate at 950 °C could not have $R_L \equiv$(Ni, Ga) eutectic alloy droplets. This is consistent with the Ni particle droplets not found at the nanobelt tips. The only other possibility was then that a $R_L \equiv$(Ni, Ga, ϑ), or

Table 16.5 List of some selected Ni/Ga phases of the NiGa alloy identified during the past years

Phase	Composition, atomic % Ni	Comments	References
NiGa	50	Cubic, stable at 1220 °C	Ingerly [91]
$NiGa_5$	16.7	Stable at 611 °C	Okamoto [90]
Ni_3Ga_7	29–30	A solid solution of phases	Jones [89]
Ni_2Ga_3	40.9	Trigonal, stable at 895 °C	Ingerly [91]
Ni_3Ga_4	42.4–43.4	Stable at 961 °C	Okamoto [90]
Ni_3Ga_2	60	Decomposes at $T > 900$ °C	Okamoto [90]
Ni_3Ga	75	Non-stoichiometric compound alloy	Okamoto [90]
Ni_5Ga_3	62.5	Orthorhombic, A solid solution of phases	Jones [89]
Ni_5Ga_3	62.5	Stable at 741 °C	Ingerly [91]
β	42.0–69.4	–	Okamoto [90]
ε	59.0–59.5	–	Okamoto [90]
γ'-$Ni_{13}Ga_9$	59.9	Monoclinic, stable at 680 °C	Ingerly [91]
δ	63.0–66.5	–	Okamoto [90]
α'	70.0–77.0	–	Okamoto [90]
Ni_3Ga	74.8	Cubic, stable at 1212 °C	Ingerly [91]

$R_L \equiv$(Ni, Ga, N, ϑ) species solid solution was formed during the pre-nucleation stage of growth, and that it was responsible for the nanobelt growths. Recall that Ga has a melting point of 29.76 °C, and probably because of this, the solid solution was molten. Following the formation of the $R_L$species of nanoparticle, $R_S \equiv$Ga and $R_S \equiv$N vapor atoms diffused through the R_L species yielding GaN molecules. Upon supersaturation, they were crystallized into GaN nanobelts at the interface of molten nanoparticles and the solid Si substrate, which was underneath these nanoparticles.

16.8.3.4 α-Si_3N_4 Nanobelt Synthesis

Huang et al. [93] grew Si_3N_4 nanobelts using silicon powder (99 wt% pure) milled and placed on a pure graphitic carbon felt substrate. It was coated with nickel nitrate: $Ni(NO_3)_2$. The Si powder, together with the substrate was then placed in an alumina tube furnace and heated in a N_2 (purity 99.999%) flow to 1450 °C. It was held at this temperature for 3 h. For the sake of comparison, the whole experiment was repeated under identical growth conditions, but without the presence of catalyst in the furnace. XRD pattern and the FTIR spectrum revealed that the product was composed of only α-Si_3N_4, and that only a few short nanowires grew during deposition without the presence of $Ni(NO_3)_2$. It implied that the presence of $Ni(NO_3)_2$ was key to the α-Si_3N_4 nanobelt growth. The following reactions probably took place during the nanobelt growth:

$$2Ni(NO_3)_2 \rightarrow 2NiO + 4NO_2 + O_2(g) \tag{16.6}$$

$$2NiO + C \rightarrow 2Ni(s) + CO_2(g) \tag{16.7}$$

We believe that Ni released from the reaction (16.7) acted as nanoparticle for growth. During the pre-nucleation stage of growth, Ni atoms created Ni islands on the carbon felt substrate. Note that the N_2 vapor was highly reactive at the growth temperature of 1450 °C. Based on transformation equation (16.8), stated in the following, plenty of Si powder was converted to Si vapor at this temperature:

$$Si(s) \rightarrow Si(g). \tag{16.8}$$

R_S≡Si and R_S≡N vapor atoms served as source species for α-Si_3N_4 nanobelt growth during the pro-nucleation stage of growth. Even though the reactions (16.6) and (16.7) took place during the pre-nucleation stage of growth tending to consume C and $Ni(NO_3)_2$, there were residual C and $Ni(NO_3)_2$ left in the growth chamber. There were also by-products and contaminants present in the chamber.

Let us now examine the state of the Ni catalyst nanoparticle and the possibility of the formation of (Ni, Si) eutectic alloy, which was required for the VLS growth. According to Dutra et al. [94] and Ram and Bhan [95], Ni and Si undergo eutectic transformation producing β_3-Ni_3Si alloy of a monoclinic structure. At 1143 °C, this monoclinic structure nevertheless suffers a polymorphic transformation to β_2-Ni_3Si and finally to β_1-Ni_3Si at 1035 °C. Gaudet et al. [96] observed several highly textured phases created during the thermally induced reaction of a Ni film with Si substrate. They identified a metastable hexagonal compound with $a = 3.8$ Å and $c = 4.9$ Å referred to as θ phase in the phase diagram. It was normally stable above 825 °C. According to the phase diagram, it had a composition ranging from Ni/Si ratios of 2:1 (33 at.% Si) to 3:2 (40 at.% Si). The lattice parameters of the θ phase ranged from $a = 3.836$ Å and $c = 4.948$ Å for 37.5 at.% Si to $a = 3.802$ Å and $c = 4.863$ Å for 43 at.% Si. The discussions made above suggest that no stable eutectic R_L≡(Ni, Si) phase could be created on the nanoparticle surface at the growth temperature of 1450 °C. The alternative would therefore be the possibility of a R_L species solid solution mediating the growth. This solid solution on the nanoparticle surface could be a combination of more than one Ni-Si phase; it could be R_L≡(Ni, Si, C, N, O, ϑ) species. Based on the descriptions in Chap. 12, the most viable growth mechanism would then be the VQS mechanism.

16.8.4 The R_L Species Responsible for Nanobelt Growths

The R_L species responsible for the growth of some nanowires and nanobelts [78, 83, 86, 93, 97–108] are listed in Table 16.6. As stated earlier, these R_L species were created on the nanoparticle surface during the pre-nucleation stage of growths. And

Table 16.6 The physicochemical processes, the R_S source species, and the R_L nanoparticle species (e.g., cluster, alloy and/or solid solution) that lead to the formation of nanowires (NWs) and/or nanobelts (NBs)

No.	Material NW (NB)	Process (temp, °C)	Species on the nanoparticle surf	R_L species	Characteristics of the R_L species	References
1	SnO_2 (NW)	Thermal evaporation (850)	Au, Sn, O, N_2	R_L≡(Au, Sn)	Eutectic alloy	Mondal [78]
2	SnO_2 (NB)	Thermal evaporation (1000)	Au, Sn, O, N_2	R_L≡(Au, Sn, O)	Non-eutectic alloy, Solid solution	Mondal [78]
3	ZnO (NB)	Thermal evaporation (700)	$ZnCl_2$, O_2	R_L≡(Zn, Cl, O)	Cluster, Solid solution	Zhang [83]
4	In_2O_3 (NB)	Thermal evaporation (1400)	In, O	R_L≡(In, O)	Cluster, Solid solution	Kong [97]
5	CdO (NB)	Thermal evaporation (1000)	Cd, O	R_L≡(Cd, O)	Cluster, Solid solution	Wang [98]
6	GaN (NB)	Thermal evaporation (950)	Ga, Ni, NH_3	R_L≡(Ni, Ga, N)	Non-eutectic alloy, Solid solution	Xiang [86]
7	Sn_3O_4 NB)	Carbothermal reduction (1200)	SnO_2, N_2, C, O	R_L≡(Sn, C, N, O), R_L≡(SnO_{2-z}) suboxide	Cluster, Solid solution	Berengue [100]
8	α-Si_3N_4 (NB)	Thermal CVD (1450)	Si, N_2, $Ni(NO_3)_2$	R_L≡(Si, N, Ni, O)	Non-eutectic alloy, Solid solution	Huang [93]
9	SiO_2 (NB)	Thermal evaporation (1300)	SiO, Si clusters	R_L≡(Si, O)	Cluster, Solid solution	Zhang [101]
10	AlN (NB)	Vapor–solid process (1200)	Al, NH_3, $AlCl_3$	R_L≡(Al, Cl, N)	Non-eutectic alloy, Solid solution	Xie [102]
11	ZnO (NB)	Thermal evaporation (850–860)	Zn, Sn, O	R_L≡(Sn, Zn, O)	Non-eutectic alloy, Solid solution	Wang [103]

(continued)

Table 16.6 (continued)

No.	Material NW (NB)	Process (temp, °C)	Species on the nanoparticle surf	R_L species	Characteristics of the R_L species	References
12	α-Al_2O_3 (NB)	Thermal evaporation (1350)	Al, $A_{l2}O_3$	$R_L \equiv$(Al, Cl, N), $R_L \equiv$(Al_2O) suboxide	Non-eutectic alloy, Solid solution	Zhang [104]
13	CdS (NB)	Thermal evaporation (850)	Au, In, CdS	$R_L \equiv$(Au, Cd, In, S)	Non-eutectic alloy, Solid solution	Ma [105]
14	NiSi (NB)	Thermal evaporation (900)	Ni, N_2, $SiHCl_3$	$R_L \equiv$(Ni, Si, Cl) $R_L \equiv$(Ni_ySi_z)	Cluster, alloy, Solid solution	Zhang [106]
15	SnS_2 (NB)	Hydrothermal process (180)	CS_2, NH_3, $SnCl_4$	$R_L \equiv$(Sn, C, S, Cl) $R_L \equiv$(SnO_{2-z}) suboxide	Cluster, alloy, Solid solution	Ma [107]
16	InN (NB)	Guided stream CVD (650)	a-SiN,Au, In	$R_L \equiv$(Au, Si, In)	Solid solution	Hu [108]

essentially, all of them are non-eutectic alloys, clusters, or solid solutions. Physico-chemical growth processes, the catalysts, if any, and the source species present on the nanoparticle surface during growth are also listed in Table 16.6. The R_L species for growths in several entries of this table resulted from the presence of oxygen on the nanoparticle surface. In fact, almost all growth chambers have intentionally or unintentionally introduced oxygen. And although not shown, the R_L species for all nanobelt growths might have oxygen content. These could as well be reactive, amorphous,porous, and molten suboxides. The entry 6 shows that the presence of Ni and N on the substrate surface for GaN nanobelt growth modified the R_L species to R_L≡(Ni, Ga, N). Entry 1 shows Au-catalyzed SnO_2 nanowire growth carried out at ~750 to 850 °C. The (Au, Sn) alloy constituting the R_L species becomes the molten $Au_{0.8}Sn_{0.2}$eutectic alloy at $T = 278$ °C. This R_L species was however a molten R_L≡$Au_{0.86}Sn_{0.14}$ alloy at the growth temperature of 750 °C. And hence, the VLS type mechanism was probably involved for the said nanowire growth at 750 °C. Entry 2 shows that, due to increase in O at a higher temperature of ~1000 °C, the alloy was transformed to R_L≡(Au, Sn, O), which is a non-eutectic alloy suitable for nanobelt growths. As indicated in entries 14 and 15, Cl played decisive role in establishing the final composition of the R_L species; it was transformed to solid solution by the Cl atoms. Huang et al. [93] observed many long α-Si_3N_4 nanobelts formed by the catalytic action of $Ni(NO_3)_2$. Very few such nanobelts could, on the other hand, grow without the catalytic action of $Ni(NO_3)_2$. As shown in entry 8, this nickel nitride was decomposed in the growth environment yielding R_L≡(Si, N, Ni, O) solid solution for α-Si_3N_4 nanobelt growth. We indicated in several occasions, but not mentioned here that phase transformations take place during growth and multiple phases may simultaneously participate in yielding R_L species for growth. So, the R_L species mentioned here may not always be the one solely responsible for growth. We indicated in entries 7, 12, 14, and 15 that highly reactive, yet amorphous, porous, molten suboxide, or subsilicide may also effectively serve as the R_L species. Note that clusters, non-eutectic alloys, non-stoichiometric oxides and solid solutions are not solid; as described in Sect. 16.12; they are rather quasiliquid (quasisolid). Following our discussions in Chap. 12, various nanobelts were thus formed during the pro-nucleation stage of growth by their mediation by the VQS mechanism.

16.9 SFLS and SoLS Mechanisms Are the Special Cases of the VQS Mechanism

16.9.1 Catalyst-Mediated Si and Ge Nanowires Grown by the SFLS Mechanism

Hanrath [109] synthesized Ge nanowires in supercritical cyclohexane or hexane using MET≡Au. The (Au, Ge) binary system has a eutectic temperature of 361 °C for the Ge content of 28 at.% in the R_L≡(Au, Ge) alloy. The Ge nanowires thus grown were largely defect-free. HRTEM showed defects only in less than 5% of the nanowires investigated. This implied that the R_L species for the growth had hardly any contaminant. Au-catalyzed Ge nanowire growth at a temperature of 370 °C and a pressure of 10 MPa using diphenylgermane was obviously by the VLS mechanism. Au-catalyzed Ge nanowire growth at 400 °C and 15 MPa using tetraethylgermane was also by the VLS mechanism. It was evident from the $R_L \equiv Au_{0.72}Ge_{0.28}$ alloy formed at the nanowire tip. With adequate supply of Ge, the eutectic composition of the (Au, Ge) alloy at 400 °C should have been $Au_{0.70}Ge_{0.30}$. Instead it was $Au_{0.72}Ge_{0.28}$ simply because the $R_L \equiv$(Au, Ge) alloy could achieve stability not at 400 °C, but at a lower temperature of ~361 °C. The VLS-type growth of Ge nanowires in supercritical fluid demonstrates that this growth also can take place only under very specific conditions of the R_L species.

Tuan et al. [110] used colloidal nickel nanoparticles to synthesize catalyst-mediated Si nanowires in an organic solvent. The reaction temperatures for the growth were 400–520 °C, and the pressures were from 14.3 to 23.4 MPa. These temperatures and pressures were well above the critical point of the solvent. The silicon precursors, i.e., arylsilane, alkylsilane, and trisilane were decomposed to silicon by Ni nanoparticles. Thermal decomposition of these precursors in organic solvents was otherwise found to be very difficult. The thermal stability of the Si–C and Si–Si bonds in alkylsilane and trisilane was just too strong to be broken even by Au nanoparticles. Ni nanoparticles were the only exceptions. These were confirmed by solid-phase alloying of Si in the Ni, which yielded (Ni, Si) solid solution at a temperature far below the eutectic temperature of 864 °C. Nash and Nash [111] suggested that the (Ni, Si) phase at $T < 800$ °C was not eutectic, peritectic, polymorphic, peritectoid, or eutectoid. We believe that gradual increase in temperature during growth accompanied phase changes in the (Ni, Si) binary alloy. These phase changes also led to changes in the solubility of Si in Ni and hence in SECINI and ξ_m. And at 460 °C, the solid solubility of Si in Ni reached its peak value. It was high enough to yield $R_L \equiv$(Ni, Si) solid solution of almost the highest value ρ_{cmod0} of the porosity ρ_c and almost largest magnitude $\alpha_{amoreff0}$ of the effective amorphicity $\alpha_{amoreff}$ (see Chap. 15). Optimal surface roughness, highest HET concentration, and greatest surface melting compatible to $\alpha_{amoreff}$ lead SECINI to approach SECINI0 and ξ_m to approach unity, e.g., 1. Both SECINI and ξ_m were then gradually decreased with further increase in temperature. This is why nanowires were not formed at $T <$ 400 °C before SECINI became SECINI0 and $\xi_m \approx 1$; and curly Si nanowires riddled with defects were formed at temperature between 400 and 450 °C. Also, low yields of nanowires with poor crystallinity and a large amount of particulate by-products were observed at $T > 500$ °C. Notably, very few poor-quality nanowires were produced at 520 °C, and no nanowires were produced at $T > 580$ °C because both SECINI and ξ_m were gradually lowered from their peak values at about 400 °C. Further, Ni nanoparticles were found [112] to promote the formation of high-quality crystalline Ge nanowires in organic solvents at growth temperature of approximately 410 °C, which is 352 °C below the lowest eutectic temperature of 762 °C of the $R_L \equiv$(Ni, Ge) binary alloy. EDS measurements of the $R_L \equiv$(Ni, Ge) alloy found at the tip of most of the nanowires indicated that the alloy composition varied from $Ni_{0.33}Ge_{0.67}$

to $Ni_{0.26}Ge_{0.74}$. These alloys were non-eutectic; they were essentially solid solutions, and not solid; they were significantly different from the eutectic alloy of composition $Ni_{0.37}Ge_{0.63}$.

16.9.2 Catalyst-Free Si and Ge Nanowires Grown by the SFLS Mechanism

Yuan and Tuan [113] reported the catalyst-free synthesis of single-crystalline silicon nanowires on a SiO_z-coated Si substrate in supercritical benzene. No metal catalysts were employed for growth. The nanowire synthesis was very sensitive to reaction temperature. Only a small amount of spherical particles with diameters of hundreds of nanometers were grown at temperatures $T < 400$ °C. Nanowires could however be produced at temperatures 400 °C $\leq T \leq$ 440 °C; they were curly and short exhibiting poor crystalline quality. Straight and long nanowires could be synthesized at a temperature of 460 °C; and only micrometer-sized Si particles were formed at temperatures exceeding 500 °C. The surface energy of Si is 1140 mJ/m^2. But the surface energy of C is 75–150 mJ/m^2, and the surface energy of SiO_2 is 287 mJ/m^2. HRTEM images of the as-synthesized Si nanowires showed that these nanowires were single crystalline, but coated with a thin (<2 nm), rough, amorphous layer composed of silicon dioxide and carbon. So, as usual, the low surface energy materials sheathed the high surface energy nanowires. It was confirmed by EDS micrographs. It suggested that the R_L species responsible for nanowire growths was $R_L \equiv$(Si, O, C, ϑ), and that it was a solid solution. The discussions made above demonstrate that the R_L species both for catalyst-free and catalyst-mediated nanowires growths via the solution phase technique gradually changed in composition and phases with change in temperature. At low temperature, (a) it was solid with SECINI far off SECINI0, $\xi_m \approx 0$, and no solubility of Si in Ni; (b) at moderately low temperature, it was nearly solid with SECINI still far off SECINI0, $\xi_m \ll 1$, and low solubility of Si in Ni; (c) at moderately high temperature, it was nearly molten with SECINI not yet close to SECINI0, $\xi_m < 1$, and solubility of Si in Ni moderately high; (d) at an optimally high temperature, it was fully molten with SECINI $\approx$ SECINI0, $\xi_m \approx 1$, and solubility of Si in Ni high. We argue that the $R_L \equiv$(Si, O, C, ϑ) species was solid solution suited for most efficient growth of nanowires by the VQS mechanism.

Ag_2S, Cu_2S, and Ag_2Se nanoparticles are generally mesoporous [114, 115]. Nanopore dimension in these nanoparticles can generally be smaller than 2 nm. Depending on the process of synthesis, nanopore dimensions can even be smaller—in the subnanomater range. This corroborates with the observation by Wang et al. [115], who attributed the exceptional catalytic ability of the Ag_2S, Cu_2S, and Ag_2Se catalyst nanoparticles to their structural characteristics, namely high density of vacancies and high mobility of silver or copper(I) cations in the rigid sublattice of Se^{2-} or S^{2-} ions. We believe the basics of the SoSS mechanism lie in the formation of solid solution and the diffusion/transport through this solid solution. The SoSS mechanism, like

SoLS mechanism, mediates growth at relatively low temperatures (typically 100–230 °C). The chalcogenite nanoparticles for them are superionic conductors. They have a high density of cation vacancies, which promotes adsorption of foreign species onto nanoparticle surface causing the R_L species to be cluster or solid solution. These foreign species may also be the R_S species yielding alloyed R_L species.

Being cluster or solid solution, these R_L species in both catalyst-free and catalyst-mediated growths, stated above, are generally non-stoichiometric; they are amorphous, semi-amorphous, or amorphous-like (see Chap. 12). They have coarsened surface accompanied by pits, vacancies, dislocations, etc., which give rise to nanopores. The R_L species may have humps, hillocks, protrusions, mounds, etc. Sharp edges of these humps, hillocks, protrusions, mounds, etc., undergo size-dependent mesoscopic effect (see Figs. 12.1a, b, 12.2a, b of Chap. 12). The R_L species consequently becomes quasiliquid (quasisolid) having solid structure together with molten (semimolten) islands. Hence, the nanowire growths by both the SFLS and SoSS mechanisms follow the tenets of the VQS mechanism.

16.10 SLS Mechanism is a Special Case of the VQS Mechanism

16.10.1 Demonstration with Si Nanowire Growths

Poulose et al. [116] employed MET≡Au to produce Si nanowires on Si(100) substrate. But no siliconnanowire was grown at the (Au, Si) eutectictemperature $T_E = 363$ °C, which corresponds to the $R_L \equiv Au_{0.81}Si_{0.19}$ eutectic alloy. Instead siliconoxidenanowires were grown at $T = 1100$ °C. The EDX analysis indicated that the R_L species for growth had gold, silicon, andoxygen: R_L≡(Si, Au, O). But the nanowires contained only silicon and oxygen. The R_L species, as observed in the experiment, were pockets of the gold silicide alloy phase at the gold-silica interface. Nanowire growth was found to occur from pits generated on/near the R_L species surface. This confirmed that the R_L species was not liquid as required by the SLS mechanism; it was rather solid solution exhibiting coarsened surface. The only source of silicon for nanowires was silicon substrate. Had the R_L≡(Au, Si) eutectic alloy been responsible for their growths at $T = 1100$ °C, based on the liquidus line of the Au/Si binary phase diagram, the Si content in the R_L≡(Au, Si) alloy would then be ~59.7 at.%. But Si released from the solid silicon substrate at $T = 1100$ °C was not sufficient; vapor pressure of Si derived from the solid silicon substrate was too low to permit it. Hence, the Si content in the R_L≡(Au, Si) alloy was lower than 59.7 at.%. It was a cluster, a solid solution, or a non-eutectic alloy, but not a eutectic alloy needed for growth by the SLS mechanism.

16.10.2 Demonstration with Indium Oxide Nanowire Growth

The In_2O_3 nanowire growth by Chen et al. [117] using $R_L \equiv$(Cu, In) alloy compound was claimed to be carried out by the SLS mechanism. Cu and Ni form a continuous series of solid solutions. The lattice spacing of the face-centered cubic solid solution thus formed varies almost linearly between Cu and In. For this, the atomic concentration is 3.6153×10^{-8} cm for Cu, and the atomic concentration is 3.5238 $\times$ 10^{-8} cm for Ni. There occurred change in phases during annealing of the (Cu, In) compound prior to growth. As the temperature was gradually increased, the (Cu, In) compound was transformed to $Cu_{11}In_9$ and In; $Cu_{11}In_9$ was transformed next to the thermal equilibrium R_L species: $R_L \equiv Cu_{16}In_9$, releasing In. This means In was driven out of the (Cu, In) compound during annealing at a relatively lower temperature of 550 °C. $R_L \equiv Cu_{16}In_9$ has a stable η phase at this temperature. It was not a liquid phase as required by the SLS mechanism; it was rather a cluster (solid solution). It served as catalyst for nanowire growth. In atoms diffused out of it through nanopores and reacted with O atoms yielding In_2O_3 molecules. Upon supersaturation, these molecules were nucleated into In_2O_3nanowires [117].

16.10.3 Demonstration with Silicon Nitride Nanowire Growths

α-Si_3N_4 nanowire growths by Liu et al. [118] on $Ni(NO_3)_2$-coated Si substrate were described earlier. Considering that these nanowires by Liu et al. [118] were sheathed with SiO_z amorphous layer, and that $Ni(NO_3)_2$ coating of the Si substrate contained O elements, we argue that the catalytic bases at the bottom of nanowires had also O elements. So, the catalytic bases might have been clusters, alloys, and/or solid solutions of Si, Ni, N, and O elements implying that the R_L species responsible for growth was $R_L \equiv$(Si, Ni, N, O). The presence of SiO_z sheath indicates that, being very highly reactive, it might have played also a role in the nucleation of nanowires. The $R_L$species had coarsened surface with pits and sharp-edged hillocks. They might have contained other impurity atoms. Based on binary phase diagrams [4], the eutectic alloy $NiSi_2$ is formed at 993 °C, and the non-eutectic $Ni_{0.3}Si_{0.7}$ alloy is formed at the growth temperature of 1150 °C. Due to the presence of N and O vapors, the composition of the non-eutectic alloy of the R_L species might have been different from the eutectic $Ni_{0.3}Si_{0.7}$ alloy.

16.10.4 Demonstration with Si Covered by a Layer of Ni

In an experiment by Yan et al. [119], the Si substrate was covered with a thin layer of Ni eventually creating $R_L \equiv$(Ni, Si) alloy. The Ni/Si binary phase diagram [4] suggests that the eutectic point for the $R_L \equiv (Ni, Si)$ alloy is 993 °C. However, nanowire growth took place at a temperature just above 930 °C on nanoparticles exhibiting $R_L \equiv Ni_2Si$ non-eutectic alloy species. TEM investigation showed that these nanoparticles were visible between the substrate surface and the amorphous Si nanowires confirming that these nanowires were actually grown on these nanoparticles, and the growth was done by the VQS mechanism.

16.10.5 *Inference*

The illustrations cited above demonstrate that the R_L species for the SLS nanowire growth by various experiments are not liquid; they are not either solid. Instead they are non-eutectic solid solutions. Depending on the growth conditions, they could as well be non-eutectic liquid solution. While in solid solution, they had the following features characteristic of NP2 nanoparticles:

1. disturbed, disordered, amorphous (semi-amorphous) surface with or without grains, and grain boundaries,
2. rough, coarsened surface,
3. high concentration of HETs instrumental for precursor decomposition at a temperature lower than that required for thermal decomposition,
4. voids, vacancies, dislocations, etc., and hence hillocks, pits, sharp edges, etc.,
5. porous and quasiliquid (quasisolid) (e.g., $0 < \xi_m < 1$) state,
6. polarity.

The nanowires grew therefore by the VQS mechanism. The SLS mechanism was a special case of the VQS mechanism. The R_L species for the SLS growth could rarely exhibit eutectic molten alloy with $\xi_m \approx 1$.

16.11 OAG Mechanism is a Special Case of the VQS Mechanism

Various investigations described in the OAG growth chapter (e.g., Chap. 10) suggest that the OAG mechanism has not yet been fully understood. The very observation that prompted the possibility of the OAG mechanism to be an important growth mechanism was the significant increases in the yield of Si nanowire product due to the presence of SiO_2 in the Si source. The Si nanowire product obtained using 50% SiO_2 and 50% Si was 30 times larger than that obtained using a metal-containing target. Park et al. [120] employed non-stoichiometric SiO_z, InO_z, and AsO_z, ($z \approx 1$) substrate nanoparticles (e.g., SUBSANOs), but not metal (e.g., not Au) nanoparticles (e.g., METANOs), for the growth of InAs nanowires. Based on the OAG mechanism, it is not clear why they could grow nanowires at 580 °C on only SiO_z, but not on InO_z, and AsO_z. Based on the OAG mechanism, it is not also clear why Lan et al. [121]

could observe that oxide-assisted growth rate of InN nanowires by employing Au and In_2O_3 was much higher than the oxide-assisted growth rate by employing only In_2O_3. In both cases, InN nanowires were produced on the In_2O_3 nanoparticle surface. Liu et al. [69, 70] grew CNTs at 900 °C by making use of non-stoichiometric SiO_z ($1 \leq z < 2$) formed on Si substrate. Liu et al. [69] noted that sputter-deposited SiO_2, but not thermally grown SiO_2, could serve as SUBSANOs for CNT growths. As stated earlier, sputter-deposited SiO_2 is structurally disordered having the composition of non-stoichiometric SiO_z; it has high-energy sites (HETs) and plausibly dipole moment. It corresponds to NP2 type nanoparticles. Thermally grown SiO_2 is, on the other hand, stoichiometric, crystalline, and non-amorphous; it does not exhibit HETs. It corresponds to NP1 type nanoparticles. All these explain probably why there occurred a significant increase in the yield of Si nanowire product due to the presence of SiO_2 in the Si source. They suggest that there was a reaction: $Si + SiO_2 \rightarrow 2SiO$ yielding highly reactive SiO, and that this SiO is non-stoichiometric having the composition of solid solution of NP2 type nanoparticle.

We explore further. A close look of Table 13.1 of Chap. 13 would indicate that the materials at nanowire tips examined after nanowire growths were solid solutions; they were disturbed and disordered. For example, Tang et al. [122] found a solid solution of R_L≡(In, P, O) at the tip of InP nanowires. Likewise, Lan et al. [121] observed a solid solution of R_L≡ (Au, In, O) at the tips of InN nanowires, and Shi et al. [123] found R_L≡(Ga_zO) at the tip of GaP and GaAs nanowires, respectively. All of these nanowires were believed to be grown by the OAG mechanism. The R_L≡(Ga_zO) was found to be amorphous in the same way as R_L≡(In, P, O) and R_L≡(Au, In_2O_3) were found to be solid solutions. Also, R_L≡(Ga_zO) served the same purpose as did the R_L≡(In, P, O) and R_L≡(Au, In_2O_3) solid solutions for nanowire growth. The R_L≡(Au, In_2O_3) solid solution could actually be R_L≡(Au, In, O) solid solution. As discussed earlier (see Chap. 12), solid solution exhibits disturbed, disordered, amorphous (semi-amorphous, amorphous-like) lattice structure, and coarsened surface. This surface is porous and quasiliquid (quasisolid) with $0 \leq \xi_m \leq 1$; it belongs to the NP2 type nanoparticles. It satisfies the requirements for nanomaterials growths by the VQS mechanism. It demonstrates that, just not any oxide surface, but the ones with disturbed, disordered, amorphous (semi-amorphous, amorphous-like) surface lattice structure, coarsening, dipole moment, porosity, HETs, and surface melting (see Figs. 12.1a, b, 12.2a, b of Chap. 12) catalyze nanomaterial growth. This explains why SiO_z, but not InO_z and AsO_z ($z \approx 1$) could catalyze InAs nanowires at 580 °C [120]. Although they are all non-stoichiometric, the degree of non-stoichiometry for $z \approx 1$ and at T=580 °C is different in different oxide.This non-stoichiometry strongly depends on specific value of z between 1 and 2. This is the reason of why SiO_z ($z \approx 1$) was porous at 580 °C, but InO_z and AsO_z ($z \approx 1$) were non-porous at 580 °C. This explains also why InN nanowire growth using Au and In_2O_3 was much higher than InN nanowire growth just by using In_2O_3 [121]. The R_L species surface produced for the said InN nanowire growth was far more stressed and hence much more porous with both Au and In_2O_3 and with just In_2O_3.

Key to the selective area epitaxial growth mechanism [124] is the presence of an oxide mask layer of well-defined nanopores. The R_S species diffuse through these nanopores to the L/S interface for supersaturation prior to layer-by-layer nanowire growth. This mask blocks the nanowire growth in the opaque mask areas exhibiting no nanopores. At low temperatures, oxides are LXOs; but at high temperatures, they are HXOs (see Table 10.2 of Chap. 10). SUBSANOs, which are made of HXOs, may be regarded as mask layers. The nanopores of this mask layer depend on temperature. The size of the nanopores increases generally with increase in temperature. At a certain temperature, this size becomes comparable to those of the $R_S{\equiv}X$ and $R_S{\equiv}Y$ species or X_mY_n molecules formed from the $R_S{\equiv}X$ and $R_S{\equiv}Y$ species. They all diffuse through nanopores to the L/S interface for supersaturation before nanomaterial (nanowire, nanobelt, etc.) growths. Liu et al. [70] found SiO_z to be amorphous by itself and hence porous. We explained earlier why Park et al. [120] could not grow InAs nanowires at 580 °C by employing InO_z and AsO_z catalysts. To reiterate, it could plausibly because of one or more of the following reasons:

1. The nanopores in the R_L species of InO_z and AsO_z SUBSANOs at the growth temperature of 580 °C were too small for the R_S species to diffuse to the L/S interface.
2. The effective amorphicity $\alpha_{amoreff}$ of the R_L species of InO_z and AsO_z SUBSANOs at the growth temperature of 580 °C was too small as compared to the peak effective amorphicity $\alpha_{amoreff0}$ of them.
3. The surface roughness of the R_L species of InO_z and AsO_z SUBSANOs at the growth temperature of 580 °C was too small to have high concentration of highly effective HETs.

Kim et al. [50] found that nanowire growth rate increased with increase in temperature because the diffusion of the R_S species through the nanopores of the HXO SUBSANO to the L/S interface for supersaturation increased with increase in temperature.

16.12 SCG Mechanism is a Special Case of the VQS Mechanism

16.12.1 SCG Mechanism Based on Substrate Surface Disorder

Dhara and Giri [125] synthesized Si nanowires by the SCG mechanism. For the growth of these Si nanowires, commercial Si(100) wafer was first thermally oxidized at 900 °C in pure O_2 atmosphere for one hour. A thin Si layer (about 300 nm) was then deposited on it at room temperature using RF magnetron sputtering. This sputter deposition of the Si layer during the pre-nucleation stage of growth was carried out with a RF power of 80 W and an Ar pressure of 0.01 mbar for one hour. After the sputter deposition, the coated substrates were heat treated for one hour in an inert Ar gas atmosphere. Si nanowires grew at 1000 °C on nanoparticles thus created on substrate. These nanoparticles had disturbed (disordered) lattice structure. And it is consistent with the findings of Stan et al. [126] and Pernis et al. [127]. Stan et al. noted that AlN thin film deposited by sputtering had local defect structure within the crystallites induced by ion bombardment during sputtering. Using high-resolution XRD and TEM micrographs, Parnis et al. [127] observed agglomerates of point defects in SiGe films formed under ion bombardment prior to growth. Xu et al. [128] performed InN nanowire growth using indium and ammonia at 650–850 °C. As confirmed by the EDX analysis, a wetting layer of In-rich $In_{1-z}N_z$ was formed during the pre-nucleation stage of growth. Note that $In_{1-z}N_z$ is a solid solution of In and N atoms: $R_L \equiv$(In, N); this is why the observed nanoparticles were amorphous and no indium droplets were found at the nanowire tips during or after growth. Bundles of InN nanowires protruding out of $R_L \equiv$(In, N), however, suggested that the R_L species were most likely in the liquid [quasiliquid, quasisolid) state.

16.12.2 SCG Mechanism Based on Substrate Surface Roughness

Cuscuná et al. [129] noted that the surface roughness measured by AFM during SCG growth of Si nanowires was primarily related to the amorphous ridges located at the grain boundaries of the grains, which we believe, served as nanoparticles. Obviously, the rough nanoparticle surface had both pits and hillocks detected by SEM and AFM. The surface roughness resulted from surface disturbance and disorder and had heights and hollows. Surface material of the nanoparticle was presumably solid solution, which was the key parameter contributing to surface roughness and also to nanowire growth. This is justified in light of finding by Zi et al. [130] that thermally created pits on III-V substrates were crucial for the self-catalyzed nanowire growth. This is justified based on prior studies [131, 132] that edges and corners of platelets and the tips of hillocks are crucial for the self-catalyzed growth.

16.12.3 SCG Mechanism Based on Substrate Surface Porosity

Koblmüller et al. [133] performed self-catalytic growth of vertically aligned, free-standing InAs nanowires on Si (111) substrates by solid source MBE. During the pre-nucleation stage of growth, an ultrathin amorphous SiO_z mask was formed on Si(111) substrate. SEM, high-resolution XRD, and HRTEM analyses showed that this SiO_z mask facilitated epitaxial InAs nanowire growth. SEM images of InAs nanowires indicated direct relationship of the vertical directionality of the nanowires to the underlying Si(111) substrate. Interestingly, the nucleation of the nanowires was found to occur directly at the interface to the substrate surface and the mask layer.

This implied that the 3-nm thick SiO_z mask layer generated SiO_z nanoparticles, and that these nanoparticles were not only disordered and amorphous, but also had nanopores; the In and As source species diffused through these nanopores before reacting together to form InAs supersaturated species and finally nanowires on the nanoparticle surface.

16.12.4 Inferences

The experimental observations stated above demonstrate that nanoparticles exhibiting R_L species for SCG growth are generally a solid solution; it is amorphous (semi-amorphous or amorphous-like). It has coarsened surface accompanied by pits, vacancies, dislocations, etc., which give rise to nanopores. It also has humps, hillocks, protrusions, mounds, etc. (see Chap. 12). Sharp edges of these humps, hillocks, protrusions, mounds, etc. (see Fig. 12.1a, b), undergo size-dependent mesoscopic effect. The R_L species is consequently quasiliquid (quasisolid). Hence, following the tenets of the VQS mechanism, the nanowire growth is catalyzed by the quasiliquid (quasisolid) R_L species.

16.13 Boron Nitride Nanotube Growths Are by the VQS Mechanism

16.13.1 BNNTs and CNTs Compared

As we know, boron nitride nanotubes (BNNTs) possess similar tubular structure [134, 135] as carbon nanotubes (CNTs). And hence, both BNNTs and CNTs exhibit some identical properties, such as excellent mechanical stability and high thermal conductivity. CNTs are though semimetallic and semiconducting, while BNNTs are insulating. BNNTs have been synthesized by methods such as arc discharge, ball-milling, CVD, and laser ablation. Among them, CVD has been adopted quite late to synthesize BNNTs. Nevertheless, it offers better controllability of growth parameters, experimental setup, precursors, catalysts, and temperature, and hence ensures better quality of nanomaterials. Although CVD is essentially the same for the growths of CNT and BNNT, it differs in the types of starting materials, complexity of the precursors, and growth conditions for them. For example, vapor carbon sources are usually used for the growth of CNTs, but liquid or solid boron and boron nitride sources are used for the growth of BNNTs (Table 16.7).

Table 16.7 Characteristics of some BNNT growths by various techniques using various nitrogen sources

No.	Substrate	Catalyst	Nitrogen Source	Growth temp (°C)	Technique	Suggested R_L species formed for growth	References
1	Carbon fibers	$B_4N_3O_2H$	$C_3N_6H_6$	1200	CVD	B–N–O together with Si, Al, Ca yielded $R_L \equiv$ (B, O, Si, Al, Ca) species	Ma [137, 138]
2	Fe	–	BN powder	600	PEPLD	BN reacted with Fe yielding $R_L \equiv$(Fe, B, N) species	Wang [140]
3	Si	SiO_2	NH_3	1175	OAG	SiO_2 is transformed into $R_L \equiv SiO_z$, $z < 2$. B_2O_2 and NH_3 landing on SiO_z reacted yielding BN molecules which diffused through SiO_z	Gan [141]
4	Si	MgO, Ni, Fe	NH_3	1200	CVD	Together with B_2O_2 and NH_3, (1) MgO yielded $R_L \equiv$(Mg, B, N, O) species, (2) FeO yielded $R_L \equiv$(Fe, B, N, O) species, (3) Ni yielded $R_L \equiv$(Ni, B, N, O) species	Lee [142]
5	Si	FeAl	NH_3	900	PECVD	In presence of Fe, B_2H_6, NH_3 and O_2 yielded $R_L \equiv$(Fe, B, N, H, O) species	Su [143]
6	Si	Ferrocene	N	1450	Facile	In the presence of $Fe(C_5H_5)_2$, H_3BNH_3 yielded $R_L \equiv$(Fe, C, B, N, H) species, Fe and C of this may be in the form of Fe_3C	Zhong [144]
7	Si	Ni	NH_3	800	PECVD	The gas mixture B_2H_6–NH_3–H_2 on Ni yielded $R_L \equiv$(Ni, B, N, H) species	Guo [145]

(continued)

Table 16.7 (continued)

No.	Substrate	Catalyst	Nitrogen Source	Growth temp (°C)	Technique	Suggested R_L species formed for growth	References
8	$LiNO_3$ on BN bar	FeO, MgO	NH_3	1280–1320		(B, MgO, FeO) on $LiON_3$ yielded B_xO_y and the $R_L \equiv$ (B, Mg, Fe, O) species, B_xO_y reacted with NH_3 yielding BN molecules which diffused through the R_L species	Matveev [146]

Although not specifically shown as component of the R_L species, the R_L species may contain O and other contaminants, no matter if they are introduced intentionally or unintentionally in the chamber

16.13.2 Key Features of BNNT Growths

Boron (B) has a melting temperature of 2076 °C. But BN has a melting temperature of 2973 °C. Boron forms eutectic alloy with a number of metals. BNNT syntheses mediated by metal catalysts have therefore been tried. A list of precursors, substrates, temperature for growth and catalyst nanoparticles used for the growth is presented in Table 16.8. According to binary phase diagrams, there can be a number of R_L≡(MET, B) eutectic alloys formed with MET and B atoms. Characteristics of some of them are depicted in Table 16.9. An interesting feature of BNNT syntheses, unlike CNT syntheses, is that METANOs just for being of certain suitable shape, size, and composition are not generally employed for the METANO-mediated BNNT growths. As apparent from Fig. 16.7, instead several metals, metal oxides or metallic compounds are allowed to react in the growth chamber during the pre-nucleation stage of growth. The end products of these reactions yield together an alloy, cluster, or solid solution suitable to serve as catalyst nanoparticles for the growth. Figure 16.8 demonstrates that, depending on temperature and the abundance of the metal and B, many of the (MET, B) binary alloys can have multiple eutectic compositions. The synthesis of BNNTs by various techniques takes place at a temperature between 600 and 1500 °C. The R_L species for the CVD synthesis of BNNTs may be eutectic or non-eutectic alloy. If non-eutectic, the BNNT synthesis may follow the VQS mechanism. We repeat. Nanoparticles are formed during the very first step of the pre-nucleation stage of growth; BNNTs grow on these nanoparticles during the pro-nucleation stage of growths. This growth may require reactions between two or more precursors. Following the formation of nanoparticles, R_L species are created. This R_L species on the nanoparticle surface may again be a eutectic alloy, a non-eutectic alloy, a solid solution, and/or a non-stoichiometric compound of disturbed, disordered, porous, amorphous (semi-amorphous), and coarsened lattice structure. The formation of the R_L species is followed by the surface diffusion of the R_S species onto the nanoparticle surface. Obviously, the R_S species are generated from their precursors on the nanoparticle of the R_L species surface. The immediate consequence of this is the generation of a shell, or of both the shell and hill at the nanoparticle's peripheral surface. Obviously,

Table 16.8 List of Precursor(s), substrates, growth temperatures, and the catalyst nanoparticles employed for some representative catalyst-mediated boron nitride nanotube growths

No.	Precursor(s)	Substrate	Temperature (°C)	Catalyst
1	B, MgO, FeO	Si/SiO_2	1200	FeO, MgO
2	B, MgO, FeO	Si/SiO_2	1100–1200	MgO, Ni, Fe
3	B, NH_4NO_3, Fe_2O_3	–	1300	Fe_2O_3
4	B, MgO, FeO	Si/SiO_2	1200–1400	Mg, Fe
5	B, MgO, FeO	–	1200	Mg, Fe
6	B, Fe_2O_3	–	900–1400	Fe

The data are reproduced from [134]

Table 16.9 Eutectic alloys formed during metal-catalyzed BN nanotube growths and the characteristics of the R_L≡(MET, B) species of these eutectic alloys

MET	Eutectic alloy 1		Eutectic alloy 2		Eutectic alloy 3		Eutectic alloy 4	
	T_E (°C)	MET atomic %	T_E (°C)	MET atomic %	T_E (°C)	MET atomic %	T_E (°C)	MET atomic %
Fe	1159	0.84	1391	0.67	1484	0.37	–	–
Ni	1002	0.50	1089	0.70	1014	0.59	1060	0.84
Co	1118	0.82	1246	0.65	1333	0.39	–	–
Cr	1628	0.85	1890	0.66	2059	0.44	1867	0.15
Mo	1872	0.13	2357	0.35	2257	0.70	2164	0.79

the shell, and the hill, if any, may be a eutectic alloy, a non-eutectic alloy, a solid solution, and/or a non-stoichiometric compound of disturbed, disordered, porous, amorphous (semi-amorphous), and coarsened lattice structure. Pro-nucleation stage begins at the end of the pre-nucleation stage. And the most important function of this stage is the bulk diffusion of the R_S≡B and R_S≡N species through the R_L species of the shell to the liquid/solid interface, and the reaction (interaction) of the R_S≡B and R_S≡N species to yield BN molecules during the diffusion through the R_L species. If the BN molecules are otherwise formed on the R_L species surface, they must diffuse through the R_L species for the nucleation and growth of the BNNTs. All these are evident from a series of investigations [136–147] carried out during the past years. Salient features of some of them are listed in Table 16.7. The VQS mechanism followed for BNNT growths observed in some of them is elaborated in the following.

16.13.3 BNNT Growths by Using Borazine

Lourie et al. [136] used borazine ($B_3H_6N_3$) as boron nitride source for BN nanotube growth on silicon-wafer substrate. This substrate had a 250 nm thermally grown SiO_2 passivation layer. The substrate, together with borazine and the catalyst, namely nickel boride (e.g., NiB or Ni_2B) was placed at the center of a tube furnace reactor for growth. The substrate and catalyst were exposed to the borazine-containing carrier gas flow for ~30 min at a temperature of 1000–1100 °C. BNNTs, as white deposits, were observed on the substrate by optical microscopy. These were grown by the base growth mode. The melting point of nickel boride is 1230 °C, which is much higher than the growth temperature of 1000–1100 °C. Yet base growth of multiwalled BNNTs required diffusion of the R_S≡B and R_S≡N species through the R_L species of the shell [148, 149]. And it took place at 1100 °C during the pro-nucleation stage of growth. We argue that this R_L species was actually a R_L≡(Ni, B, N) solid solution rather than R_L≡(Ni, B, N) eutectic alloy. For this to be a molten eutectic alloy at $T \approx$ 1100 °C (see Table 16.8), it would be R_L≡(Ni, B) and the atomic % of Ni in it would be high, about 70. The very shapes of the bulbous tips would then be hemispherical. The very irregular shapes of these tips as was apparent from enlarged TEM images suggest that it did not happen. This means the R_L species was a non-eutectic alloy solid solution of B, Ni, and N and had quasiliquid (quasisolid) morphology needed for the growth by the VQS mechanism.

16.13.4 BNNT Growths by Using B–N–O Powder

Ma et al. [137] synthesized BNNTs using B–N–O powder generated from melamine diborate ($C_3N_6H_6$•$2H_3BO_3$). This powder was rapidly heated in the N_2 atmosphere to produce multiwalled BNNTs. It was found that the BNNT tips were composed of B–N–O clusters. Some of the BNNT tips also had Si, Al, and Ca. B–N–O clusters were not stable, and hence, might have trapped Si, Al, and Ca thus becoming more disordered and amorphous. They appeared as bulbous (spherical) with minimum surface energy required for their stability. Ma et al. suggested that the said clusters played central role in the BNNT growths. And it proved to be justified when Ma et al. [138] grew subsequently BNNTs partially filled with amorphous B–N–O mixture directly on α-Al_2O_3 nanoparticles. Recall that nanotubes are grown on peripheral shells which have disturbed (disordered) amorphous structure. The experiments by Ma et al. [137] demonstrated that the R_L species of these shells created during the pre-nucleation stage of growth were composed of solid solutions. They were R_L≡(B, N, O) or R_L≡(B, N, O, Si, Al, Ca), with quasiliquid (quasisolid) morphology. The R_S≡B and R_S≡N species underwent bulk (volume) diffusion through them to the liquid/solid interface for BNNT nucleation and growth during the pro-nucleation stage of growth.

16.13.5 BNNT Growths by Using Amorphous Boron

Kim et al. [139] performed the growth of BNNT from amorphous boron formed on the surface of the iron nanoparticles. These iron nanoparticles originated from stainless steel vessel. Kim et al. [139] noted that BNNT growth was actually governed by BN clusters on Fe nanoparticles. We argue that these clusters were components of the R_L species and were actually R_L≡(Fe, B, N) solid solutions. Crystalline boron powder was thus transformed into amorphous R_L species: R_L≡(Fe, B, N) solid solution on the Fe nanoparticle surface promoted BNNT growths in the annealing phase. Amorphous nanoshells at the periphery of the iron nanoparticles were responsible for the growths. The long milling time was crucial for high yield of BNNTs as it promoted the nitration process between boron and NH_3 and thus facilitated the formation of the R_L≡(Fe, B, N) species responsible for the BNNT synthesis. The annealing temperature of 1200 °C was probably optimal, which insured the highest conversion rate from nitration products to BNNTs.

16.13.6 Effects of Processing Parameters on BNNT Growths

Su et al. [143] studied the effect of processing parameters (such as plasma power, gas flow ratio, and process temperature) on high-yield and high-purity BNNTs, which were grown at 900 °C from a mixture of the feed gases: B_2H_6, NH_3, and O_2, respectively. The growth was conducted with Ar as carrier gas and $Fe_{1\text{-}z}Al_z$ as catalyst. During growth, several active species (such as H_2, BH_2, BNH_2, B_2H_2, NH, NH_2) were detected by QMS analysis. The O_2 additives helped enhance the dehydrogenation process even on the surface of the catalyst or in the atmosphere. While Fe catalyst can be a thin film, $Fe_{1-z}Al_z$ alloy can be a non-eutectic alloy, solid solution.

Increased plasma power increased the density of B and N radicals. Bombarded B, N, and O atoms were adsorbed onto (Fe, Al) solid solution creating a more effective (Fe, Al, B, N, O) catalyst solid solution. The $R_L \equiv$(Fe, Al, B, N, O) species, thus created, had increased surface disorder, optimal effective surface amorphicity, and optimal surface coarsening needed for increased BNNT growth plausibly by the VQS mechanism.

16.13.7 Effects of Laser Irradiation on BNNT Growths

Yu et al. [147] synthesized BNNT by laser irradiation of a mixture of $Ni_{1-z}Co_z$ catalyst and h-BN powder. BNNTs of 1–3 layers and of diameter 1.5–8 nm were realized. $Ni_{1-z}Co_z$ catalyst was nested at the tips (ends) of some of the BNNTs. The laser irradiation converted the $Ni_{1-z}Co_z$ catalyst surface into $R_L \equiv$(Ni, Co, B, N) species, which is a solid solution. Note that $Ni_{1-z}Co_z$ catalyst itself is a (Ni, Co) non-eutectic alloy solid solution. But it has lower amorphicity and porosity; it also has lower concentration of HETs than that of (Ni, Co, B, N) cluster or solid solution. As a result, it created high density of the R_S species released from their precursors. This density was large enough to generate both the shell and the hill at the edge of the catalyst nanoparticle (see Sects. 4.3 and 4.4 of Chap. 4). Single-walled BNNTs were consequently produced. The $Ni_{1-z}Co_z$ catalyst played a role in synthesizing high-quality single-walled BNNTs.

16.14 Conclusions

16.14.1 Key Conclusions

There is one and only one universal growth mechanism, called the αQS growth mechanism. All the known and widely employed growth mechanisms are the special cases and different variants of this growth mechanism. Three parameters: ξ_m, SECINI=SECINI0, and $\alpha_{amoreff0}$ introduced in this investigation may probably very effectively define the state-of-the-art condition of a growth mechanism. Unless the nanomaterials growths are mediated by droplet (ξ_m=1), for example, by the VLS mechanism, or by solid (ξ_m=0), for which the probability is small, these growths are carried out, in general, by the VQS mechanism. And Knudsen diffusion of the R_S species through nanopores created in the porous catalyst nanoparticle plays a crucial role in the growths. This implies that nanopore radius and nanopore density are important for most of the growths. The R_L species of nanoparticle must be liquid, quasiliquid or quasisolid in order for supersaturation, nucleation, crystallization and growth at the liquid/solid, quasiliquid/solid or quasisolid/solid interface. If the nanopore radius and the nanopore density remain unaltered during growth, then nanomaterial growth rate and consequently the nanomaterial length should increase linearly with increase in the precursor amount of the source species. If the precursor amount is large enough to provide unabated supply of the source species during growth, this nanomaterial length should increase linearly with also the increase in growth time. Hedrich et al. [150] studied ZnO nanowhisker length as function of the supplied precursor amount and growth time. The growth was believed to be by the VS mechanism. The results are shown in Figs. 16.5 and 16.6. Interestingly, the variations of nanowhisker length as function of precursor amount and growth time were always linear. These would not happen without the growth being via diffusion through droplet or molten nanopores.

We summarize. Efficient nanomaterials growths would require the following:

1. Ability of the nanoparticle surface to attract precursor(s) of the R_S species and optimally high concentration of R_S species released on the R_L species surface;
2. Ability of the nanoparticle surface to decompose the precursor(s) of the R_S species at the growth temperature which may be much lower than the thermal decomposition temperature of the precursor(s);
3. Optimal and selective level of absorption and adsorption of the precursor(s) and of the source species and of desorption of their by-products on the nanoparticle surface. This would happen if the sticking coefficients of them (e.g., precursor(s) and the R_S species) are high;
4. High concentration of grain boundaries and nanopores of (1) appropriately large radius and (2) straight smooth cross section in the R_L species;
5. High, smooth, uninterrupted Knudsen diffusion of the R_S species through the nanopores and grain boundaries of the R_L species;

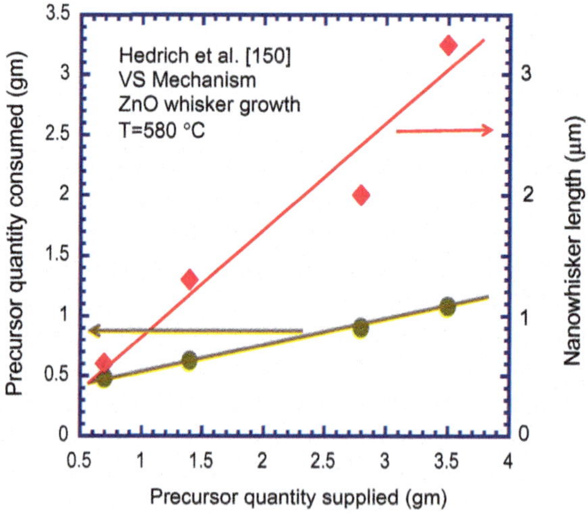

Fig. 16.5 Precursor amount consumed and nanowhisker length obtained as function of the amount of the zinc acetylacetonate hydrate powder precursor supplied during the growth of ZnO nanowhiskers. The plots were obtained with experimental data by Hedrich et al. [150]

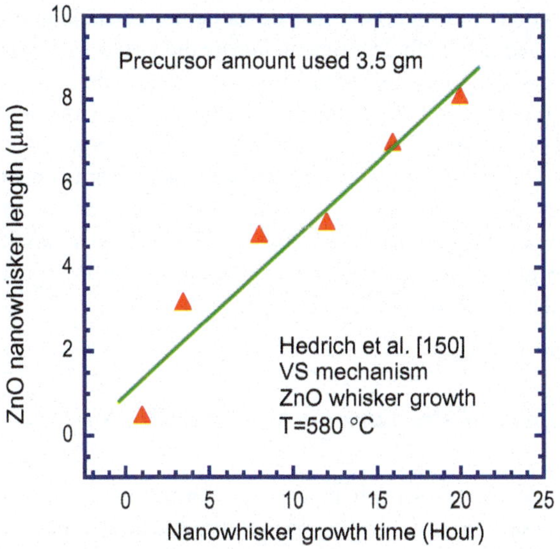

Fig. 16.6 Nanowhisker length obtained as function of time of nanowhisker growth using a fixed amount of zinc acetylacetonate hydrate powder precursor. The plots were obtained with experimental data by Hedrich et al. [150]

6. An optimally large chemical potential difference between the R_S species in the vapor phase and the R_S species in the solid phase;
7. Melting (semi-melting) of the droplet or nanopores and/or grain boundaries, particularly of the sharp edges of the pits, hillocks, etc., formed on the coarsened nanoparticle surface.

16.14.2 Sticking of the Source Species and of the Precursors of the Source Species

Needless to say, sticking of the precursors and of the R_S source species on the nanoparticle surface is key to nanomaterials growths. This sticking must follow adsorption of the most desired species and desorption of the most undesired species released or landed on the nanoparticle surface. As Pendulic [151] suggested, both the adsorption and desorption are frequently and rather completely controlled by defects and disorders on the nanoparticle surface. He noted that, in many instances, even low level of defect concentrations can give rise to profound changes in adsorption and desorption. Kennou et al. [152] studied the adsorption of Cs on Si(100) surfaces by making use of LEED, AES, EELS, and TDS measurements under identical experimental conditions on a flat and a 5° vicinal surface exhibiting steps. The influence of the steps was determined by comparing the measured Cs adsorption on the flat and the stepped surface. They found that Cs adsorbs preferentially on step sites.

And this adsorption on the stepped surface had a higher sticking coefficient, larger binding energy, and lower dipole moment than on the flat surface. Raschke and Höfer [153] noted that step, defect, and terrace sites all contribute to adsorption, and that the sticking coefficient increases almost linearly with increase in temperature. All these observations highlight the very novelty of the VQS mechanism detailed in Chaps. 12–14. They raise, at the same time, serious concerns about the validity of the VS and the VSS mechanism. Recall that the VS mechanism advocates growth on flat, crystalline substrate surface, and the VSS mechanism advocates growth on flat, crystalline metal surface. Sticking coefficient of the R_S species and of their precursors on flat, crystalline surface is supposedly small.

16.14.3 Impact of Pressure on Nanoparticle Surface

Precursor pressure may be as much important as temperature in dictating the nanomaterials growth modes. Also, depending on pressure and temperature, a metal nanoparticle (METANO) for growth may be eutectically or non-eutectically molten even before it attains the eutectic temperature T_E. Tian et al. [154] performed Ni-catalyzed hafnium carbide nanowire growth at $T = 1025$ °C, which is ~75 °C below the $R_L \equiv$(Ni, Hf) eutectic temperature. Recall that Kodambaka et al. [2] performed Au-catalyzed Ge nanowire growth earlier at a temperature below the eutectic temperature. They noted that the Ge_2H_6 precursor should preferably have some minimal pressure. This pressure would be essential for stabilizing the liquid (quasiliquid, quasisolid) state of the R_L species of nanoparticles at a temperature $T < T_E$. Kodambaka et al. observed that the catalyst droplet tended to solidify if the Ge_2H_6 precursor pressure during growth was reduced to a value lower than a certain optimal value. With the reduction of this pressure, the droplets could easily solidify at $T < T_E$.

The fact that the precursor pressure influences the droplet state during VLS growth does not rule out the influence of the Gibbs–Thomson effect on this growth. It is true because the pressure can also impact the surface energy, for example, through a change in hydrogen coverage. In situ imaging by Kodambaka et al. [2] confirmed the effect of Ge_2H_6 pressure on surface energy. A change in this surface energy modifies the force balance at the triple phase line. And this line governs the steady-state droplet shape and nanowire diameter [155]. The shape of droplets during nanowire growth as a function of Ge_2H_6 pressure indicated that a reduction in the Ge_2H_6 pressure, even if small, could lead to a decrease in the droplet height, but an increase in the droplet diameter leading to a change in the contact angle ϕ (see Chap. 4). The droplet shape and hence the angle ϕ therefore varied with Ge_2H_6 pressure, and this was due to the impact of pressure on surface energy. The fact that pressure affected the droplet state implies that the pressure plays a critical role in dictating the growth mechanism. This also implies that an increase in the precursor pressure during, for example, the VLS growth can affect the liquid droplet created at $T = T_E$. This liquid droplet may be transformed into a quasiliquid droplet ($\xi_m < 1$) or even a solidified droplet ($\xi_m \approx 0$). Yu et al. [156] performed diamond anvil cell synchrotron

X-ray study on liquid $Ga_{0.858}In_{0.142}$eutectic alloy. The study elucidated the pressure-induced crystallization of the eutectic liquid alloy at a pressure of about 3.4 GPa. Even when decompressed, the high-pressure crystallization phase remained stable until the pressure went down to 2.3 GPa at which it transformed back into the liquid state. *Ab initio* molecular dynamics calculations confirmed the low-pressure crystallization process and elucidated the structure, dynamic, and electronic properties for the liquid-to-crystalline phase transition. Yu et al. [156] concluded that pressure-induced phase transformation resulted from changes in local atomic packing of the nearest neighbors. The analysis reported above suggests that, under appropriately high precursor pressure, the mechanism for growth at a temperature $T = T_E$ may change from the VLS mechanism to the VQS mechanism,

Recall that the surface melting is an important element of the VQS mechanism; see Sect. 12.5 of Chap. 12. If the precursor pressure is not maintained during growth by the VQS mechanism, and it is increased beyond a certain limit, the surface melting of the R_L species may disappear. The nanoparticle may, as a result, be gradually transformed from the NP2 type nanoparticle to the NP1 type nanoparticle. And the nanomaterial growth via the VQS mechanism may cease. *An essential requirement of nanomaterials growths by the VQS mechanism is therefore to maintain a chamber pressure, which is optimal, and which insures that the surface melting of the nanoparticle up to a depth of* δ_{amor} (see Figs. 12.3b and 12.4b of Chap. 12) *remains intact during the growths.*

16.14.4 Steps Involved in Growth Kinetics

There are three steps involved in the growth kinetics. They must ensure the efficient nanomaterials growths spelled out above. The three steps together must establish stable nanomaterials growth rates. We reiterate the growth kinetics. Step 1 is the decomposition reaction step at the vapor–liquid, vapor-quasiliquid (vapor-quasisolid) interface. This step introduces R_S species into the R_L species. Step 2 is for the diffusion of the R_S species through the R_L species; the R_S species thus travel through a distance δ_{amor} until this species reaches the L/S (QL/S, QS/S) interface. Step 3 is for the supersaturation and crystallization of the R_S species at the L/S (QL/S, QS/S) interface. The growth rate would typically vary from much lower than 1 nm/s to several nm/s. Usually, δ_{amor} (see Figs. 12.3b and 12.4b of Chap. 12) is only a few nm. The diffusion of the R_S species through the R_L species may hence be fast, and the transport of the R_S species through the R_L species may not be a limiting factor in the kinetics of growth. The growth rate may hence be determined by the steady state in which the rate for Step 1 and the rate for Step 3 would be equal and hence in equilibrium. Both of them can be proportional to the surface area yielding a stable growth rate.

Importantly, the flux of the precursor gas molecules or ions of the R_S species on the R_L species surface can be directly proportional to the pressure of the precursor gas molecules. Any increase in the precursor partial pressure during Step 1 would cause an increase in the available R_S species on the R_L species surface. Unless the R_S species are excessively crowded and have a retarding effect, they would lead to an increase in the resulting growth rate. This is corroborated with experiments by Dailey et al. [157] and by Jeong et al. [158]. An optimal partial pressure during growth by the VQS mechanism must serve two purposes: (1) It must keep the surface melting intact, and (2) it must help maintain reasonably high growth rate. An optimal partial pressure should be equivalent to an optimal chemical potential difference between the R_S species in the vapor phase and the R_S species in the solid phase. The concentration of the R_S species, and probably also the concentration gradient of the R_S species, in the molten (semimolten) R_L species would be high if this chemical potential difference is high. Supersaturation of the R_S species would consequently also be high, which would lead to a high level of crystallization rate in Step 3. And it would take place at the L/S (QL/S, QS/S) interface.

16.14.5 Competitive Role of Temperature and Pressure During Growths

Unlike monometallic nanoparticles, bimetallic nanoparticles lack long-range crystal order (see Sects. 3.6 and 3.7 of Chap. 3). Based on Sect. 3.6 of Chap. 3 and Sect. 12.9 of Chap. 12, bimetallic nanoparticles may be classified as NP2 nanoparticles; these are most suitable as METANOs for nanomaterials growths by the VQS mechanism. In view of this, vertically aligned CNT growths by Pint et al. [159] employing $Fe_{1-z}Mo_z$ bimetallic nanoparticles would be the VQS growths. Effects of temperature and gas pressure on these growths as observed by Pint et al. would shed some light on the VQS mechanism. Pint et al. [159] noted that the CNT heights as function of temperature (precursor pressure kept very low, only 1.4 Torr) were dependent on the Mo mole fraction z in the $Fe_{1-z}Mo_z$ nanoparticle alloy. The CNT heights as function of precursor pressure (temperature kept at 600 °C and 750 °C, respectively) were also dependent on the Mo mole fraction z in the $Fe_{1-z}Mo_z$ nanoparticle alloy. While the CNT heights at 750 °C gradually increased with increase in precursor pressure, the CNT heights at 600 °C gradually decreased with increase in precursor pressure. Apparently, this is puzzling, because an increase in precursor pressure led to an increase in the $R_S{\equiv}C$ species on the $Fe_{1-z}Mo_z$ nanoparticle surface. But an increase in temperature led to an increase in the diffusion of the $R_S{\equiv}C$ species through the nanopores of the R_L species of the $Fe_{1-z}Mo_z$ nanoparticle surface. At 750 °C, the Knudsen diffusivity of the $R_S{\equiv}C$ species through the nanopores of the R_L species of the $Fe_{1-z}Mo_z$ nanoparticle was high. So, it led the CNT heights to increase almost linearly with increase in precursor pressure. Gradually increased $R_S{\equiv}C$ species resulting from gradual increase in precursor pressure could all diffuse through the nanopores of the R_L species without interruption. And they could participate in CNT growth at the relatively high temperature of 750 °C. This did not happen at 600 °C because the $R_S{\equiv}C$ diffusivity at this temperature was low. And as a result, gradually increased $R_S{\equiv}C$ species resulting from increase in precursor pressure could not all diffuse through the nanopores of the R_L species. They could not therefore participate in CNT growth at the relatively low temperature of 600 °C. The overcrowded R_S species blocked the nanopores a few nanosecond after the initial beginning of diffusion through the nanopores. The investigation by Pint et al. [159] thus demonstrates that an optimal level of growth temperature together with an optimal level of precursor pressure is central to the effective nanomaterials growths by the VQS mechanism.

Let us be more specific. Metal-assisted growth by the VQS mechanism may take place at a temperature $T < T_E$, $T = T_E$, or $T > T_E$. Any efficient growth at all of these temperature regimes must ensure the requirements (a)–(d) spelled out earlier in this section (viz. Sect. 16.14). However, if the growth temperature is relatively low, e.g., $T < T_E$, the diffusivity of the R_S species would be relatively low, and the flux of the R_S species on the R_L species surface would not be large enough to yield overcrowded R_S species. If this surface is overcrowded, it would block nanopores causing reduction in growth rate. If the growth temperature is moderate, e.g., $T \approx T_E$, the diffusivity of the R_S species would be moderately large. And in this paradigm,

1. *The flux of the R_S species on the R_L species surface must be such that the liquid droplet is transformed into quasiliquid (quasisolid) cluster or solid solution*;
2. *The concentration of the R_S species on the R_L species surface should be such that they diffuse through the R_L species without interruption.*

High rate of nanomaterials growths can be realized under these circumstances. Also,

1. *If the growth temperature is high, e.g., $T > T_E$, the diffusivity of the R_S species would be large. To cope with it, the pressure of the flux of the R_S species on the R_L species surface must be such that the R_L species is quasiliquid (quasisolid) and not solid.*
2. *Also, the concentration and concentration gradient of the R_S species on the R_L species surface and inside the RL species should be high. There should also be a motive force. And hence the R_S species should diffuse through the R_L species without any interruption by this high concentration of the R_S species. High growth rate of nanomaterials can be achieved under these circumstances.*

References

1. C.J. Adkins, *Equilibrium Thermodynamics* (Cambridge University Press, Cambridge, 1983)
2. S. Kodambaka, J. Tersoff, F.M. Ross, Ge nanowire growth below the eutectic temperature. Science **316**, 729–732 (2007)
3. C.-C. Chen, C.-C. Yeh, C.-H. Chen, M.-H. Yu, H.-L. Liu, J.-J. Wu, K.-H. Chen, L.-C. Chen, J.-Y. Peng, Y.-F. Chen, J. Am. Chem. Soc. **123**, 2791–2798 (2001)
4. T.B. Massalski (ed.), *Binary Alloy Phase Diagrams*, vol. 3, 2nd edn. (American Society of Metals, Metals Park, 1986)

5. Y. Osada, H. Nakayama, M. Shindo, T. Odaka, Y. Ogata, Growth and structure of silicon fibers. J. Electrochem. Soc. **126**, 31–38 (1979)
6. K.A. Dick, K. Deppert, L.S. Karlsson, L.R. Wallenberg, L. Samuelson, W. Seifert, A new understanding of Au-assisted growth of III-V semiconductor nanowires. Adv. Funct. Mater. **15**, 1603–1610 (2005)
7. L.C. Campos, M. Tonezzer, A.S. Ferlauto, V. Grillo, R. Magalhães-Paniago, S., Oliveira, L.O. Ladeira, R.G. Lacerda, Vapor-solid-solid growth mechanism driven by epitaxial match between solid AuZn alloy catalyst particle and ZnO nanowire at low temperature. Adv. Mater. **20**, 1499–1504 (2008)
8. K. Kang, D.A. Kim, H.-S. Lee, C.-J. Kim, J.-E. Yang, M.-H. Jo, Low-temperature deterministic growth of Ge nano-wires using Cu solid catalysts. Adv. Mater. **20**, 4684–4690 (2008)
9. T.I. Kamins, R.S. Williams, Y. Chen, Y.-L. Chang, Y.A. Chang, Chemical vapor deposition of Si nanowires nucleated by $TiSi_2$ islands on Si. Appl. Phys. Lett. **76**, 562 (2000)
10. N. Han, F. Wang, A.T. Hui, J.J. Hou, G. Shan, F. Xiu, T.-F. Hung, J.C. Ho, Facile synthesis and growth mechanism of Ni-catalyzed GaAs nanowires on non-crystalline substrates. Nanotechnology **22**, 285607 (2011)
11. M. He, I. Minus, P. Zhou, S.N. Mohammad, J.B. Halpern, R. Jacobs, W.L. Sarney, L. Salamanca-Riba, R.D. Vispute, Growth of large-scale GaN nanowires and tubes by direct reaction of Ga with ammonia. Appl. Phys. Lett. **77**, 3731–3733 (2000)
12. M. He, P. Zhou, S.N. Mohammad, G.L. Harris, J.B. Halpern, R. Jacobs, W.L. Sarney, L. Salamanca-Riba, Growth of GaN nanowires by direct reaction of Ga with ammonia. J. Cryst. Growth **231**, 357–365 (2001)
13. A.M.S. El Ahl, M. He, P. Zhou, L. Salamanca-Riba, F. Felt, H. Shaw, A.K. Sharma, M. Jah, D. Lakins, T. Steiner, S.N. Mohammad. A systematic study of effects of growth conditions on the (nano-, meso-, micro) size and (1-, 2-, 3-dimensional) shape of GaN single crystals grown by direct reaction of Ga with ammonia. J. Appl. Phys. **94**, 7749 (2003)
14. M. He, S.N. Mohammad, Novel chemical vapor deposition technique for the synthesis of high-quality single-crystal nanowires and nanotubes. J. Chem. Phys. **124**, 064714 (2006)
15. M. He, M.E.E. Fahmi, S.N. Mohammad, InAs nanowires and whiskers grown by reaction of indium with GaAs. Appl. Phys. Lett. **82**, 3749 (2003)
16. M. He, S.N. Mohammad, Structural characteristics of single-crystal nanowires grown by self-catalytic chemical vapor deposition method. J. Vac. Sci. Technol. B **25**, 1909 (2007)
17. M. He, S.N. Mohammad, Novelty of self-catalytic nanowire growth: a case study with InN nanowires. J. Vac. Sci. Technol. B **25**, 940 (2007)
18. M. He, A. Motayed, S.N. Mohammad, Phase separations of single-crystal nanowires grown by self-catalytic chemical vapor deposition method. J. Chem. Phys. **126**, 064704 (2007)
19. J.N. Pratt, J.M. Bird, Solid electrolyte cell studies of solid nickel-gallium alloys. J. Phase Equil. **14**, 465–472 (1993)
20. P. Feschotte, P. Eggimann, Les systèms binaries cobalt-gallium et nickel-gallium: étude compare. J. Less-Common Met. **63**, 15 (1979). (in French)
21. E. Hellner, Das System Nickel-Gallium, Z. Metallkunde. **41**, 480–484 (1950)
22. W.X. Yuan, Z.Y. Qiao, H. Ipser, G. Eriksson, Thermodynamic assessment of the Ni-Ga system. J. Phase Equil. **25**, 68–74 (2004)
23. C.C. Lee, C.Y. Wang, G. Matijasevic, Au-In bonding below the eutectic temperature. IEEE Trans. Compo. Hybrid Manufact. Technol. **16**, 311–316 (1993)
24. Y. Cai, S.K. Chan, I.K. Sou, Y.F. Chan, D.S. Su, N. Wang, The size-dependent growth direction of ZnSe nanowires. Adv. Mater. **18**, 109–114 (2006)
25. S. Hoffmann, R. Sharma, C.T. Wirth, F. Cervantes-Sodi, C. Ducati, T. Kasama, R.E. Dunin-Borkowski, J. Drucker, P. Bennett, J. Robertson, Ledge-flow-controlled catalyst interface dynamics during Si nanowire growth. Nat. Mater. **7**, 372–375 (2008)
26. Z. Liu, V. Srot, P.A. van Aken, J.C. Yang, M. Rühle, Microstructure characterization of iron catalyst assisted SiC nanowires. Micros. Microanal. **13**(Suppl 2), 754–755 (2007)

27. T.I. Kamins, X. Li, R.S. Williams, Growth and structure of chemically vapor deposited Ge nanowires on Si subs-trates. Nano Lett. **4**, 503–506 (2004)
28. C. Chêze, L. Geelhaar, A. Trampert, O. Brandt, H. Riechert, Collector phase transitions during vapor-solid-solid nucleation of GaN nanowires. Nano Lett. **10**, 3426–3431 (2010)
29. A. Hiraki, K. Shuto, S. Kim, W. Kammura, M. Iwami, Room temperature interfacial reaction in Au-semiconductor systems. Appl. Phys. Lett. **31**, 611–612 (1977)
30. B. Pecz, R. Veresegyhazy, G. Radnoczi, A. Barna, I. Mojzes, O. Geszri, G. Vincze, Cross-sectional transmission electron microscopic study of Au/GaP and Au/InP contacts. J. Appl. Phys. **70**, 332–336 (1991)
31. V. Malina, Z. Sroubek, I. Mojzes, R. Veresegyhazy, B. Pecz, Interaction of thin gold films with GaP during heat treatment in a vacuum. Semicond. Sci. Technol. **2**, 428–436 (1987)
32. A. Piotrowska, E. Kaminska, A. Bercz, J. Adamczewska, A. Turos, Gold-based ohmic contacts on III–V compounds: thermally induced reactions between metallization and the semiconductor compound. Thin Solid Films **130**, 231–236 (1985)
33. A. Barcz, E. Kaminska, A. Piotrowska, Fundamental and practical aspects of alloying encapsulated gold-based contacts to GaAs. Thin Solid Films **149**, 251–260 (1987)
34. J. Gyulai, J.W. Mayer, V. Rodriguez, A.Y.C. Yu, H.J. Gopen, Alloying behavior of Au and Au-Ge on GaAs. J. Appl. Phys. **42**, 3578–3585 (1971)
35. J.-C. Harmand, G. Patriarche, N. Péré-Laperne, M.-N. Mérat-Combes, L. Travers, F. Glas, Analysis of vapor-liquid-solid mechanism for Au-assisted GaAs nanowire growth. Appl. Phys. Lett. **87**, 203101–203103 (2005)
36. C.J. Lee, J. Park, Y. Huh, J.Y. Lee, Temperature effect on the growth of carbon nanotubes using thermal chemical vapor deposition. Chem. Phys. Lett. **343**, 33–38 (2001)
37. M. Ritchel, A. Leonhardt, D. Elefant, S. Oswald, B. Büchner, Rhenium-catalyzed growth of carbon nanotubes. J. Phys. Chem. C **111**, 8414–8417 (2007)
38. S. Esconjaureguia, C.M. Whelan, K. Maex, The reasons why metals catalyze the nucleation and growth of carbon nanotubes and other carbon nanomorphologies. Carbon **47**, 659–669 (2009)
39. D. Takagi, H. Kobayashi, H. Hibino, S. Suzuki, Y. Homma, Mechanism of gold-catalyzed carbon material growth. Nano Lett. **8**, 832–835 (2008)
40. D. Takagi, Y. Homma, H. Hibino, S. Suzuki, Y. Kobayashi, Single-walled carbon nanotube growth from highly activated metal nanoparticles. Nano Lett. **6**, 2642–2645 (2006)
41. C.J. Lee, J. Park, J.M. Kim, Y. Huh, J.Y. Lee, K.S. No, Low-temperature growth of carbon nanotubes by thermal chemical vapor deposition using Pd, Cr, and Pt as co-catalyst. Chem. Phys. Lett. **327**, 277–283 (2000)
42. F.Z. Bouanis, L. Baraton, V. Huc, D. Pribat, C.S. Cojocaru, High-quality single-walled carbon nanotubes synthesis by hot filament CVD on Ru nanoparticle catalyst. Thin Solid Films **519**, 4594–4597 (2011)
43. Z. Sadeghian, A.M. Rashidi, Synthesis, optimization and characterization of multiwalled carbon nanotubes produced by spray pyrolysis of hexane. Mater. Sci. Technol. **26**, 1191–1196 (2010)
44. S.G. Sundaresan, A.V. Davydov, M.D. Vaudin, I. Levin, J.E. Maslar, Y.-L. Tian, M.V. Rao, Growth of Silicon carbide nanowires by a microwave heating-assisted physical vapor transport process using group VIII metal catalysts. Chem. Mater. **19**, 5531–5537 (2007)
45. A.-F. Lehlooh, S.M. Fayyad, S.H. Mahmood, Mössbauer spectroscopy study of Fe–Si solid solution prepared by mechanical milling. Hyperfine Interact. **139**(140), 335–344 (2002)
46. W.-C. Hou, L.-Y. Chen, W.-C. Tang, F.C.N. Hong, Control of seed detachment in Au-assisted GaN nanowire growths. Cryst. Growth Des. **11**, 990–994 (2011)
47. N. Wang, Y. Cai, R.Q. Zhang, Growth of nanowires. Mater. Sci. Eng. R **60**, 1–51 (2008)
48. C. Liu, H.M. Cheng, Carbon nanotubes: controlled growth and applications. Mater. Today **16**, 19–28 (2013)
49. H. Liu, D. Takagi, S. Chiashi, T. Chokan, Y. Homma, Investigation of catalytic properties of Al_2O_3 particles in the growth of single-walled carbon nanotubes. J. Nanosci. Nanotechnol. **10**, 4068–4073 (2010)

50. B.-S. Kim, T.-W. Koo, J.-H. Lee, D.S. Kim, Y.C. Jung, S.W. Hwang, B.L. Choi, E.K. Lee, J.M. Kim, D. Whang, Catalyst-free growth of single-crystal silicon and germanium nanowires. Nano Lett. **9**, 864–869 (2009)
51. R. Rao, D. Liptak, T. Cherukuri, B.I. Yakobson, B. Maruyama, In situ evidence for chirality-dependent growth rates of individual carbon nanotubes. Nat. Mater. **11**, 213–216 (2012)
52. S. Kumar, I. Levchenko, K. Ostrikov, J.A. McLaughlin, Plasma-enabled, catalyst-free growth of carbon nanotubes on mechanically-written Si features with arbitrary shape. Carbon **50**, 325–329 (2012)
53. J.H. Lin, C.-S. Chen, H.-L. Ma, C.-W. Chang, C.-Y. Hsu, H.-W. Chen, Self-assembling of multi-walled carbon nanotubes on a porous carbon surface by catalyst-free chemical vapor deposition. Carbon **46**, 1619–1623 (2008)
54. J.H. Lin, C.S. Chen, M.H. Rümmeli, A. Bachmatiuk, Z.Y. Zeng, H.L. Ma, B. Büchner, H.W. Chen, Growth of carbon nanotubes catalyzed by defect-rich graphite surfaces. Chem. Mater. **23**, 1637–1639 (2011)
55. J.-H. Lin, C.-S. Chen, M.H. Rummeli, Z.-Y. Zeng, Self-assembly formation of multi-walled carbon nanotubes on gold surfaces. Nanoscale **2**, 2835–2840 (2010)
56. D. Takagi, Y. Kobayashi, Y. Homma, Carbon nanotube growth from diamond. J. Am. Chem. Soc. **131**, 6922–6923 (2009)
57. B. Kruszka, A.P. Terzyk, M. Wiśniewski, P.A. Gauden, M. Szybowicz, Synthesis of carbon nanotubes and nanotube forests on copper catalyst. Mater. Res. Express **1**, 035040 (2014)
58. E.S. Kukovitskyz, S.G. Lvov, Increased carbon chemical vapor deposition and carbon nanotube growth on metal substrates in confined spaces. ECS J. Solid St. Sci. Technol. **2**, M1–M8 (2013)
59. C. Du, N. Pan, CVD growth of carbon nanotubes directly on nickel substrate. Mater. Lett. **59**, 1678–1682 (2005)
60. C. Masarapu, B. Wei, Direct growth of aligned multiwalled carbon nanotubes on treated stainless steel substrates. Langmuir **23**, 9046–9049 (2007)
61. M. Kurk, M. Jaroniec, Y. Bereznitski, Adsorption study of porous structure development in carbon blacks. J. Colloid Int. Sci. **182**, 282–288 (1996)
62. M. Kusunoki, T. Suzuki, T. Hirayama, N.A. Shibata, A formation mechanism of carbon nanotube films on SiC (0001). Appl. Phys. Lett. **77**, 531–533 (2000)
63. M.H. Rümmeli, E. Borowiak-Palen, T. Gemming, T. Pichler, M. Knupfer, M. Kalbác, L. Dunsch, O. Jost, S.R.P. Silva, W. Pompe, B. Büchner, Nano Lett. **5**, 1209–1215 (2005)
64. J. Li, J.B. Li, Y.C. Yin, Y.J. Chen, S.F. Bi, Water-assisted chemical vapor deposition synthesis of boron nitride nanotubes and their photoluminescence property. Nanotechnology **24**, 365605 (2013)
65. N. Muradov, F. Smith, A.T. Raissi, Catalytic activity of carbons for methane decomposition reaction. Catal. Today **102–103**, 225–233 (2005)
66. N. Muradov, Catalysis of methane decomposition over elemental carbon. Catalysis Commun. **2**, 89–94 (2001)
67. H.W. Liang, M. Lu, D.J. Shen, B.H. Li, Z.Z. Zhang, C.X. Shan, J.Y. Zhang, X.W. Fan, G.T. Du, Growth of vertically aligned single crystal ZnO nanotubes by plasma-molecular beam epitaxy. Solid-St. Commun. **137**, 182–186 (2006)
68. S. Huang, Q. Cai, J. Chen, Y. Qian, L. Zhang, Metal-catalyst-free growth of single-walled carbon nanotubes on substrates. J. Am. Chem. Soc. **131**, 2094–2095 (2009)
69. B. Liu, W. Ren, L. Gao, S. Li, S. Pei, C. Liu, C. Jiang, H.-M. Cheng, Metal-catalyst-free growth of single-walled carbon nanotubes. J. Am. Chem. Soc. **131**, 2082–2083 (2009)
70. B. Liu, D.-M. Tang, C. Sun, C. Liu, W. Ren, F. Li, W.-J. Yu, L.C. Yin, L. Zhang, C. Jiang, H.-M. Cheng, Importance of oxygen in the metal-free catalytic growth of single-walled carbon nanotubes from SiO_x by a vapor-solid-solid mechanism. J. Am. Chem. Soc. **133**, 197–199 (2011)
71. S.A. Steiner, T.F. Baumann, B.C. Bayer, R. Blume, M.A. Worsley, W.J. MoberlyChan, E.L. Shaw, R. Schlogl, A.J. Hart, S. Hofmann, B.L. Wardle, Nanoscale zirconia as a nonmetallic catalyst for graphitization of carbon and growth of single- and multiwall carbon nanotubes. J. Am. Chem. Soc. **131**, 12144–12154 (2009)

72. U.P. Gomes, D. Ercolani, N.V. Sibirev, M. Gemmi, V.G. Dubrovskii, F. Beltram, L. Sorba, Catalyst-free growth of InAs nanowires on Si(111) by CBE. Nanotechnology **26**, 415604 (2015)
73. C.Y. Nam, D. Tham, J.E. Fischer, Effect of the polar surface on GaN nanostructure morphology and growth orientation. Appl. Phys. Lett. **85**, 5676–5678 (2004)
74. X. Wen, S. Wang, Y. Ding, Z.L. Wang, S. Yang, Controlled growth of large-area, uniform, vertically aligned arrays of α-Fe_2O_3 nanobelts and nanowires. J. Phys. Chem. B **109**, 215–220 (2005)
75. J. Song, X. Wang, E. Riedo, Z.L. Wang, Systematic study on experimental conditions for large-scale growth of aligned ZnO nanwires on nitrides. J. Phys. Chem. B **109**, 9869–9872 (2005)
76. S.H. Dalal, D.L. Baptista, K.B.K. Teo, R.G. Lacerda, D.A. Jefferson, W.I. Milne, Controllable growth of vertically aligned zinc oxide nanowires using vapor deposition. Nanotechnology **17**, 4811–4818 (2006)
77. J.-S. Lee, K. Park, S. Nahm, S.-W. Kim, S. Kim, Ga_2O_3 nanomaterials synthesized from ball-milled GaN powders. J. Cryst. Growth **244**, 287–295 (2002)
78. S.P. Mondal, S.K. Ray, J. Ravichandran, I. Manna, Temperature dependent growth and optical properties of SnO_2 nanowires and nanobelts. Bull. Mater. Sci. Indian Acad. Sci. **33**, 357–364 (2010)
79. M.O. Orlandi, E.R. Leite, R. Aguiar, J. Bettini, E. Longo, Growth of SnO nanobelts and dendrites by a self-catalytic VLS process. J. Phys. Chem. B **110**, 6621–6625 (2006)
80. Z.-G. Chen, F. Li, G. Liu, Y. Tang, H. Cong, G.Q. Lu, H.-M. Cheng, Preparation of high purity ZnO nanobelts by thermal evaporation of ZnS. J. Nanosci. Nanotechnol. **6**, 704–707 (2006)
81. C.H. Bartholomew, Mechanisms of catalyst deactivation. Appl. Catal. A: General **212**, 17–60 (2001)
82. K. Vanheusden, W.L. Warren, C.H. Seager, D.K. Tallant, J.A. Voigt, B.E. Gnade, Mechanisms behind green photoluminescence in ZnO phosphor powders. J. Appl. Phys. **79**, 7983 (1996)
83. J. Zhang, W. Yu, L. Zhang, Fabrication of semiconducting ZnO nanobelts using a halide source and their photoluminescence properties. Phys. Lett. A **299**, 276–281 (2002)
84. D. Vogtenhuber, R. Podloucky, J. Redinger, E.L.D. Hebenstreit, W. Hebenstreit, U. Diebold, Ab initio and experimental studies of chlorine adsorption on the rutile TiO_2 (110) surface. Phys. Rev. B **65**, 125411 (2002)
85. W. Gao, T.A. Baker, L. Zhou, D.S. Pinnaduwage, E. Kaxiras, C.M. Friend, Chlorine adsorption on Au(111): chlorine overlayer or surface chloride? J. Am. Chem. Soc. **130**, 3560–3565 (2008)
86. X. Xiang, C. Cao, H. Zhu, Catalytic synthesis of single-crystalline gallium nitride nanobelts. Solid State Commun. **126**, 315–318 (2003)
87. I. Katayama, S. Igi, Z. Kozuka, Thermodynamic study of solid Ni-Ga alloys by E.M.F. measurements using the solid electrolyte, Trans. JIM **15**, 447 (1974).
88. W.B. Pearson, A nickel-gallium superlattice (Ni_3Ga). Nature **173**, 364 (1954)
89. M.J. Jones, *Baccalaureate Degree Thesis* (Pennsylvania State University, Chemical Engineering Dept, 2014)
90. H. Okamoto, Ga-Ni (gallium-nickel). J. Phase Equilib. Diffus. **31**, 575 (2010)
91. D.B. Ingerly, D. Swenson, C.-H. Jan, Y.A. Chang, Phase equilibriums of the Ga-Ni-As ternary system. J. Appl. Phys. **80**, 543–550 (1996)
92. U. Rizal, B.P. Swain, Raman characterization of gallium nitride nanowires deposited by chemical vapor deposition, in *Advances in Power Systems and Energy Management*, ed. by A. Garg, A.K. Bhoi, P. Sanjeevikumar, K.K. Kamani (Springer, Singapore, 2016), p. 47
93. J. Huang, S. Zhang, Z. Huang, Y. Wen, M. Fang, Y. Liu, Catalyst-assisted synthesis and growth mechanism of ultra-long single crystal a-Si_3N_4 nanobelts with strong violet–blue luminescent properties. CrystEngComm **14**, 7301–7305 (20112)
94. A.T. Dutra, P.L. Ferrandini C.A.R. Costa, M.C. Goncalves, R. Caram, Growth and solid/solid transformation in a Ni–Si eutectic alloy. J. Alloys Compd. **399**, 202–207 (2005)
95. R.P. Ram, S. Bhan, On the Structure of Ni_3Si ($Beta_2$) and Ni_3Si ($Beta_3$), Zeit. Metallkd. (Int. J. Mater. Res.) **66**, 521–524 (1975)

96. S. Gaudet, C. Coia, P. Desjardins, C. Lavoie, Metastable phase formation during the reaction of Ni films with Si(001): the role of texture inheritance. J. Appl. Phys. **107**, 093515 (2010)
97. X.Y. Kong, Z.L. Wang, Structures of indium oxide nanobelts. Solid State Commun. **128**, 1–4 (2003)
98. Z.L. Wang, Functional oxide nanobelts: materials, properties and potential applications in nanosystems and biotechnology. Annu. Rev. Phys. Chem. **55**, 159–196 (2004)
99. Z.W. Pan, Z.R. Dai, Z.L. Wang, Lead oxide nanobelts and phase transformation induced by electron beam irradiation. Appl. Phys. Lett. **80**, 309–311 (2002)
100. O.M. Berengue, R.A. Simon, A.J. Chiquito, C.J. Dalmaschio, E.R. Leite, H.A. Guerreiro, F.E.G. Guimarães, Semiconducting Sn_3O_4 nanobelts: growth and electronic structure. J. Appl. Phys. **107**, 033717 (2010)
101. Z.Y. Zhang, X.L. Wu, L.L. Xu, J.C. Shen, G.G. Siu, P.K. Chu, Synthesis, growth mechanism, and light-emission properties of twisted SiO_2 nanobelts and nanosprings. J. Chem. Phys. **129**, 164702 (2008)
102. T. Xie, M. Ye, Z. Jiang, Y. Qin, Y.-C. Wu, G.-W. Meng, L.-D. Zhang, Chloride assisted growth of aluminum nitride nanobelts and their enhanced dielectric responses. Chin. J. Chem. Phys. **21**, 586–590 (2008)
103. X. Wang, Y. Ding, C.J. Summers, Z.L. Wang, Large-scale synthesis of six-nanometer-wide ZnO nanobelts. J. Phys. Chem. B **108**, 8773–8777 (2004)
104. Y. Zhang, R. Li, X. Zhou, M. Cai, X.-L. Sun, Hierarchical Al_2O_3 nanobelts and nanowires: morphology control and growth mechanism. Cryst. Growth Des. **9**, 4230-4234 (2009)
105. R.M. Ma, L. Dai, H.B. Huo, W.Q. Yang, G.G. Qin, P.H. Tan, C.H. Huang, J. Zheng, Synthesis of high quality *n*-type CdS nanobelts and their applications in nanodevices. Appl. Phys. Lett. **89**, 203120 (2006)
106. H.-L. Zhang, F. Li, C. Liu, H.-M. Cheng, The facile synthesis of nickel silicide nanobelts and nanosheets and their application in electrochemical energy storage. Nanotechnology **19**, 165606 (2008)
107. D. Ma, W. Zhang, Q. Tang, R. Zhang, W. Yu, Y. Qian, Large-scale hydrothermal synthesis of SnS_2 nanobelts. J. Nanosci. Nanotechnol. **5**, 806–809 (2005)
108. M.-S. Hu, W.-M. Wang, T.T. Chen, L.-S. Hong, C.-W. Chen, C.-C. Chen, Y.-F. Chen, K.-H. Chen, L.-C. Chen, Sharp infrared emission from single-crystalline indium nitride nanobelts prepared using guided-stream thermal chemical vapor deposition. Adv. Funct. Mater. **16**, 537–541 (2006)
109. T. Hanrath, Germanium Nanowires: Synthesis, Characterization, and Utilization. *Doctoral Thesis* (University of Texas, Austin, 2004)
110. H.-Y. Tuan, D.C. Lee, T. Hanrath, B.A. Korgel, Catalytic solid-phase seeding of silicon nanowires by nickel nanocrystals in organic solvents. Nano Lett. **5**, 681–684 (2005)
111. P. Nash, A. Nash, The Ni−Si (nickel-silicon) system. Bull. Alloy Phase Diagram **8**, 6–14 (1987)
112. H.-Y. Tuan, D.C. Lee, T. Hanrath, B.A. Korgel, Germanium nanowire synthesis: an example of solid-phase seeded growth with nickel nanocrystals. Chem. Mater. **17**, 5705–5711 (2005)
113. F.-W. Yuan, H.-Y. Tuan, Supercritical fluid-solid growth of single-crystalline silicon nanowires: an example of metal-free growth in an organic solvent. Cryst. Growth Des. **10**, 4741–4745 (2010)
114. D. Wang, T. Xie, Q. Peng, Y. Li, Ag, Ag_2S and Ag_2Se nanocrystals: synthesis, assembly, and construction of mesoporous structures. Am. Chem. Soc. **130**, 4016–4022 (2008)
115. J. Wang, K. Chen, M. Gong, B. Xu, Q. Yang, Solution-solid-solid mechanism: superionic conductors catalyze nanowire growth. Nano Lett. **13**, 3996–4000 (2013)
116. M. Poulose, O.K. Varghese, C.A. Grimes, Synthesis of gold-silica composite nanowires through solid-liquid-solid phase growth. J. Nanosci. Nanotechnol. **3**, 341–346 (2003)
117. G. Chen, L. Wang, X. Sheng, H. Liu, X. Pi, Y. Zhang, D. Li, D. Yang, Growth of In_2O_3 nanowires catalyzed by Cu via a solid–liquid–solid mechanism. Nanoscale Res. Lett. **5**, 898–903 (2010)

118. H. Liu, Z. Huang, J. Huang, M. Fang, Y.-G. Liu, X. Wu, X. Hu, S. Zhang, Novel, low-cost solid-liquid-solid process for the synthesis of α-Si_3N_4 nanowires at lower temperatures and their luminescence properties. Sci. Report **5**, 17250 (2015)
119. H.F. Yan, Y.J. Xing, Q.L. Hang, D.P. Yu, Y.P. Wang, J. Xu, Z.H. Xi, S.Q. Feng, Growth of amorphous silicon nanowires via a solid–liquid–solid mechanism. Chem. Phys. Lett. **323**, 224–228 (2000)
120. H.D. Park, S.M. Prokes, M.E. Twigg, R.C. Cammarata, A.-C. Gaillot, Si-assisted growth of InAs nanowires. Appl. Phys. Lett. **89**, 223125 (2006)
121. Z.H. Lan, W.M. Wang, C.L. Sun, S.C. Shi, C.W. Hsu, T.T. Chen, K.H. Chen, C.C. Chen, Y.F. Chen, L.C. Chen, Growth mechanism, structure and IR photoluminescence studies of indium nitride nanorods. J. Cryst. Growth **269**, 87–94 (2004)
122. C. Tang, Y. Bando, Z. Liu, D. Goldberg, Synthesis and structure of InP nanowires and nanotubes. Chem. Phys. Lett. **376**, 676–678 (2003)
123. W.S. Shi, Y.F. Zheng, N. Wang, C.S. Lee, S.T. Lee, Synthesis and microstructure of gallium phosphide nanowires. J. Vac. Sci. Technol. B **19**, 1115–1118 (2001)
124. J. Motohisa, J. Noborisaka, J. Takeda, M. Inari, T. Fukui, Catalyst-free selective-area MOVPE of semiconductor nanowires on (111)B oriented substrates. J. Cryst. Growth **272**, 180–185 (2004)
125. S. Dhara, P.K. Giri, Self-catalytic growth of horizontal and straight Si nanowires on Si substrates using a sputter deposition technique. Solid State Commun. **150**, 1923–1927 (2010)
126. G.E. Stan, I. Pasuk, A.C. Galca, A. Dinescu, Highly textured (001) AlN nanostructured thin films synthesized by reactive magnetron sputtering for saw and fbar applications. Dig. J. Nanomater. Biostruct. **5**, 1041–1054 (2010)
127. D. Parnis, E. Zolotoyabko, W.D. Kaplan, M. Eisenberg, N. Mosleh, F. Meyer, C. Schwebel, Structural disorder in SiGe films grown epitaxially on Si by ion beam sputter deposition. Thin Solid Film **294**, 64–68 (1997)
128. G. Xu, Z. Li, J. Baca, J, Wu, Probing nucleation mechanism of self-catalyzed InN nanostructures. Nanoscale Res. Lett. **5**, 7–13 (2010)
129. M. Cuscuná, A. Convertino, L. Mariucci, G. Fortunato, L. Felisari, G. Nicotra, C. Spinella, A. Pecora, F. Martelli, Low-temperature, self-catalyzed growth of Si nanowires. Nanotechnology **21**, 255601 (2010)
130. Y. Zi, S. Suslov, C. Yang, Understanding self-catalyzed epitaxial growth of III–V nanowires toward controlled synthesis. Nano Lett. **17**, 1167–1173 (2017)
131. F. Gao, Z. Gu, Melting temperature of metallic nanoparticles, in *Handbook of Nanoparticles*, ed. by M. Aliofkhazraei (ed). (Springer, Cham, 2016), pp. 661–690
132. F. Jabeen, V. Grillo, S. Rubini, F. Martelli, Self-catalyzed growth of GaAs nanowires on cleaved Si by molecular beam epitaxy. Nanotechnology **19**, 275711 (2008)
133. G. Koblmüller, S. Hertenberger, K. Vizbaras, M. Bichler, F. Bao, J.-P. Zhang, G. Abstreiter, Self-induced growth of vertical free-standing InAs nanowires on Si(111) by molecular beam epitaxy. Nanotechnology **21**, 365602 (2010)
134. P. Ahmad, M.U. Khandaker, Z.R. Khan, Y.M. Amin, Synthesis of boron nitride nanotubes via chemical vapor deposition: a comprehensive review, RSC Adv. **5**, 35116–35137 (2015); see also, J.H. Kim, T.V. Pham, J.H. Hwang, C.S. Kim, M.J. Kim, Boron nitride nanotubes: synthesis and applications, Nano Converg. **5**, 17 (2018)
135. S. Kalay, Z. Yilmaz, O. Sen, M. Emanet, E. Kazanc, M. Çulha, Synthesis of boron nitride nanotubes and their applications. Beilstein J. Nanotechnol. **6**, 84–102 (2015)
136. O.R. Lourie, C.R. Jones, B.M. Bartlett, P.C. Gibbons, R.S. Ruoff, W.E. Buhro, CVD growth of boron nitride nanotubes. Chem. Mater. **12**, 1808–1811 (2000)
137. R. Ma, Y. Bando, T. Sato, K. Kurashima, Growth, morphology, and structure of boron nitride nanotubes. Chem. Mater. **13**, 2965–2971 (2001)
138. R. Ma, Y. Bando, T. Sato, K. Kurashima, Thin boron nitride nanotubes with unusual large inner diameters. Chem. Phys. Lett. **350**, 434–440 (2001)
139. J. Kim, S. Lee, Y.R. Uhm, J. Jun, C.K. Rhee, G.M. Kim, Synthesis and growth of boron nitride nanotubes by a ball milling–annealing process. Acta Mater. **59**, 2807–2813 (2011)

140. J. Wang, V.K. Kayastha, Y.K. Yap, Z. Fan, J.G. Lu, Z. Pan, I.N. Ivanov, A.A. Puretzky, D.B. Geohegan, Low temperature growth of boron nitride nanotubes on substrates. Nano Lett. **5**, 2528–2532 (2005)
141. Z.W. Gan, X.X. Ding, Z.X. Huang, X.T. Huang, C. Cheng, C. Tang, S.R. Qi, Growth of boron nitride nanotube film in situ. Appl. Phys. A **81**, 527–529 (2005)
142. C.H. Lee, M. Xie, V. Kayastha, J. Wang, Y.K. Yap, Patterned growth of boron nitride nanotubes by catalytic chemical vapor deposition. Chem. Mater. **22**, 1782–1787 (2010)
143. C.-Y. Su, Z.-Y. Juang, K.-F. Chen, B.-M. Cheng, F.-R. Chen, K.-C. Leou, C.-H. Tsai, Selective growth of boron nitride nanotubes by the plasma-assisted and iron-catalytic CVD methods. J. Phys. Chem. **113**, 14681–14688 (2009)
144. B. Zhong, X. Huang, G. Wen, H. Yu, X. Zhang, T. Zhang, H. Bai, Large-scale fabrication of boron nitride nanotubes via a facile chemical vapor reaction route and their cathodoluminescence properties. Nanoscale Res Lett. **6**, 36 (2011)
145. L. Guo, R.N. Singh, Selective growth of boron nitride nanotubes by plasma-enhanced chemical vapor deposition at low substrate temperature. Nanotechnology **19**, 065601 (2009)
146. A.T. Matveev, K.L. Firestein, A.E. Steinman, A.M. Kovalskii, O.I. Lebedev, D.V. Shtansky, D. Golberg, Boron nitride nanotube growth via boron oxide-assisted chemical vapor transport-deposition process using $LiNO_3$ as a promoter. Nano Res. **8**, 2063–2207 (2015)
147. D.P. Yu, X.S. Sun, C.S. Lee, I. Bello, S.T. Lee, H.D. Gu, K.M. Leung, G.W. Zhou, Z.F. Dong, Z. Zhang, Synthesis of boron nitride nanotubes by means of excimer laser ablation at high temperature. Appl. Phys. Lett. **72**, 1966–1968 (1998)
148. S.N. Mohammad, Systematic investigation of the growth mechanisms for the synthesis of the conventional, doped, and bamboo-shaped nanotubes, primarily the carbon nanotubes. Carbon **75**, 133–148 (2014)
149. S. Ganji, Hill model for the base growths and tip growths of doped and undoped carbon nanotubes. J. Nanosci. Nanotechnol. **18**, 7623–7640 (2018)
150. C. Hedrich, S. Haugg, L. Pacarizi, K.P. Furlan, R.H. Blick, R. Zierold, Low-temperature vapor-solid growth of ZnO nanowhiskers for electron field emission. Coatings **9**, 698 (2019)
151. K.D. Rendulic, The influence of surface defects on adsorption and desorption. Appl. Phys. A **47**, 55–62 (1988)
152. S. Kennou, K. Kamaratos, S. Ladas, C.A. Papageorgopolous, The influence of steps on the adsorption of Cs on Si(100). Surf. Sci. **216**, 462–471 (1989)
153. M.B. Raschke, U. Höfer, Influence of steps and defects on the dissociative adsorption of molecular hydrogen on silicon surfaces. Appl. Phys. B **68**, 649–655 (1999)
154. S. Tian, H. Li, Y. Zhang, S. Zhang, Y. Wang, J,.-C. Ren, X. Qiang, Single-crystalline hafnium carbide nanowire growth below the eutectic temperature by CVD 384, 44–49 (2013)
155. F.M. Ross, J. Tersoff, M.C. Reuter, Sawtooth faceting in silicon nanowires. Phys. Rev. Lett. **95**, 146104 (2005)
156. Q. Yu, A.S. Ahmad, K. Ståhl, X.D. Wang, Y. Su, K. Glazyrin, H.P. Liermann, H. Franz, Q.P. Cao, D.X. Zhang, J.Z. Jiang, Pressure-induced structural change in liquid GaIn eutectic alloy. Sci. Report **7**, 1139 (2017)
157. J.W. Dailey, J. Taraci, T. Clement, Vapor-liquid-solid growth of germanium nanostructures on silicon. J. Appl. Phys. **96**, 7556 (2004)
158. H. Jeong, T.E. Park, H.K. Seong, M. Kim, U. Kim, H.J. Choi, Growth kinetics of silicon nanowires by platinum- assisted vapor–liquid–solid mechanism. Chem. Phys. Lett. **467**, 331–334 (2009)
159. C.L. Pint, N. Nicholas, S.T. Pheasant, J.G. Duque, A. Nicholas G. Parra-Vasquez, G. Eres, M. Pasquali, R.H. Hauge, Temperature and gas pressure effects in vertically aligned carbon nanotube growth from Fe-Mo catalyst. J. Phys. Chem. C **112**, 14041–14051 (2008)

Chapter 17
Conclusions

Abstract In conclusion, the integral elements of the book are the kinetics and mechanism of the syntheses of X_mY_n nanomaterials. The book is as much about the existing knowledge of the nanomaterials syntheses, as about new knowledge and research pertaining to nanomaterials syntheses. The unified character of the growth phenomena described in the book should further our understanding of the correlations between material properties and growth mechanisms. Several design rules and guidelines have emerged from investigations detailed in the book. These design rules, if followed, should advance our understanding of the growth kinetics and growth mechanism.

17.1 General Conclusions

During the past decade, nanotechnology has emerged as a highly multidisciplinary field, drawing from disciplines such as applied physics, materials science, colloidal science, device physics, supramolecular chemistry, biological sciences, and even mechanical engineering, chemical engineering, and electrical engineering. The focus of this book has been the synthesis of X_mY_n nanomaterials needed to nurture the scientific and engineering activities in these disciplines and necessary to develop nanoscale nanotechnologies never realized before. The essential elements of the book are the kinetics and mechanism of the syntheses of X_mY_n nanomaterials. Both of them are needed to discern the new and novel characteristics of these materials. The book is as much about the existing knowledge of the nanomaterials syntheses, and the kinetics and mechanism of these syntheses, as about the evolving new knowledge and research pertaining to nanomaterials syntheses, and the kinetics and mechanism of these syntheses. The X_mY_n nanomaterials include, but not limited to nanowires, carbon nanotubes, semiconductor nanotubes, nanobelts, nanorings, nanodots , including quantum dots and quantum rings, grapheme, and nanofibers.

Nanoscience encompasses the understanding of the fundamental physics, chemistry, biology, and technology of the nanometer-scale nanomaterials (e.g., nanocrystals). Nanotechnology, on the other hand, is broadly classified into three groups,

S. N. Mohammad, *Synthesis of Nanomaterials*, Springer Series in Materials Science 307, https://doi.org/10.1007/978-3-030-57585-4_17

namely, nano-enabled tools, nanoparticles, and nano-enabled drugs. The successful development of this technology relies on the availability of technologically feasible new nanomaterials. In this context, the growth mechanisms are central because the characteristics and functionalities of new nanomaterials are largely dictated by the mechanisms by which they are grown, and also by the corresponding manifestations.

The understanding of nanomaterials syntheses is noticeably complicated by the variety of experimental techniques used for growths, wide differences in the behavior of FECA nanoparticles employed for the growths, the materials characteristics of the supports, the materials characteristics of the substrates, the composition and behavior of the precursors, and importantly the end effects of the growth conditions. Too many different techniques used for the growths often lead to too many different observations to allow in-depth understanding of growths. Too many mechanisms to explain these growths have therefore emerged. Almost always new mechanisms resulted from the old and existing mechanisms not being able to explain the results of new growths. Not all of these mechanisms, but only the major ones of them employed widely to explain growths constitute the heart of this book.

The book elaborates all major nanomaterials synthesis routes including the CVD and MBE technique, has presented an overview of FECA nanoparticles and has set forth an account of pre-synthesis and pro-synthesis events before venturing to articulate features of nanomaterials syntheses and the fundamentals of the mechanisms of these syntheses via vapor-phase, liquid-phase, and solid-phase processes. Basic ideas underlying the foundation and operation of all possible mechanisms and the strengths and weaknesses of these mechanisms have been elucidated. Again, the growth mechanisms include all well-known mechanisms widely employed for growths. These are, for example, the vapor-phase mechanisms such as the VLS, VSS, VS, and OAG mechanisms, the solution and supercritical fluid-based mechanisms such as SoLS and SFLS mechanisms, the solid-phase mechanism such as the SLS mechanism, and the self-catalytic growth mechanism. Among various mechanisms, the VLS and the VSS mechanisms are perhaps the most widely employed MET-mediated mechanisms, while the VS mechanism is perhaps the most widely employed MET-free mechanism. SFLS and SoLS mechanisms are also the MET-mediated growth mechanisms successful in producing many different nanowires, including the III-V and II-VI nanowires of broad diameter distributions and properties. The novelty of these mechanisms as parallel to VLS growth mechanism has been detailed. It has been noted that nanowire growths by the SFLS and SoLS mechanisms are governed, not only by the catalyst nanoparticle, but also by the precursor from which this nanoparticle is obtained, and also by the liquid solution in which the nanoparticle remains immerged during growth. The details of the growths carried out by both the MET-mediated and the MET-free mechanisms have been discussed in some details. Features of these growth mechanisms have been broadly examined. As they demonstrate, nucleations in the MET-mediated (METANO based) mechanisms are different from those in the MET-free (SUBSANO based) mechanisms. Various attributes of the vapor-phase, solution-phase, and solid-phase growths as demonstrated by existing literature are listed in Table 17.1. Various attributes of

Table 17.1 Various attributes of the vapor-phase, solution-phase, and solid-phase growths as demonstrated by existing literature; L/S is the abbreviation of liquid/solid; specific growth mechanisms are given in the parenthesis

Feature	Vapor-phase growth	Solution-phase growth	Solid-phase growth
Medium in which R_S species is released and then adsorbed on nanoparticle surface	Vapor	Solution	Solid
Structure of FECA nanoparticle surface before growth	Solid (VLS, VS, VSS)	Solid	Solid (SLS)
Structure of FECA nanoparticle surface during growth	Molten metal alloy (VLS); crystalline solid metal (VSS); crystalline solid substrate (VS)	Molten metal alloy (SoLS, SFLS); crystalline solid metal (SoSS, SFSS)	Molten metal alloy (SLS)
Condition of FECA nanoparticle surface during growth	Liquid (VLS); solid (VS, VSS)	Liquid (SoLS, SFLS); solid (SoSS, SFSS)	Liquid (SLS)
Nanoparticle state in which R_S species diffuse through the R_L species of it	Liquid (VLS); solid (VS, VSS)	Liquid (SoLS, SFLS); solid (SoSS, SFSS)	Liquid (SLS)
Interface at which supersaturation and nucleation take place	L/S (VLS), undefined (VS, VSS)	L/S (SoLS, SFSS); undefined (SoSS, SFSS)	L/S (SLS)

the vapor-phase, solution-phase, and solid-phase growths as demonstrated by analyses, experimental evidences, and investigations carried out in Chaps. 5–16 are, however, listed in Table 17.2. They suggest that the vapor-phase, solution-phase, and the solid-phase growths all have some common traits. This is quite evident from comparison of various entries of Tables 17.1 and 17.2. A close look of Table 17.2 may indicate that the only difference in the vapor-phase, solution-phase, and the solid-phase growths is that the R_S ($R_S \equiv X$ and $R_S \equiv Y$) source species are released, for example, from their precursors and then adsorbed on the FECA nanoparticle surface in vapor medium during vapor-phase growth, in solution medium during solution-phase growth, and in solid medium during solid-phase growth. Otherwise, the $R_S \equiv X$ and $R_S \equiv Y$ species are essentially the same, no matter in which medium they are generated and which precursors are used for their release.

Based on descriptions in the book, despite all the difficulties, the vapor deposition techniques, in particular, have emerged as versatile and capable of producing high-quality nanocrystal structures from the bottom up. These techniques appear to have potential to demonstrate competitive approaches for producing technologically promising nanostructures. Undeniably, detailed understanding of the nucleation and

Table 17.2 Various attributes of the vapor-phase, solution-phase, and solid-phase growths as demonstrated to be true based on analysis, investigation, experimental evidences, etc., performed in the present study

Feature	Vapor-phase growth (Source species)	Solution-phase growth (Source species)	Solid-phase growth (Source species)
Medium in which R_S species is released and then adsorbed on nanoparticle surface	Vapor [R_S≡X, R_S≡Y]	Solution [R_S≡X, R_S≡Y]	Solid [R_S≡X, R_S≡Y]
Structure of FECA nanoparticle surface before growth	Solid	Solid	Solid
Structure of FECA nanoparticle surface during growth	Alloy, cluster, or solid solution	Alloy, cluster, or solid solution	Alloy, cluster, or solid solution
Condition of FECA nanoparticle surface during growth	Quasiliquid (quasisolid), porous	Quasiliquid (quasisolid), porous	Quasiliquid (quasisolid), porous
Nanoparticle state in which R_S species diffuse through the R_L species of it	Quasiliquid (quasisolid)	Quasiliquid (quasisolid)	Quasiliquid (quasisolid)
Nanoparticle state in which R_S species are streamlined and react inside the R_L species of it	Quasiliquid (quasisolid)	Quasiliquid (quasisolid)	Q uasiliquid (quasisolid)
Interface at which supersaturation, crystallization and nucleation take place	QL/S, or QS/S	QL/S, or QS/S	QL/S, or QS/S

QL/S is for quasiliquid/solid, and QS/S is for quasisolid/solid; quasiliquid (quasisolid) state may span between ξ_m=0 and ξ_m=1

growths by the vapor deposition processes is however still lacking. An important element of the book is how to overcome these problems so that critical knowledge for achieving fine control of the nanomaterials morphology and properties and for improving the reliability and reproducibility of the growths by these techniques can be realized. This is immensely important for an evolution of the nanomaterials products from the laboratory-based synthesis to the industrial-level manufacturing.

The book elaborates the major problems that arise in finding a specific mechanism suitable for the growth of many different types of X_mY_n nanomaterials, such as nanowires, nanotubes, nanodots, nanobelts, and grapheme. For example, while many nanowire growths are believed to be mediated by the VLS mechanism, many other nanowire growths are believed to be mediated by the VSS mechanism, and many other nanowire growths are believed to be mediated by the VS mechanism. Almost none of the carbon nanotube growths are believed to be mediated by the VLS mechanism. Although these carbon nanotube growths are believed to be done by the VSS and VS mechanisms, the intricate details of these growths by these mechanisms cannot be explained in details. The major drawbacks of these two mechanisms emanate from two major problems. It is not clear how nanomaterials source species and their precursors stick firmly to the crystalline nanoparticle surfaces. The sticking coefficient of these species on crystalline solid surfaces is marginally small. It is not also evident how vapor source species are nucleated into solid nanomaterial(s) without going through supersaturation at the liquid/solid interface. Nanobelts are also widely believed to be synthesized by the VSS and VS mechanisms. Yet the details of the nanobelt syntheses by the VS and VSS mechanisms could never be explained and established. A rational unification of the growth mechanisms has become exceedingly challenging. The problems are compounded by insufficient controls over the techniques and by the absence of knowledge of various possible properties of these techniques. It is nevertheless necessary to achieve versatility of a suitable deposition mechanism. And if achieved, it should provide not only more opportunities and flexibilities in exploring nanomaterials morphologies and orientation with controlled properties, but also can vastly advance the complex and complicated growth phenomena that lead to variance in nanomaterials engineering. It has been determined that an in-depth understanding of the eutectic, non-eutectic, and non-stoichiometric alloying behaviors of FECA nanoparticles (both METANO and SUBSANO) is crucial for vapor-phase, liquid-phase, and solid-phase mechanisms. Defect control is also vital for nanomaterials applied in electronics and optoelectronics. An understanding of the growth kinetics can greatly facilitate the synthesis of nanomaterials suitable for complex heterostructure devices with controlled compositions and distributions.

17.2 Concerns About the VSS and the VS Mechanisms Reaffirmed

We go back to the VSS and the VS mechanisms. Unfortunately, these mechanisms do not appear to have the fundamental physicochemical base crucially required for understanding the nanomaterials growths by them. It is shrouded with confusions that the VSS growth at a temperature T lower or higher than the binary alloy eutectic temperature T_E takes place via the crystalline solid metal nanoparticle. It happens while the VLS growth at $T = T_E$ takes place via the liquid alloy catalyst nanoparticle. Also, the growth rate increases with temperature. Bulk diffusivity of the R_S species is always faster through liquid than through solid. Yet, due to the absence of strong fundamental base, the observed VSS growth rate of nanowires at $T < T_E$ comparable to the VLS growth rate of the same nanowires at $T = T_E$ cannot be comprehended. For the same reason, it is not known why the VSS growth rate is marginally small under some experimental condition, but dramatically high under some other experimental conditions. It is puzzling that such high growth rates by the VSS mechanism could experimentally be accomplished despite the sticking coefficients of the R_S source species on crystalline metal nanoparticles is very small.

Parallel to VSS mechanism, SoSS and SFSS mechanisms have been proposed for the growths of X_mY_n nanomaterials in solution. The SoSS (SFSS) mechanism differs from the SoLS and SFLS mechanisms in that solid, superionic silver, or copper chalcogenide (e.g., Ag_2S, Cu_2S, and Ag_2Se) nanoparticles, rather than low-melting metallic nanoparticles, are used to catalyze the nanomaterials growths. These growths are carried out at temperatures typically 100–230 °C lower than the growths by the corresponding SFLS and SoLS mechanism. The growths by the SoSS (SFSS) mechanism are therefore by solid catalyst nanoparticles, rather than by the eutectic liquid nanoparticles, as in the SFLS and SoLS mechanisms. It has been argued in the text (see Chaps. 12–16) that, during the pre-nucleation stage of growth by all mechanisms, including the SoSS (SFSS) mechanism, the FECA metal nanoparticles (e.g., METANOs) undergo surface treatment and/or surface functionalization, and react with the R_S≡X and/or R_S≡X species. The kinetic processes, prior to METmediated low-temperature pro-nucleation stage of growth of X_mY_n nanomaterials, thus lead to the formation of R_L species as R_L≡(MET, X, Y) metastable alloys or R_L species as R_L≡(MET, X) metastable alloys if R_S≡Y species is volatile. These alloys generally have a significant content of both FECA and X. They can be cluster or solid solution. If some contaminant ϑ and oxygen atoms are present, they may even be R_L≡(MET, X, ϑ), R_L≡(MET, X, Y), R_L≡(MET, X, Y, ϑ), and/or R_L≡(MET, X, Y, ϑ, O) alloy, cluster or solid solution. There may also be the possibility of R_L≡(oxygenated MET) or R_L≡(MET oxide) species. There may as well be R_L≡(EMNO) species. The R_L species, thus created, exhibit surface disorder, surface amorphicity, surface porosity, surface coarsening, and surface melting, and it may have polar characteristics. Similar situations arise for growths by the MET-free VLS mechanism. The R_L species for these growths may not though have traces of MET element(s).

17.3 Concerns About Catalyst Droplets During Growths Affirmed

While growing Ga-assisted Ge nanowires in solution, Pertl et al. [1] found high content of Ga in Ge nanowires. While exploring the cause of it, the R_L≡(Ga, Ge) alloy was found to have a significant content of both MET≡Ga and X≡Ge. And, as a result, Ge nanowires were contaminated with Ga when the R_S≡Ge species, diffusing through the R≡(Ga, Ge) alloy, cluster or solid solution prior to nucleation and growth knocked out Ga atoms from the R_L≡(Ga, Ge) species. Good examples of similar other R_L species are R_L≡(Sn, Ge) alloys [2–4] and R_L≡(Bi, Ge) alloys [5] formed during low-temperature growths of Ge nanowires catalyzed by Sn and Bi nanoparticles, respectively. Wang et al. [6] found that these nanoparticles are mesoporous even before undergoing pre-nucleation stage of growth. The pore sizes of these nanoparticles are in the range of 2–50 nm. We believe it was due to low melting point of the MET≡Ga, Sn, and Bi, and because R_L≡(Ga, Ge) alloy, R_L≡(Sn, Ge) alloy, and R_L≡(Bi, Ge) alloy were very unstable at the growth temperature(s). They point to the inherent flaws of highly molten R_L species at growth temperature(s) much higher than the melting point of the MET(s) during nanomaterials growths.

Alekberov et al. [7] performed first-principles calculations to study the cation and anion monovacancies in β-Ag_2S. They found that the neutral vacancies in this β-Ag_2S have the lowest formation energies. And also due to low formation energies, Ag vacancies in β-Ag_2S are high-concentration dominant defects. These are acceptor-like defects. Also, the Ag atoms in them are not rigidly fixed at their sites. This means the formation of additional defects may be common in Ag_2S crystal. Ballikaya et al. [8] noted that the nano-Ag_2Se films possess multiphase nanoscale features. They all appear to support our view that the catalyst nanoparticles for SoSS (SFSS) growths are not solid, but alloy, cluster, or solid solution.

17.4 The Key Elements of the Book

The most important goal of the book has been to try to establish a common—a general platform for the mechanism and kinetics, and for the synthesis of nanomaterials—just not one nanomaterial, but all nanomaterials including the ones already realized and the ones not yet realized. To accomplish this goal, an essential element of the book has been to demonstrate that, unlike the VS and the VSS mechanisms, the VLS mechanism does have fundamental chemico-physical base for nanomaterials growths. VQS mechanism has similarly fundamental chemico-physical base (see Chaps. 12–16) for nanomaterials growths. In fact, the study reported in the book suggests that the VSS and VS mechanisms are actually two different specific variants of the VQS mechanism. It also highlights a deep-rooted flaw of the VLS mechanism. Notably, the VLS growth is accomplished by using (MET, X) eutectic alloy formed at the eutectic temperature T_E. And the formation of this eutectic alloy at the eutectic temperature T_E is very sensitive to the presence of contaminants. It is sensitive also to the mole fractions of MET and X. Even if all conditions are satisfied, the (MET, X) alloy may not be eutectic at the eutectic temperature T_E if the (MET, X) alloy contains contaminant(s). Further, the VLS mechanism may not explain MET-mediated growths at $T < T_E$ and $T > T_E$. VLS mechanism may not also explain MET-mediated non-eutectic growth at $T = T_E$.

To achieve the stated goal, it has, in general, been necessary that we capture unified character of the growth phenomena and further our understanding of the correlations between material properties and growth mechanisms. This is crucial in order to guide the design and development of a universal growth mechanism urgently needed for nanomaterials-based new technology development. To fulfill this goal, attention in the book was turned to the comprehensive investigation of the VQS mechanism, which was discovered a decade ago in our laboratory. Careful investigations of various growth behaviors have been presented to assess the potential and applicability of this mechanism. For this, two different avenues have been followed. One of them involves metal nanoparticle, and the other one is free from metal nanoparticle. The nanoparticles have been classified as NP1 and NP2 nanoparticles, and also as METANOs (e.g., metal -based FECA nanoparticles) and SUBSANOs (e.g., substrate-based FECA nanoparticles). And within the framework of them, various features of the VQS mechanism, namely the concept of quasiliquid (quasisolid) nanoparticle surface, have been extensively discussed. It has been determined that surface amorphicity of the nanoparticle surface, surface coarsening of nanoparticle surface, surface looseness and porosity of nanoparticle surface, melting (semi-melting) of nanoparticle surface, nanoparticle structure and morphology, phase transition(s), and phase separation(s) during nanomaterials synthesis at the growth temperature T, the generation of dipole moments, and the creation of high-energy sites (HETs) on the nanoparticle surface are the integral elements of the VQS mechanism. Importantly, these elements are not arbitrary. They are rather judiciously planned and controlled; they are interrelated, and even interdependent. They can, for example, be derived from the fundamental viewpoint of the effective surface amorphicity α_{amoreff} taking into consideration α_{amoreff0} of the nanoparticle surface. We believe α_{amoreff} would have some relevance to SECINI, and α_{amoreff0} would have some relevance to SECINI0, both introduced earlier in Chap. 3. And to find and establish all these would be challenging. Among various elements, HETs may be noteworthy. They have paramount importance as tools in decomposing the source-species precursors at growth temperatures, which may be much lower than the thermal decomposition temperatures of these precursors. It has been argued that the FECA nanoparticle (both METANO and SUBSANO) surface (and possibly subsurface) exhibiting looseness should be (1) a eutectic alloy, a non-eutectic alloy, a cluster, or a solid solution. It should be (1) amorphous, semi-amorphous, amorphous-like, (2) paracrystalline, (3) stepped, (4) influenced by synergy, and/or (5) inundated with defects and dislocations. For further illumination of the VQS mechanism, many different means to control nanomaterials growths on surfaces, and the large-scale order and novel properties of the resulting nanomaterials have been presented. The processes for the SUBSANO-based growths by the VQS mechanism include HETs, semi-melting of the surface, and the nanopores created on the nanoparticle surface. These processes for the MET-mediated growth by the VQS mechanism also include the HETs, semi-melting of the surfaces, and nanopores created on the nanoparticle surfaces. They incorporate as well the surface energy-mediated exchange of materials on the nanoparticle surface. The deeper insights of various physical and chemical processes contributing to the VQS mechanism may consequently have been unraveled. All these reveal the fundamental base of the VQS mechanism and demonstrate that this VQS mechanism is founded on one single mechanistic platform common for nanowire, nanotube, nanodot, nanofiber, nanobelt, nanosheet, graphene, and other nanomaterials growths.

The need of a rational choice of the substrates, and of the materials structures (e.g., support and/or FECA nanoparticles) formed on these substrates has been discussed. The importance of surface functionalization of the FECA nanoparticle (particularly, METANO) caused by surface treatment and/or induced by charge transfer, defect transfer, and others from the support and/or substrate has also been explained.

While examining the generality of the VQS mechanism, attempts have been made to determine if the (1) vapor-phase mechanisms (e.g., the VLS, VSS, VS, OAG, SLS , and SCG mechanisms), the solution-phase mechanisms (e.g., SFLS, SoLS, SSS mechanisms), and the solid-phase mechanism (e.g., SLS mechanism) are all special cases of the VQS mechanism. In fact, it has been established that all of these mechanisms, including the VSS mechanism, for both the low-temperature ($T < T_E$) and high-temperature ($T > T_E$) growths are special cases of the METANO-based VQS mechanism. It has been established that, indeed, the VS, OAG , and the self-catalytic growth mechanisms are all different variants of the SUBSANO-based VQS mechanism. Various presentations have shown that the VLS, VSS, and VS mechanisms have very narrow domains of applications. Various illustrations have demonstrated that the carbon nanotubes, boron nitride nanotubes, and semiconductor nanobelts are all grown by the VQS mechanism, and not by the VSS or VS mechanism. The growth behaviors not explained by most of the existing mechanisms (see Chaps. 5–11) have been examined. It has been noted that they all can be explained by the VQS mechanism.

Table 17.1 has been prepared based on experimental evidences and first-principle simulation results. This table shows that the only differences among the vapor-phase, solution-phase, and the solid-phase growth mechanisms are the media of release of the R_S species from their precursor(s) and their adsorption on the FECA nanoparticle surface. All other features for them appear to be common and relevant to the ones proposed for the VQS mechanism. This is true even if FECA nanoparticles are in stationary condition (1) in vapor medium in the vapor-phase mechanisms, (2) in floating condition in liquid medium in solution-phase mechanism, and (3) under stationary condition in solid medium in the solid-phase mechanism.

Some very unique features of the VQS mechanism have been manifested from the model calculations and also from the analyses of the role of surface energy in the METANO-mediated growths. Recall that the surface energy of FECA nanoparticles is sensitive to the FECA nanoparticle dimension. Taking this surface energy into account, it has been explained probably for the first time why Si and Ge nanowire growths are far more successful than other nanowire growths. It has been explained probably for the first time also why carbon nanotube growth rates with METANOs from Fe, Ni, and Co are much higher than the carbon nanotube growth rates with METANOs from other metals.

The manifestation made above suggests that the unified growth mechanism (called αQS mechanism) put forth in Chap. 16 may eventually pave the way of the development of atomic-scale nanomaterials with excellent control, for example, of doping, orientation, growth rate, and morphology. The said development would involve complex nanomaterials structures, such as heterostructures and superlattice structures with clean and abrupt interfaces, branched nanomaterials, nanomaterials with bridges, and complete nanomaterials devices. They would all be achieved during the growth stage without the use of complicated nanofabrication technique(s). The excellent control over material composition and morphology, along with the relative ease of synthesis will allow the fabrication of large-scale, homogeneous nanomaterials with uniform dimensions (e.g., length and diameter) at critical locations, such as doped locations. Deep understanding of the said unified mechanism (e.g., αQS mechanism) will also enable the integration of different materials and functions in the same system with atomic-scale precision. Graphene is a single layer of sp^2 bonded carbon atoms. It constitutes a fascinating new field of immense technological potential. Just one atom thick, and yet harder than diamond, stronger than steel, and lighter than paper, it has heralded a new era in science and technology. However, it has yet to have a commercial breakthrough which depends squarely on the basic understanding of the mechanism of its growth. We believe αQS mechanism is the true mechanism of this growth [9].

17.5 Design Rules and Guidelines

A few design rules and guidelines have emerged from investigations that tended to embellish the book. They are listed as follows:

1. Nanomaterials growths on a nanoparticle surface are primarily dictated by the R_L species formed on the nanoparticle surface. So, the formation of a nanomaterial of desired shape, size, composition, morphology, and dimension is crucially governed by the R_L species (see Sect. 3.4 of Chap. 3). This R_L species may have many different shapes, sizes, and compositions. The shape, size, and composition of this R_L species may be different from the shape, size, and composition of the nanoparticle on which it is created. The shape and size of the nanomaterial formed on this R_L species may be the same as the shape and size of this R_L species. To be specific, the shape and size of the nanomaterial formed on the R_L species may be the same as the shape and size of the R_L species, but not necessarily the same as the shape and size of the nanoparticle on which the R_L species is created. We cite an example, the nanoparticle for the growth of a nanobelt may have circular cross-section. But the R_L species formed on it

may be thin and rectangular. It may be formed along a certain porous, amorphized grain boundary between two crystalline (near crystalline) grains. It may be dictated by directional anisotropy and charge polarization. The shape and size of the nanobelt formed on this R_L species may be the same as the shape and size of the said R_L species itself.

2. Nanoparticles should not be excessively unstable and molten eutectic alloy. Such nanoparticles would lack the ability to prevent the migration of the MET species from the droplet into the X_mY_n nanomaterials that it mediates to grow.
3. If the nanoparticles are clusters or solid solutions in composition, they should not have any component overly molten (for example, Ga of the GaGe solid solution) at the growth temperature. As described above [1–6], nanomaterials would be contaminated with MET atoms when the R_S species diffusing, for example, through the $R_L \equiv$(MET, X), $R_L \equiv$(MET, X, Y), or $R_L \equiv$(MET, X, Y, ϑ) species alloy, cluster, or solid solution prior to nucleation and growth knock out MET atoms and even ϑ contaminants into the growing X_mY_n nanomaterial.
4. If the nanoparticles are clusters or solid solutions in composition, they should, at the growth temperature, preferably have molten nanopores in an otherwise solid, amorphous R_L species composition. These nanoparticles should be stable at the growth temperature, and yet porous enough for the diffusion of the R_S species to the QL/S or QS/S interface for supersaturation, nucleation, and growth of the X_mY_n nanomaterial.
5. If the nanoparticle is a BNP: METANO$\equiv$(MET1)$_{1-z}$(MET2)$_z$, then MET1 and MET2 of this BNP may react with the $R_S \equiv$X species creating a core–shell structure with (MET1)$_{1-z}$(X)$_z$ alloy in the core and (MET2)$_{1-z}$(X)$_z$alloy ($z \neq z'$) in the shell [10]. The alloy forming the shell and primarily mediating the X_mY_n nanomaterials growth should not be overly molten and hence unstable. It should rather be stable, quasiliquid (quasisolid) and have nanopores and grain boundaries of desired dimension and concentration.
6. The nanopores transporting the R_S($R_S \equiv$X, $R_S \equiv$Y) species to the QL/S or QS/S interface must be straight and have proper dimension needed for the said transport. They must not be bent and twisted. The straight, molten (semimolten) grain boundaries would serve, as well, as the channels for the transport of the R_S species to the QL/S, QS/S interface.
7. The growth temperature and chamber pressure must simultaneously be so tuned that the R_L species of the nanoparticle is quasiliquid (quasisolid) implying that nanopores and grain boundaries of the R_L species are suitable enough (as stated above) for the diffusion of the R_S species through the R_L species. Again, both of them must be tuned to ensure smooth, uninterrupted diffusion of the R_S species through the R_L species
8. Unless tapering is desired, the vapor source deposition on the sidewalls must not be dominant due to the reduction in the temperature-dictated precursor flux, the effective diffusion length of the R_S species adatoms, and also the equilibrium vapor source pressure on the top nanoparticle surface. Several factors related to nanoparticle govern the radial growth of X_mY_n nanomaterials. If the growth is mediated by a relatively unstable droplet and the peripheral areas of the growth

front of this droplet is in the vicinity of vapor, liquid, and solid (generally called tristate), these peripheral areas are very susceptible to sticking of the vapor species and hence to the deposition on the radial direction. If on the other hand, the growth is mediated by a relatively stable nanoparticle, which (a) is smooth (not rough) across the radial surface (but not across the top axial surface), (b) is solid [suppose, with $\xi_m \approx 0.5$ and $\alpha_{amor} = \alpha_{amoreff0}$], and (c) exhibits nanopores of optimal dimension and density, then the radial growth as compared to the axial growth would be marginal. This is largely because the peripheral areas of the nanoparticle would be least susceptible to sticking of the vapor species and hence to the deposition on the radial direction. On the other hand, axial growth would be large. It would, of course, depend on temperature and pressure. It would depend, as well, on the background gases, such as H_2, which must be controlled. Recall that this H_2 is known to passivate the top axial surface of the nanoparticle thus suppressing the decomposition of the precursor(s) of the R_S species on this surface and also the adsorption of the R_S(R_S≡X and R_S≡Y) species onto this surface. It is also known to reduce the surface roughness of the top axial surface of nanoparticle. It thus allows the radial growth to dominate over the axial growth.

9. Surface treatment must be conducted in such a judicious manner that the effective surface amorphicity $\alpha_{amoreff}$ of the R_L species surface would equal $\alpha_{amoreff0}$. Following relationships spelled out in Chap. 15, all parameters, such as porosity, surface melting, surface roughness, dipole moment, and HETs would commensurate with the condition of $\alpha_{amore} = \alpha_{amoreff0}$; they would correspond to the optimal conditions for growths.
10. Surface treatment of FECA nanoparticle must be conducted in such a judicious manner that its R_L species surface has pre-determined number of grains and grain boundaries. For example, the R_L species surface for nanowire growth may have many grains and grain boundaries, and defects and disorders all throughout the grains and grain boundaries. But these defects and disorders must commensurate with the condition of $\alpha_{amore} = \alpha_{amoreff0}$ and be on the entire R_L species surface of desired shape and size. The R_L species surface for nanotube growth may have many grains and grain boundaries, and defects and disorders all throughout the grains and grain boundaries. But they must commensurate with $\alpha_{amore} = \alpha_{amoreff0}$ and be only in the peripheral shell. The R_L species surface for nanobelt growth may have grains and grain boundaries. Also, while the grains should be free from defect, disorder, and amorphicity, at least one of the grain boundaries may have defects and disorders, including dopant atoms. These defects, disorders, and dopant atoms may be on the said grain boundary and/or in areas adjacent to it. So, this grain boundary appears as a thickened straight stripe exhibiting the characteristics of the R_L species (see Chap. 4). It should be along a certain specific direction dictated, for example, by charge polarization. Also, the said defect and disorder must commensurate with the condition of $\alpha_{amore} = \alpha_{amoreff0}$. All these would ensure nanowire growth over the entire nanoparticle surface, nanotube growth over the peripheral shell of the nanoparticle surface, and nanobelt growth over the thickened grain boundary of

the nanoparticle surface. If the R_L species of a nanoparticle has defect-free grain boundaries separating porous, amorphous defected grains, several nanowires may be grown on the grains of the R_L species of the nanoparticle.

11. The nanomaterials growths may suffer from the desorption of the $R_S{\equiv}X$ and/or the $R_S{\equiv}Y$ species. The immediate consequence of this desorption may be long incubation time. We cite an example. Zettler et al. [11] studied GaN nanowire growths on Si substrate by the MBE technique. They noted that the growth time for the growths of a certain length of nanowires at 785 and 815 °C was ~4 h. However, the growth time for the same length of nanowires at 835 °C was 7.5 h. The delay in growth was attributed to increased incubation time. An increase in temperature accompanied an increase in desorption of the $R_S{\equiv}Ga$ species due to decrease in sticking coefficient of the $R_S{\equiv}Ga$ species on the Si substrate surface. In general, the higher the temperature, the higher is the kinetic energy of the $R_S{\equiv}X$ and the $R_S{\equiv}Y$ species, and the lower is the sticking coefficient of these species. One outright remedy of large incubation time for the growth of X_mY_n nanomaterials due, for example, to large desorption of the $R_S{\equiv}X$ species would be to conduct this growth in the $R_S{\equiv}X$-rich environment. The second remedy would be to increase the sticking coefficient of the $R_S{\equiv}X$ species on the substrate surface. There can be several different means to achieve this goal. Some of them may be related to surface treatment and/or surface functionalization described in Chap. 14. We exemplify one of them pertaining to the formation of a thin support (buffer) layer on the substrate. This support layer should have thermal expansion coefficient significantly different from that of the substrate. Suppose, the substrate is Si and the support is AlN. As listed in Table 12.1 of Chap. 12, the thermal expansion coefficient of Si is 2.56×10^{-6}/K, but the thermal expansion coefficient of AlN is 5.63×10^{-6}/K. Such a large difference in thermal expansion coefficients results in significant tensile stress in the AlN support layer and a compressive stress in the underlying Si interfacial region. Obviously, this tensile stress in the AlN support layer increases with increase in temperature. The immediate consequence of it is the formation of cracks needed for relaxing the compressive stress in various local regions of it. The islands of disordered lattice containing patches of Al, N, and AlN are consequently formed on the Si substrate surface. Also, there may occur a stress gradient, which leads to the migration of Si atom from the compressed regions to the stress-free cracks and islands. The migration of Al and N into the Si lattice is not also ruled out. The island composition thus resembles the one of clusters and solid solutions. The island size increases with temperature. The island composition is rough and possesses electronegativity. It is attractive enough to enhance the sticking coefficient of the R_S species on the support-covered substrate surface even at high growth temperature. The probability of desorption of the R_S species from the substrate surface thus dramatically decreases leading to increase in the nanomaterial growth.
12. As elaborated in Chap. 4, there must be a limit to the choice of FECA metal creating a droplet for the catalysis of nanomaterial growth. This FECA metal should have such a composition that it does not yield a droplet of excessively

large contact angle ϕ. In general, the larger the contact angle ϕ, the larger is the solid–liquid interface energy γ_{SL}, and the larger is the instability of the droplet.

References

1. P. Pertl, M.S. Seifner, C. Herzig, A. Limbeck, M. Sistani, A. Lugstein, S. Barth, Solution-based low-temperature synthesis of germanium nanorods and nanowires. Monatshefte Für Chemie - Chemical Monthly **149**, 1315–1320 (2018)
2. S. Barth, M.S. Seifner, J. Bernardi, Microwave-assisted solution–liquid–solid growth of $Ge_{1-z}Sn_z$ nanowires with high tin content. Chem. Commun. **51**, 12282–12285 (2015)
3. M.S. Seifner, F. Biegger, A. Lugstein, J. Bernardi, S. Barth, Microwave-assisted $Ge_{1-z}Sn_z$ nanowire synthesis: precursor species and growth regimes. Chem. Mater. **27**, 6125–6130 (2015)
4. K. Ramasamy, P.G. Kotula, A.F. Fidler, M.T. Brumbach, J.M. Pietryga, S.A. Ivanov, $Ge_{1-z}Sn_z$ alloy nanocrystals: a first step toward solution-processed group IV photovoltaics. Chem. Mater. **27**, 4640–4649 (2015)
5. K. Tabatabaei, H. Lu, B.M. Nolan, X. Cen, C.E. McCold, X. Zhang, R.L. Brutchey, K. van Benthem, J. Hihath, S.M. Kauzlarich, Bismuth doping of germanium nanocrystals through colloidal chemistry. Chem. Mater. **29**(17), 7353–7363 (2017)
6. D. Wang, T. Xie, Q. Peng, Y. Li, Ag, Ag_2S, and Ag_2Se nanocrystals: synthesis, assembly, and construction of mesoporous structures. J. Am. Chem. Soc. **130**(12), 4016–4022 (2008)
7. O. Alekberov, Z. Jahangirli, R. Paucar, S. Huseynova, N. Abdulzade, A. Nakhmedov, K. Wakita, N. Mamedov, Band structure and vacancy formation in β-Ag_2S: Ab-initio study. Phys. Status Solidi C **12**, 672–675 (2015)
8. S. Ballikaya, M. Sertkol, Y. Oner, T.P. Bailey, C. Uher, Fracture structure and thermoelectric enhancement of Cu_2Se with substitution of nanostructured Ag_2Se. Phys. Chem. Chem. Phys. **21**, 13569–13577 (2019)
9. S.N. Mohammad, VQS (vapor-quasiliquid-solid, vapor-quasisolid-solid) mechanism lays down general platform for the syntheses of graphene by chemical vapor deposition. J. Appl. Phys. **120**, 214305 (2016)
10. K.-L. Wu, Y. Chou, C.-C. Su, C.-C. Yang, W.-I. Lee, Y.-C. Chou, Controlling bottom-up rapid growth of single crystalline gallium nitride nanowires on silicon. Sci. Report **7**, 17942 (2017)
11. J.K. Zettler, C. Hauswald, P. Corfdir, M. Musolino, L. Geelhaar, H. Riechert, O. Brandt, S. Fernández-Garrido, High-temperature growth of GaN nanowires by molecular beam epitaxy: toward the material quality of bulk GaN. Cryst. Growth Des. **15**, 4104–4109 (2015)

Appendix A
Effective Amorphicity of FECA Nanoparticle Surface

Abstract In this appendix, the amorphicity α_{amor} of the FECA nanoparticle surface has been described. It has been noted that the surface amorphicity α_{amor} has an effective value called the effective surface amorphicity $\alpha_{amoreff}$. While the surface amorphicity α_{amor} gradually increases from 0 to 1 with increase in the surface disorder, the effective surface amorphicity $\alpha_{amoreff}$ gradually increases, reaches a peak, and then decreases with increase in the surface disorder. A model for the effective surface amorphicity $\alpha_{amoreff}$ has been presented. Although empirical, it is highly flexible and most suitable for the study of nanoparticle surface characteristics and also the nanomaterial growths.

A.1 Introduction

Numerous physicochemical variables govern the shapes, sizes, and surface disorder of FECA (METANO and SUBSANO) nanoparticles. The surface structure and the characteristic impact of this structure of nanoparticles (both METANOs and SUBSANOs), for the effective catalytic activities of these nanoparticles, have been elaborated in Chap. 14, Sect. 14.2. Our discussions in Chaps. 12–14 demonstrate that well-thought-out and optimal structural disturbance (disorder) of nanoparticle surface leads to the most important design features for new opportunities for nanoparticle design. Surface disorder and anisotropy generated naturally or artificially on the nanoparticle surface must therefore be embraced as useful design parameters. Among them, anisotropy of the nanoparticle surface can have significant impact on the surface active sites. Uncontrollable imperfection and anisotropy can though be *a priori* challenging. In this appendix, we describe the tuning of the disorder and anisotropy of nanoparticle surfaces. Surface disorder gives rise to surface amorphicity α_{amor}. This amorphicity must be such that it nurtures and not impedes the catalytic activities of nanoparticle. This means it must be the constructive amorphicity and not

S. N. Mohammad, *Synthesis of Nanomaterials*, Springer Series in Materials Science 307, https://doi.org/10.1007/978-3-030-57585-4

the destructive amorphicity. And depending on surface treatment (plasma treatment, acid treatment, aqua regia treatment, water vapor treatment, thermal annealing, sputtering, etc.), there can be surface functionalization of nanoparticle surface induced, for example, by substrate and/or support. There can also be structural vibrations and fluctuations of the nanoparticle surface. All of them can give rise to several competing factors that can govern the effectiveness of the amorphicity of the nanoparticle surface. To be more specific, there can be an effective surface amorphicity α_{amoreff0} of the nanoparticle surface. The end result of all these would be: most favorable and optimal (1) surface looseness, surface vacancy, and particularly surface porosity ρ_c resulting from the surface amorphicity α_{amor} ($0 \leq \alpha_{\text{amor}} \leq 1$); (2) surface roughness; (3) high-energy sites (HETs) resulting from unsaturated dangling bonds of the amorphous nanoparticle surface; (4) surface melting; (5) dipole moment; and consequently (6) enhanced diffusivity of the $R_S{\equiv}X$ and $R_S{\equiv}Y$ species through the nanoparticle surface. In other words, they all commensurate with the optimal effective surface amorphicity α_{amoreff0}.

A.2 The Role of the R_S Species in Creating Effective Surface Amorphicity

A.2.1 What is effective surface amorphicity?

We noted earlier that the surface amorphicity α_{amor} of the nanoparticle surface can be between 0 and 1. This surface would be semi-amorphous if the amorphicity α_{amor} is, for example, between 0.0 and 0.1. The number of the R_S species generated by HETs can be very large. In fact, it can be larger than that capable of diffusing through the nanopores of the porous nanoparticle surface. There is a certain optimal limit of the amorphicity α_{amor}. And beyond this limit, the increasing number of the R_S species released from their precursor by HETs or some other means can be increasingly blocked at the surface. A parameter defined by

$$u_T = \left[\frac{\alpha_{\text{amor}}}{1 - \alpha_{\text{amor}}}\right], \tag{A.1}$$

would probably best quantify the characteristics of the amorphicity and probably of the said blockade.

Note that there can be increase in surface amorphicity due to increase in surface treatment, surface functionalization, and surface vibration. And this increase will be different for different nanoparticles of a different shape, size, morphology, and material composition. It would, for example, be different for a SUBSANO than for a METANO. Under identical condition(s), it can increase more rapidly for some nanoparticle than for some other nanoparticle. The barrier to diffusion of the R_S

species through nanoporous nanoparticle surface would obviously be different for different amorphicities of the same or different nanoparticles. And it would happen due to increasing blockade of nanopores by the excessive accumulation of the R_S species on the nanoparticle surface. There can thus be an effective amorphicity $\alpha_{amoreff}$ exhibiting one of the four following characteristics: (a) It will increase slowly, reach a peak, and then decrease slowly with increase in amorphicity α_{amor}. (b) It will increase rapidly, reach a peak, and then decrease rapidly with increase in amorphicity α_{amor}. (c) It will increase slowly, reach a peak, and then decrease rapidly with increase in amorphicity α_{amor}. (d) It will increase rapidly, reach a peak, and then decrease slowly with increase in amorphicity α_{amor}. All these may probably be accomplished by w_T defined in terms of the parameters η_0 and ν, and expressed as follows:

$$w_T = \eta_0 u_T^{\nu}. \tag{A.2}$$

If $\Im_1$ and $\Im_2$ represent the effects described above, then $\Im_1$ may be given by

$$\Im_1 = \sin(w_T). \tag{A.3}$$

Similarly, $\Im_2$ may be given by

$$\Im_2 = \cosh(w_T). \tag{A.4}$$

Temperature should play a critical role in dictating the effective amorphicity $\alpha_{amoreff}$ and porosity ρ_c of the nanoparticle surface. Both of them should increase with increase in temperature, which is corroborated with experiments available in the literature. These experiments showed that an increase in temperature leads to an increase in porosity. If T_0 is a weighing parameter: suppose T_0=300 °C, and b is a parameter that quantifies the effect of $\Im_4$ given by

$$\Im_4 = \left(\frac{T}{T_0}\right)^b. \tag{A.5}$$

Taking into consideration all the effects described above, the effective amorphicity $\alpha_{amoreff}$ would then be given by

$$\alpha_{amoreff} = \Im_4\left(\frac{\Im_1}{\Im_2}\right), \tag{A.6}$$

A.2.2 *Nanopore radius*

Note that $\alpha_{amoreff}$ depends on the parameters ν, b, and η_0. The pore radius r_c of the nanopore may then be given by

$$r_c = r_{cn}\alpha_{\text{amoreff}} = r_{cn}\Im_4\left(\frac{\Im_1}{\Im_2}\right), \tag{A.7}$$

where r_{cn} is a constant. Unless otherwise stated, we may assume $r_{cn} = 0.15$ nm. It may be noted that, based on (A.7), the nanopore radius is dependent on the effective amorphicity α_{amoreff}, and that the nanopore radius r_c gradually decreases for α_{amor} increasing beyond $\alpha_{\text{amor}} = \alpha_{\text{amoreff0}}$. Due to excessive blockade, the nanopore radius becomes almost zero, for example, for $\alpha_{\text{amor}} > 0.80$.

A.3 Novelty of Surface Amorphicity

We believe the analytical formula (A.6) developed for effective surface amorphicity is very important. Although empirical, it appears to explain most, if not all, experimental results pertaining of the amorphicity effect of nanomaterial growths. Figure A.1a shows the variation of α_{amoreff} with u_T, which is composed of α_{amor}; it is for $\eta_0 = 1$, $\nu = 1$, and $b = 1$. It demonstrates that an increase in α_{amor} from 0 to 0.8 leads to a gradual increase in u_T. It accompanies also a gradual increase in α_{amoreff} until α_{amoreff} attains a peak at $\alpha_{\text{amor}} = \alpha_{\text{amoreff0}}$, and then attains a gradual decrease in α_{amoreff}. The rapid increase in u_T beyond where α_{amoreff} attains the peak α_{amoreff0} highlights the fact that the negative effect of α_{amor} increases almost exponentially for $\alpha_{\text{amor}} > \alpha_{\text{amoreff0}}$. Figure A.1b shows the variations of $\Im_1$ and $\Im_2$ with α_{amor}. Note that α_{amoreff} is determined by the competitive behavior of $\Im_1$ and $\Im_2$. $\Im_1$ slowly increases and then decreases with increase in α_{amor}. In contrast, $\Im_2$ remains almost constant and then rapidly increases with increase in α_{amor}. As a result, α_{amoreff} increases, reaches a peak at α_{amoreff0}, and then decreases with further increase in α_{amor}. Figure A.2 shows the impact of the parameter b on the effective amorphicity α_{amoreff} for three different values of α_{amor}. As apparent from Fig. A.2, the effective amorphicity α_{amoreff} increases almost exponentially with increase in the parameter b. This increase is the highest for α_{amor} close to α_{amoreff0}. Figure A.3 shows the effect of temperature T on the effective amorphicity α_{amoreff} for two different values of the parameter b. It demonstrates that the effective amorphicity α_{amoreff} increases rapidly, but not exponentially with increase in the temperature T. This increase is though larger for larger value of the parameter b.

Figure A.4a depicts the effect of ν on the effective amorphicity α_{amoreff}. One can see that there are peaks in the plots of α_{amoreff} as function of α_{amor} and that these peaks remain unchanged even when the curvature of these plots changes with change in ν. It has a good implication. It demonstrates a condition in which α_{amoreff} as function of α_{amor} increases rapidly, reaches a peak at $\alpha_{\text{amoreff}} = \alpha_{\text{amoreff0}}$, and then decreases slowly. However, the increase and decrease are different for different values of ν. It is different, for example, for $\nu = 0.6$ than for $\nu = 1.2$. Figure A.4a also demonstrates that, for certain value of ν, such as $\nu = 1.2$, α_{amoreff}, as function of α_{amor}, can increase slowly, reach a peak at $\alpha_{\text{amor}} = \alpha_{\text{amoreff0}}$, and then decrease rapidly with increase in

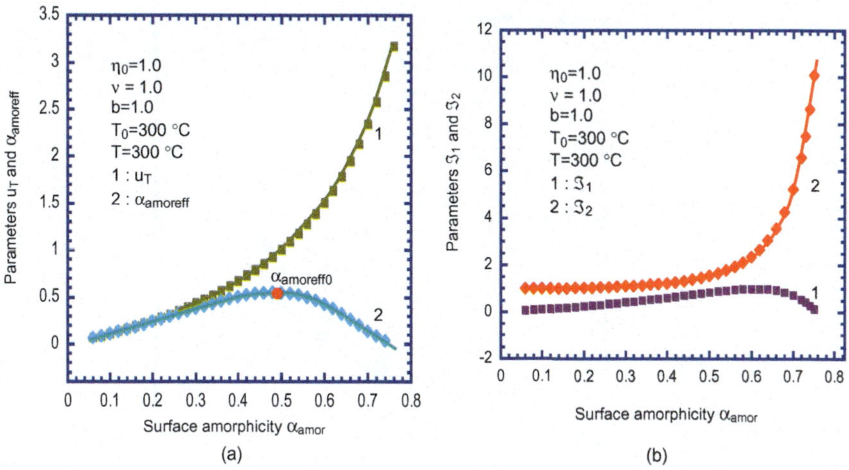

Fig. A.1 **a** Variation of the parameter u_T and of the effective surface amorphicity α_{amoreff} with the surface amorphicity α_{amor}; **b** variation of the parameter $\Im_1$ and of the parameter $\Im_2$ with the surface amorphicity α_{amor}

α_{amor}; Fig. A.4b also shows the variation of α_{amoreff} with α_{amor}; however, it is for $\nu = 1$ and $b = 1$. It shows the impact of η_0 on the effective amorphicity α_{amoreff}. It demonstrates that the peaks, as well as the curvature of the plots of α_{amoreff} as function of α_{amor}, remain unchanged even when there occurs change in η_0. It suggests that an increase in η_0 leads to a shift in the peak of the α_{amoreff} versus α_{amor} plots to larger α_{amor}. Figure A.4c is for $\nu = 1$, $\eta_0 = 1$, and $b = 1$. It shows the effect of temperature on the effective amorphicity α_{amoreff}. It demonstrates that α_{amoreff} gradually increases with increase in temperature T. The rate of this increase is higher for larger b than for smaller b. Figure A.4a–c suggests that, although (A.6) is empirical, as function of α_{amor}, ν, η_0, and b, it is quite flexible. Figure A.5 shows the amorphicity α_{amor} and the effective amorphicity α_{amoreff}, both as function of the nanoparticle surface disorder. There are four pink curves, namely curves 1, 2, 3, and 4, all for the effective amorphicity α_{amoreff}. Based on discussions made above, the amorphicity α_{amor} can be between 0 and 1. But the effective amorphicity α_{amoreff} can be larger than unity. In fact, depending, for example, on temperature T, it can be much larger than 1.0 (see Figs. A.2 and A.3). However, no matter how large or small the effective amorphicity α_{amoreff} is, it increases with increase in the amorphicity α_{amor}, reaches a peak, and then decreases with further increase in amorphicity α_{amor}. It is very small for $\alpha_{\text{amor}} \approx 0$ and $\alpha_{\text{amor}} \approx 1$. All these are reflected from Fig. A.5.

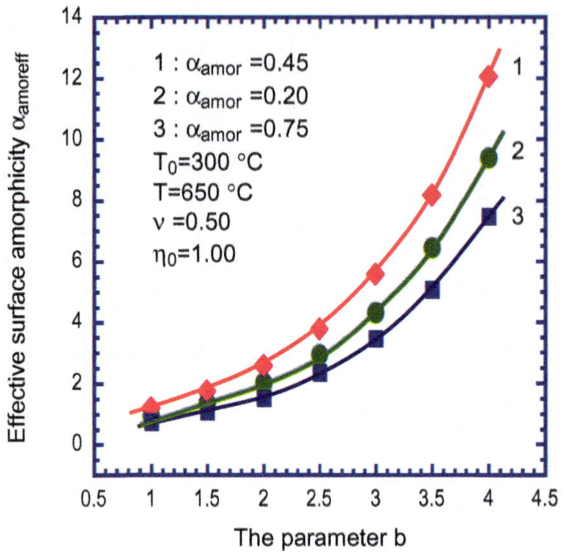

Fig. A.2 Variation of the effective surface amorphicity $\alpha_{amoreff}$ with the parameter b for three different values of the surface amorphicity α_{amor}

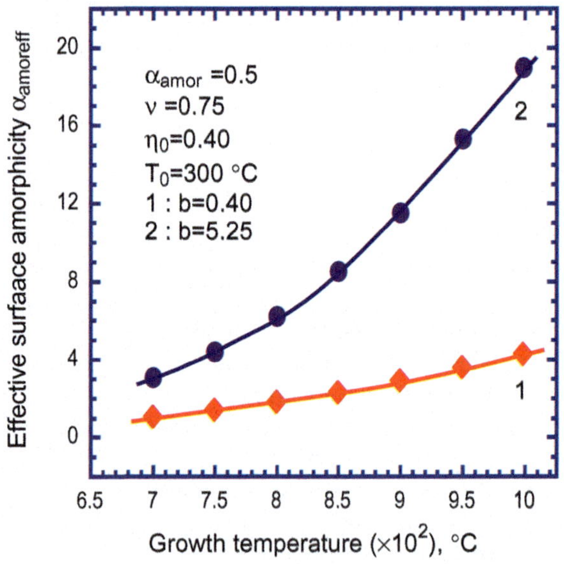

Fig. A.3 Variation of the effective surface amorphicity $\alpha_{amoreff}$ with the growth temperature T for two different values of the parameter b; the surface amorphicity α_{amor} for this variation is 0.5

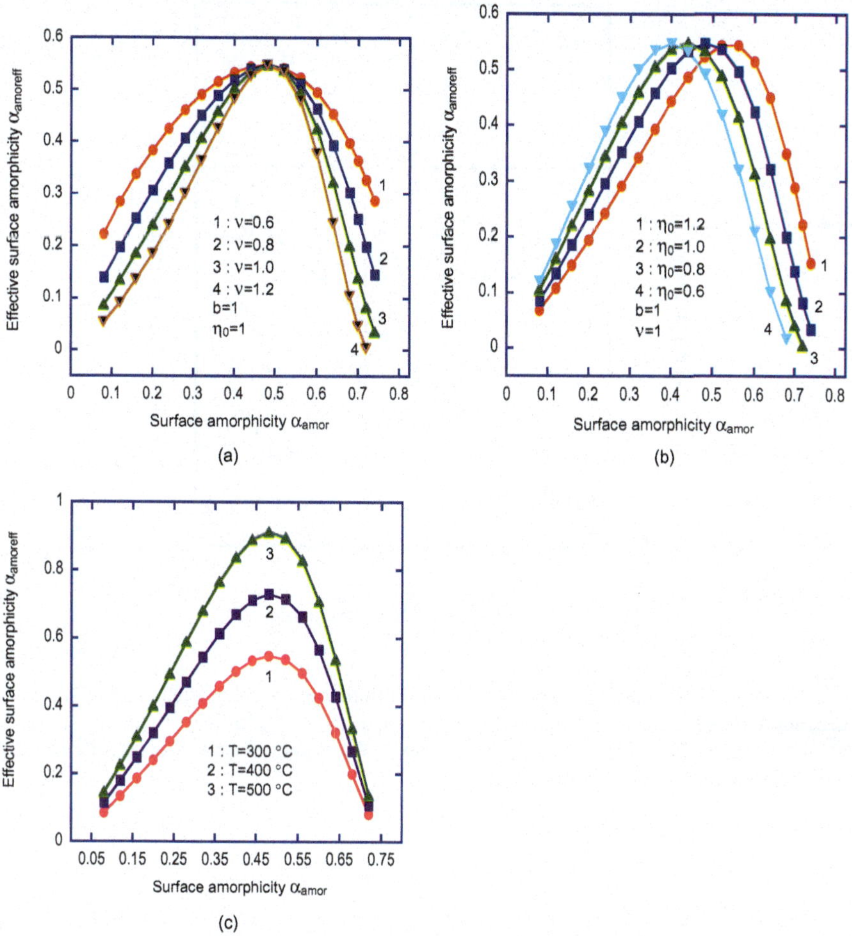

Fig. A.4 Variation of the effective surface amorphicity α_{amoreff} with the surface amorphicity α_{amor} for **a** four different values of the parameter ν, **b** four different values of the parameter η_0, and **c** three different values of the temperature T

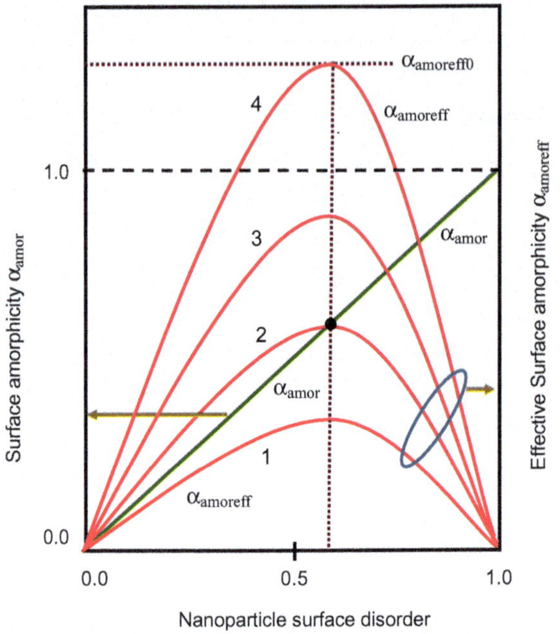

Fig. A.5 Schematic diagram showing the variation of the effective surface amorphicity α_{amoreff} and the surface amorphicity α_{amor} with the surface disorder of the nanoparticle surface. Curves 1, 2, 3, and 4 are all for the effective surface amorphicity α_{amoreff}, but for four different temperatures of increasing values. Note that α_{amoreff0} is the peak of the α_{amoreff}; there is α_{moreff0} for each of the four curves (namely, curves 1 to 4) for α_{amoreff}.

Index

A

Ab initio, 36, 343

Activation, 53, 58, 72, 73, 90, 110, 111, 115, 116, 167, 193, 255, 262, 263, 272, 278, 310, 312, 370

Adatom, 60–62, 85, 113, 115, 132, 256, 291, 304, 341, 342, 423

Adhesive, 53, 60

Advantage, 16–18, 22, 23, 25, 34, 37, 38, 79, 107, 144, 148, 179, 268, 271, 348, 373

Alloy, 4, 34, 40, 41, 43, 45–47, 59, 69–71, 73–85, 88, 89, 91–94, 96, 101–104, 107–109, 112, 113, 124, 139, 140, 142, 145, 146, 148, 150, 153, 160–164, 166–169, 183, 187, 197, 202, 209, 210, 212, 215, 220, 223, 224, 227, 233, 234, 239, 240, 242, 245, 258–263, 265, 268, 269, 271, 272, 274, 279, 280, 283, 289, 297, 303, 307, 308, 310, 311, 315, 327, 328, 351–361, 363, 366, 372, 378–388, 395–399, 403, 404, 415, 416, 418–420, 423

AlN, 8, 39, 125, 131, 132, 200, 201, 216, 244, 295, 373, 381, 391, 425

Amorphicity, 30, 40–42, 55, 66, 132, 181, 188, 196, 207, 208, 210, 212, 217, 223–225, 228, 229, 231, 232, 236, 238, 239, 253–256, 269, 270, 280, 282, 284, 289, 290, 297, 298, 307, 309, 315, 321, 323–329, 331, 334–339, 341–344, 351–355, 363, 369, 371, 376, 384, 390, 399, 418, 420, 424, 427–434

Amorphous, 9, 18, 28, 29, 37, 38, 55, 77, 78, 101, 103, 106, 108, 123, 127, 128, 154, 159, 163, 165–169, 179, 180, 183, 188–192, 196, 199, 203, 208, 210, 212, 214, 215, 217, 218, 221, 222, 224, 225, 227, 231, 232, 234, 236, 254–256, 259, 279, 283, 284, 289–292, 296–298, 300, 301, 306–309, 312, 313, 324, 327, 330, 333, 351, 359, 360, 362, 363, 370–373, 375–377, 383, 385, 387–392, 395, 397, 398, 420, 423, 428

Amphoteric, 207, 238

Angle, 53, 62–65, 67, 71, 72, 88, 93, 107, 245, 246, 360, 402, 426

Anisotropic, 53, 59, 123, 132

Annealing, 21, 28, 35, 38, 61, 72, 74, 77, 82, 106, 132, 160, 161, 168, 169, 178, 188, 191, 192, 195, 196, 199, 210, 212, 214, 216, 217, 225, 229, 230, 234, 297, 301–303, 353, 360, 361, 366, 372, 378, 387, 398, 428

Appendix, 30, 41, 55, 188, 208, 212, 223, 224, 253, 256, 284, 290, 295, 323, 325–327, 334, 335, 340, 427

Arc discharge, 13, 14, 19, 20, 392

Atomic, 2, 9, 13, 17, 27, 33, 34, 38, 40, 41, 44, 45, 59, 67, 69, 75, 77, 78, 83–85, 88, 91, 102, 103, 106, 107, 109, 124, 127, 150, 162, 166, 181, 183, 184, 215, 218, 220, 227, 232, 240, 254, 259, 264, 265, 267, 271, 273, 274, 279, 293, 297, 299, 309, 311, 314, 323, 327, 329, 356, 357, 360, 361, 366, 375, 378–380, 383, 386, 387, 396, 397, 403, 422

S. N. Mohammad, *Synthesis of Nanomaterials*, Springer Series in Materials Science 307, https://doi.org/10.1007/978-3-030-57585-4

Au, 6, 23, 28, 42–44, 46–48, 58, 63, 67, 71–76, 79–88, 92, 94, 95, 101–106, 108, 111, 112, 124, 127, 128, 131, 134, 142, 145, 146, 149, 150, 153, 161, 165–167, 169, 176, 183, 184, 187, 209, 215, 227, 239, 246, 253, 257–260, 263–266, 268, 269, 271–274, 281, 292, 295, 298, 302, 303, 315, 356, 357, 359–361, 363, 365, 366, 368–370, 377, 381–384, 386, 389, 402

B

Barrier, 53, 58, 67, 81, 87, 245, 253, 261–263, 268, 272, 274–276, 278, 310, 312, 428

Bimetallic, 27, 43–45, 225, 296, 310, 311, 404

Binary, 2, 43, 69, 71, 74, 76, 78, 82, 83, 88, 92, 108, 109, 124, 150, 153, 164, 165, 197, 260, 271, 279, 280, 327, 328, 356, 358, 359, 363, 364, 366, 383, 384, 386, 387, 395, 418

BN, 1, 8, 20, 42, 194, 216, 227, 244, 291, 306, 329, 331–333, 347, 393–395, 397–399

BNP, 45, 46, 423

Boron nitride nanotubes (BNNTs), 2, 19, 20, 194, 227, 308, 369, 392, 393, 395, 397–399, 421

Buffer, 37, 39, 82, 200, 201, 295–297, 425

C

Carbon, 1, 2, 4, 6, 8, 9, 18–20, 38, 42, 46, 48, 58, 66, 67, 89, 90, 105–107, 114–116, 123, 127–129, 134, 154, 174, 212, 215, 216, 224, 225, 232, 243, 259, 273–275, 277–281, 283, 284, 292, 296–298, 308–310, 312, 313, 327, 328, 331–333, 339, 341, 342, 364, 368–370, 375, 379, 380, 385, 392, 393, 422

Carbon Nanotube (CNT), 1, 2, 6, 9, 18–20, 24, 37, 42, 46, 48, 54, 69, 77, 89, 90, 104, 106, 114, 115, 121, 127–129, 134–136, 180, 181, 195, 201, 224, 226, 227, 231, 243, 253, 262, 273, 274, 276–281, 283, 284, 292, 296–298, 301–305, 309, 310, 313, 321, 331, 335, 336, 338–343, 363, 364, 367–371, 373, 389, 392, 395, 404, 405, 413, 417, 421, 422

Catalyst, 15, 16, 20, 27, 30, 38, 40, 42, 43, 48, 53, 54, 62, 64, 65, 67, 69, 75, 79, 81, 88–90, 101–105, 107, 109, 113–116, 123, 126–128, 131, 134–136, 139, 142, 146, 147, 149, 153, 154, 159, 163, 164, 166–169, 174, 179, 184, 185, 188, 193, 194, 201, 207, 226, 231, 236–238, 242, 243, 245, 246, 253, 257–260, 271, 272, 275, 280, 281, 283, 289, 292, 294–299, 302–307, 310, 312, 313, 315, 330, 353–355, 357, 358, 362, 364, 366, 367, 371, 373, 376–380, 383, 385, 387, 392–395, 397–399, 402, 414, 418, 419

Chalcogenite, 153, 386

Characteristics, 1, 2, 9, 15, 16, 30, 41, 53, 54, 61, 62, 64, 65, 71, 74, 79–82, 92, 101, 108, 113, 123, 129–131, 176, 207–209, 225, 226, 228, 238, 256, 261, 262, 268, 269, 276, 283, 289, 292, 295, 307, 315, 330, 347, 352–354, 358, 361, 365, 368, 371, 381, 382, 385, 393, 395, 396, 413, 414, 418, 424, 427, 429

Charge, 45, 61, 93, 95, 96, 147, 163, 210, 223, 224, 237–239, 299–302, 304, 305, 309, 312, 313, 421, 423, 424

Chemical Beam Epitaxy (CBE), 13, 14, 17, 23, 112

Chemical Vapor Deposition (CVD), 13–17, 21, 23, 24, 32, 34, 42, 48, 56–59, 81, 85, 91, 94, 103, 105, 106, 108, 111, 114–116, 124, 125, 127, 128, 134, 139, 144, 153, 180, 184, 190, 191, 196, 210, 224, 226, 227, 234, 243, 257–259, 272, 277, 281, 283, 292, 293, 296–298, 303, 308, 356, 358, 365, 367, 368, 381, 382, 392, 393, 395, 414

Cluster, 29, 41, 61, 77, 91, 141, 166, 169, 170, 174, 175, 189, 196, 200, 209, 210, 212, 223, 234, 240, 242, 246, 260, 283, 289, 291–293, 298, 301, 307, 308, 310–312, 315, 328, 351–353, 355, 363, 381, 386, 387, 395, 405, 416, 418–420, 423

Coarsened, 15, 28, 29, 37, 38, 66, 161, 188, 189, 192, 195, 202, 208, 210, 212, 214, 215, 221, 222, 234, 255, 256, 289–291, 298, 309–312, 351, 362, 371, 372, 386–389, 392, 395, 397, 401

Coarsening, 41, 47, 188, 189, 196, 207, 208, 212, 214, 215, 217, 218, 220, 225, 231, 284, 311, 353–355, 363, 369, 373, 389, 399, 418, 420
Composition, 1, 2, 7, 14, 16, 17, 23, 27, 28, 35, 38, 40, 41, 44, 45, 47, 54, 74, 75, 82–84, 87, 88, 91, 93, 107, 129, 130, 140, 151, 159, 161–166, 173, 184, 185, 196, 203, 209, 220, 223, 226, 227, 229, 235, 237, 240, 242, 245, 253, 255, 256, 259, 274, 289, 292, 296–298, 302, 303, 308, 310–312, 315, 323, 351–353, 355, 356, 358, 361, 363–366, 369, 371, 372, 375, 378–380, 383–385, 387, 389, 395, 414, 417, 422, 423, 425, 428
Concern, 88, 121, 402, 418, 419
Contact, 53, 60, 62–65, 71, 72, 93, 107, 148, 245, 310, 402, 426
Contaminant, 41, 74, 77, 82, 83, 87, 94, 96, 132, 162, 163, 167, 196, 202, 209, 215, 223, 227, 254, 255, 260, 261, 263, 292, 307, 352, 353, 360, 369, 373, 375–377, 380, 384, 394, 418, 420, 423
Contamination, 88, 93, 94, 123, 126, 154, 179, 190, 254, 353
Controversy, 108
Core, 36, 45, 53, 55, 65, 66, 81, 163, 164, 166, 180–184, 222, 243, 278–281, 306, 323, 360, 423
Criteria, 69, 78, 90, 91
Criterion, 64, 91–94, 96
Crystal, 8, 16, 23, 59, 60, 71, 82, 106, 130, 131, 144, 146, 147, 150, 154, 190, 191, 201, 202, 207, 218, 219, 225, 264, 265, 281, 308, 314, 327, 366, 378, 404, 419
Crystallinity, 34, 38, 145, 148, 184, 282, 315, 384
Crystallographic, 59, 69, 89, 103, 129, 144, 150, 154, 201, 314

D

Defect, 53, 87, 107, 126, 180, 181, 190, 214, 216, 217, 231, 232, 237, 245, 272, 289, 298–302, 305, 307, 309–311, 324, 371, 384, 391, 401, 417, 419–421, 424
Deposition, 13–16, 18–21, 23, 24, 34, 37, 70, 71, 80, 91–93, 122, 124, 130, 147, 154, 174, 181, 191, 200, 224, 232, 256, 298, 311, 314, 379, 390, 415, 417, 423, 424
Design, 53, 148, 413, 420, 422, 427
Diagram, 3, 7, 31, 43, 45, 56, 57, 69–71, 74–76, 78, 82, 88, 102, 108, 109, 122, 124, 140, 141, 143, 160, 164–166, 175, 192, 203, 210, 214, 241, 260, 262, 270, 271, 280, 300, 325, 356–359, 361, 364, 366, 380, 386, 387, 395, 434
Dick, 113, 132, 357, 361, 362
Diffusion, 4, 6, 15, 16, 24, 41, 55, 58, 66, 70, 71, 80, 81, 84, 87–90, 92–94, 96, 102–106, 109, 110, 113, 115, 122, 124, 127, 129, 132, 141, 142, 153, 160, 161, 166, 167, 170, 183, 187, 200, 202, 203, 207, 217, 232, 234, 237, 239, 254–256, 273, 274, 279–281, 289, 299, 301, 304, 306, 308, 310–312, 321, 326–333, 335–337, 340, 343, 344, 354, 357, 358, 362–364, 369, 370, 372, 375, 378, 385, 390, 395, 397–400, 403–405, 423, 428
Diffusivity, 4, 19, 85, 87, 113, 143, 148, 232, 233, 281, 321, 326–328, 330–332, 334–336, 338, 404, 405, 418, 428
Dipole, 41, 61, 62, 163, 207, 225, 238, 239, 282, 284, 302, 315, 352–354, 389, 402, 420, 424, 428
Disadvantage, 16, 22, 24, 180
Discrepancy, 279, 362
Disorder, 28, 29, 37–39, 58, 66, 77, 106, 123, 124, 126, 132, 161, 166, 169, 181, 188, 189, 192, 195, 202, 208, 210, 212, 214–220, 222–225, 229, 231, 238, 255, 282, 290–293, 295, 298, 300, 302, 306, 308, 311, 312, 324, 327, 351, 353, 354, 360, 362–364, 369–373, 388, 389, 391, 392, 395, 397–399, 401, 418, 424, 425, 427, 431, 434
Disturbance, 61, 132, 208, 218, 222, 238, 239, 282, 302, 305, 354, 391, 427
Dopant, 17, 163, 179, 223, 254, 255, 261, 299, 360, 424
Doping, 1, 14, 17, 147, 179, 225, 237, 306, 351, 422
Droplet, 4, 40, 46–48, 53, 59, 60, 62–64, 69–75, 77, 80–85, 87, 89, 91–96, 107, 108, 113, 122, 132, 139, 142, 145, 153, 160, 162, 165, 168, 169, 183,

189, 190, 194, 196, 201, 202, 209, 217, 221, 233, 245, 246, 256, 260, 271, 281, 302, 349, 353–355, 357, 363, 366, 378, 391, 399, 402, 405, 419, 423–426
Dynamics, 36, 47, 69, 79, 90, 102, 106, 125, 169, 220, 223, 228, 254, 343, 361, 403

E
Energy, 3, 8, 17, 18, 23, 33, 34, 36, 38, 41, 45, 47, 58–67, 71, 72, 79, 84, 90, 91, 93–95, 108, 110, 115, 116, 128–130, 180, 182, 184, 189, 192, 193, 201, 207, 215, 219, 224, 233, 234, 243, 244, 253, 255, 260–265, 267, 269, 272, 274, 276, 279, 281, 296, 299, 300, 306, 312, 314, 321–323, 373, 385, 398, 402, 421, 425, 426, 428
Energy Dispersive X-ray (EDX), 184, 203, 256, 259, 260, 302, 354, 360, 361, 364, 375, 386, 391
Epitaxy, 13, 14, 17, 23, 121, 190, 199
Ethylene, 128, 230, 283, 309, 365, 368
Eutectic, 4, 41, 43, 46, 69–79, 81–84, 88, 91, 92, 96, 101, 103, 104, 108, 109, 113, 122, 139, 140, 142, 145–147, 150, 153, 160, 161, 163–167, 169, 202, 209, 220, 233, 239, 240, 242, 245, 254–261, 263, 265, 268, 269, 271, 272, 279, 280, 351–360, 363–366, 378, 380, 381, 383–388, 395–397, 402, 403, 418–420, 423
Evidence, 110, 207, 224, 228, 415, 416, 421
Exchange, 253, 261–263, 265, 268–270, 272–276, 278, 421

F
FECA nanoparticles (FECANOs), 27, 29, 41, 61, 208, 214, 221, 236, 253, 289, 290, 321, 322, 347
Foreign Element Catalytic Agents (FECAs), 4, 15–17, 27–29, 41, 43, 44, 46–48, 53, 54, 61–63, 69, 71, 76–80, 84, 87–90, 101, 102, 104–113, 115, 121, 122, 124, 130, 147, 149, 151, 159–161, 173, 176, 187, 190, 207–209, 212, 214, 218, 220–222, 224, 236–243, 245, 256, 266, 272, 277, 290, 321–323, 347, 414–416, 418, 420, 421, 424, 425, 427
Functionalization, 27, 44, 47, 53, 207, 209, 236, 289, 295, 298, 299, 307, 311, 418, 421, 425, 428

G
GaAs, 2, 8, 24, 43, 80–82, 84, 87, 103, 108, 110–113, 150, 175–177, 182, 190, 192, 194, 195, 198, 200, 201, 244, 259, 291, 307, 357, 358, 360–362, 366, 389
GaN, 2, 4, 8, 18, 38, 39, 42, 112, 113, 126, 130, 132, 175–177, 190, 191, 195, 197, 200, 201, 203, 216, 232, 234, 244, 291, 295, 297, 308, 311, 324, 329, 354, 355, 359–362, 366, 373, 374, 377–379, 381, 383, 393, 425
Ge, 2, 3, 8, 16, 37, 38, 42, 43, 46, 76, 84, 85, 92, 95, 103–105, 107–113, 123, 142, 144–150, 153, 175, 177, 178, 180, 209, 216, 239, 240, 244, 253, 257, 258, 264, 265, 267, 269, 271–274, 293, 296, 329, 357, 358, 360, 363, 383, 384, 402, 419, 421
Generality, 347, 421
Germanium, 109, 145
Gold, 28, 34, 67, 144, 146, 148, 166, 183, 216, 290, 292, 302, 374, 386
Gomes, 130, 132, 133, 372
Grain, 58, 77, 161, 187, 195, 196, 210, 212, 214, 218, 220, 231, 254, 256, 260, 281, 291, 297, 298, 311, 351, 360, 362, 372, 388, 391, 400, 424
Grain boundary, 77, 161, 187, 195, 196, 203, 210, 212, 214, 220, 225, 231, 238, 256, 260, 291, 351, 362, 372, 388, 423, 424
Graphene, 1, 2, 4, 8, 9, 20, 89, 101, 190, 218, 313, 421, 422
Guidelines, 413, 422

H
HET, 238, 239, 284, 291, 324, 326, 338, 384
Heterointerface, 61, 299, 301, 302
Heterostructures, 14, 17, 107, 151, 153, 182, 197, 417, 422
Heuting, 212
High-Energy Site (HETs), 200, 224, 389, 420
Hill, 66, 67, 239, 278–280, 306, 324, 395, 397, 399
Hillock, 15, 39, 161, 187, 189, 191, 192, 194–196, 210, 212, 216, 217, 220,

221, 234, 297, 312, 313, 352, 353, 386–388, 391, 392, 401
Hofmann, 47, 81, 106, 195, 226, 360
Hypothesis, 90, 135, 159, 161, 162

I
Illustration, 105, 106, 129, 164, 210, 215, 218, 220, 225, 237, 290, 292, 295, 296, 298, 354, 357, 359, 362, 372, 388, 421
InAs, 8, 23, 40, 69, 86, 87, 110–113, 121, 132, 133, 142, 175–177, 187, 190–193, 199, 200, 217, 244, 302, 362, 373, 388, 390–392
Incubation, 69, 72–75, 77, 341, 425
InN, 8, 124, 176, 177, 244, 259, 308, 359, 382, 389, 391
Interaction, 18, 44, 53, 54, 207, 224, 228, 261, 306, 310, 397
Interface, 6, 39, 44, 45, 55, 59–64, 70, 71, 79, 80, 87, 89, 92–94, 102, 104, 107, 109, 110, 113, 122, 124, 142, 148, 161, 162, 168, 180, 182, 189, 190, 201, 202, 216, 217, 219, 221, 222, 234, 236, 240, 245, 255, 261, 262, 291, 301, 305, 306, 308, 314, 353, 357, 362, 363, 372, 375, 376, 379, 386, 390, 391, 397, 398, 403, 404, 415–417, 423, 426
Interstitial, 45, 226, 298, 311, 312
Island, 38, 127, 295, 372, 425

K
Kinetics, 1, 19, 34, 38, 72, 80, 139, 145, 151, 163, 315, 403, 413, 417, 419
Knudsen, 4, 66, 273, 289, 311, 312, 321, 326–331, 333–336, 338, 340, 343, 344, 364, 399, 400, 404
Kodambaka, 47, 48, 113, 353, 357, 402

L
Laser, 14, 18, 19, 21, 30, 34, 179, 194, 196, 197, 225, 230, 258, 290, 392, 399
Lifetime, 27, 48, 54, 60, 88, 134, 135, 283
Liquid, 4–6, 15, 30, 32, 33, 40, 41, 46, 47, 59, 60, 62, 63, 69–71, 75, 78, 83, 84, 88, 89, 94, 102, 104, 106–108, 113, 124, 127, 140, 142, 144, 145, 165, 168, 184, 191, 196, 208–210, 220–222, 232–235, 242, 246, 256, 260, 308, 314, 348, 349, 352, 355, 356, 363, 366, 375, 386–388, 391, 392, 402, 403, 405, 414, 415, 417, 418, 421, 424
Liquid/solid (L/S), 6, 70, 110, 162, 182, 189, 222, 306
Liquidus, 71, 74, 75, 82, 166, 356, 358, 364, 386
Low, 14, 17, 19–22, 24, 27, 38, 67, 72, 76, 80, 82, 83, 85, 88, 91, 93, 108–110, 113, 115, 127, 130, 139, 142–145, 147, 148, 150, 153, 161, 168, 174, 180, 196, 209, 214, 220, 224, 226, 227, 233, 234, 236, 239, 240, 259, 269, 271, 274, 280, 313, 324, 326, 341, 347, 355, 357, 361, 363, 364, 366, 367, 372, 374, 376–378, 384–386, 390, 401, 404, 405, 419

M
Macrostate, 4–6, 71, 348, 349
Material, 2, 16, 18–21, 28, 29, 33, 37, 39–41, 83, 92, 103, 104, 108, 113, 128, 130, 143, 147, 154, 163, 173, 174, 178, 182, 190, 202, 210, 212, 216, 224, 230, 236, 238, 240, 244, 256, 258, 262, 268, 270, 274, 289, 295, 299, 300, 307–309, 314, 315, 322, 347, 348, 351, 355, 359, 363, 377, 381, 382, 391, 413, 420, 422, 428
Mechanism, 1, 2, 4–7, 16, 41, 42, 55, 57, 64, 69–72, 75, 77–81, 83, 87, 89–91, 101–110, 113, 121–123, 125–127, 129, 130, 132–134, 139, 140, 143, 144, 146–151, 153, 159, 161, 163–165, 167–169, 171, 173, 174, 177, 179–185, 187, 188, 190, 193, 194, 197, 199, 201–203, 207, 212, 217, 221, 224–229, 232, 233, 236, 240, 242, 243, 245, 246, 253, 255, 256, 259, 277, 282, 289, 290, 293, 298, 307, 315, 321, 322, 329, 347–351, 355–358, 360–364, 366, 367, 369–378, 380, 383–390, 392, 395, 397, 399, 400, 402–405, 413, 414, 417–422
Melting, 6, 27, 40, 41, 46, 85, 88, 91, 92, 96, 108, 109, 139, 141, 142, 147, 150, 151, 153, 154, 165, 169, 181, 184, 188, 189, 191, 196, 202, 207, 209, 210, 212, 214, 218–221, 225, 236, 239, 240, 242, 243, 254, 255, 258, 263, 265, 267, 269, 272–274, 282,

308, 312, 315, 352–356, 358, 363–365, 368, 369, 373, 375–377, 379, 384, 389, 395, 397, 401, 403, 404, 418–420, 424, 428
Membrane, 81, 148, 176, 218, 314, 322
MET, 4, 46, 47, 62, 70, 71, 73, 74, 76–80, 82, 88, 91–96, 101, 103, 104, 106, 108, 109, 139–142, 145, 147, 150, 153, 154, 159–163, 165, 167, 169, 188, 189, 202, 208, 209, 220, 223, 225, 232, 233, 239, 240, 246, 253–256, 258–263, 268–274, 289, 290, 347, 349–353, 356, 357, 359–361, 363–366, 368, 373, 375, 377, 383, 386, 395, 396, 414, 418–421, 423
Metal, 4, 14, 20–23, 27, 29, 31, 32, 34, 38, 40, 42–48, 59, 60, 64, 69–71, 76–80, 84, 87–90, 101, 105–109, 121–125, 127, 128, 130, 134, 135, 139, 140, 142, 147, 149, 151, 153, 154, 159–161, 173, 174, 179, 181–185, 187, 188, 190, 196, 207, 209, 217, 221, 223–228, 236–240, 243, 253, 259, 264, 266, 271, 275, 276, 281–283, 289, 290, 293–299, 301, 303, 307–310, 312, 313, 315, 321, 347, 350, 354, 363, 367, 376, 385, 388, 395, 402, 405, 415, 418, 420, 422, 425
Metallo-organic, 55, 139, 142
Metalorganic, 17, 23, 140
METANO, 15, 27–31, 34, 35, 40, 41, 44, 53, 54, 57, 59, 61–63, 66, 71, 73, 74, 79, 84, 89, 91–93, 122, 130, 135, 140, 145, 188, 207, 208, 214, 223, 225, 237, 238, 240, 253, 255, 256, 258, 260–265, 269, 271–276, 278–281, 289, 290, 304, 311, 313, 322, 324, 336, 347, 349, 351–353, 356, 360, 362–365, 395, 402, 404, 414, 417, 418, 420–422, 427
MOCVD, 23, 40, 82, 112, 197, 308
Model, 2, 201, 219, 253, 262, 263, 321, 322, 343, 344, 363, 377, 421, 427
Molecular Beam Epitaxy (MBE), 13, 14, 16–18, 23, 24, 38, 39, 73, 74, 81, 82, 103, 112, 126, 190, 195, 200, 259, 295, 308, 414, 425
Molten, 40, 41, 59, 62, 63, 70, 71, 74, 77, 78, 80, 82, 83, 88–94, 107, 108, 141, 142, 150, 160, 161, 166, 181, 183, 184, 188, 189, 192, 200, 202, 208–210, 212, 214, 220–222, 232, 234–236, 242, 243, 245, 246, 259, 263, 268, 274, 281, 315, 351–353, 355–357, 362–364, 366, 372, 375–377, 379, 383, 385, 386, 397, 404, 415, 419, 423
Moment, 41, 61, 62, 163, 207, 225, 236, 238, 239, 282, 284, 302, 315, 352–354, 389, 402, 420, 424, 428
Morphology, 1, 2, 6, 7, 16, 17, 19, 22, 28, 34, 35, 41, 44, 54, 66, 84, 89, 106, 107, 129, 180, 188, 197, 201, 207, 209, 218, 221, 223, 228, 242, 245, 253, 255, 282, 283, 289, 295, 302, 303, 314, 315, 351–353, 355, 369, 373, 397, 398, 417, 422, 428
MWCNT, 2, 20, 42, 67, 104, 106, 115, 116, 127, 128, 215, 224, 276–281, 292, 310, 340, 368–370

N

Nanobelts, 1–3, 8, 30, 31, 121, 129–131, 210, 295, 314, 347, 373–381, 383, 390, 413, 417, 421–424
Nanocolumns, 39
Nanocrystals, 1, 2, 4, 6, 7, 14, 17, 21, 38, 59, 70, 79, 80, 84, 144, 146, 154, 220, 243, 282, 290, 315, 348, 350, 413, 415
Nanodots, 1–4, 6, 8, 23, 30, 64, 66, 101, 234, 295, 413, 417
Nanofibers, 1, 2, 4, 6, 8, 19, 20, 89, 90, 101, 105, 121, 259, 292, 295, 303, 413, 421
Nanomaterials, 1–8, 13–19, 21, 24, 27, 28, 30, 35, 37, 38, 40, 42, 44, 46–48, 53–55, 57, 59, 61, 62, 64–66, 69–71, 75, 76, 78, 79, 83, 90–93, 101, 107, 121–123, 126, 129, 130, 132, 139, 140, 143, 144, 148, 159, 173, 182, 187, 196, 197, 203, 207–209, 212, 220, 224–226, 228, 231, 232, 234, 236, 238–243, 245, 246, 253–256, 262, 281–283, 289–293, 295, 298, 303, 306, 310–315, 321, 322, 324, 329, 330, 334, 335, 344, 347–352, 354, 356, 357, 364, 367, 373, 374, 389, 390, 392, 399–405, 413, 414, 417–423, 425, 427, 430
Nanoparticle, 4, 6, 13, 15–18, 20–23, 25, 27–38, 40–48, 53–55, 57–64, 66, 67, 69, 71, 77–80, 84, 88–92, 94, 102, 104–111, 113–115, 121, 123, 124, 127–130, 139–146, 148–151, 153, 154,

159–163, 166, 167, 170, 177, 181–183, 187–189, 191, 192, 194–201, 203, 207–210, 212, 214–228, 231–233, 235–243, 245, 246, 253–255, 259, 262–265, 268–271, 273, 274, 276, 278–284, 289–292, 294, 296–299, 301, 303–314, 321–325, 327, 334, 336, 338, 339, 341–344, 347, 348, 351–353, 355, 357, 358, 363–366, 371, 373, 375–386, 388–392, 395, 398–404, 414–424, 427–429, 431, 434
Nanopore, 39, 54, 58, 59, 77, 170, 188, 189, 200, 202, 203, 210, 212, 214, 216–218, 222, 227, 230, 235, 236, 238, 239, 261, 263, 273, 274, 284, 290, 299, 314, 322–325, 327, 329–337, 342–344, 352, 362–364, 370, 372, 373, 385–387, 389, 390, 392, 399, 400, 404, 405, 421, 423, 424, 428–430
Nanoribbon, 2
Nanoring, 1, 2, 30, 65, 66, 306, 413
Nanoscience, 3, 413
Nanosheet, 2, 295, 303, 421
Nanotube, 1–4, 6, 8, 18, 19, 23, 30, 31, 42, 48, 53, 55, 57, 58, 65, 66, 89, 92, 93, 96, 101, 107, 121, 127, 129, 134, 187, 190, 191, 194, 197, 217, 221, 222, 231, 233, 234, 237, 238, 243, 274, 275, 278–280, 291, 295–298, 301–303, 306, 308, 310, 313, 314, 323, 324, 330, 331, 338, 340, 343, 344, 347, 364, 367, 370, 398, 413, 417, 421, 422, 424
Nanowire, 1–4, 6, 8, 16, 18, 19, 23, 24, 30, 31, 37–40, 42–44, 46–48, 53, 55, 56, 58, 59, 62–66, 69–73, 75–88, 90–96, 101–105, 107–111, 113, 115, 121, 123–126, 129–134, 139–154, 159–171, 173–185, 187–203, 217, 221, 226, 227, 232–234, 238–243, 245, 246, 253, 255–260, 268, 269, 271–273, 281, 282, 291–293, 295, 297, 302, 303, 306–308, 311, 313–315, 322, 324, 330, 332, 338, 354–364, 366, 371, 374, 379–381, 383–392, 402, 413, 414, 417–419, 421, 424, 425
Novel, 6, 21, 27, 53, 127, 283, 295, 298, 413, 420
Novelty, 27, 139, 187, 197, 207, 253, 289, 344, 402, 414, 430
NP1, 30, 207, 225, 238, 253, 290, 292, 293, 348, 367, 370, 371, 389, 403, 420
NP2, 30, 207, 225, 226, 238, 239, 253, 290–293, 321, 348, 352, 367, 370, 371, 373, 388, 389, 403, 404, 420
Nucleation, 14, 15, 20, 30, 35, 38, 39, 44, 53, 55–57, 59, 60, 69, 72, 84, 85, 87, 90, 92, 94, 103, 113, 115, 123, 124, 129, 132, 141, 142, 151, 153, 160, 162, 168, 169, 173, 174, 179, 181, 184, 190, 196, 200, 201, 212, 217, 226, 228, 231, 232, 234, 236, 239, 243, 245, 254, 255, 259, 298, 304, 306, 310, 313, 324, 349–351, 362, 370, 372, 378, 387, 391, 397, 398, 414–416, 419, 423

O

Ostwald ripening, 16, 27, 46, 47
Oxide, 5, 6, 20–23, 39, 43, 44, 72, 81, 113, 123–125, 127–130, 140, 163, 167–169, 173–177, 179–184, 187, 193, 195, 200, 202, 215–218, 225, 226, 230, 236, 238, 243, 244, 254–256, 259, 282, 291, 293–304, 309, 313–315, 347, 367, 369, 371, 373, 386, 387, 389, 390, 395, 418
Oxide-Assisted Growth (OAG), 5, 6, 111, 112, 123, 163, 173–185, 202, 203, 256, 259, 315, 347, 388, 389, 393, 414, 421
Oxygen, 16, 20, 22, 47, 48, 91, 124, 125, 162, 163, 165–167, 174, 184, 196, 201, 202, 223, 224, 227, 236, 237, 254, 255, 259, 282, 294, 303, 309, 315, 360, 369, 371, 374–377, 383, 386, 418

P

Paracrystalline, 289, 307, 309, 420
Permeability, 321, 326, 329, 331–333
Phase, 5, 14–17, 21–23, 30, 43, 45, 53–55, 60, 69, 71, 73–78, 82, 83, 87, 88, 91, 102, 107–109, 122, 124, 126, 127, 139, 140, 143, 148–151, 153, 159, 163–165, 182–184, 200, 201, 207, 209, 210, 212, 217–220, 223, 226–228, 232, 234, 240, 245, 246, 253–258, 260–263, 271, 282, 289, 299, 301, 308, 315, 347–349, 351–361, 363, 364, 366, 378–380, 383–387,

395, 398, 401–404, 414–417, 420, 421
Photoluminescence, 126, 180
Physical Vapor Deposition (PVD), 13, 14, 21, 32, 257
Plasma, 14, 16, 20, 28, 34, 35, 38, 58, 59, 126, 210, 215, 224, 225, 232, 259, 277, 279, 281, 282, 290, 297, 298, 303, 308–311, 324, 348, 349, 362, 367–370, 372, 398, 399, 428
Polar, 61, 126, 297, 376, 418
Polarity, 19, 238, 297, 388
Porosity, 66, 207, 212, 216, 218, 225, 226, 230–237, 239, 255, 256, 263, 269, 280, 282, 284, 297, 309, 315, 321–329, 331–336, 338, 341, 352–355, 363, 367, 369, 373, 384, 389, 391, 399, 418, 420, 424, 428, 429
Porous, 28, 29, 41, 54, 58, 59, 66, 104, 161, 162, 169, 170, 199, 201, 202, 210, 212, 214, 217, 218, 221, 222, 224, 226, 228, 231, 232, 235, 236, 256, 279, 283, 290, 291, 296, 297, 307–310, 315, 326–329, 332, 352, 353, 362, 364, 366–372, 375–377, 383, 388–390, 395, 397, 399, 416, 423, 428
Precursor, 15, 16, 21–23, 27, 28, 34, 41, 43, 45, 48, 55, 57–59, 67, 69, 70, 74, 75, 77, 82, 83, 85, 89–91, 93, 94, 101–103, 105, 106, 109, 110, 112–116, 122, 123, 127, 128, 134, 135, 139–146, 148, 151–154, 175, 180, 192, 196, 200, 210, 212, 215, 223, 224, 226, 228, 238, 239, 243, 254, 259, 261, 276, 277, 283, 289, 291, 293, 302, 306, 308, 309, 312, 313, 321, 324, 335, 341–343, 348, 349, 351, 352, 354, 356, 359, 360, 364, 369, 370, 384, 388, 392, 395, 400–405, 414, 415, 417, 420, 421, 423, 424, 428
Pre-nucleation, 38, 53, 55, 57–59, 61, 67, 71, 141, 159, 161, 191, 201, 202, 212, 215, 224, 239, 253, 255, 259, 279, 292, 297, 303, 351, 370, 375, 377–380, 390, 391, 395, 397, 398, 418, 419
Pressure, 14–16, 18–21, 23–25, 35, 41, 47, 48, 60, 69, 71, 74, 77, 78, 81–86, 91, 96, 109, 110, 112, 113, 122, 123, 125, 128, 129, 133, 143, 144, 146, 150, 154, 165, 167, 178, 191, 192, 195–197, 201, 208–210, 223, 226, 229, 232, 233, 235, 245, 254, 255, 261, 263, 271, 276, 322, 327–329, 351, 353, 354, 356, 357, 361, 363, 366, 373, 374, 384, 386, 390, 402–405, 423, 424
Pro-nucleation, 24, 55, 67, 71, 141, 142, 208, 279, 370, 375–377, 380, 395, 397, 398, 418
Pulsed Laser Deposition (PLD), 13, 14, 18

Q

Quantum dots, 1, 2, 13, 18, 30, 66, 413
Quantum rings, 30, 66, 413
Quasiliquid, 4–6, 41, 55, 59, 83, 192, 207–210, 212, 214, 218, 220–222, 232–234, 243, 245, 253, 256, 263, 269, 274, 280, 289, 306, 314, 348, 349, 352–355, 362, 363, 372, 383, 389, 391, 397, 398, 402, 416, 420, 423
Quasisolid, 4–6, 41, 59, 192, 207–210, 212, 214, 218, 221–223, 232–235, 243, 245, 256, 263, 269, 274, 280, 306, 314, 348, 349, 352, 354, 355, 362, 363, 372, 383, 388, 389, 391, 397, 398, 405, 416, 420, 423

R

Rate, 15, 17, 18, 21, 23, 24, 40, 46–48, 59, 69, 74, 78, 80–83, 85–88, 90, 93, 101, 103, 108–116, 124, 126, 128–130, 132, 134–136, 149, 151–153, 161, 165, 166, 169–171, 173, 181, 190, 191, 193, 196, 197, 201, 234, 238, 253, 276–278, 280–282, 284, 291, 301, 304, 305, 310, 321, 329, 330, 335, 336, 338–344, 357, 359, 362–364, 367, 369, 373, 375, 377, 389, 390, 398, 400, 403–405, 418, 422, 431
Reaction, 15–17, 19–23, 33–35, 46, 53, 54, 59, 60, 67, 78, 84, 87, 91, 96, 110, 129, 134, 139, 142–147, 149–151, 154, 161, 167, 173, 174, 177, 180, 182, 190, 196, 198, 200, 210, 223, 224, 228, 234, 254, 256, 261–263, 306, 309, 312, 329, 331, 359, 374–377, 379, 380, 384, 385, 389, 395, 397, 403
Reactivity, 19, 35, 36, 41, 113, 181, 239, 241, 325, 326, 354

R_L species, 27, 40–42, 44, 46, 47, 55, 59, 60, 71, 91–93, 96, 101, 102, 106, 108, 109, 113, 122, 132, 141, 154, 160, 161, 163, 166, 169, 182, 188, 196, 202, 203, 208, 210, 212, 214, 220–224, 226, 227, 231, 233–236, 240–242, 253–256, 258–260, 269, 272, 274, 279, 282–284, 289–293, 297, 304, 307–310, 315, 321–329, 331–334, 344, 347, 349–354, 358–364, 366, 369–373, 375–377, 379–388, 390–395, 397, 398, 400, 403–405, 418, 419, 422–424
Rough, 124, 195, 215, 218, 221, 351, 370, 385, 388, 391, 424, 425
Roughness, 41, 66, 124, 154, 187, 194–196, 202, 210, 215, 228–230, 236, 255, 291, 315, 321, 330, 332–336, 338, 354, 371, 384, 390, 391, 424, 428
Rule, 71, 314, 402, 413, 422

S
Screw, 123, 125, 207, 208, 282, 298, 311, 315
SECINI, 41, 71, 73, 77, 79, 83, 208, 253, 254, 256, 290, 347, 348, 353, 354, 357, 362, 366, 384, 385, 420
SECINI0, 41, 71, 73, 74, 77, 79, 82, 83, 208, 253, 254, 256, 290, 348, 354, 362, 366, 384, 385, 420
Seed, 24, 53, 59–62, 108, 122, 127, 128, 139, 141, 143–146, 148, 149, 154, 165, 188–191, 200, 203, 258, 366
Selection, 23, 59, 69, 78, 79
Selective Area Epitaxy (SAE), 187, 190, 194, 198, 199
Self-Catalytic Growth (SCG), 5, 6, 55, 111, 112, 123, 187–190, 193, 194, 196–203, 217, 315, 347, 390–392, 414, 421
SEM, 19, 38, 125, 128, 129, 168, 191, 195, 196, 198, 215, 234, 276, 281, 297, 369, 374, 377, 391
Semiconductor, 1, 2, 8, 17, 19, 42, 43, 71, 103, 121, 127, 129–131, 140, 143, 147, 148, 153, 182, 243, 245, 281, 295, 300, 347, 373, 413, 421
Semiliquid, 181, 208, 220, 234, 235
Semimolten, 40, 59, 80, 92, 107, 108, 161, 184, 188, 189, 192, 200, 202, 208–210, 212, 221, 222, 234, 236, 355, 363, 377, 386, 423
Semisolid, 208, 209
Sensor, 3, 14, 168, 373
Sheath, 19, 81, 173, 177, 180, 182–184, 387
Shell, 9, 34, 44, 45, 48, 53, 55, 58, 64–67, 112, 147, 163, 164, 166, 177, 180–184, 209, 222, 223, 238, 239, 243, 275, 278–281, 306, 323, 324, 352, 353, 360, 370, 395, 397–399, 423, 424
SiC, 4, 42, 112, 123, 125, 127, 215, 216, 244, 257, 259, 267, 360, 364, 369
SiGe, 37, 38, 42, 296, 391
Silicon (Si), 2–4, 6, 8, 16, 17, 23, 28, 37–40, 42–44, 46–48, 54, 63, 64, 69, 71–88, 92, 94, 95, 101, 103, 104, 107–113, 121, 123, 124, 126, 128, 130–133, 142, 144, 146–150, 153, 154, 161, 163–171, 173–185, 187, 188, 190, 193, 195, 196, 199–202, 215–217, 221, 227, 229, 230, 232, 234, 238–240, 243, 244, 246, 253, 257–259, 263–265, 267–269, 271–273, 277, 281, 282, 291–293, 295–297, 301–304, 310, 315, 324, 329, 330, 332, 336–338, 356–360, 362, 364, 365, 367–369, 371–373, 376–391, 393, 395, 397, 398, 401, 421, 425
Single-Walled Carbon Nanotubes (SWCNTs), 2, 20, 42, 48, 66, 67, 90, 106, 127, 128, 150, 215, 217, 220, 228, 239, 254, 276–279, 281, 292, 297, 309, 313, 332, 368, 369, 371
SiO_2, 22, 28, 37–39, 42, 54, 81, 103, 121, 123, 124, 127, 128, 131, 173–177, 179, 181–185, 190, 193, 200, 201, 216, 243, 244, 267, 291–293, 295, 297, 299, 301, 302, 304, 313, 365, 367, 369–371, 381, 385, 388, 389, 393, 395, 397
Sol-gel, 13, 21–23, 27, 30, 32, 34, 45, 297
Solid-Liquid-Solid (SLS), 5, 6, 55, 159–164, 166–171, 234, 347, 349, 386–388, 414, 415, 421
Solubility, 19, 59, 76, 78, 88, 102, 109, 111, 113, 142, 149, 227, 240, 255, 262, 263, 265, 268, 269, 271–274, 276, 278–281, 312, 363, 366, 375, 378, 384, 385
Solution, 4, 5, 21, 22, 27, 29, 34, 40, 41, 43, 45–47, 53, 55, 59–61, 63, 76, 77, 88, 91, 103, 105, 106, 139–141, 143–151, 153, 160, 163, 166, 168,

169, 189, 196, 200, 202, 209, 210, 212, 215, 223, 226–228, 232, 240, 242, 243, 259, 260, 262, 283, 289, 291–293, 297, 298, 307, 308, 312, 314, 315, 328, 347–354, 356, 358–361, 363, 366, 370, 373, 375, 377–389, 391, 392, 395, 397–399, 405, 414–416, 418–421, 423, 425
Solution-Liquid-Solid (SoLS), 5, 6, 43, 55, 139–142, 144, 146–148, 151, 153, 242, 245, 347, 349, 383, 386, 414, 415, 418, 421
Solution-Solid-Solid (SoSS), 5, 139, 153, 245, 246, 347, 385, 386, 415, 418, 419
Solvothermal, 13, 14, 19
Stage, 24, 38, 53, 55, 57–59, 61, 67, 69, 71, 79, 80, 89, 141, 142, 145, 146, 148, 159, 161, 181, 190, 191, 196, 201, 202, 208, 212, 215, 223, 224, 239, 253, 255, 259, 279, 292, 298, 303, 341, 351, 360, 370, 375–380, 390, 391, 395, 397, 398, 418, 419, 422
Stepped, 289, 307, 309, 401, 402, 420
Sticking, 41, 255, 261, 306, 307, 400–402, 417, 418, 424, 425
Strength, 1, 2, 7, 9, 13, 24, 45, 54, 57, 101, 107, 121, 139, 143, 147, 148, 159, 173, 187, 223, 312, 414
Stress, 16, 34, 37–39, 94, 96, 216, 225, 255, 261, 264, 295, 296, 301, 308, 353, 360, 377, 425
Structure, 1, 4, 6–9, 13, 14, 16, 19, 23, 27, 30, 34, 37, 38, 41, 44, 45, 53, 54, 58–61, 64, 67, 74, 79, 82, 84, 89, 106, 115, 121, 123, 128–132, 140, 146–148, 150, 154, 163, 164, 180, 184, 188, 190–192, 196, 199, 201, 202, 207–210, 212, 214, 216–218, 221, 222, 224, 227, 235–237, 242, 253, 255, 260, 264, 265, 278, 281, 283, 289, 290, 292–295, 298–302, 307–312, 322, 332, 349, 353, 359, 360, 364, 368, 370, 373, 378, 380, 386, 389, 391, 392, 395, 397, 398, 403, 415, 416, 420–423, 427
SUBSANO, 15, 27–31, 37–41, 53, 54, 57, 59, 61–63, 66, 121, 122, 130, 132, 134, 135, 140, 184, 185, 188, 196, 207, 208, 214, 223–225, 236, 237, 240, 253, 289, 290, 292–299, 301–304, 307–315, 321, 322, 324, 347, 349, 351, 352, 362, 367, 369–373, 388–390, 414, 417, 420, 421, 427
Substrate, 6, 15–18, 20–24, 27–32, 35–41, 44, 53–55, 57, 60, 61, 65, 71, 72, 74, 79–82, 91–93, 101–105, 111, 112, 121–128, 130, 132–134, 142, 147, 159–171, 175–177, 179, 180, 181, 183–185, 188–190, 192–202, 207, 212–217, 220, 224, 231, 232, 234, 254, 255, 259, 271, 277, 283, 289–299, 302–311, 315, 332, 347, 349–351, 353, 359, 360, 362, 364, 365, 367–372, 374, 376–380, 383, 385–391, 393–395, 397, 402, 414, 415, 420, 421, 425, 428
Sulfur, 173, 184, 216, 236, 315, 376
Supercritical, 4–6, 33, 34, 43, 59, 139, 140, 143, 144, 146, 148–150, 153, 348, 383–385, 414
Supercritical Fluid-Liquid-Solid (SFLS), 5, 6, 43, 55, 139–141, 143–149, 151, 347, 383, 385, 386, 414, 415, 418, 421
Supercritical Fluid-Solid-Solid (SFSS), 5, 139, 153, 154, 347, 415, 418, 419
Superiority, 126, 146, 187, 197, 202, 245
Supersaturation, 60, 69–71, 92, 110, 113, 124, 142, 161, 162, 167, 169, 181, 182, 196, 234, 240, 255, 259, 289, 304, 313, 314, 349, 357, 362, 363, 372, 375–377, 379, 387, 390, 403, 404, 415–417, 423
Support, 16, 27, 28, 41, 44, 53–55, 57, 61, 74, 89, 90, 93, 102, 114, 175, 207, 208, 224, 227, 228, 236–239, 254, 290, 293–299, 301, 305–307, 315, 326, 327, 414, 419, 421, 425, 428
Surface, 1, 7, 15–18, 21–24, 27–30, 34–42, 44–47, 53–62, 64–67, 70–74, 77–80, 84, 88–96, 101, 102, 106, 108–110, 113, 115, 122–130, 132, 134, 135, 139–144, 148, 154, 160–167, 170, 177, 179–185, 187–189, 191–196, 198–202, 207–226, 228–240, 243, 244, 253–256, 258–270, 272–276, 278–284, 289–315, 321–336, 338, 339, 341–344, 348, 351–356, 359–364, 367–373, 375, 377, 380, 383–392, 395, 397–405, 415–418, 420–425, 427–434
Surfaces of the Functional Substrates (SFS), 294–296, 298, 299, 301, 302
Synergy, 44, 289, 307, 310, 420

Synthesis, 1–4, 6–8, 13–19, 21, 22, 24, 25, 30, 32–35, 37, 38, 42, 43, 45, 46, 53–55, 57, 59, 64, 66, 71, 109, 127, 130, 135, 139, 141–146, 148–150, 153, 154, 159, 163, 168, 173, 194, 208, 225, 243, 253, 259, 289, 292, 314, 324, 373–376, 385, 395, 398, 413, 414, 417, 419, 420, 422

T

Tamman, 212, 214
TEM, 84, 87, 89, 124, 125, 128, 129, 142, 154, 166, 181, 191, 195, 200, 234, 245, 256, 272, 281, 297, 302, 355, 371, 374, 388, 391, 397
Temperature, 2, 4, 8, 15–25, 27, 28, 34, 35, 38, 39, 41, 43, 46–48, 59, 60, 69–88, 90–94, 96, 101–110, 113–115, 122–125, 127–130, 132–136, 139, 141–143, 145–151, 153, 154, 159–161, 163–170, 174, 176–178, 180, 181, 183, 184, 187, 188, 190–197, 199–202, 209, 210, 212, 214, 216–221, 223, 224, 226–230, 233–237, 239, 240, 242, 245, 254–256, 258–263, 265, 267–269, 271–276, 278, 280, 281, 291–295, 299, 302, 303, 308, 309, 312–315, 321, 327, 331–334, 336–338, 340–342, 347, 348, 351–364, 366–368, 372–380, 383–388, 390, 392, 395, 397, 398, 400, 402–405, 418–420, 423–425, 429–434
Template, 23, 30, 111, 148, 194, 289, 306, 313–315
Thermodynamic, 6, 60, 69, 73, 79, 80, 85, 90, 103, 123, 210, 255, 282, 315, 353, 356, 360, 378
Tip, 54, 62, 64, 89, 93, 104, 105, 123–125, 128, 148, 150, 162–165, 176, 181, 184, 203, 208, 234, 236, 241, 242, 246, 256–259, 278, 283, 293, 306, 307, 355, 359, 364, 384, 389
Transformation, 21, 23, 77, 107, 207, 212, 226, 232, 234, 253–255, 263, 289, 358, 380, 403
Transition, 30, 42, 60, 71, 207, 218, 223, 227, 255, 264, 299, 358, 360, 403, 420
Treatment, 29, 30, 34, 53, 184, 209, 210, 212, 215, 223–225, 227, 228, 230, 234, 279, 289, 290, 292, 293, 295, 298, 301, 307, 309–311, 335, 361, 362, 372, 418, 421, 424, 425, 428

V

Vapor, 5, 6, 13–16, 18, 21, 23, 30, 32, 34, 38, 53, 55, 62, 69–71, 78–80, 91–94, 101, 109, 122, 124, 125, 130, 134, 149, 151, 163, 169, 174, 175, 177, 178, 181, 182, 184, 189, 191, 196, 212, 219, 225, 232–234, 243, 259, 281, 283, 284, 315, 348–350, 369, 374–377, 379, 380, 386, 392, 401, 404, 414, 415, 417, 421, 423, 424, 428
Vapor Liquid Solid (VLS), 4–6, 16, 24, 41, 44, 46, 55, 63, 64, 69–73, 75–85, 87–91, 101, 107–111, 113, 121, 122, 132, 139, 140, 144, 146, 148, 150, 151, 159, 161–163, 169, 173, 180, 181, 184, 188, 190, 202, 203, 227, 232, 233, 236, 240, 245, 246, 255, 259, 271, 347, 353, 355–357, 360, 362–364, 366, 373, 374, 378, 380, 383, 384, 399, 402, 403, 414, 415, 417–421
Vapour Quasiliquid Solid (VQS), 6, 55, 207, 212, 221, 224–229, 232, 236, 240, 242, 243, 245, 246, 253, 255, 256, 282, 283, 289, 290, 293, 298, 307, 315, 321, 322, 347, 349, 355–358, 361–364, 366, 367, 370–373, 375–377, 380, 385, 386, 388, 389, 392, 395, 397, 399, 402–405, 419–421
Vapor Solid Solid (VSS), 5, 6, 55, 83, 101–111, 113, 122, 126, 139, 151, 153, 188, 232, 233, 236, 245, 246, 277, 347, 349, 353, 357, 361–364, 367, 372, 402, 414, 415, 417–419, 421
Vapor Solid (VS), 5, 6, 55, 112, 121–127, 129, 130, 132–134, 347, 349, 350, 367, 369–373, 400, 402, 414, 417–419, 421

W

Water, 19, 22, 28, 34, 94, 123, 134, 225, 283, 284, 293, 369, 371, 428
Water-assisted, 48, 121, 134, 135
Weakness, 1, 7, 13, 24, 101, 121, 139, 159, 173, 180, 187, 202, 357, 414

X

X-ray, 19, 23, 102, 123, 181, 184, 227, 308, 403
X-ray Diffraction (XRD), 19, 89, 302, 308, 374, 378, 379, 391

Z

ZnO, 19, 23, 37, 42, 84, 88, 102, 103, 112, 113, 123, 129–131, 174, 216, 227, 244, 291, 297, 299, 311, 329, 332, 357, 362, 370, 374, 376, 377, 381, 400

The manufacturer's authorised representative in the EU is Springer Nature Customer Service Centre GmbH, Europaplatz 3, 69115 Heidelberg, Germany. If you have any concerns regarding our products, please contact ProductSafety@springernature.com

Printed and bound by CPI Group (UK) Ltd, Croydon, CR0 4YY
30/09/2026
02225454-0013